Molecular Biology and Biotechnology
Fourth Edition

This book is dedicated to the memory of
Christopher J. Dean
who died shortly after producing his chapter for this volume.

Molecular Biology and Biotechnology
Fourth Edition

Edited by

John M. Walker and Ralph Rapley
University of Hertfordshire, Hatfield, UK

RS•C
ROYAL SOCIETY OF CHEMISTRY

ISBN 0-85404-606-2

A catalogue record for this book is available from the British Library

Published by The Royal Society of Chemistry,
Thomas Graham House, Science Park, Milton Road, Cambridge CB4 0WF, UK

For further information see our web site at www.rsc.org

Typeset by Paston PrePress Ltd, Beccles, Suffolk
Printed in Great Britain by Athenaeum Press Ltd, Gateshead, Tyne and Wear

Preface

One of the exciting aspects of being involved in the field of molecular biology is the ever accelerating rate of progress, both in the development of new methodologies and the practical applications of these methodologies. Indeed, such developments led to the idea of the first edition of *Molecular Biology and Biotechnology* and subsequent editions have reflected the fast moving nature of the area. To keep pace with the ever expanding technological changes we have increased the basic molecular biology content of the book from one to two chapters in this latest edition. In recent years the development of the World Wide Web has been exponential and now provides an essential source of information and access to databases for the molecular biologist. We therefore considered it both appropriate and timely to include a new chapter devoted to the subject of Bioinformatics. Other chapter titles remain the same as the previous edition but this should not mask the significant updating of the content of these chapters in response to major developments in each area. Indeed, in order to reflect research developments, the majority of these chapters have required a total re-write rather than simple updating.

PCR (only introduced as stand-alone chapter in the last edition) is firmly established as a day-to-day tool and its revolutionary effect on the field is evidenced by its inclusion in chapters throughout the book. Molecular biology continues to profoundly affect progress in areas such as plant biotechnology, food technology (especially the contentious area of genetically modified foods), vaccine development, use and application of monoclonal antibodies, clinical treatment and diagnosis, the production of transgenic plants and animals, and many other areas of research relevant to the pharmaceutical industry. All these areas have been fully updated in this edition. In addition, we continue to ensure that biotechnology is not just considered at the gene level and full consideration continues to be given to aspects of large-scale production and manufacturing with chapters on fermentation technology, downstream processing and the applications of immobilised biocatalysts.

Our continued intention is that this book should primarily have a teaching function. As such, the book should prove of interest both to undergraduates studying for biological or chemical qualifications and to postgraduate and other scientific workers who need a sound introduction to this ever rapidly expanding area.

John M. Walker
Ralph Rapley

Contents

Chapter 15 Bioinformatics **405**

Peter M. Woollard

Contributors

J.R. Adair, *Cambridge, UK*

G.F. Bickerstaff, *Department of Biological Sciences, University of Paisley, Paisley PA1 2BE, UK*

V.C. Bugeja, *Department of Biosciences, University of Hertfordshire, College Lane, Hatfield, Hertfordshire AL10 9AB, UK*

M.F. Chaplin, *School of Applied Science, South Bank University, 103 Borough Road, London SE1 0AA, UK*

B.P.G. Curran, *School of Biological Sciences, Queen Mary and Westfield College, University of London, Mile End Road, London E1 4NS, UK*

P. Debenham, *University Diagnostics, South Bank Technopark, 90 London Road, London SE1 6LN, UK*

C.J. Dean, *formerly of McElwain Laboratories, Institute of Cancer Research, UK*

E. Green, *South Thames Regional Genetics Centre, Division of Medical and Molecular Genetics, Guys Hospital, St Thomas Street, London SE1 9RT, UK*

P.M. Hammond, *Centre for Applied Microbiology and Research, Porton Down, Salisbury, Wiltshire SP4 0JG, UK*

M.G.K. Jones, *Western Australian State Agricultural Biotechnology Centre, Murdoch University, Loneragan Building, Perth 6150, Western Australia*

M. Mackett, *CRC Department of Molecular Biology, Molecular Genetics Section, Paterson Institute of Cancer Research, Christie Hospital NHS Trust, Manchester M20 9BX, UK*

P.D. Martin, *University Diagnostics, South Bank Technopark, 90 London Road, London SE1 6LN, UK*

J.J. Mullins, *The Molecular Physiology Laboratory, The Wilkie Building, University of Edinburgh Medical School, Teviot Place, Edinburgh EH8 9AG, UK*

L.J. Mullins, *The Molecular Physiology Laboratory, The Wilkie Building, University of Edinburgh Medical School, Teviot Place, Edinburgh EH8 9AG, UK*

E.J. Murray, *Roche Products, Broadwater Road, Welwyn Garden City, Hertfordshire AL7 3AY, UK*

R.K. Pawsey, *School of Applied Science, South Bank University, 103 Borough Road, London SE1 0AA, UK*

R. Rapley, *Department of Biosciences, University of Hertfordshire, College Lane, Hatfield, Hertfordshire AL10 9AB, UK*

T.K. Sawyer, *ARIAD Pharmaceuticals, 26 Landsdowne Street, Cambridge, Massachusetts 02139, USA*

M.D. Scawen, *Centre for Applied Microbiology and Research, Porton Down, Salisbury, Wiltshire SP4 0JG, UK*

R.J. Slater, *Department of Biosciences, University of Hertfordshire, College Lane, Hatfield, Hertfordshire AL10 9AB, UK*

P.F. Stanbury, *Department of Biosciences, University of Hertfordshire, College Lane, Hatfield, Hertfordshire AL10 9AB, UK*

J. M. Walker, *Department of Biosciences, University of Hertfordshire, College Lane, Hatfield, Hertfordshire AL10 9AB, UK*

D.R. Williams, *Department of Biosciences, University of Hertfordshire, College Lane, Hatfield, Hertfordshire AL10 9AB, UK*

P.M. Woollard, *Genomics Unit, UK Genetics, GlaxoWellcome, Gunnells Wood Road, Stevenage, Hertfordshire SG1 2NY, UK*

CHAPTER 1

Fermentation Technology

PETER F. STANBURY

1 INTRODUCTION

Microorganisms are capable of growing on a wide range of substrates and can produce a remarkable spectrum of products. The relatively recent advent of *in vitro* genetic manipulation has extended the range of products that may be produced by microorganisms and has provided new methods for increasing the yields of existing ones. The commercial exploitation of the biochemical diversity of microorganisms has resulted in the development of the fermentation industry and the techniques of genetic manipulation have given this well-established industry the opportunity to develop new processes and to improve existing ones. The term fermentation is derived from the Latin verb *fervere*, to boil, which describes the appearance of the action of yeast on extracts of fruit or malted grain during the production of alcoholic beverages. However, fermentation is interpreted differently by microbiologists and biochemists. To a microbiologist the word means any process for the production of a product by the mass culture of microorganisms. To a biochemist, however, the word means an energy-generating process in which organic compounds act as both electron donors and acceptors, that is, an anaerobic process where energy is produced without the participation of oxygen or other inorganic electron acceptors. In this chapter fermentation is used in its broader, microbiological context.

2 MICROBIAL GROWTH

The growth of a microorganism may result in the production of a range of metabolites but to produce a particular metabolite the desired

1

organism must be grown under precise cultural conditions at a particular growth rate. If a microorganism is introduced into a nutrient medium that supports its growth, the inoculated culture will pass through a number of stages and the system is termed batch culture. Initially, growth does not occur and this period is referred to as the lag phase and may be considered a period of adaptation. Following an interval during which the growth rate of the cells gradually increases, the cells grow at a constant, maximum rate and this period is referred to as the log or exponential phase, which may be described by the equation

$$dx/dt = \mu x \tag{1}$$

where x is the cell concentration (mg ml^{-1}), t is the time of incubation (h), and μ the specific growth rate (h^{-1}). On integration equation (1) gives

$$x_t = x_0 e^{\mu t} \tag{2}$$

where x_0 is the cell concentration at time zero and x_t is the cell concentration after a time interval, t h.

Thus, a plot of the natural logarithm of the cell concentration against time gives a straight line, the slope of which equals the specific growth rate. The specific growth rate during the exponential phase is the maximum for the prevailing conditions and is described as the maximum specific growth rate, or μ_{max}. Equations (1) and (2) ignore the facts that growth results in the depletion of nutrients and the accumulation of toxic by-products and thus predict that growth continues indefinitely. However, in reality, as substrate (nutrient) is exhausted and toxic products accumulate, the growth rate of the cells deviates from the maximum and eventually growth ceases and the culture enters the stationary phase. After a further period of time, the culture enters the death phase and the number of viable cells declines. This classic representation of microbial growth is illustrated in Figure 1. It should be remembered that this description refers to the behaviour of both unicellular and mycelial (filamentous) organisms in batch culture, the growth of the latter resulting in the exponential addition of viable biomass to the mycelial body rather than the production of separate, discrete unicells.

As already stated, the cessation of growth in a batch culture may be due to the exhaustion of a nutrient component or the accumulation of a toxic product. However, provided that the growth medium is designed such that growth is limited by the availability of a medium component,

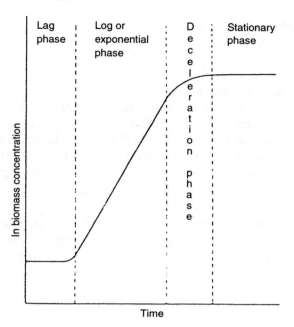

Figure 1 *Growth of a 'typical' microorganism under batch culture conditions*
(Reproduced with permission from P. F. Stanbury, A. Whitaker and S. J. Hall,
'Principles of Fermentation Technology', Pergamon Press, Oxford, 1995)

growth may be extended by addition of an aliquot of fresh medium to the
vessel. If the fresh medium is added continuously, at an appropriate rate,
and the culture vessel is fitted with an overflow device, such that culture is
displaced by the incoming fresh medium, a continuous culture may be
established. The growth of the cells in a continuous culture of this type is
controlled by the availability of the growth limiting chemical component
of the medium and, thus, the system is described as a chemostat. In this
system a steady-state is eventually achieved and the loss of biomass via
the overflow is replaced by cell growth. The flow of medium through the
system is described by the term dilution rate, D, which is equal to the rate
of addition of medium divided by the working volume of the culture
vessel. The balance between growth of cells and their loss from the
system may be described as

$$dx/dt = growth - output$$

or

$$dx/dt = \mu x - Dx$$

Under steady-state conditions,

$$\mathrm{d}x/\mathrm{d}t = 0$$

and, therefore, $\mu x = Dx$ and $\mu = D$.

Hence, the growth rate of the organisms is controlled by the dilution rate, which is an experimental variable. It will be recalled that under batch culture conditions an organism will grow at its maximum specific growth rate and, therefore, it is obvious that a continuous culture may be operated only at dilution rates below the maximum specific growth rate. Thus, within certain limits, the dilution rate may be used to control the growth rate of a chemostat culture.

The mechanism underlying the controlling effect of the dilution rate is essentially the relationship between μ, specific growth rate, and s, the limiting substrate concentration in the chemostat, demonstrated by Monod[1] in 1942:

$$\mu = \mu_{max}s/(K_s + s) \tag{3}$$

where K_s is the utilization or saturation constant, which is numerically equal to the substrate concentration when μ is half μ_{max}. At steady-state, $\mu = D$, and, therefore,

$$D = \mu_{max}\bar{s}/(K_s + \bar{s})$$

Where $\bar{s}$ is the steady-state concentration of substrate in the chemostat, and

$$\bar{s} = K_sD/(\mu_{max} - D) \tag{4}$$

Equation (4) predicts that the substrate concentration is determined by the dilution rate. In effect, this occurs by growth of the cells depleting the substrate to a concentration that supports that growth rate equal to the dilution rate. If substrate is depleted below the level that supports the growth rate dictated by the dilution rate the following sequence of events takes place:

(i) The growth rate of the cells will be less than the dilution rate and they will be washed out of the vessel at a rate greater than they are being produced, resulting in a decrease in biomass concentration.

[1] J. Monod, 'Recherches sur les Croissances des Cultures Bacteriennes', Herman and Cie, Paris, 1942.

(ii) The substrate concentration in the vessel will rise because fewer cells are left in the vessel to consume it.

(iii) The increased substrate concentration in the vessel will result in the cells growing at a rate greater than the dilution rate and biomass concentration will increase.

(iv) The steady-state will be re-established.

Thus, a chemostat is a nutrient-limited self-balancing culture system that may be maintained in a steady-state over a wide range of sub-maximum specific growth rates.

Fed-batch culture is a system that may be considered to be intermediate between batch and continuous processes. The term fed-batch is used to describe batch cultures that are fed continuously, or sequentially, with fresh medium without the removal of culture fluid. Thus, the volume of a fed-batch culture increases with time. Pirt[2] described the kinetics of such a system as follows. If the growth of an organism were limited by the concentration of one substrate in the medium the biomass at stationary phase, x_{max}, would be described by the equation:

$$x_{max} = YS_R$$

where Y is the yield factor and is equal to the mass of cells produced per gram of substrate consumed and S_R is the initial concentration of the growth limiting substrate. If fresh medium were to be added to the vessel at a dilution rate less than μ_{max} then virtually all the substrate would be consumed as it entered the system:

$$FS_R = \mu(X/Y)$$

where F is the flow rate and X is the total biomass in the vessel, *i.e.* the cell concentration multiplied by the culture volume.

Although the total biomass (X) in the vessel increases with time the concentration of cells, x, remains virtually constant; thus $dx/dt = 0$ and $\mu = D$. Such a system is then described as quasi-steady-state. As time progresses and the volume of culture increases, the dilution rate decreases. Thus, the value of D is given by the expression

$$D = F/(V_0 + F_t)$$

where F is the flow rate, V_0 is the initial culture volume, and t is time. Monod[1] kinetics predict that as D falls residual substrate concentration

[2] S. J. Pirt, 'Principles of Microbe and Cell Cultivation', Blackwell, Oxford, 1975.

should also decrease, resulting in an increase in biomass. However, over the range of growth rates operating the increase in biomass should be insignificant. The major difference between the steady-state of the chemostat and the quasi-steady-state of a fed-batch culture is that in a chemostat D (hence, μ) is constant whereas in a fed-batch system D (hence, μ) decreases with time. The dilution rate in a fed-batch system may be kept constant by increasing, exponentially, the flow rate using a computer-control system.

3   APPLICATIONS OF FERMENTATION

Microbial fermentations may be classified into the following major groups:[3]

(i)   Those that produce microbial cells (biomass) as the product.
(ii) Those that produce microbial metabolites.
(iii) Those that produce microbial enzymes.
(iv) Those that modify a compound which is added to the fermentation – the transformation processes.
(v)   Those that produce recombinant products.

3.1   Microbial Biomass

Microbial biomass is produced commercially as single cell protein (SCP) for human food or animal feed and as viable yeast cells to be used in the baking industry. The industrial production of bakers' yeast started in the early 1900s and yeast biomass was used as human food in Germany during the First World War. However, the development of large-scale processes for the production of microbial biomass as a source of commercial protein began in earnest in the late 1960s. Several of the processes investigated did not come to fruition owing to political and economic problems but the establishment of the ICI Pruteen process for the production of bacterial SCP for animal feed was a milestone in the development of the fermentation industry.[4] This process utilized continuous culture on an enormous scale (1500 m^3) and is an excellent example of the application of good engineering to the design of a microbiological process. However, the economics of the production of SCP as animal feed were marginal, which eventually led to the discontinuation of the

[3] P. F. Stanbury, A. Whitaker and S. J. Hall, 'Principles of Fermentation Technology', 2nd Edn, Pergamon Press, Oxford, 1995.
[4] D. H. Sharp, 'Bioprotein Manufacture—A Critical Assessment', Ellis Horwood, Chichester, 1989, Chapter 4, p. 53.

Pruteen process. The technical expertise gained from the Pruteen process assisted ICI in collaborating with Rank Hovis MacDougall on a process for the production of fungal biomass to be used as human food.[5] A continuous fermentation process for the production of *Fusarium graminearum* biomass (marketed as Quorn®) was developed utilizing a 40 m³ air-lift fermenter. This process was based on sound economics and has proved to be a major economic success.

3.2 Microbial Metabolites

The kinetic description of batch culture may be rather misleading when considering the product-forming capacity of the culture during the various phases, for, although the metabolism of stationary phase cells is considerably different from that of logarithmic ones, it is by no means stationary. Bu'Lock *et al.*[6] proposed a descriptive terminology of the behaviour of microbial cells which considered the type of metabolism rather than the kinetics of growth. The term 'trophophase' was suggested to describe the log or exponential phase of a culture during which the sole products of metabolism are either essential to growth, such as amino acids, nucleotides, proteins, nucleic acids, lipids, carbohydrates, *etc.* or are the by-products of energy-yielding metabolism such as ethanol, acetone and butanol. The metabolites produced during the trophophase are referred to as primary metabolites. Some examples of primary metabolites of commercial importance are listed in Table 1.

Bu'Lock *et al.* suggested the term 'idiophase' to describe the phase of a culture during which products other than primary metabolites are synthesized, products which do not have an obvious role in cell metabolism. The metabolites produced during the idiophase are referred to as the secondary metabolites. The interrelationships between primary and secondary metabolism are illustrated in Figure 2, from which it may be seen that secondary metabolites tend to be synthesized from the intermediates and end-products of primary metabolism. Although the primary metabolic routes shown in Figure 2 are common to the vast majority of microorganisms, each secondary metabolite would be synthesized by very few microbial taxa. Also, not all microbial taxa undergo secondary metabolism; it is a common feature of the filamentous fungi and bacteria and the sporing bacteria but it is not, for example, a feature of the Enterobacteriaceae. Thus, although the

[5] A. P. J. Trinci, *Mycol. Res.*, 1992, **96**, 1.
[6] J. D. Bu'Lock, D. Hamilton, M. A. Hulme, A. J. Powell, D. Shepherd, H. M. Smalley and G. N. Smith, *Can. J. Microbiol.*, 1965, **11**, 765.

Table 1   *Some examples of microbial primary metabolites and their commercial significance*

Primary metabolite	Producing organism	Commercial significance
Ethanol	*Saccharomyces cerevisiae*	'Active ingredient' in alcoholic beverages
Citric acid	*Aspergillus niger*	Various uses in food industry
Glutamic acid	*Corynebacterium glutamicum*	Flavour enhancer
Lysine	*Corynebacterium glutamicum*	Feed additive
Polysaccharides	*Xanthamonas* spp.	Applications in food industry; enhanced oil recovery

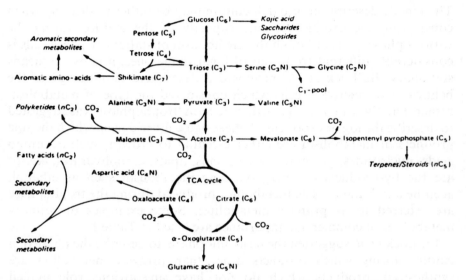

Figure 2   *The inter-relationships between primary and secondary metabolism*
(Reproduced with permission from W. B. Turner, 'Fungal Metabolites', Academic Press, 1971)

taxonomic distribution of secondary metabolism is far more limited than that of primary metabolism, the range of secondary products produced is enormous. The classification of microbial products into secondary and primary metabolites should be considered as a convenient, but in some cases, artificial system. To quote Bushell,[7] the classification should not be allowed to act as a conceptual straitjacket, forcing the reader to consider all products as either primary or secondary metabolites. It is sometimes difficult to categorize a product as primary or secondary, and the kinetics

[7] M. E. Bushell, in 'Principles of Biotechnology', ed. A. Wiseman, Chapman and Hall, New York, 1988, p. 5.

Table 2 *Some examples of microbial secondary metabolites and their commercial significance*

Secondary metabolite	Commercial significance
Penicillin	Antibiotic
Cephalosporin	Antibiotic
Streptomycin	Antibiotic
Griseofulvin	Antibiotic (anti-fungal)
Pepstatin	Treatment of ulcers
Cyclosporin A	Immunosuppressant
Gibberellin	Plant growth regulator
Lovastatin	Cholesterol synthesis inhibitor

of production of certain compounds may change, depending on the growth conditions employed.

At first sight it may seem anomalous that microorganisms produce compounds which do not appear to have any metabolic function and are certainly not by-products of catabolism as are, for example, ethanol and acetone. However, many secondary metabolites exhibit antimicrobial properties and, therefore, may be involved in competition in the natural environment;[8] others have, since their discovery in idiophase cultures, been demonstrated to be produced during the trophophase where, it has been claimed, they act in some form of metabolic control.[9] Although the physiological role of secondary metabolism continues to be the subject of considerable debate its relevance to the fermentation industry is the commercial significance of the secondary metabolites. Table 2 summarizes some of the industrially important groups of secondary metabolites.

The production of microbial metabolites may be achieved in continuous, as well as batch, systems. The chronological separation of trophophase and idiophase in batch culture may be studied in continuous culture in terms of dilution rate.[10–12] Secondary metabolism will occur at relatively low dilution rates (growth rates) and, therefore, it should be remembered that secondary metabolism is a property of slow-growing, as well as stationary, cells. The fact that secondary metabolites are produced by slow-growing organisms in continuous culture indicates

[8] A. L. Demain, *Search*, 1980, **11**, 148.
[9] I. M. Campbell, *Adv. Microb. Physiol.*, 1984, **25**, 2.
[10] S. J. Pirt, *Chem. Ind. (London)*, May 1968, 601.
[11] S. J. Pirt and R. C. Righelato, *Appl. Microbiol.*, 1967, **15**, 1284.
[12] L. H. Christensen, C. M. Henriksen, J. Nielson, J. Villadsen and M. Egel-Mitani, *J. Biotechnol.*, **42**, 95.

Table 3   *Some examples of the repression of sec-*
              ondary metabolism by medium compo-
              nents

Medium component	Repressed secondary metabolite
Glucose	Penicillin
Glucose	Actinomycin
Glucose	Neomycin
Glucose	Streptomycin
Phosphate	Candicidin
Phosphate	Streptomycin
Phosphate	Tetracycline

that primary metabolism is continuing in idiophase-type cells. Thus, secondary metabolism is not switched on to remove an accumulation of metabolites synthesized entirely in a different phase; synthesis of the primary metabolic precursors continues through the period of secondary biosynthesis.

The control of the onset of secondary metabolism has been studied extensively in batch culture and, to a lesser extent, in continuous culture. The outcome of this work is that a considerable amount of information is available on the interrelationships between the changes occurring in the medium and the cells at the onset of secondary metabolism and the control of the process. Primary metabolic precursors of secondary metabolites have been demonstrated to induce secondary metabolism, for example, tryptophan in alkaloid[13] biosynthesis and methionine in cephalosporin biosynthesis.[14] On the other hand, medium components have been demonstrated to repress secondary metabolism, the earliest observation being that of Saltero and Johnson[15] in 1953 of the repressing effect of glucose on benzyl penicillin formation. Carbon sources that support high growth rates tend to support poor secondary metabolism and Table 3 cites some examples of this situation. Phosphate sources have also been implicated in the repression of secondary metabolism, as exemplified in Table 3.

Therefore, it is essential that repressing nutrients should be avoided in media to be used for the industrial production of secondary metabolites or that the mode of operation of the fermentation maintains the potentially repressing components at sub-repressing levels, as discussed in a later section of this chapter.

[13] J. F. Robers and H. G. Floss, *J. Pharmacol. Sci.*, 1970, **59**, 702.
[14] K. Komatsu, M. Mizumo and R. Kodaira, *J. Antibiot.*, **28**, 881.
[15] F. V. Saltero and M. I. Johnson, *Appl. Microbiol.*, 1953, **1**, 2.

3.3 Microbial Enzymes

The major commercial utilization of microbial enzymes is in the food and beverage industries[16] although enzymes do have considerable application in clinical and analytical situations, as well as their use in washing powders. Most enzymes are synthesized in the logarithmic phase of batch culture and may, therefore, be considered as primary metabolites. However, some, for example the amylases of *Bacillus stearothermophilus*,[17] are produced by idiophase cultures and may be considered as equivalent to secondary metabolites. Enzymes may be produced from animals and plants as well as microbial sources but the production by microbial fermentation is the most economic and convenient method. Furthermore, it is now possible to engineer microbial cells to produce animal or plant enzymes, as discussed in Section 3.5.

3.4 Transformation Processes

As well as the use of microorganisms to produce biomass and microbial products, microbial cells may be used to catalyse the conversion of a compound into a structurally similar, but financially more valuable, compound. Such fermentations are termed transformation processes, biotransformations, or bioconversions. Although the production of vinegar is the oldest and most well-established transformation process (the conversion of ethanol into acetic acid), the majority of these processes involve the production of high-value compounds. Because microorganisms can behave as chiral catalysts with high regio- and stereospecificity, microbial processes are more specific than purely chemical ones and make possible the addition, removal, or modification of functional groups at specific sites on a complex molecule without the use of chemical protection. The reactions that may be catalysed include oxidation, dehydrogenation, hydroxylation, dehydration and condensation, decarboxylation, deamination, amination, and isomerization. The anomaly of the transformation process is that a large biomass has to be produced to catalyse, perhaps, a single reaction. The logical development of these processes is to perform the reaction using the purified enzyme or the enzyme attached to an immobile support. However, enzymes work more effectively within their microbial cells, especially if co-factors such as reduced pyridine nucleotide need to be

[16] D. J. Jeenes, D. A. MacKenzie, I. N. Roberts and D. B. Archer, in 'Biotechnology and Genetic Engineering Reviews', ed. M. P. Tombs, Intercept, Andover, 1991, Vol. 9, Chapter 9, p. 327.
[17] A. B. Manning and L. L. Campbell, *J. Biol. Chem.*, 1961, **236**, 2951.

regenerated. A compromise is to employ resting cells as catalysts, which may be suspended in a medium not supporting growth or attached to an immobile support. The reader is referred to Goodhue *et al.*[18] for a detailed review of transformation processes.

3.5 Recombinant Products

The advent of recombinant DNA technology has extended the range of potential microbial fermentation products. It is possible to introduce genes from higher organisms into microbial cells such that the recipient cells are capable of synthesizing foreign (or heterologous) proteins. Examples of the hosts for such foreign genes include *Escherichia coli, Saccharomyces cerevisiae* and other yeasts as well as filamentous fungi such as *Aspergillus niger* var. *awamori.* Products produced in such genetically manipulated organisms include interferon, insulin, human serum albumin, factor VIII and factor IX, epidermal growth factor, bovine somatostatin and bovine chymosin. Important factors in the design of these processes include the secretion of the product, minimization of the degradation of the product, and the control of the onset of synthesis during the fermentation, as well as maximizing the expression of the foreign gene. These aspects are considered in detail in references 19, 20 and 21.

4 THE FERMENTATION PROCESS

Figure 3 illustrates the component parts of a generalized fermentation process. Although the central component of the system is obviously the fermenter itself, in which the organism is grown under conditions optimum for product formation, one must not lose sight of operations upstream and downstream of the fermenter. Before the fermentation is started the medium must be formulated and sterilized, the fermenter sterilized, and a starter culture must be available in sufficient quantity and in the correct physiological state to inoculate the production fermenter. Downstream of the fermenter the product has to be purified and further processed and the effluents produced by the process have to be treated.

[18] C. T. Goodhue, J. P. Rosazza and G. P. Peruzzutti, in 'Manual of Industrial Microbiology and Biotechnology', ed. A. L. Demain and A. Solomons, American Society for Microbiology, Washington, DC, 1986, p. 97.
[19] J. R. Harris, 'Protein Production by Biotechnology', Elsevier, London, 1990.
[20] A. Wiseman, 'Genetically-engineered Proteins and Enzymes from Yeast: Production and Control', Ellis Horwood, Chichester, 1991.
[21] R. C. Hockney, *Trends in Biotechnology*, **12**, 456.

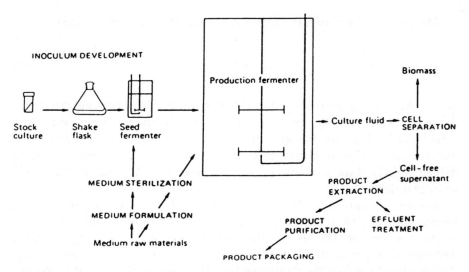

Figure 3 *A generalized, schematic representation of a fermentation process*
(Reproduced with permission from P. F. Stanbury, A. Whitaker and S. J. Hall,
'Principles of Fermentation Technology', Pergamon Press, Oxford, 1995)

4.1 The Mode of Operation of Fermentation Processes

As discussed earlier, microorganisms may be grown in batch, fed-batch,
or continuous culture, and continuous culture offers the most control
over the growth of the cells. However, the commercial adoption of
continuous culture is confined to the production of biomass and, to a
limited extent, the production of potable and industrial alcohol. The
superiority of continuous culture for biomass production is overwhelming, as may be seen from the following account, but for other microbial
products the disadvantages of the system outweigh the improved process
control which the technique offers.

Productivity in batch culture may be described by the equation[3]

$$R_{batch} = (x_{max} - x_0)/(t_i + t_{ii}) \tag{5}$$

where R_{batch} is the output of the culture in terms of biomass concentration per hour, x_{max} is the maximum cell concentration achieved at
stationary phase, x_0 is the initial cell concentration at inoculation, t_i, is
the time during which the organism grows at μ_{max} and t_{ii} is the time
during which the organism is not growing at μ_{max} and includes the lag
phase, the deceleration phase, and the periods of batching, sterilizing and
harvesting.

The productivity[3] of a continuous culture may be represented as

$$R_{\text{cont}} = D\bar{x}(1 - t_{\text{iii}}/T) \qquad (6)$$

where R_{cont} is the output of the culture in terms of cell concentration per hour, t_{iii} is the time period prior to the establishment of a steady-state and includes time for vessel preparation, sterilization and operation in batch culture prior to continuous operation, T is the time period during which steady-state conditions prevail, and $\bar{x}$ is the steady-state cell concentration.

Maximum output of biomass per unit time (*i.e.* productivity) in a chemostat may be achieved by operating at the dilution rate giving the highest value of $D\bar{x}$, this value being referred to as D_{max}. Batch fermentation productivity, as described by equation (5), is an average for the total time of the fermentation. Because $dx/dt = \mu x$, the productivity of the culture increases with time and, thus, the vast majority of the biomass in a batch process is produced near the end of the log phase. In a steady-state chemostat, operating at, or near, D_{max} the productivity remains constant, and maximum, for the whole fermentation. Also, a continuous process may be operated for a very long time so that the non-productive period, t_{iii} in equation (6), may be insignificant. However, the non-productive time element for a batch culture is a very significant period, especially as the fermentation would have to be re-established many times during the running time of a comparable continuous process and, therefore, t_{ii} would be recurrent.

The steady-state nature of a continuous process is also advantageous in that the system should be far easier to control than a comparable batch one. During a batch fermentation, heat output, acid or alkali production, and oxygen consumption will range from very low rates at the start of the fermentation to very high rates during the late logarithmic phase. Thus, the control of the environment of such a system is far more difficult than that of a continuous process where, at steady-state, production and consumption rates are constant. Furthermore, a continuous process should result in a more constant labour demand than a comparable batch one.

A frequently quoted disadvantage of continuous systems is their susceptibility to contamination by foreign organisms. The prevention of contamination is essentially a problem of fermenter design, construction, and operation and should be overcome by good engineering and microbiological practice. ICI recognized the overwhelming advantages of a continuous biomass process and overcame the problems of contamination by building a secure fermenter capable of very long periods of aseptic operation, as described by Smith.[22]

The production of growth-associated by-products, such as ethanol, should also be more efficient in continuous culture. However, continuous brewing has met with only limited success and UK breweries have abandoned such systems owing to problems of flavour and lack of flexibility.[23] The production of industrial alcohol, on the other hand, should not be limited by the problems encountered by the brewing industry and continuous culture should be the method of choice for such a process. The adoption of continuous culture for the production of biosynthetic (as opposed to catabolic) microbial products has been extremely limited. Although, theoretically, it is possible to optimize a continuous system such that optimum productivity of a metabolite should be achieved, the long-term stability of such systems is precarious, owing to the problem of strain degeneration. A consideration of the kinetics of continuous culture reveals that the system is highly selective and will favour the propagation of the best-adapted organism in a culture. Best-adapted in this context refers to the affinity of the organism for the limiting substrate at the operating dilution rate. A commercial organism is usually highly mutated such that it will produce very high amounts of the desired product. Therefore, in physiological terms, such commercial organisms are extremely inefficient and a revertant strain, producing less of the desired product, may be better adapted to the cultural conditions than the superior producer and will come to dominate the culture. This phenomenon, termed by Calcott[24] as contamination from within, is the major reason for the lack of use of continuous culture for the production of microbial metabolites.

Although the fermentation industry has been reluctant to adopt continuous culture for the production of microbial metabolites, very considerable progress has been made in the development of fed-batch systems.[25,26] Fed-batch culture may be used to achieve a considerable degree of process control and to extend the productive period of a traditional batch process without the inherent disadvantages of continuous culture described previously. The major advantage of feeding a medium component to a culture, rather than incorporating it entirely in the initial batch, is that the nutrient may be maintained at a very low

[22] R. L. Smith, *Philos. Trans. R. Soc. London., Ser. B*, 1980, **290**, 341.
[23] B. H. Kirsop, in 'Topics in Enzyme and Fermentation Biotechnology', ed. A. Wiseman, Ellis Horwood, Chichester, 1982, p. 79.
[24] P. H. Calcott, 'Continuous Culture of Cells', CRC Press, Boca Raton, Fl., 1981, Vol. 1, p. 13.
[25] A. Whitaker, *Process Biochem.*, 1980, **15**(4), 10.
[26] T. Yamane and S. Shimizu, *Adv. Biochem. Eng./Biotechnol.*, 1984, **30**, 147.

concentration during the fermentation. A low (but constantly replenished) nutrient level may be advantageous in:

   (i) Maintaining conditions in the culture within the aeration capacity of the fermenter.
   (ii) Removing the repressive effects of medium components such as rapidly used carbon and nitrogen sources and phosphate.
   (iii) Avoiding the toxic effects of a medium component.
   (iv) Providing a limiting level of a required nutrient for an auxotrophic strain.

The earliest example of the commercial use of fed-batch culture is the production of bakers' yeast. It was recognized as early as 1915 that an excess of malt in the production medium would result in a high rate of biomass production and an oxygen demand which could not be met by the fermenter.[27] This resulted in the development of anaerobic conditions and the formation of ethanol at the expense of biomass. The solution to this problem was to grow the yeast initially in a weak medium and then add additional medium at a rate less than the organism could use it. It is now appreciated that a high glucose concentration represses respiratory activity, and in modern yeast production plants the feed of molasses is under strict control based on the automatic measurement of traces of ethanol in the exhaust gas of the fermenter. As soon as ethanol is detected the feed rate is reduced. Although such systems may result in low growth rates, the biomass yield is near that theoretically obtainable.[28]

The penicillin fermentation provides a very good example of the use of fed-batch culture for the production of a secondary metabolite.[29] The penicillin process is a two-stage fermentation; an initial growth phase is followed by the production phase or idiophase. During the production phase glucose is fed to the fermentation at a rate which allows a relatively high growth rate (and therefore rapid accumulation of biomass) yet maintains the oxygen demand of the culture within the aeration capacity of the equipment. If the oxygen demand of the biomass were to exceed the aeration capacity of the fermenter anaerobic conditions would result and the carbon source would be used inefficiently. During the production phase the biomass must be maintained at a relatively low growth rate

[27] G. Reed and T. W. Nagodawithana, 'Yeast Technology', 2nd Edn, Avi, Westport, 1991.
[28] A. Fiechter, in 'Advances in Biotechnology, 1, Scientific and Engineering Principles', ed. M. Moo-Young, C. W. Robinson and C. Vezina, Pergamon Press, Toronto, 1981, p. 261.
[29] J. M. Hersbach, C. P. Van der Beek and P. W. M. Van Vijek, in 'Biotechnology of Industrial Antibiotics', ed. E. J. Vandamme, Marcel Dekker, New York, 1984, p. 387.

and, thus, the glucose is fed at a low dilution rate. Phenylacetic acid is a precursor of the penicillin molecule but it is also toxic to the producer organism above a threshold concentration. Thus, the precursor is also fed into the fermentation continuously, thereby maintaining its concentration below the inhibitory level.

5 THE GENETIC IMPROVEMENT OF PRODUCT FORMATION

Owing to their inherent control systems, microorganisms usually produce commercially important metabolites in very low concentrations and, although the yield may be increased by optimizing the cultural conditions, productivity is controlled ultimately by the organism's genome. Thus, to improve the potential productivity, the organism's genome must be modified and this may be achieved in two ways: by (i) classical strain improvement by mutation and selection and (ii) the use of recombination.

5.1 Mutation

Each time a microbial cell divides there is a small probability of an inheritable change occurring. A strain exhibiting such a changed characteristic is termed a mutant and the process giving rise to it, a mutation. The probability of a mutation occurring may be increased by exposing the culture to a mutagenic agent such as UV light, ionizing radiation, and various chemicals, for example nitrosoguanidine, nitrous acid and caffeine. Such an exposure usually involves subjecting the population to a mutagen dose which results in the death of the vast majority of the cells. The survivors of the mutagen exposure may then contain some mutants, the vast majority of which will produce lower levels of the desired product. However, a very small proportion of the survivors may be improved producers. Thus, it is the task of the industrial geneticist to separate the desirable mutants (the superior producers) from the very many inferior types. This approach is easier for strains producing primary metabolites than it is for those producing secondary metabolites, as may be seen from the following examples.

The synthesis of a primary microbial metabolite (such as an amino acid) is controlled such that it is only produced at a level required by the organism. The control mechanisms involved are the inhibition of enzyme activity and the repression of enzyme synthesis by the end product when it is present in the cell at a sufficient concentration. Thus, these mechanisms are referred to as feedback control. It is obvious that a

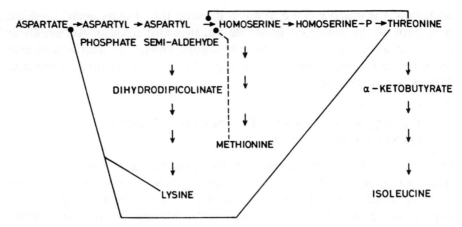

Figure 4   *The control of biosynthesis of lysine in* Corynebacterium glutamicum: *Biosynthetic route* →; *Feedback inhibition* —●; *Feedback repression* – – –●

good 'commercial' mutant should lack the control systems so that 'overproduction' of the end product will result. The isolation of mutants of *Corynebacterium glutamicum* capable of producing lysine will be used to illustrate the approaches which have been adopted to remove the control systems.

The control of lysine synthesis in *C. glutamicum* is illustrated in Figure 4 from which it may be seen that the first enzyme in the pathway, aspartokinase, is inhibited only when both lysine and threonine are synthesized above a threshold level. This type of control is referred to as concerted feedback control. A mutant which could not catalyse the conversion of aspartic semialdehyde into homoserine would be capable of growth only in a homoserine-supplemented medium and the organism would be described as a homoserine auxotroph. If such an organism were grown in the presence of very low concentrations of homoserine the endogenous level of threonine would not reach the inhibitory level for aspartokinase control and, thus, aspartate would be converted into lysine which would accumulate in the medium. Thus, a knowledge of the control of the biosynthetic pathway allows a 'blueprint' of the desirable mutant to be constructed and makes easier the task of designing the procedure to isolate the desired type from the other survivors of a mutation treatment.

The isolation of bacterial auxotrophs may be achieved using the penicillin enrichment technique developed by Davis.[30] Under normal culture conditions an auxotroph is at a disadvantage compared with the

[30] B. D. Davis, *Proc. Natl. Acad. Sci. USA*, 1949, **35**, 1.

parental (wild-type) cells. However, penicillin only kills growing cells and, therefore, if the survivors of a mutation treatment were cultured in a medium containing penicillin and lacking the growth requirement of the desired mutant only those cells unable to grow would survive, *i.e.* the desired auxotrophs. If the cells were removed from the penicillin broth, washed, and resuspended in a medium containing the requirement of the desired auxotroph then the resulting culture should be rich in the required type. Nakayama *et al.*[31] used this technique to isolate a homoserine auxotroph of *C. glutamicum* which produced 44 g l^{-1} lysine.

An alternative approach to the isolation of mutants which do not produce controlling end products (*i.e.* auxotrophs) is to isolate mutants which do not recognize the presence of controlling compounds. Such mutants may be isolated from the survivors of a mutation treatment by exploiting their capacity to grow in the presence of certain compounds which are inhibitory to the parental types. An analogue is a compound which is similar in structure to another compound and analogues of primary metabolites are frequently inhibitory to microbial cells. The toxicity of the analogue may be due to any of a number of possible mechanisms; for example, the analogue may be incorporated into a macromolecule in place of the natural product, resulting in the production of a defective compound, or the analogue may act as a competitive inhibitor of an enzyme for which the natural product is a substrate. Also, the analogue may mimic the control characteristics of the natural product and inhibit product formation despite the fact that the natural product concentration is inadequate to support growth. A mutant which is capable of growing in the presence of an analogue inhibitory to the parent may owe its resistance to any of a number of mechanisms. However, if the toxicity were due to the analogue mimicking the control characteristics of the normal end product, then the resistance may be due to the control system being unable to recognize the analogue as a control factor. Such analogue-resistant mutants may also not recognize the natural product and may, therefore, overproduce it. Thus, there is a reasonable probability that mutants resistant to the inhibitory effects of an analogue may overproduce the compound to which the analogue is analogous. Sano and Shiio[32] made use of this approach in attempting to isolate lysine-producing mutants of *Brevibacterium flavum*. The control of lysine formation in *B. flavum* is the same as that illustrated in Figure 4 for *C. glutamicum*. Sano and Shiio demonstrated that the lysine analogue

[31] K. Nakayama, S. Kituda and S. Kinoshita, *J. Gen. Appl. Microbiol.*, 1961, **7**, 41
[32] K. Sano and I. Shiio, *J. Gen. Appl. Microbiol.*, 1970, **16**, 373.

S-(2-aminoethyl)cysteine (AEC) only inhibited growth completely in the presence of threonine, which suggests that AEC combined with threonine in the concerted inhibition of aspartokinase and deprived the organism of lysine and methionine. Mutants were isolated by plating the survivors of a mutation treatment onto agar plates containing both AEC and threonine. A relatively high proportion of the resulting colonies were lysine overproducers, the best of which produce more than 30 g l^{-1}. A fuller account of the isolation of amino acid and nucleotide producing strains may be found in reference 3.

Thus, a knowledge of the control systems may assist in the design of procedures for the isolation of mutants overproducing primary metabolites. The design of procedures for the isolation of mutants overproducing secondary metabolites has been more difficult owing to the fact that far less information was available on the control of production and, also, that the end products of secondary metabolism are not required for growth. Thus, many current industrial strains have been selected using direct, empirical, screens of the survivors of a mutation treatment for productivity rather than cultural systems which give an advantage to potential superior producers. A programme typical of the industry in the 1950s to 1970s is illustrated in Figure 5. The throughput of such programmes has been increased by miniaturizing[3] the systems. Small volumes of liquid media in tubes or microtitre plates have been used coupled with the use of robots to automate the process. However, as more information has accumulated on the biosynthesis and control of secondary metabolites, directed selection approaches have also been used, thus reducing the empirical nature of the screens. Mutants capable of producing increased levels of secondary metabolite precursors have been isolated by techniques similar to those used for the improvement of primary metabolite producers. For example, analogue-resistant mutants have been isolated giving improved yields of pyrrolnitrin, candicidin and cephamycin.[3] Relief of carbon repression has been achieved on of mutants resistant to 2-deoxyglucose, a glucose analogue.[33] Further examples of selection methods for the isolation of improved secondary metabolite-producing strains are given in reference 3.

5.2 Recombination

Hopwood[34] defined recombination as any process which helps to generate new combinations of genes that were originally present in

[33] D. A. Hodgson, *J. Gen. Microbiol.*, 1982, **128**, 2417.
[34] D. A. Hopwood in 'Genetics of Industrial Micro-organisms', ed. O. K. Sebec and A. J. Laskin, American Society of Microbiology, Washington, DC, 1979, p. 1.

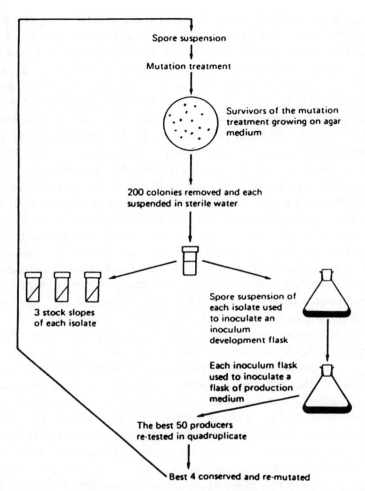

Figure 5 *A strain improvement programme for a secondary metabolite producing culture*
(Reproduced with permission from P. F. Stanbury, A. Whitaker and S. J. Hall,
'Principles of Fermentation Technology', 2nd Edn, Pergamon Press, Oxford,
1995)

different individuals. Compared with the use of mutation techniques for
the improvement of industrial strains the use of recombination was fairly
limited in the early years of improvement programmes. However,
techniques are now widely available which allow the use of recombination as a system of strain improvement. *In vivo* recombination may be
achieved in the asexual fungi (*e.g. Penicillium chrysogenum*, used for the
commercial production of penicillin) using the parasexual cycle.[35] The

[35] K. D. Macdonald and G. Holt, *Sci. Prog.*, 1976, **63**, 547.

technique of protoplast fusion has increased greatly the prospects of combining together characteristics found in different production strains. Protoplasts are cells devoid of their cell walls and may be prepared by subjecting cells to the action of wall-degrading enzymes in isotonic solutions. Cell fusion, followed by nuclear fusion, may occur between protoplasts of strains which would not otherwise fuse and the resulting fused protoplast may regenerate a cell wall and grow as a normal cell. Protoplast fusion has been achieved using filamentous fungi, yeasts, streptomycetes and bacteria. For example, Tosaka *et al.*[36] improved the rate of glucose consumption (and therefore lysine production) of a high lysine producing strain of *B. flavum* by fusing it with another *B. flavum* strain which was a non-lysine producer but consumed glucose at a high rate. Among the fusants one strain exhibited high lysine production with rapid glucose utilization. Chang *et al.*[37] used protoplast fusion to combine the desirable properties of two strains of *P. chrysogenum* producing penicillin V into one producer strain. Lein[38] has described the use of protoplast fusion for the improvement of penicillin production in the procedures used by Panlabs Inc. and DeWitt *et al.*[39] reviewed the technique for the improvement of actinomycete processes.

In vitro recombination has been achieved by the techniques of *in vitro* recombinant DNA technology discussed elsewhere in this book. Although the most well publicised recombinants achieved by these techniques are those organisms which synthesize foreign products (see Section 3.5), very considerable achievements have been made in the improvement of strains producing conventional products. The efficiency of *Methylomonas methylotrophus*, the organism used in the ICI Pruteen process, was improved by the incorporation of a plasmid containing a glutamate dehydrogenase gene from *E. coli*.[40] The manipulated organism was capable of more efficient ammonia metabolism which resulted in a 5% improvement in carbon conversion. However, the strain was not used on the large-scale plant due to problems of scale-up.

In vitro DNA technology has been used to amplify the number of copies in a critical pathway gene (or operon) in a process organism. Although gene amplification is not an example of recombination it is best

[36] O. Tosaka, M. Karasawa, S. Ikeda and Y. Yoshii, Abstracts of 4th International Symposium on Genetics of Industrial Microorganisms, 1982, p. 61.
[37] L. T. Chang, D. T. Terasaka and R. P. Elander, *Dev. Ind. Microbiol.*, 1982, **23**, 21.
[38] J. Lein, in 'Overproduction of Microbial Metabolites', ed. Z. Vanek and Z. Hostalek, Butterworths, Boston, 1986, p. 105.
[39] J. P. DeWitt, J. V. Jackson and T. J. Paulus, in 'Fermentation Process Development of Industrial Organisms', ed. J. O. Neway, Marcel Dekker, New York, 1989, Chapter 1, p. 1.
[40] J. D. Windon, M. J. Worsey, E. M. Pioli, D. Pioli, P. T. Barth, K. T. Atherton, E. C. Dart, D. Byrom, K. Powell and P. J. Senior, *Nature (London)*, 1980, **287**, 396.

considered in the context of DNA manipulative techniques. Threonine production by *E. coli* has been improved by incorporating the entire threonine operon of a threonine analogue-resistant mutant into a plasmid which was then introduced back into the bacterium. The plasmid copy number in the cell was approximately 20 and the activity of the threonine operon enzymes was increased 40 to 50 times. The organism produced 30 g l^{-1} threonine compared with the 2–3 g l^{-1} of the non-manipulated strain.[41] Miwa *et al.*[42] utilized similar techniques in constructing an *E. coli* strain capable of synthesizing 65 g l^{-1} threonine.

The application of the techniques of genetic manipulation to the improvement of *C. glutamicum* (see Section 5.1) was hindered by the availability of a suitable vector. However, vectors have been constructed and considerable progress has been made in the improvement of amino acid fermentations.[3] Threonine, histidine and phenylalanine production have been improved using gene amplification techniques. In these examples the cloned genes were mutant forms which were resistant to feedback control and had been obtained using the conventional mutagenesis/screening systems described in Section 5.1. Thus, the *in vitro* DNA techniques have built upon the achievements of conventional strain improvement.

Phenylalanine has become a very important fermentation product because it is a precursor in the manufacture of the sweetener, aspartame. Backman *et al.*[43] described the rationale used in the construction of an *E. coli* strain capable of synthesizing commercial levels of phenylalanine. *Escherichia coli* was chosen as the producer because of its rapid growth, the availability of DNA manipulative techniques, and the extensive genetic database. Several of the phenylalanine genes are subject to control by the repressor protein of the *tyr* R gene. *In vitro* techniques were used to generate *tyr* R mutations and introduce them into the production strain. The promoter of the *phe* A gene, was replaced to remove repression and attenuation control. As an alternative to the traditional technique of generating a tyrosine auxotroph an excision vector carrying the *tyr* A gene was incorporated into the chromosome. The vector is excised from the chromosome at a slightly increased temperature. Thus, auxotrophy may be induced *during* the fermentation by careful temperature manipulation, thus allowing tyrosine limitation

[41] V. G. Debabof, in 'Overproduction of Microbial Products', ed. V. Krumphanzl, B. Sikyta and Z. Vanek, Academic Press, London, 1982, p. 345.
[42] K. Miwa, S. Nakamori and H. Momose, Abstracts of 13th International Congress of Microbiology, Boston, USA, 1982, p. 96.
[43] K. Backman, M. J. O'Connor, A. Maruya, E. Rudd, D. McKay, R. Balakrishnan, M. Radjai, V. Di Pasquantonio, D. Shoda, R. Hatch and K. Venkatasubramanian, *Ann. N. Y. Acad. Sci.*, 1990, **589**, 16.

to be imposed after the growth phase and at the beginning of the production phase. However, the final step in the genetic manipulation of the organism was the traditional step of isolating an analogue-resistant mutant to relieve the feedback inhibition of DAHP synthase by phenylalanine.

The application of *in vitro* recombinant DNA technology to the improvement of secondary metabolite formation is not as developed as it is in the primary metabolite field. However, considerable advances have been made in the genetic manipulation of the streptomycetes[44] and the filamentous fungi[45] and a number of different strategies have been devised for cloning secondary metabolite genes.[46] The first such genes which were cloned were those coding for resistance of the producer organism to its own antibiotic.[47] Complete antibiotic synthesizing pathways have now been cloned and the genes for antibiotic biosynthesis have been shown to be clustered together on the chromosome in both prokaryotes and eukaryotes.[48] Baltz[49] cited the application of recombinant DNA technology to the development of improved strains for the production of tylosin, pristinamycin and daptomycin and the discovery of several global and pathway specific secondary metabolism regulatory genes has opened the way to a new means of increasing yields.

6 CONCLUSIONS

Thus, microorganisms are capable of producing a wide range of products, a range which has been increased by the techniques of *in vitro* recombinant DNA technology to include mammalian products. Improved productivity may be achieved by the optimization of cultural conditions and the genetic modification of the producer cells. However, a successful commercial process for the production of a microbial metabolite depends as much upon chemical engineering expertise as it does on that of microbiology and genetics.

[44] K. F. Chater, *Microbiology*, 1998, **144**, 727.
[45] C. A. M. J. J. van den Hondel and P. J. Punt, in 'Applied Molecular Genetics of Fungi', ed. J. F. Peberdy, C. E. Caten, J. E. Ogden and J. W. Bennett, Cambridge University Press, Cambridge, 1991, p. 1.
[46] I. S. Hunter and S. Baumberg, in 'Microbial Products: New Approaches', ed. S. Baumberg, I. S. Hunter and P. M. Rhodes. Society for General Microbiology Symposium, Cambridge University Press, Cambridge, 1989.
[47] I. S. Hunter in 'Fermentation Microbiology and Biotechnology' ed. E. M. T. El-Mansi and C. F. A. Bryce, Taylor and Francis, London, 1999, p. 121.
[48] J. Thompson, T. Kieser, J. M. Ward and D. A. Hopwood, *Gene*, 1982, **20**, 51.
[49] R. H. Baltz, *Trends in Microbiology*, 1998, **6**(2), 76.

CHAPTER 2

Molecular Analysis and Amplification Techniques

RALPH RAPLEY

1 INTRODUCTION

Numerous developments in molecular biology and biotechnology have provided the tools to effectively analyse and manipulate nucleic acids. The development of these technologies has, however, relied on a number of crucial discoveries. Many of these have been in the isolation and characterization of numerous DNA manipulating enzymes such as DNA polymerase, DNA ligase, reverse transcriptase, *etc*. However, perhaps the most important was the isolation and application of a number of enzymes that enabled the reproducible digestion of DNA. These enzymes, termed restriction endonucleases, provided not only a turning point for the analysis of DNA, but also the potential to develop recombinant DNA technology (see Chapter 3).

1.1 Enzymes Used in Molecular Biology

Restriction endonucleases recognize certain DNA sequences, usually 4–6 base-pairs (bp) in length, and cleave them in a defined pattern. The nucleotide sequences recognized are palindromic or of an inverted repeat nature, reading the same in both directions on each strand.[1] When cleaved they leave a flush (blunt-ended) or staggered (cohesive-ended) fragment, depending on the particular enzyme, as indicated in Figure 1. An important property of staggered ends is that DNA from different sources digested by the same restriction endonuclease will be comple-

[1] H. O. Smith and K. W. Wilcox, *J. Mol. Biol.*, 1970, **51**, 379.

HindII Resriction Enzyme

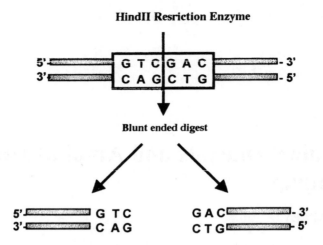

Blunt ended digest

Blunt ended restriction fragments produced

EcoRI Restriction enzyme

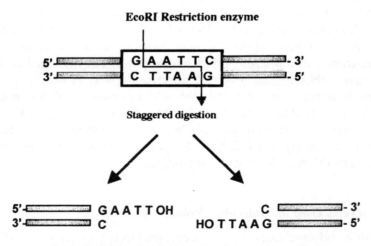

Staggered digestion

Sticky/cohesive ended restriction fragments produced

Figure 1   *The types of DNA fragments resulting from digestion with restriction endo-*
nuclease enzymes that leave blunt ends (upper) or sticky/cohesive ends (lower)

mentary (or 'sticky') and so will anneal to each other. The annealed
strands are held together only by hydrogen bonding between complementary bases on opposite strands. Covalent joining of ends on each of
the two strands may be brought about by the enzyme DNA ligase. This is
widely exploited in molecular biology to enable the construction of

Table 1   *A selection of four, six and eight nucleotide recognition sequence targets for specific restriction endonucleases. Note in the first panel the ability of some enzymes to provide blunt ends whilst others may provide cohesive or sticky ends*

Name	Recognition sequence	Digestion products		

Four nucleotide recognition sequence

	↓			
*Hae*III	5'-GGCC-3'	5'-GG	CC-3'	Blunt-end digestion
	3'-CCGG-5'	3'-CC	GG-5'	
*Hpa*II	5'-CCGG-3'	5'-C	CGG-3'	Cohesive-end digestion
	3'-GGCC-5'	3'-GGC	C-5'	

Six nucleotide recognition sequence

	↓			
*Bam*HI	5'-GGATTC-3'	5'-G	GATCC-3'	
	3'-GGCC-5'	3'-CCTAG	G-5'	
*Eco*RI	5'-GAATTC-3'	5'-G	AATCC-3'	
	3'-CTTAAG-5'	3'-CTTAA	G-5'	
*Hind*III	5'-AAGCTT-3'	5'-A	AGCTT-3'	
	3'-TTCGAA-5'	3'-TTCGA	A-5'	

Eight nucleotide recognition sequence

	↓			
*Not*I	5'-GCGGCCGC-3'	5'-GC	GGCCGC-3'	
	3'-CGCCGGCG-5'	3'-CGCCGG	CG-5'	

recombinant DNA, *i.e.* the joining of DNA fragments from different sources. Approximately 500 restriction enzymes have been characterized and these recognize over 100 different target sequences (see Table 1). A number of these, termed isoschizomers, recognize different target sequences but produce the same staggered ends or overhangs. A number of other enzymes have proved to be of value in the manipulation of DNA and are indicated at appropriate points within the text.

2   EXTRACTION AND SEPARATION OF NUCLEIC ACIDS

2.1   DNA Extraction Techniques

The use of DNA for analysis or manipulation usually requires that it is isolated and purified to a certain extent. DNA is recovered from cells by the gentlest possible method of cell rupture to prevent the DNA from fragmenting by mechanical shearing. This is usually in the presence of EDTA which chelates the Mg^{2+} ions needed by the enzymes that degrade DNA, termed DNase. Ideally, cell walls, if present, should be digested enzymatically, *e.g.* the lysozyme treatment of bacteria, and the cell membrane should be solubilized using detergent. If physical disruption is necessary, it should be kept to a minimum, and should involve cutting or squashing of cells, rather than the use of shear forces. Cell disruption (and most subsequent steps) should be performed at 4°C, using glassware and solutions which have been autoclaved to destroy DNase activity.[2]

After release of nucleic acids from the cells, RNA can be removed by treatment with ribonuclease (RNase) which has been heat treated to inactivate any DNase contaminants; RNase is relatively stable to heat as a result of its disulfide bonds, which ensure rapid renaturation of the molecule on cooling. The other major contaminant, protein, is removed by shaking the solution gently with water-saturated phenol, or with a phenol/chloroform mixture, either of which will denature proteins but not nucleic acids. Centrifugation of the emulsion formed by this mixing produces a lower organic phase, separated from the upper aqueous phase by an interface of denatured protein. The aqueous solution is recovered and deproteinized repeatedly until no more material is seen at the interface. Finally, the deproteinized DNA preparation is mixed with two volumes of absolute ethanol, and the DNA allowed to precipitate out of solution in a freezer. After centrifugation, the DNA pellet is redissolved in a buffer containing EDTA to inactivate any DNases present. This solution can be stored at 4°C for at least a month. DNA solutions can be stored as frozen, although repeated freezing and thawing tends to damage long DNA molecules by shearing. A flow diagram summarizing the extraction of DNA is shown in Figure 2. The procedure described above is suitable for obtaining total cellular DNA. If the DNA from a specific organelle or viral particle is needed, it is best to isolate the organelle or virus before extracting its DNA, since the recovery of a particular type of DNA from a mixture is usually rather difficult. Where a high degree of purity is required, DNA may be subjected to density gradient ultracentrifugation through caesium chloride which is particularly useful for the preparation plasmid DNA. It is possible to check the integrity of the DNA by agarose gel electrophoresis and determine the concentration of the DNA by using the fact that 1 absorbance unit (A) equates to 50 μg ml^{-1} of DNA. Hence

$$50 \times A_{260} = \text{concentration of DNA sample } (\mu\text{g ml}^{-1})$$

[2] J. Sambrook, E. M. Fritsch, and T. Maniatis, 'Molecular Cloning: A Laboratory Manual', 2nd Edn, Cold Spring Harbor, NY, 1989.

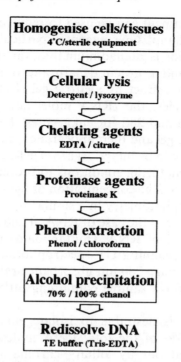

Figure 2 *Flow diagram indicating the general steps involved in DNA extraction*

The identification of contaminants may also be undertaken by scanning UV-spectrophotometry from 200 nm to 300 nm. A ratio of absorbance at 260 nm:absorbance at 280 nm of approximately 1.8 indicates that the sample is free of protein contamination, which absorbs strongly at 280 nm.

2.2 RNA Extraction Techniques

The methods used for RNA isolation are very similar to those described above for DNA; however, RNA molecules are relatively short, and therefore less easily damaged by shearing, so cell disruption can be rather more vigorous.[3] RNA, however, is very vulnerable to digestion by RNases which are present endogenously in various concentrations in certain cell types and exogenously on fingers. Gloves should be worn and a strong detergent should be included in the isolation medium to immediately denature any RNases. Subsequent deproteinization should be particularly rigorous, since RNA is often tightly associated with

[3] P. Jones, J. Qiu and D. Rickwood, 'RNA Isolation and Analysis', Bios Scientific Publishers, Oxford, 1994.

proteins. DNase treatment can be used to remove DNA, and RNA can be precipitated by ethanol. One reagent in particular which is commonly used in RNA extraction is guanadinium thiocyanate; this is both a strong inhibitor of RNase and a protein denaturant. It is possible to check the integrity of an RNA extract by agarose gel electrophoresis. The most abundant RNA species, the rRNA molecules, are 23S and 16S for prokaryotes and 18S and 28S for eukaryotes. These appear as discrete bands on the agarose gel and their appearance indicates that other RNA components are likely to be intact. Agarose gel electrophoresis is usually carried out under denaturing conditions to prevent secondary structure formation in the RNA. The concentration of the RNA may be estimated by using UV-spectrophotometry in a similar way to that used for DNA (Section 2.1). However in the case of RNA at 260 nm, 1 absorbance unit equates to 40 μg ml^{-1} of RNA. Contaminants may also be identified in the same way by scanning UV-spectrophotometry, although, in the case of RNA, a 260 nm : 280 nm absorbance ratio of approximately 2 would be expected for a sample containing little or no contaminating protein.

In many cases it is desirable to isolate eukaryotic mRNA which constitutes only 2–5% of the total RNA molecules in cellular RNA. This may be carried out by affinity chromatography on oligo(dT)-cellulose columns. At high salt concentrations, the mRNA containing poly(A) tails binds to the complementary oligo(dT) molecules of the affinity column, and so mRNA will be retained; all other RNA molecules can be washed through the column by further high salt solution. Finally, the bound mRNA can be eluted using a low concentration of salt. Nucleic acid species may also be subfractionated by more physical means such as electrophoretic or chromatographic separations based on differences in nucleic acid fragment sizes or physicochemical characteristics.[4]

3 ELECTROPHORESIS OF NUCLEIC ACIDS

In order to analyse DNA by size, electrophoresis in agarose or polyacrylamide gels is usually undertaken. Electrophoresis may be used analytically or preparatively, and can be qualitative or quantitative.[5] Large fragments of DNA such as chromosomes may also be separated by a modification of electrophoresis termed pulsed field gel electrophoresis (PFGE).[6] The easiest and most widely applicable method is electro-

[4] R. Rapley and D. L. Manning, 'RNA Isolation and Characterisation Protocols', Humana Press, Totowa, NY, 1998.
[5] P. Jones, 'Gel Electrophoresis of Nucleic Acids', Wiley, Chichester, UK, 1995.
[6] D. C. Schwartz and C. R. Cantor, *Cell*, 1984, **37**, 67.

phoresis in horizontal agarose gels, as indicated in Figure 3. This is followed by staining of the DNA with the dye ethidium bromide. This dye binds to DNA by in insertion between stacked base pairs, termed intercalation, and exhibits strong orange/red fluorescence when illuminated with ultraviolet light. Alternative stains to ethidium bromide, such as CyberGreen or Gelstar, are also available. These have similar sensitivities but are less hazardous to use.

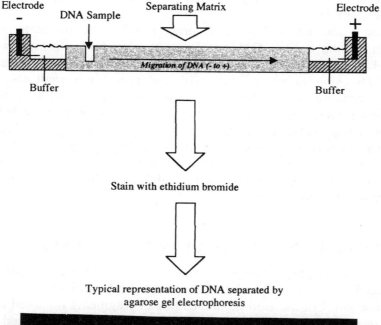

Stain with ethidium bromide

Typical representation of DNA separated by agarose gel electrophoresis

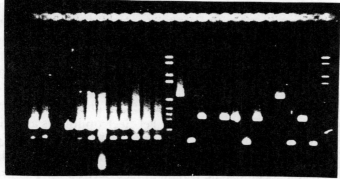

Figure 3   *Apparatus used in a typical gel electrophoresis set up. Following the completion of electrophoresis the gel is removed and stained with an intercalating dye (ethidium bromide); application of UV light to the gel reveals the point to which the DNA has migrated (indicated in this Polaroid photograph)*

In general electrophoresis is used to check the purity and intactness of a DNA preparation or to assess the extent of a enzymatic reaction during for example the steps involved in the cloning of DNA. For such checks 'mini-gels' are particularly convenient as they need little preparation, use small samples and provide results quickly. Agarose gels can be used to separate molecules larger than about 100 bp. For higher resolution or for the effective separation of shorter DNA molecules polyacrylamide gels are the preferred method. In recent years a number of acrylic gels have been developed which may be used as an alternative to agarose and polyacrylamide.

When electrophoresis is used preparatively, the fragment of gel containing the desired DNA molecule is physically removed with a scalpel. The DNA is then recovered from the gel fragment in various ways. This may include crushing with a glass rod in a small volume of buffer, using agarase to digest the agrose leaving the DNA, or by the process of electroelution. In this method the piece of gel is sealed in a length of dialysis tubing containing buffer, and is then placed between two electrodes in a tank containing more buffer. Passage of an electrical current between the electrodes causes DNA to migrate out of the gel piece, but it remains trapped within the dialysis tubing and can therefore be recovered easily.

4   RESTRICTION MAPPING OF DNA FRAGMENTS

Restriction mapping involves the size analysis of restriction fragments produced by several restriction enzymes both individually and in combination. Comparison of the lengths of fragments obtained allows their relative positions within the DNA fragment to be deduced. Any mutation which creates, destroys or moves the recognition sequence for a restriction enzyme leads to a restriction fragment polymorphism (RFLP).[7] An RFLP can be detected by examining the profile of restriction fragments generated during digestion. Conventionally, this required the purification of the original starting DNA sample before digestion with single or multiple restriction enzymes. The resultant fragments were then size-separated by gel electrophoresis and visualized by staining with ethidium bromide. Routine RFLP analysis of genomic DNA samples generally also involves hybridization with labelled gene probes to detect a specific gene fragment (see Section 5). The first useful RFLP was described for the detection of sickle cell anaemia. In this case

---

[7] E. K. Green, in 'Molecular Biomethods Handbook', ed. R. Rapley and J. M. Walker, Humana Press, Totowa, NY, 1998.

a difference in the pattern of digestion with the restriction endonuclease HhaI could be identified between DNA samples from normal individuals and patients with the disease. This polymorphism was later shown to be the result of a single base substitution in the gene for β-globin which changed a codon, GAG, specific for the amino acid glutamine, to GTG, which encodes valine.

RFLPs may arise by a number of different means which alter the relative position of restriction endonuclease recognition sequences (Figure 4). In general most polymorphisms are randomly distributed throughout a genome, however there are certain regions where a particularly high concentration of polymophisms exist. These are termed hypervariable regions and have been found in regions flanking structural genes from several sources. These were first identified as differences in the numbers of short repeated sequences, termed mini-satellites. These occurred within the genomes from different individuals, as evidenced by RFLP analysis using specific gene probes. Several types

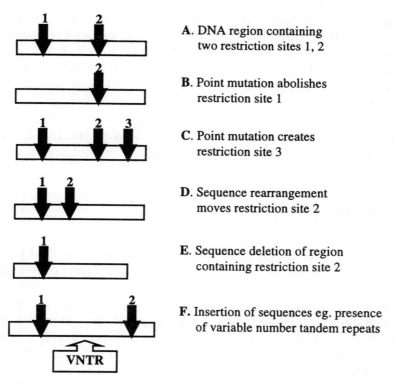

A. DNA region containing two restriction sites 1, 2

B. Point mutation abolishes restriction site 1

C. Point mutation creates restriction site 3

D. Sequence rearrangement moves restriction site 2

E. Sequence deletion of region containing restriction site 2

F. Insertion of sequences eg. presence of variable number tandem repeats

Figure 4 *The influence of alterations in DNA to the generation of patterns in restriction fragment length polymorphisms*

of mini-satellite sequences such as variable numbers of tandem repeats (VNTR) have now been described. Their detection has consequently allowed the development of the technique of genetic fingerprinting (see Chapter 11).

5 NUCLEIC ACID BLOTTING AND HYBRIDIZATION

Electrophoresis of DNA restriction fragments allows separation based on size to be carried out, however it provides no indication as to the presence of a specific, desired fragment among the complex sample. This can be achieved by transferring the DNA from the intact gel onto a piece of nitrocellulose or nylon membrane placed in contact with it. This provides a more permanent record of the sample since DNA will begin to diffuse out of a gel that is left for a few hours. First the gel is soaked in alkali to render the DNA single stranded. It is then transferred to the membrane so that the DNA becomes bound to it in exactly the same pattern as that originally on the gel.[8] This transfer, named a Southern blot after its inventor Ed Southern, is usually performed by drawing large volumes of buffer by capillary action through both gel and membrane, thus transferring DNA from the gel to the membrane. Alternative methods are also available for this operation such as electrotransfer or vacuum assisted transfer. Both are claimed to give a more even transfer and are much more rapid, although they do require equipment more expensive than the capillary transfer system. Transfer of the DNA from the gel to the membrane allows the membrane to be treated with a labelled DNA gene probe. This single-stranded DNA probe will hybridize under the right conditions to complementary single-stranded DNA fragments immobilised onto the membrane.

5.1 Hybridization and Stringency

The conditions of hybridization are critical if the process is to take place effectively. This is usually referred to as the stringency of the hybridization, and conditions are peculiar to each individual gene probe and for each sample of DNA. Two of the most important components are the temperature and the salt concentration. Higher temperatures and low salt concentrations, termed high stringency, provide a favourable environment for perfectly matched probe and template sequences, whereas reduced temperatures and high salt concentrations, termed low stringency, allow the stabilization of mismatches in the duplex. In addition,

---

[8] E. M. Southern, *J. Mol. Biol.*, 1975, **98**, 503.

inclusion of denaturants such as formamide allow the hybridization temperatures to be reduced without affecting the stringency.

A series of post-hybridization washing steps with a salt solution such as SSC, containing sodium citrate and sodium chloride, is then carried out to remove any unbound probe and control the binding of the duplex. The membrane is developed either using autoradiography if the probe is radiolabelled or by a number of non-radioactive methods. The precise location of the probe and its target may be then visualized. The steps involved in Southern blotting are indicated in Figure 5.

It is also possible to analyse DNA from different species or organisms by blotting the DNA and then using a gene probe representing a protein or enzyme from one of the organisms. In this way it is possible to search for related genes in different species. This technique is generally termed Zoo blotting.

An analogous process of nucleic acid blotting can be used to transfer RNA separated by gel electrophoresis onto membranes similar to those used in Southern blotting. This process, termed Northern blotting, allows the identification of specific mRNA sequences of a defined length by hybridization to a labelled gene probe.[9] With this technique it is possible not only to detect specific mRNA molecules but it may also be used to quantify the relative amounts of a specific mRNA present in a tissue or sample. It is usual to separate the mRNA transcripts by gel electrophoresis under denaturing conditions since this improves resolution and allows a more accurate estimation of the sizes of the transcripts. The format of the blotting may be altered from transfer from a gel to direct application to slots on a specific blotting apparatus containing the nylon membrane. This is termed slot or dot blotting and provides a convenient means of measuring the abundance of specific mRNA transcripts without the need for gel electrophoresis; it does not, however, provide information regarding the size of the fragments.

Hybridization techniques are essential to many molecular biology experiments, and the format of the hybridization may be altered to improve speed sensitivity and throughput. One interesting alternative is termed surface plasmon resonance (SPR). This is an optical system based on differences between incident and reflected light in the presence or absence of hybridization, its main advantage being that the kinetics of hybridization can be undertaken in real time and without a DNA label. A further exciting method for hybridization uses arrays of single-stranded DNA molecules tethered to small hybridization chips. Hybrid-

---

[9] J. C. Alwine, D. J. Kemp and G. R. Stark, *Proc. Natl. Acad. Sci. USA*, 1977, **74**, 5350.

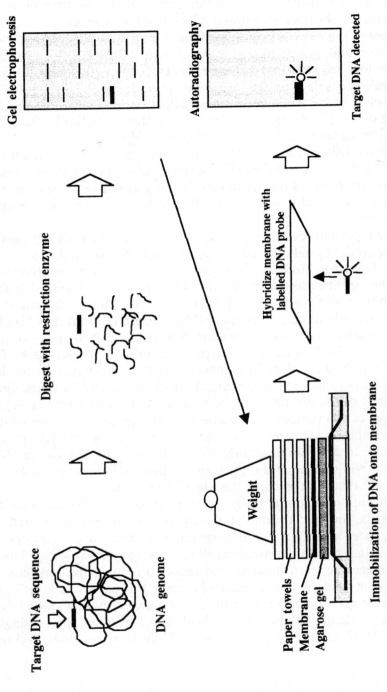

Gel electrophoresis

Autoradiography

Target DNA detected

Digest with restriction enzyme

Hybridize membrane with labelled DNA probe

Weight

Paper towels
Membrane
Agarose gel

Immobilization of DNA onto membrane

Target DNA sequence

DNA genome

Figure 5 *Typical steps involved in Southern blotting and the detection of a specific fragment of DNA (denoted by heavy black line) by hybridisation with a specific complementary labelled gene probe*

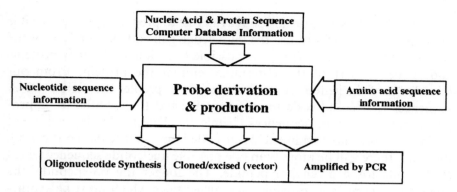

Figure 6 *Sources and nature of information required in the design and synthesis of gene probes*

ization to a DNA sample is detected by computer, allowing DNA mutations to be quickly and easily analysed (see Chapter 3).

6 PRODUCTION OF GENE PROBES

The availability of a gene probe is essential in many molecular biology techniques and yet in many cases its production is one of the most difficult steps, as indicated in Figure 6. The information needed to produce a gene probe may come from many sources, but with the development and sophistication of genetic databases this is usually one of the first stages.[10] There are a number of genetic databases throughout the world and it is possible to search these over the internet and identify particular sequences relating to a specific gene or protein. In some cases it is possible to use related proteins from the same gene family to gain information on the most useful DNA sequence. Similar proteins or DNA sequences but from different species may also provide a starting point with which to produce a so-called heterologous gene probe. Although in some cases probes are already produced and cloned it is possible, armed with a DNA sequence from a DNA database, to chemically synthesize a single-stranded oligonucleotide probe. This is usually undertaken by computer-controlled gene synthesizers which link dNTPs together based on a desired sequence. It is essential to carry out certain checks before probe production to determine that the probe is unique, is not able to self-anneal or that it is self-complementary, all of which may compromise its use.

[10] M. Aquino de Muro, in 'Molecular Biomethods Handbook', ed. R. Rapley and J. M. Walker, Humana Press, Totowa, NY, 1998.

Where little DNA information is available to prepare a gene probe it is possible in some cases to use the knowledge gained from analysis of the corresponding protein. Thus it is possible to isolate and purify proteins and sequence part of the *N*-terminal end of the protein. From our knowledge of the genetic code, it is possible to predict the various DNA sequences that could code for the protein, and then synthesize appropriate oligonucleotide sequences chemically. Due to the degeneracy of the genetic code most amino acids are coded for by more than one codon, therefore there will be more than one possible nucleotide sequence which could code for a given polypeptide. The longer the polypeptide, the greater the number of possible oligonucleotides which must be synthesized. Fortunately, there is no need to synthesize a sequence longer than about 20 bases, since this should hybridize efficiently with any complementary sequences and should be specific for one gene. Ideally, a section of the protein should be chosen which contains as many tryptophan and methionine residues as possible, since these have unique codons and there will therefore be fewer possible base sequences which could code for that part of the protein. The synthetic oligonucleotides can then be used as probes in a number of molecular biology methods.

6.1   DNA Gene Probe Labelling

An essential feature of a gene probe is that it can be visualized by some means. In this way a gene probe that hybridizes to a complementary sequence may be detected and the desired sequence identified from a complex mixture. There are two main ways of labelling gene probes: this has traditionally been carried out using radioactive labels, but the use of non-radioactive labels is gaining in popularity.[11,12] Perhaps the most used radioactive label is 32-phosphorus (^{32}P), although for certain techniques 35-sulfur (^{35}S) and tritium (^{3}H) are used. These labels may be detected by the process of autoradiography where the labelled probe molecule, bound to sample DNA located, for example, on a nylon membrane, is placed in contact with an X-ray sensitive film. Following exposure the film is developed and fixed as a black and white negative which reveals the precise location of the labelled probe and therefore the DNA to which it has hybridized. A comparison of labelling strategies is indicated in Table 2.

[11] G. H. Keller and M. M. Manak, eds., 'DNA Probes' 2nd Edn, Stockton Press, New York, 1993.
[12] T. P. McCreery and T. R. Barrette, in 'Molecular Biomethods Handbook', ed. R. Rapley and J. M. Walker, Humana Press, Totowa, NY, 1998.

Table 2 *A general comparison of methods used in the labelling of nucleic acids*

Labelling method	Enzyme	Probe type	Specific activity
5′ end labelling	Alkaline phosphatase Polynucleotide kinase	DNA	Low
3′ end labelling	Terminal transferase	DNA	Low
Nick translation	DNase I DNA polymerase	DNA	High
Random hexamer	DNA polymerase I	DNA	High
PCR	*Taq* DNA polymeras	DNA	High
Riboprobes (cRNA)	RNA polymerase	RNA	High

6.2 Non-radioactive DNA Labelling

Non-radioactive labels are increasingly being used to label DNA gene probes. Until recently, radioactive labels were more sensitive than their non-radioactive counterparts, however developments in non-radioactive labels have led to similar sensitivities which, when combined with their improved safety, have led to their greater acceptance.

The labelling systems are termed either direct or indirect. Direct labelling allows an enzyme reporter such as alkaline phosphatase to be coupled directly to the DNA. Although this may alter the characteristics of the DNA gene probe it offers the advantage of rapid analysis since no intermediate steps are needed. Indirect labelling is presently more popular. This technique relies on the incorporation of a nucleotide which has a label attached. At present three of the most used labels are biotin, fluorescein and digoxygenin. These molecules are covalently linked to nucleotides using a carbon spacer arm of 7, 14 or 21 atoms. Specific binding proteins may then be used as a bridge between the nucleotide and a reporter protein such as an enzyme. For example, biotin incorporated into a DNA fragment is recognized with a very high affinity by the protein streptavidin. This may be either coupled or conjugated to a reporter enzyme molecule such as alkaline phosphatase or horseradish peroxidase (HRP). This is usually used to convert a colourless substrate into a coloured insoluble compound and also offers a means of signal amplification. Alternatively labels such as digoxygenin incorporated into DNA sequences may be detected by monoclonal antibodies, again conjugated to reporter molecules including alkaline phosphatase. Thus, rather than a detection system relying on autoradiography, which is necessary for radiolabels, a series of reactions resulting in either a colour, light or chemiluminescence reaction takes place. This has important practical implications since autoradiography may take 1–3 days whereas colour and chemiluminescent reactions take minutes. In addition, non-

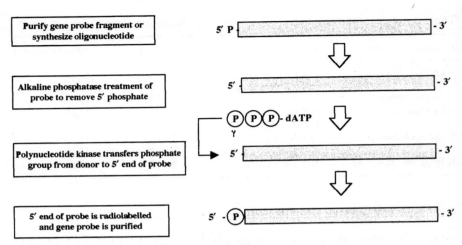

Figure 7   *Steps involved in the preparation of a probe labelled at the 5' end*

radiolabelling and detection minimize the potential health and safety hazards encountered when using radiolabels.

6.3   End Labelling of DNA

The simplest form of labelling DNA is by 5' or 3' end labelling.[13] 5' end labelling involves a phosphate transfer or exchange reaction where the 5' phosphate of the DNA to be used as the probe is removed and in its place a labelled phosphate, usually ^{32}P, is added. This is usually carried out by two enzymes. The first, alkaline phosphatase, is used to remove the existing phosphate group from the DNA. Following removal of the released phosphate from the DNA, a second enzyme, polynucleotide kinase, is added which catalyses the transfer of a ^{32}P labelled phosphate group to the 5' end of the DNA (see Figure 7). The newly labelled probe is then purified, usually by chromatography through a sephadex column, and may be used directly.

Labelling the 3' end of the DNA molecule is slightly less complex. Here, a new labelled (*e.g.* [32]PadATP or biotin) deoxyribonucleoside triphosphate (dNTP) is added to the 3' end of the DNA by the enzyme terminal transferase, as indicated in Figure 8. Although this is a simpler reaction, a potential problem exists because a new nucleotide is being added to the existing sequence which alters the complete sequence of the DNA. This may affect hybridization to its target sequence. End-labelling methods also suffer from the fact that only one label is added to the DNA

---

[13] J. A. Matthews and L. L. Kricka, *Anal. Biochem.*, 1988, **169**, 1.

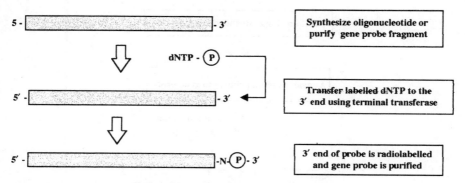

Figure 8   *Steps involved in the preparation of a probe labelled at the 3′ end*

so they are of a lower activity in comparison to methods which incorporate labels along the length of the DNA.

6.4   Random Primer Labelling of DNA

The DNA to be labelled is first denatured and then placed under renaturing conditions in the presence of a mixture of many different random sequences of hexamers or hexanucleotides.[14] These hexamers will, by chance, bind to the DNA sample wherever they encounter a complementary sequence and so the DNA will rapidly acquire an approximately random sprinkling of hexanucleotides annealed to it. Each of the hexamers can act as a primer for the synthesis of a fresh strand of DNA catalysed by DNA polymerase since it has an exposed 3′ hydroxyl group, as seen in Figure 9. The Klenow fragment of DNA polymerase is used for random primer labelling because it lacks a 5′–3′ exonuclease activity. This is prepared by cleavage of DNA polymerase with subtilisin, giving a large enzyme fragment which has no 5′ to 3′ exonuclease activity, but which still acts as a 5′ to 3′ polymerase. Thus when the Klenow enzyme is mixed with the annealed DNA sample in the presence of dNTPs, including at least one which is labelled, many short stretches of labelled DNA will be generated. In a similar way to random primer labelling, the polymerase chain reaction may also be used to incorporate radioactive or non-radioactive labels.

6.5   Nick Translation Labelling of DNA

A traditional method of labelling DNA is by the process of nick

---

[14] A. P. Feinberg and B. Vogelstein, *Anal. Biochem.*, 1983, **132**, 6.

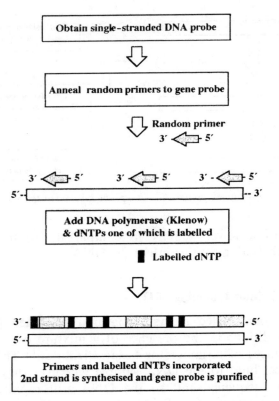

Figure 9   *Steps involved in the preparation of an internally labelled probe by the use of random hexanucleotides*

translation.[15] Low concentrations of DNase I are used to make occasional single-strand nicks in the double-stranded DNA that is to be used as the gene probe. DNA polymerase then fills in the nicks, using an appropriate dNTP, at the same time making a new nick to the 3' side of the previous one. In this way the nick is translated along the DNA. If labelled dNTPs are added to the reaction mixture, they will be used to fill in the nicks (Figure 10). In this way the DNA can be labelled to a very highly specific activity.

The methods of DNA labelling are summarized in Table 2.

7   THE POLYMERASE CHAIN REACTION

There have been a number of key developments in molecular biology techniques, but one that has had the most impact in recent years has been

[15] P. W. J. Rigby, M. Dieckman, C. Rhodes, and P. Berg, *J. Mol. Biol.*, 1977, **113**, 237.

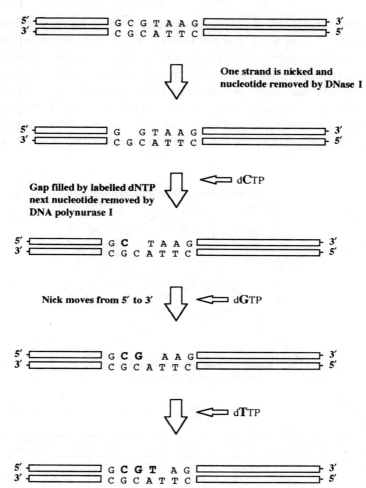

Figure 10 *Steps involved in the preparation of an internally labelled probe by the use of nick translation*

the polymerase chain reaction, or PCR. One of the reasons for the adoption of the PCR is the elegant simplicity of the reaction and relative ease of the practical steps.[16] It is frequently one of the first techniques used when analysing DNA, and it has opened up the analysis of cellular and molecular processes to those outside the field of molecular biology.

The PCR is used to amplify a precise fragment of DNA from a complex mixture of starting material, usually termed the template DNA, and in many cases requires little DNA purification. It does require

[16] R. K. Saiki, S. J. Scharf, F. Faloona, K. B. Mullis, G. T Horn, H. A. Erlich and N. Arnheim, *Science*, 1985, **230**, 1350.

knowledge of the DNA sequence information which flanks the fragment of DNA to be amplified (target DNA). From this information two oligonucleotide primers can be chemically synthesized, each complementary to a stretch of DNA to the 3′ side of the target DNA, one oligonucleotide for each of the two DNA strands (see Figure 11). Further reagents required for the PCR include a DNA polymerase, each of the four nucleotide dNTP building blocks of DNA in equimolar amounts (50–200 μM) and a buffer appropriate for the enzyme. This is

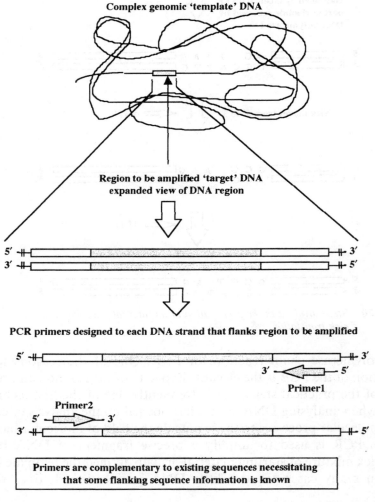

Figure 11 *Simplified scheme of the relationship of complex genomic DNA, region to be amplified and potential primer binding sites*

usually optimised for Mg^{2+} concentration (0.5–5 mM). This is a critical component of the PCR, affecting not only the enzyme, but also primer/template binding and effective incorporation of dNTPs which together form a soluble complex. The PCR may be thought of as a technique analogous to the DNA replication process that takes place in cells since the outcome is the same – the generation of new complementary DNA stretches based upon the existing ones. It is also a technique that has, in many cases, replaced the traditional DNA cloning methods since it fulfils the same function, the production of large amounts of DNA from limited starting material, but achieves this in a fraction of the time needed to clone a DNA fragment. Although not without its drawbacks, the PCR is a remarkable development which is changing the approach of many scientists to the analysis of nucleic acids and continues to have a profound impact on core biosciences and biotechnology.

7.1 Stages and Components of the PCR

The PCR consists of three defined sets of times and temperatures termed steps: (i) denaturation, (ii) annealing and (iii) extension, as shown in Figure 12. Each of these steps is repeated 30–40 times, termed cycles. In

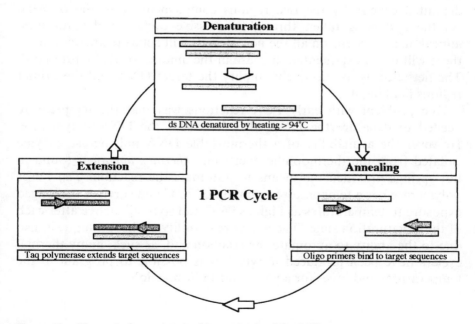

Figure 12 *The typical steps involved in one cycle of the PCR*

the first cycle the double-stranded template DNA is denatured by heating the reaction to above 90°C. The region within the complex DNA which is to be specifically amplified (target DNA) is made accessible. The reaction mixture is then cooled to 40–60°C. The precise temperature is critical and each PCR system has to be defined and optimized. Reactions that are not optimized may give rise to other DNA products in addition to the specific target, or may not produce any amplified products at all. The annealing step allows the hybridization of the two oligonucleotide primers, which are present in excess, to bind to their complementary sites that flank the target DNA. The annealed oligonucleotides act as primers for DNA synthesis, since they provide a free 3' hydroxyl group for DNA polymerase. The DNA synthesis step is termed extension and is carried out by a thermostable DNA polymerase, most commonly *Taq* DNA polymerase.[17]

DNA synthesis proceeds from both 3' ends of the primers until the new strands have been extended along and beyond the target DNA. It is important to note that, since the new strands extend beyond the target DNA they will contain a region near their 3' ends which is complementary to the other primer. Thus, if another round of DNA synthesis is allowed to take place, the new strands, as well as the original strands, will be used as templates, as indicated in Figure 13. Most interestingly, the products obtained from the new strands will have a precise length, delimited exactly by the two regions complementary to the primers. As the system is taken through successive cycles of denaturation, annealing and extension all the new strands will act as templates and so there will be an exponential increase in the amount of DNA produced. The net effect is to selectively amplify the target DNA and the primer regions flanking it.

One problem with early PCR reactions was that the temperature needed to denature the DNA also denatured the DNA polymerase. However the availability of a thermostable DNA polymerase enzyme isolated from the thermophilic bacterium *Thermus aquaticus* found in hot springs provided the means to automate the reaction. *Taq* DNA polymerase has a temperature optimum of 72°C and survives prolonged exposure to temperatures as high as 96°C and so is still active after each of the denaturation steps. The widespread utility of the technique is also due to the ability to automate the reaction and, as such, many thermal cyclers have been produced in which it is possible to program in the temperatures and times for a particular PCR reaction.

[17] K. A. Eckert and T. A. Kunkel, *Nucleic Acids Res.*, 1990, **18**, 3739.

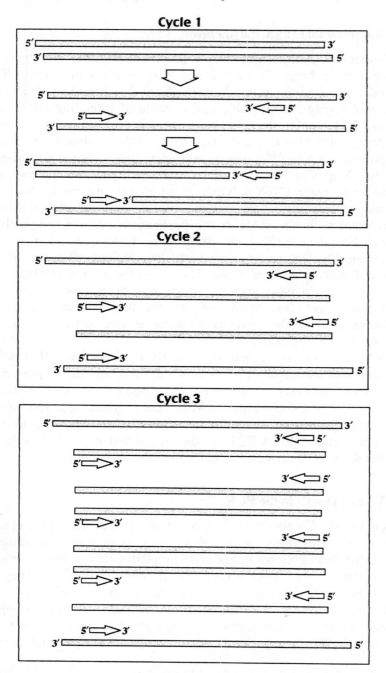

Figure 13 *The first three cycles of a PCR indicating the early events that lead to specific products and linear amplification of the original target DNA (upper and lower sequences of cycle 3)*

7.2   Thermostable DNA Polymerases

A number of thermostable DNA polymerases other than *Thermus aquaticus* have been discovered and marketed commercially. Those isolated from *Thermococcus litoralis* (Vent^{TM}), found in deep ocean floors, are highly thermostable and capable of extending templates in excess of 12 Kb pairs; they also have proofreading ability and so have a high degree of fidelity in comparison to *Taq* DNA polymerases. Further manipulation of Vent^{TM} has resulted in derivatives with greater thermostability (Deep-Vent^{TM}) and a derivative lacking the exonuclease component (exo −). *Pfu* DNA polymerase isolated from a marine bacterium also has proofreading activity and incorporates radiolabelled nucleotides and analogues efficiently. Hence it is particularly useful when producing radiolabelled gene probes or performing techniques such as cycle sequencing. One minor problem with these polymerases and another DNA polymerase isolated from *Thermotoga maritima* (UlTma^{TM}) is their 3′–5′ exonuclease activity which has been reported to cause modification and degradation of primers under initial sub-optimal conditions. This may be overcome by using various exonuclease deficient forms (exo −) of the enzyme.

One further interesting thermostable DNA polymerase is that isolated from *Thermus thermophilus* (*Tth*) which at 70°C and in the presence of Mn^{2+} is able to carry out reverse transcription reactions.[18] Following cDNA synthesis and chelation of Mn^{2+} the polymerase is able to carry out polymerization of the template. This dual activity of *Tth* DNA polymerase allows RNA-PCR to be carried out in a single tube and obviates the need for a separate cDNA synthesis reaction.

7.3   Primer Design in the PCR

The key to the PCR lies in the design of the two oligonucleotide primers. These have to not only be complementary to sequences flanking the target DNA but also must not be self-complementary or bind each other to form dimers since both prevent DNA amplification.[19,20] They also have to be matched in their GC content and have similar annealing temperatures, usually at a concentration of 1 μM. The increasing use of information from the internet and the sequences held in gene databases are useful starting points when designing primers and reaction

[18] T. W. Myers and D. H. Gelfand, *Biochemistry*, 1991, **30**, 7661.
[19] W. Rychlik, in 'Nucleic Acids Protocols Handbook', ed. R. Rapley, Humana Press, Totowa, NY, 2000.
[20] C. R. Newton and A. Graham, 'PCR', 2nd Edn, Bios Scientific Publishers, Oxford, 1997.

conditions for the PCR. A number of software packages such as Oligo, Primer, *etc.*, have allowed the process of primer design to be less troublesome. It is also possible to include more than one set of primers in a PCR. This method, termed multiplex PCR, allows the amplification of more than one product in a single reaction tube and is especially useful in molecular diagnosis of clinical disorders[21] (see Chapter 10).

It is also possible to design primers with additional sequences at their 5' end such as restriction endonuclease target sites or promoter sequences. This allows further manipulation such as cloning to be undertaken following amplification (see Chapter 3, Section 2). However, modifications such as these require that the annealing conditions be altered to compensate for the areas of non-homology in the primers.[22]

In general primers are usually designed to be between 20 and 30 bases in length. It is best to balance the melting temperature of the primer pair and to have a GC content of 40–60%. A number of factors are best avoided in primer sequences, such as the potential for secondary structure formation, primer complementarity or dimer formation and runs of purines, pyrimidines or repetitive motifs. The precise times and temperatures for a PCR may be calculated manually or by computer programs, however in practice these are usually a guide in determining the final optimal parameters for amplification.

One useful technique which may be adopted in order to amplify the desired product rather than unwanted ones is touchdown PCR.[23] Here a thermal cycler is programmed to lower the annealing temperature incrementally from a value above the expected melting temperature to a value below it. This allows optimal hybridization of the primers to the target and thus leads to amplification of the correct products over any undesired ones.

A number of PCR methods have been developed where either one or both of the primers are short and random in sequence.[24] This gives rise to arbitrary priming in genomic templates but interestingly may give rise to discrete banding patterns when analysed by gel electrophoresis. In many cases this technique may be used reproducibly to identify a particular organism or species. This is sometimes referred to as rapid amplification of polymorphic DNA (RAPDs) and has been used successfully in the detection and differentiation of a number of pathogenic strains of bacteria.

[21] J. S. Chamberlain, R. A. Gibbs, J. E. Ranier, P. N. Nguyen and C. T. Caskey, *Nucleic Acids Res.*, 1988, **16**, 11141.
[22] S. J. Scharf, G. T. Horn and H. A. Erlich, *Science*, 1986, **233**, 1076.
[23] R. H. Don, P. T. Cox, B. J. Wainwright, K. Baker and J. S. Mattick, *Nucleic Acids Res.*, **19**, 4008.
[24] J.-M. Deragon and B. S. Landry, *PCR Meths. App.*, 1992, **1**, 175.

7.4 PCR Amplification Templates

The PCR may be used to amplify DNA from a variety of sources or templates. It is also a highly sensitive technique and requires only one or two molecules for successful amplification. Unlike many manipulation methods used in current molecular biology the PCR technique is sensitive enough to require very little template preparation. Indeed the extraction of DNA from many prokaryotic and eukaryotic cells may involve a simple boiling step. However, the components of many DNA extraction techniques such as SDS, phenol, ethanol and proteinase K may adversely affect the PCR at certain concentrations.[25] In contrast, a number of reagents have been shown to improve the efficiency of amplification. In particular, the inclusion of tetramethyl ammonium chloride (TMAC) appears to improve the specificity of primer annealing whilst betaine binds and stabilizes AT sequences but destabilizes GC regions and is therefore useful in the amplification of GC-rich sequences. Nucleotide analogues such as 7-deaza-2′-deoxyguanosine triphosphate are also a useful addition when amplifying sequences with the potential to form secondary structures.

The PCR may also be used to amplify RNA, a process termed RT-PCR (reverse transcriptase-PCR) (see Figure 14 and Chapter 3). Initially a reverse transcription reaction which converts mRNA to cDNA is first carried out. This reaction normally involves the use of the enzyme reverse transcriptase although some thermostable DNA polymerases used in the PCR, *e.g. Tth*, have a reverse transcriptase activity under certain buffer conditions. This allows mRNA transcription products to be effectively analysed. It may be also be used to differentiate latent viruses (detected by standard PCR) or active viruses which replicate and thus produce transcription products which are detectable by RT-PCR. In addition, the PCR may be extended to determine relative amounts of a transcription product.[26]

7.5 Sensitivity of the PCR

The enormous sensitivity of the PCR system is also one of its main drawbacks since the very large degree of amplification makes the system vulnerable to contamination. Even a trace of foreign DNA, such as that contained in dust particles, may be amplified to significant levels and

[25] J. Bickley and D. Hopkins, in 'Analytical Molecular Biology', ed. G. C. Saunders and H. C. Parkes, RSC, Cambridge, UK, 1999.
[26] L. Raeymaekers, in 'Clinical Applications of PCR', ed. Y. M. Dennis Lo, Humana Press, Totowa, NY, 1998.

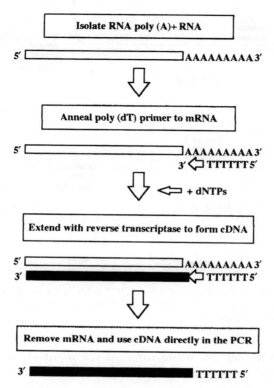

Figure 14 *Typical initial steps involved in RT-PCR including the reverse transcription step in the conversion of mRNA to cDNA*

may give misleading results. Hence cleanliness is paramount when carrying out PCR and dedicated equipment and, in some cases, laboratory areas dedicated to pre- and post-PCR stages are used. It is possible that amplified products may also contaminate the PCR although this may be overcome by UV irradiation to degrade the already amplified products so that they cannot be used as templates. A further interesting solution is to introduce uracil into the PCR by incorporating dUTP. Following analysis, the PCR products are treated with the enzyme uracil-*N*-glycosylase (UNG) which degrades the *N*-glycosidic bond in DNA containing uracil, and PCR products are fragmented and rendered useless as templates.[27]

Many traditional methods in molecular biology have now been superseded by the PCR and the applications for the technique appear to be unlimited. Some of the main techniques derived from the PCR are

[27] N. Longo, N. S. Berninger and J. L. Hartley, *Gene*, 1990, **93**, 125.

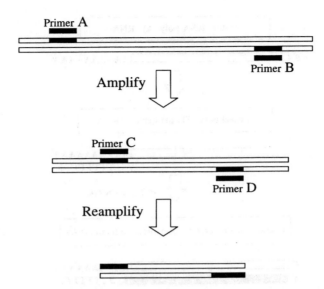

Figure 15   *Representation of a typical nested PCR where primers A and B are used to synthesise long fragments which may be subsequently used in a second PCR with primers C and D*

introduced in Chapter 10, while some of the main areas to which the PCR has been put to use are summarized in Table 3.

7.6   Modifications of the PCR

The basic PCR method has been adapted and developed in various ways in order to suit specific applications or to overcome certain problems. One such potential problem is the specificity of the reaction.[28] If mispriming occurs in a PCR due to similarities between target and non-target DNA it is possible to take an aliquot of the products and reamplify the DNA with primers internal or 3′ to the primer pair used in the original PCR. This technique, called nested PCR, also allows regions of GC-rich sequence to be amplified effectively, as indicated in Figure 15. A further method that is commonly employed to prevent mispriming in the initial stages of the PCR is termed hotstart. In this case primers can potentially bind to regions of partial complementarity to a small fraction of the DNA which will be single stranded before the denaturation stage. *Taq* DNA polymerase will then begin to synthesize DNA from these imperfectly matched duplexes even at room temperature, far below its optimum. This can be overcome by excluding a critical

[28] T. M. Haqqi, G. Sarker, C. S. David and S. S. Somer, *Nucleic Acids Res.*, 1988, **16**, 11844.

Table 3 *Selected applications and specific uses of PCR*

Field of study	Applications	Specific uses
DNA amplification	General molecular biology	Screening gene libraries
Production/labelling	Gene probe production	Use with blots/hybridization
RT-PCR	RNA analysis	Active latent viral infections
Scenes of crime	Forensic science	Analysis of DNA from blood
Microbial detection	Infection/disease monitoring	Strain typing/analysis (RAPDs)
Cycle sequencing	Sequence analysis	Rapid DNA sequencing possible
Referencing points in genome	Genome mapping studies	Sequence tagged sites (STS)
mRNA analysis	Gene discovery	Expressed sequence tags (EST)
Detection of known mutations	Genetic mutation analysis	Screening for cystic fibrosis
Quantitative PCR	Quantification analysis	5′ Nuclease (TaqMan assay)
Detection of unknown mutations	Genetic mutation analysis	Gel-based PCR methods (DGGE)
Production of novel proteins	Protein engineering	PCR mutagenesis
Retrospective studies	Molecular archaeology	Dinosaur DNA analysis
Sexing or cell mutation sites	Single-cell analysis	Sex determination of unborn
Studies on frozen sections	*In situ* analysis	Localization of DNA/RNA

RT: reverse transcriptase, RAPDs: rapid amplification polymorphic DNA, STS: sequence-tagged site, EST: expressed sequence tags, DGGE: denaturing gradient gel electrophoresis.

reagent, such as the enzyme, until the denaturation temperature has been reached. A favoured method is to use formulated wax beads (Ampliwax or PCR Gems). Here a bead of wax is layered onto the reaction mix and solidifies. The enzyme is then layered on top of the wax. Increasing the temperature in the denaturation step causes the wax to melt and allows mixing of the reagents. This is especially useful if a large number of PCRs have to be set up at any one time since all the reactions may be started simultaneously.[29]

One important application of the PCR developed in recent years is long PCR.[30] In this technique PCRs of up to 50 Kb may be produced utilizing two thermostable DNA polymerases. The key to the technique is that, in addition to *Taq* DNA polymerase, a proofreading polymerase, *e.g. Pfu* DNA polymerase, is included at a ratio of 180:1. The inclusion of proofreading activity appears to prevent premature chain termination of long products which occurs with *Taq* DNA polymerase alone. The ability to amplify long fragments of DNA has led to many potential applications, especially in the area of genome research.

7.7 Applications of the PCR

There are a number of molecular biology methods where the PCR has been used to great effect. The labelling of gene probes is one such area which has traditionally been undertaken by techniques such as nick translation. The nature of the PCR makes it an ideal method for gene probe production and labelling. PCR products may be labelled at the 5′ or 3′ end using the methods indicated in Section 6.3. This may be undertaken before the PCR by labelling the oligonucleotide primers or the resulting PCR product. However, since the PCR is essentially a two primer extension reaction, it allows the incorporation of nucleotides that have been labelled either radioactively or with a non-radioactive label such as biotin. The advantage of the PCR as a gene probe and labelling system is that it offers great flexibility and may be rapidly produced.

Another further important modification of the PCR is termed quantitative PCR.[31] This allows the PCR to be used as a means of identifying the initial concentrations of template DNA and is very useful for the measurement of, for example, a virus or an mRNA representing a protein expressed in abnormal amounts in a disease process. Early

[29] Q. Chou, M. Russell, D. E. Birch, J. Raymond and W. Bloch, *Nucleic Acids Res.*, 1992, **20**, 1717.
[30] S. Cheng, S.-Y. Cheng, P. Gravitt and R. Respess, *Nature (London)*, 1994, **369**, 684.
[31] C. A. Heid, J. Stevens, K. J. Livak and P. M. Williams, *Genome Res.*, 1996, **6**, 986.

quantitative PCR methods involved the comparison of a standard or control DNA template amplified with separate primers at the same time as the specific target DNA. These types of quantitation rely on the reaction being exponential and so any factors affecting this may also affect the result. Other methods involve the incorporation of a radiolabel through the primers or nucleotides and their subsequent detection following purification of the PCR product. An alternative automated method of great promise is the 5' exonuclease detection system or TaqMan assay.[32] Here an oligonucleotide probe is labelled with a fluorescent reporter and quencher molecule at each end. When the primers bind to their target sequence the 5' exonuclease activity of *Taq* DNA polymerase degrades and releases the reporter from the quencher. A signal is thus generated which increases in direct proportion to the number of starting molecules. Thus a detection system is able to induce and detect fluorescence in real time as the PCR proceeds. This has important implications in, for example, the rapid detection of bacterial and viral sequences in clinical samples.

One of the most useful general applications of the development of the PCR is direct PCR sequencing.[33] This traditionally involved the cloning of sequences into vectors developed for chain termination sequencing. However, the rapid accumulation of PCR products allows nucleotide sequence information to be obtained very quickly. A number of methods for direct PCR sequencing are indicated in Section 9.2. Further applications of the PCR are indicated in Table 3 and described in the respective sections in Chapters 8, 10 and 11.

8 ALTERNATIVE AMPLIFICATION TECHNIQUES

The success of the PCR process has given impetus to the development of other amplification techniques which are based on either thermal cycling or non thermal cycling (isothermal) methods.[34] In general they rely on the amplification of target sequences although there are exceptions where the signal is amplified which makes them suitable for quantitation. Primer-based techniques are, however, useful in that they may amplify more than one target in a single reaction. One of the most common alternatives, used extensively for microbial identification, is the LCR or ligase chain reaction.[35] This is a technique requiring thermal cycling and

[32] R. Okimoto and J. B. Dodgson, *Biotechniques*, 1996, **21**, 20.
[33] I. S. Bevan, R. Rapley and M. R. Walker, *PCR Meths. App.*, 1992, **4**, 222.
[34] E. Quirus and L. Gonzarlez, *Med. Clin.*, 1993, **101**, 141.
[35] F. Barany, *PCR Meths. App.*, 1992, **1**, 5.

requires a set of four oligonucleotides complementary to a target sequence containing a mutation. Two pairs of oligos on each sequence bind with a gap of one nucleotide which is sealed by a thermostable DNA ligase. This is then used as further substrate following denaturation. Any sequence containing a mutation will not allow the 3' end of the oligo to bind and so no ligation takes place as indicated in Figure 16.

A number of non thermal cycling techniques have also been devised such as the Q-β replicase system, self-sustained sequence replication (3SR) and bDNA, or branched DNA, amplification. bDNA is an interesting signal amplification method where the probe for a target sequence includes a special branched nucleotide. If the target is present the probe attaches, grows and is ultimately detected by a chemiluminescent end-point. A number of alternative amplification techniques are listed in Table 4.

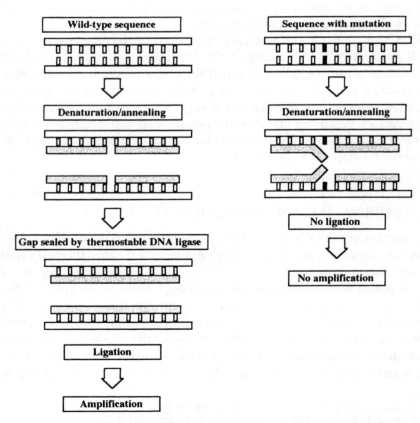

Figure 16   *Events involved in the ligase chain reaction indicating amplification of wild type sequences and non amplification of mutant sequences*

Table 4 *Selected alternative amplification techniques to the PCR including methodologies which amplify the target, e.g. DNA or RNA, or which detect the target and amplify a signal molecule bound to it*

Type of method	Assay reagents	Examples/applications
Target amplification techniques		
LCR (ligase chain reaction)	Non-isothermal employs thermostable DNA ligase	Mutation detection (clinical diagnostics)
NASBA (nucleic acid sequence based amplification)	Isothermal involving using RNA RNaseH/reverse transcriptase and T7 DNA polymerase	Viral detection, e.g. HIV
Signal amplification techniques		
Branched DNA amplification (b-DNA)	Isothermal microwell format using hybridisation of target/capture probe and signal amplification	Mutation detection

9 NUCLEOTIDE SEQUENCING OF DNA

The determination of the order or sequence of bases along a length of DNA is one of the central techniques in molecular biology. Although it is now possible to derive amino acid sequence information with a degree of reliability it is frequently more convenient and rapid to analyse the DNA coding information. The precise usage of codons, information regarding mutations and polymorphisms, and the identification of gene regulatory control sequences are also only possible by analysing DNA sequences. Two techniques have been developed for this, one based on an enzymatic method frequently termed Sanger sequencing after its developer, and a chemical method, Maxam and Gilbert, named for the same reason. At present Sanger sequencing is by far the most popular method and many commercial kits are available for its use, however, there are certain occasions such as the sequencing of short oligonucleotides where the Maxam and Gilbert method is more appropriate.

One absolute requirement for Sanger sequencing is that the DNA to be sequenced is in a single-stranded form.[36] Traditionally this demanded that the DNA fragment of interest be inserted and cloned into a specialized bacteriophage vector termed M13 which is naturally single stranded (see Chapter 3). Although M13 is still universally used, the advent of the PCR has provided the means to not only amplify a region

[36] F. Sanger, A. Nicklen and R. Coulson, *Proc. Natl. Acad. Sci. USA*, 1977, **74**, 546.

of any genome or cDNA but also very quickly generate the corresponding nucleotide sequence. This has led to an explosion in the accumulation
of DNA sequence information and has provided much impetus for gene
discovery and genome mapping (see Chapter 3).

   The Sanger method is simple and elegant, and mimics in many ways
the the natural ability of DNA polymerase to extend a growing nucleotide chain based on an existing template. Initially the DNA to be
sequence is allowed to hybridize with an oligonucleotide primer, which
is complementary to a sequence adjacent to the 3' side of DNA within a
vector such as M13 or in a PCR product. The oligonucleotide will then
act as a primer for the synthesis of a second strand of DNA, catalysed by
DNA polymerase. Since the new strand is synthesized from its 5' end,
virtually the first DNA to be made will be complementary to the DNA to
be sequenced. One of the deoxyribonucleoside triphosphates (dNTPs)
which must be provided for DNA synthesis is radioactively labelled with
^{32}P or ^{35}S, and so the newly synthesized strand will be labelled (see Figure
17).

9.1   Dideoxynucleotide Chain Terminators

The reaction mixture containing newly synthesized, labelled DNA is then
divided into four aliquots, representing the four dNTPs adenine,
cytosine, quanine and thymine (A, C, G and T). In addition to all of the
dNTPs being present in the tube containing A (A tube), an analogue of
dATP (2',3'-dideoxyadenosine triphosphate, ddATP) is added (see
Figure 18). This is similar to A but has no 3' hydroxyl group and so will
terminate the growing chain since a 5' to 3' phosphodiester linkage
cannot be formed without a 3' hydroxyl group. The situation for tube C
is identical except that ddCTP is added; similarly, the G and T tubes
contain ddGTP and ddTTP, respectively.

   Since the incorporation of ddNTP rather than dNTP is a random
event, the reaction will produce new molecules varying widely in length,
but all terminating at the same type of base. Thus four sets of DNA
sequence are generated, each terminating at a different type of base, but
all having a common 5' end (the primer). The four labelled and chain-
terminated samples are then denatured by heating and loaded next to
each other on a polyacrylamide gel for electrophoresis. Electrophoresis is
performed at approximately 70°C in the presence of urea, to prevent
renaturation of the DNA, since even partial renaturation alters the rates
of migration of DNA fragments. Very thin, long gels are used for
maximum resolution over a wide range of fragment lengths. After
electrophoresis, the positions of radioactive DNA bands on the gel are

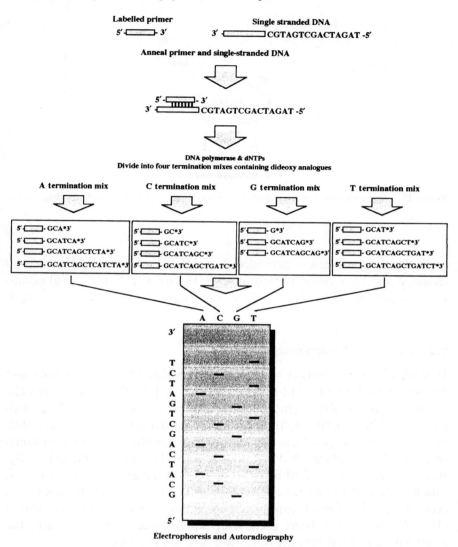

Figure 17 *Diagrammatic scheme of events undertaken in dideoxy chain termination sequencing. Here a labelled primer is annealed to the DNA sequence after which the reaction is divided into four reactions (each for A, C, G and T). Following chain termination the products are separated on a denaturing polyacrylamide gel and autoradiographed*

determined by autoradiography. Since every band in the track from the dideoxyadenosine triphosphate sample must contain molecules which terminate at adenine, and that those in the ddCTP terminate at cytosine, *etc.*, it is possible to read the sequence of the newly synthesized strand

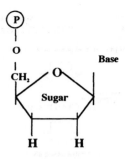

2′-3′ **Dideoxynucleotide**

Figure 18   *Representation of a typical 2′3′-dideoxy nucleoside triphosphate chain termi-*
nator

from the autoradiogram, provided that the gel can resolve differences in
length equal to a single nucleotide (see Figure 19). Under ideal conditions, sequences up to about 300 bases in length can be read from one gel.

9.2   Direct PCR Sequencing

It is possible to undertake nucleotide sequencing from double-stranded
molecules such as plasmid cloning vectors and PCR products, but the
double stranded DNA must be denatured prior to annealing with
primer. In the case of plasmid an alkaline denaturation step is sufficient,
however, for PCR products this is more problematic and a focus of much
research. Unlike plasmids, PCR products are short and reanneal rapidly,
thus preventing the annealing process. Biasing the amplification towards
one strand by using a primer ratio of 100:1 overcomes this problem to a
certain extent.[37] Denaturants such as formamide or dimethylsulfoxide
(DMSO) have also been used with some success in preventing the
reannealing of PCR strands following their separation.

It is possible to physically separate and retain one PCR strand by
incorporating a molecule such as biotin into one of the primers.
Following PCR one strand with an affinity molecule may be removed
by affinity chromatography with strepavidin leaving the complementary
PCR strand. This affinity purification provides single-stranded DNA
derived from the PCR product and, although is somewhat time-consuming, does provide high-quality single-stranded DNA for sequencing.[38]

[37] U. B. Gyllenstein and H. A. Erlich, *Proc. Natl. Acad. Sci. USA*, 1988, **85**, 7652.
[38] L. G. Mitchell and C. R. Merril, *Anal. Biochem.*, 1989, **178**, 239.

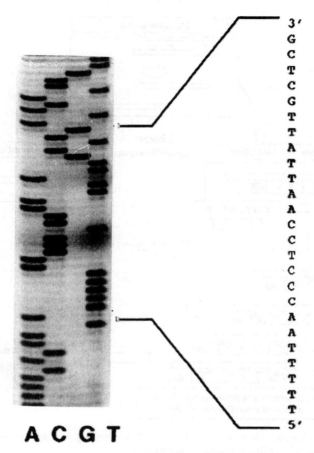

3'
G
C
T
C
G
T
T
A
T
T
A
A
C
C
T
C
C
A
A
T
T
T
T
5'

Figure 19 *A typical chain terminator sequencing pattern found after autoradiography with the sequence*
(Courtesy of Dr B.D.M. Theophilus, Birmingham Childrens Hospital, Birmingham, UK)

9.3 Cycle Sequencing

One of the most useful methods of sequencing PCR products is termed PCR cycle sequencing.[39] This is not strictly a PCR since it involves linear amplification with a single primer, as indicated in Figure 20. Approximately 20 cycles of denaturation, annealing and extension take place. Radiolabelled or fluorescently labelled dideoxynucleotides are then introduced in the final stages of the reaction to generate the chain-terminated extension products. Automated direct PCR sequencing is

[39] K. Kretz, W. Callen and V. Hedden, *PCR Meths. App.*, 1994, **3**, 107.

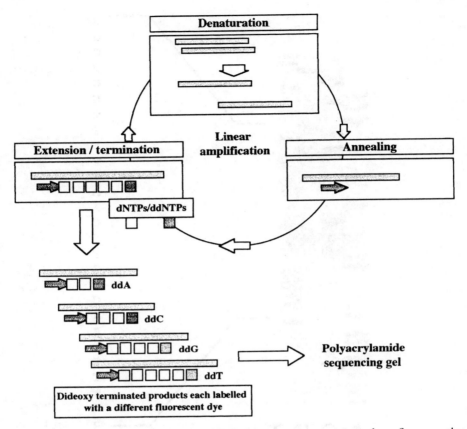

Figure 20   *Elements that make up PCR cycle sequencing reaction where fluorescently*
labelled chain terminators are used

increasingly being refined, thus allowing greater lengths of DNA to be
analysed in one sequencing run and providing a very rapid means of
analysing DNA sequences.

9.4   Automated Fluorescent DNA Sequencing

Recent advances in dye-terminator chemistry have led to the development of automated sequencing methods that involve the use of dideoxynucleotides labelled with different fluorochromes.[40] The label is
incorporated into the ddNTP and is used to carry out the chain
termination as in the standard reaction indicated in Section 9.1. The
advantage of this modification is that, since a different label is incor-

---

[40] R. K. Wilson, C. Chen, N. Avdalovic, J. Burns and L. Hood, *Genomics*, 1990, **6**, 626.

porated with each ddNTP, it is unnecessary to perform four separate reactions, and the four chain-terminated products are run on the same track of a denaturing electrophoresis gel. Each product with its base-specific dye is excited by a laser and the dye emits light at a characteristic wavelength. A diffraction grating separates the emissions which are detected by a charge-coupled device (CCD) and the sequence is interpreted by a personal computer. In addition to real-time detection, the lengths of sequence that may be analysed are in excess of 500 bp. Further improvements are likely to be made not in the sequencing reactions themselves, but in the electrophoresis of the chain terminated products. Here capillary electrophoresis may be used where liquid polymers in thin capillary tubes would substantially decrease the electrophoresis run times. The consequence of automated sequencing and the incorporation of PCR cycle sequencing has substantially decreased the time needed to undertake sequencing projects. This has given rise to the use of banks of automated robotic sequencing systems in factory style units which are now in operation in various laboratories throughout the world, especially those undertaking work for the various genome sequencing projects (see Chapter 3, Section 10).

9.5 Maxam and Gilbert Sequencing

The chemical cleavage method of DNA sequencing developed by Maxam and Gilbert is often used for sequencing small fragments of DNA such as oligonucleotides.[41] A radioactive label is added to either the 3' or the 5' end of a double-stranded DNA preparation. The strands are then separated by electrophoresis under denaturing conditions and analysed separately. DNA labelled at one end is divided into four aliquots and each is treated with chemicals which act on specific bases by methylation or removal of the base. The conditions are selected so that, on avarage, each molecule is modified at only one position along its length, and every base in the DNA strand therefore has an equal chance of being modified. Following the modification reactions, the separate samples are cleaved by piperidine, which breaks phosphodiester bonds exclusively at the 5' side of nucleotides whose base has been modified. The result is similar to that produced by the Sanger method. Each sample contains radiolabelled molecules of differing lengths, however all the labelled ends are common. The other end is cut at the same type of base. Analysis of the reaction products by electrophoresis is as described for the Sanger method.

[41] A. M. Maxam and W. Gilbert, *Proc. Natl. Acad. Sci. USA*, 1977, **74**, 560.

10 BIOINFORMATICS AND THE INTERNET

Increasingly molecular biology methods such as DNA amplification are being automated. Furthermore, sequencing technology has now reached such a level of sophistication that it is quite common for a large stretch of DNA to be sequenced and that sequence to be manipulated or stored in a computer database.[42] This has given rise to a whole new area or molecular biology, termed bioinformatics.[43] A number of large sequence facilities are now fully automated and download sequences automatically to those databases from robotic work station servers. This increase in genetic information has luckily been matched by developments in computer hardware and software.[44] There are now a large number of genetic databases which have sequence information representing a variety of organisms. The largest include GenBank at the National Institutes of Health (NIH) in the USA, EMBL at the European Bioinformatics Institute (EBI) at Cambridge, UK, and the DNA database of Japan (DDBJ) at Mishima in Japan. There are also many other databases within which specialist DNA and protein sequences are stored, all of which may be accessed over the internet.[45] A number of these important databases and internet resources are listed in Table 5.

Once a nucleotide sequence has been deduced it is possible to search an existing database for similar, homologous sequences and for generic gene or protein-coding sequences.[46] Thus it is possible to search for open reading frames, *e.g.* sequences beginning with a start codon (ATG) and continuing with a significant number of 'coding' triplets before a stop codon is reached. There are a number of other sequences that may be used to define coding sequences, and these include ribosome binding sites, splice site junctions, poly A polymerase sequences and promoter sequences that lie outside the coding regions. It is now relatively straightforward to use sequence analysis software to search a new sequence for homology within a chosen database. Software programs such as BLAST and FASTA provide the means to search for sequences of homology, and allow such sequences to be aligned, thereby giving important clues to the potential structure and function of a given DNA sequence. In some cases it is also possible to generate a graphical three-dimensional model of a putative protein encoded by a DNA sequence by

[42] E. S. Winn-Deen, *J. Clin. Ligand Assay*, 1997, **19**, 21.
[43] S. Misener and S. A. Krawetz, 'Bioinformatics: Methods and Protocols', Humana Press, Totowa, NY, 1999.
[44] T. F. Smith, *Trends Genet.*, 1998, **14**, 291.
[45] L. R. Osbourne, J. R. Lee and S. W Scherer, *Mol. Med. Today*, 1997, 370.
[46] J. W. Ficket, *Trends Genet.*, 1996, **12**, 316.

Table 5 *Nucleic acid and protein database information and resources available over the World Wide Web*

Database or resource		URL (uniform resource locator)
General DNA sequence databases		
EMBL	European genetic database	http://www.ebi.ac.uk
GenBank	US genetic database	http://nebi.nim.nih.gov
DDBJ	Japanese genetic database	http://ddbj.nig.ac.jp
Protein sequence databases		
Swiss-Prot	European protein sequence database	http://expasy.heuge.ch/sprot/ sprot-top.html
TREMBL	European protein sequence database	http://www.ebi.ac.uk/pub/databases/ trembl
PIR	US protein information resource	http://www-nbrf.gerogetown.edu/pir
Protein structure databases		
PDB	Brookhaven protein database	http://www.pdb.bnl.gov
NRL-3D	Protein structure database	http://www.gdb.org/Dan/proteins/ nrl3d.html
Genome project databases		
Human Mapping Database Johns Hopkins, USA		http://gdbwww.gdb.org
dbEST (cDNA and partial sequences)		http://www.ncbi.nih.gov
Genethon Genetic maps based on repeat markers		http://www.genethon.fr
Whitehead Institute (YAC and physical maps)		http://www.genome.wi.mit.edu

using existing sequence information. The atomic coordinates of protein structures generated from X-ray crystallography or nuclear magnetic resonance (NMR) data are also held in databases. The largest of these is the protein databank (PDB) held at Brookhaven in the USA. It is possible, although difficult at present, to predict secondary protein structures from translated nucleotide sequences. Such predicted molecular models are very complex to produce, requiring sophisticated numerical processing, however they do provide important insights into protein structure and function and are constantly being refined. Another exciting future possibility is the combination of molecular modelling with virtual reality systems, allowing real-time interaction of proteins and ligands to be observed. Whatever the means of displaying modelled proteins there is no doubt that, even now, they are extremely important in the rational design and modification of proteins and enzymes (see Chapters 14 and 18).

The major development in computing that has allowed the explosion in sequence analysis is the advent of the internet. This is a world-wide system that links numerous computers, local networks, and research, commercial and government institutions and establishments, which are all part of the World Wide Web (WWW). DNA databases and other nucleic acid sequence and protein analysis software may all be accessed over the internet, given the relevant software and authority. This is now relatively straightforward with so-called web browsers which provide a user-friendly graphical interface for sequence manipulation. Consequently the expanding and exciting new areas of bioscience research are those that analyse genome and cDNA sequence databases (genomics), and also their protein counterparts (proteomics). This type of research is sometimes referred to as 'in silico research', and there is no doubt that for both basic and biotechnological research it is as important to have internet and database access as it is to have equipment and reagents for laboratory molecular biology.

CHAPTER 3

Recombinant DNA Technology

RALPH RAPLEY

1 INTRODUCTION

Since the 1970s there have been considerable developments in the methods and techniques used to study biological processes at the molecular level. These have led to many new and powerful ways for the isolation, analysis and manipulation of nucleic acids, which, in turn, has provided insights and exciting advances in new and expanding areas of biosciences such as biotechnology, drug discovery, molecular medicine and gene therapy. The discovery of restriction endonucleases in the early 1970s was a key development which not only led to the possibility of analysing DNA more effectively, but also provided the capability to cut different DNA molecules so that they could later be joined together to create new recombinant DNA fragments, a process now termed gene cloning. The newly developed cloning strategies heralded a new and exciting era in the manipulation, analysis and exploitation of nucleic acid molecules.

Gene cloning has enabled numerous discoveries to be made and provided valuable insights into gene structure, function and regulation. Since their initial application the methods used for the production of gene libraries have been steadily refined and developed and are now seen as a cornerstone to many experiments in biochemistry and molecular biology.[1] Although the polymerase chain reaction (PCR)[2] has provided shortcuts to gene analysis there are still many cases where gene cloning techniques are not only useful but are an absolute requirement. The

[1] J. Sambrook, E. F. Fritsch and T. Maniatis, eds, 'Molecular Cloning: A Laboratory Manual', 2nd Edn. Cold Spring Harbor Laboratory, Cold Spring Harbor, NY, 1989.
[2] R. K. Saiki, S. J. Scharf, F. Faloona, K. B. Mullis, G. T. Horn, H. A. Erlich and N. Arnheim, *Science* 1985, **230**, 1350.

following provides an overview of the process of gene cloning and other methods based on recombinant DNA technology.

2 CONSTRUCTING GENE LIBRARIES

2.1 Digesting Genomic DNA Molecules

The isolation and purification of genomic DNA is the first step to many gene cloning experiments (see Chapter 6). DNA is then digested with restriction endonucleases enzymes.[3] These are the key to molecular cloning because of the specificity they have for particular DNA sequences. It is important to realize that every copy of a given DNA molecule from a specific organism will give the same set of fragments when digested with a particular enzyme. DNA from different organisms will, in general, give different sets of fragments when treated with the same enzyme. By digesting complex genomic DNA from an organism it is possible to reproducibly divide its genome into a large number of small fragments, each approximately the size of a single gene. One group of enzymes cut straight across the DNA to give flush or blunt ends whilst other restriction enzymes make staggered cuts, generating short single-stranded projections at each end of the digested DNA. These ends, termed cohesive or sticky ends, are complementary and are therefore able to base-pair with each other. In addition the phosphate groups of the DNA are always retained at the 5' end.

Over 500 enzymes, recognizing more than 200 different restriction sites, have been characterized.[4] The choice of which enzyme to use depends on a number of factors. For example, the recognition sequence of 6 base pairs (bp) will occur on average every 4096 (4^6) bases assuming a random sequence of each of the four bases. This means that digesting genomic DNA with EcoR1, which recognizes the sequence 5'-GAATTC-3', will produce fragments, each of which is on average just over 4 kb. Enzymes with 8 bp recognition sequences produce much longer fragments. Therefore very large genomes, such as human DNA, are usually digested with enzymes that produce long DNA fragments. This makes subsequent steps more manageable, since a smaller number of those fragments need to be cloned and subsequently analysed.

2.2 Ligating DNA Molecules

The DNA products with cohesive or sticky ends resulting from restric-

[3] H. O. Smith and D. Nathans, *J. Mol. Biol.*, 1973, **81**, 419.
[4] R. Fuchs and R. Blakesley, *Meth. Enzymol.*, 1983, **100**, 3.

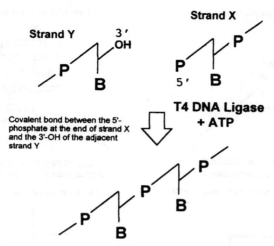

Figure 1   *Action of T4 DNA ligase in the formation of a covalent bond between two DNA strands (Y and X)*

tion enzyme digestion may be joined to any other DNA fragments treated with the same restriction enzyme. Thus, when the two sets of DNA fragments are mixed, base pairing between complementary sticky ends will result in the annealing together of fragments originally derived from different starting DNA. Pairing of fragments derived from the same starting DNA molecules, termed reannealing, will also occur, however, present day cloning vectors have in-built systems to overcome this. All the pairings between DNA fragments are transient because of the weakness of hydrogen bonding between the few bases in the sticky ends, but they can be stabilized by the use of an enzyme DNA ligase, in a process termed ligation.[5] This enzyme, usually isolated from bacteriophage T4 and hence termed T4 DNA ligase, forms a covalent bond between the 5′ phosphate at the end of one strand and the 3′ hydroxyl of the adjacent strand (Figure 1). The reaction is ATP dependent and is often undertaken at 10°C to lower the kinetic energy of molecules. This reduces the chances of base-paired sticky ends parting before they have been stabilized by ligation. However, longer reaction times are needed to compensate for the low activity of DNA ligase at this temperature. It is also possible to join blunt ends of DNA molecules, although the efficiency of this reaction is much lower than sticky-ended ligations.

The ligation of different DNA molecules also reconstructs the site of cleavage (Figure 2). Therefore recombinant molecules produced by

[5] P. Lobban and A. D. Kaiser, *J. Mol. Biol.*, 1973, **79**, 453.

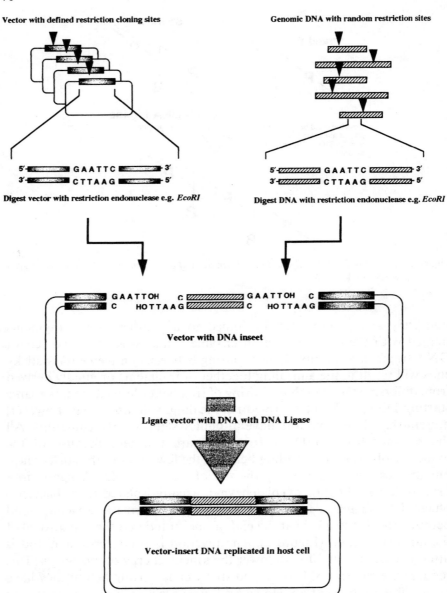

Figure 2 *Overview of the digestion and ligation process on gene cloning*

ligation of sticky ends can also be cleaved, again at the junctions, using the same restriction enzyme that was used to initially generate the fragments. In order to propagate digested DNA from an organism it is necessary to join or ligate that DNA with a specialized DNA carrier

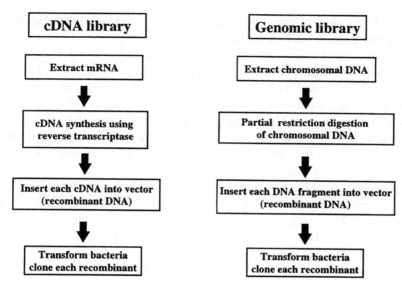

Figure 3 *Comparison of the general steps involved in the construction of genomic and cDNA libraries*

molecule, termed a cloning vector. Thus each DNA fragment is inserted by ligation into the vector DNA molecule, which allows the whole recombined DNA to be replicated indefinitely within an appropriate host cell. In this way a DNA fragment can be cloned to provide sufficient material for further detailed analysis or for further manipulation. Thus, all of the DNA extracted from an organism and digested with a restriction enzyme will result in a collection of clones. This collection of clones is known as a gene library.[6]

2.3 Considerations in Gene Library Preparation

There are two general types of gene library: a genomic library which consists of the total chromosomal DNA of an organism and a cDNA library which represents the mRNA fraction from a cell or tissue at a specific point in time (Figure 3). The choice of the particular type of gene library depends on a number of considerations. The most important of these is the final application of any DNA fragment derived from the library. If the ultimate aim is understanding the control of protein production for a particular gene or its architecture, then a genomic library must be used. However, if the goal is the production of new or

[6] D. H. Schamhart and A. C. Westerhof, *Hum. Mol. Genet.*, 1999, **8**, 1925.

modified proteins, or the determination of tissue-specific expression and timing patterns, a cDNA library is more appropriate. The main consideration in the construction of genomic or cDNA libraries is, therefore, the nucleic acid starting material. Since the genome of an organism is fixed, chromosomal DNA may be isolated from almost any cell type in order to prepare a genomic library. In contrast, however, cDNA libraries represent only the mRNA being produced from a specific cell type at a particular time in the cell's development. Thus, it is important to consider carefully the cell or tissue type from which the mRNA is to be derived in the construction of cDNA libraries.

There are a variety of cloning vectors available, many developed from naturally occurring molecules such as bacterial plasmids or bacteriophage viruses. The choice of vector also depends on whether a genomic library or cDNA library is constructed. The various types of vectors are explained in more detail in Section 3.

2.4   Genomic DNA Libraries

Genomic libraries are constructed by isolating the complete chromosomal DNA from a cell and digesting it into fragments of the desired average length with restriction endonucleases.[7] This can be achieved by partial restriction digestion with an enzyme which recognizes tetranucleotide sequences. Complete digestion with such an enzyme would produce a large number of very short fragments, as shown in Figure 4(b). However, if the enzyme is allowed to partially cleave at only a few of its potential restriction sites before the reaction is stopped, each DNA molecule will be cut into relatively large fragments, as shown in Figure 4(a). In such partial digestions the average fragment size will depend on the relative concentrations of DNA, the restriction enzyme and on the conditions and duration of incubation. It is also possible to produce fragments of DNA by physical shearing, although the ends of the fragments may need to be repaired to make them flush ended. This is achieved by using a modified DNA polymerase, Klenow polymerase, which is prepared by cleavage of DNA polymerase with subtilisin. This gives a large enzyme fragment which has no 5' to 3' exonuclease activity, but which still acts as a 5' to 3' polymerase. This will fill in any recessed 3' ends on the sheared DNA using the appropriate dNTPs.

The mixture of DNA fragments is then ligated with a vector and subsequently cloned. If enough clones are produced there will be a very

---

[7] B. D. M. Theophilus, Genomic DNA Libraries, in 'Molecular Biomethods Handbook', eds. R. Rapley and J. M. Walker, Humana Press, Totowa, NJ, 1998.

(a) **Partial DNA digestion at restriction enzyme sites (E)**

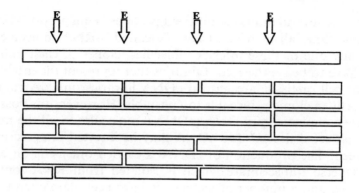

(b) **Complete DNA digestion at restriction enzyme sites (E)**

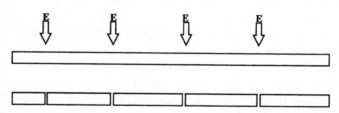

Figure 4 *Comparison of partial and complete digestion of DNA molecules at restriction enzyme sites (E)*

high chance that any particular DNA fragment such as a gene will be present in at least one of the clones. To keep the number of clones to a manageable size, fragments of about 10 kb in length are needed for prokaryotic libraries, but the length must be increased to about 40 kb for mammalian libraries. It is possible to calculate the number of clones that must be present in a gene library to give a probability of obtaining a particular DNA sequence, given by the formula:

$$N = \frac{\ln(1 - P)}{\ln(1 - f)}$$

where N is the number of recombinants, P is the probability and f is the fraction of the genome in one insert. Thus for the *Escherichia coli* DNA chromosome of 5×10^6 bp and with an insert size of 20 kb the number of clones needed (N) would be 1×10^3 with a probability of 0.99.[8]

[8] I. G. Cowell, cDNA Libraries, in 'Molecular Biomethods Handbook', eds. R. Rapley and J. M. Walker, Humana Press, Totowa, NJ, 1998.

2.5   cDNA Libraries

There may be several thousand different proteins being produced in a cell at any one time, all of which have associated mRNA molecules.[8] To identify any one of these mRNA molecules, clones of each individual mRNA have to be synthesized. Libraries that represent the mRNA in a particular cell or tissue are termed cDNA libraries. mRNA cannot be used directly in cloning since it is too unstable. However, it is possible to synthesize complementary DNA (cDNAs) molecules to all the mRNAs from the selected tissue.[9] The cDNA may be inserted into vectors and then cloned. The production of cDNA is carried out using an enzyme termed reverse transcriptase which is isolated from RNA-containing retroviruses, and a number of techniques have been developed to clone full-length cDNAs.[10]

Reverse transcriptase is an RNA-dependent DNA polymerase, and will synthesize a first strand DNA complementary to an mRNA template, using a mixture of the four deoxyribonucleoside triphosphates (dNTP).[11] There is also a requirement (as with all polymerase enzymes) for a short oligonucleotide primer to be present (Figure 5). With eukaryotic mRNA having a poly(A) tail, a complementary oligo(dT) primer may be used. Alternatively random hexamers may be used which anneal randomly to the heterogeneous mRNA in the complex. Random hexamers provide a free 3′ hydroxyl group which is used as the starting point for the reverse transcriptase. Regardless of the method used to prepare the first strand cDNA one absolute requirement is high quality undegraded mRNA (see Chapter 2, Section 2). It is usual to check the integrity of the RNA by denaturing agarose gel electrophoresis. Alternatively a fraction of the extract may be used in a cell-free translation system which, if intact mRNA is present, will direct the synthesis of proteins represented by the mRNA molecules in the sample (Section 5.4).

Following the synthesis of the first DNA strand, a poly(dC) tail is added to its 3′ end, using terminal transferase and dCTP. This will also position a poly(dC) tail on the poly(A) of mRNA. Alkaline hydrolysis is then used to remove the RNA strand, leaving single-stranded DNA which can be used to direct the synthesis of a complementary DNA strand. The second-strand synthesis requires an oligo(dG) primer, base-paired with the poly(dC) tail, which is catalysed by the Klenow fragment

[9] U. Gubler and B. J. Hoffman, *Gene* 1983, **25**, 263.
[10] H. Okayama and P. Berg, *Molecular Cellular Biol.*, 1982, **2**, 161.
[11] R. W. Old and S. B. Primrose, in 'Principles of Gene Manipulation', 5th Edn. Blackwell Science, Oxford, 1994.

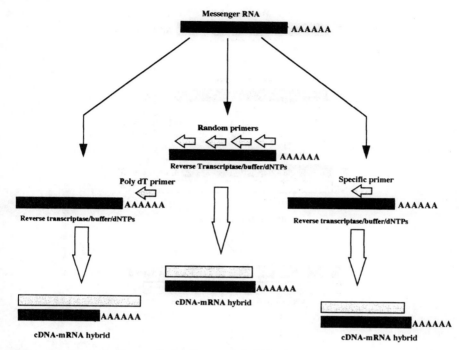

Figure 5 *Strategies for producing first-strand cDNA from mRNA*

of DNA polymerase I. The final product is double-stranded DNA, one of the strands being complementary to the mRNA.

One further method of cDNA synthesis involves the use of RNaseH. Here the first strand cDNA is carried out as above with reverse transcriptase but the resulting mRNA-cDNA hybrid is retained. RNaseH is then used at low concentrations to nick the RNA strand. The resulting nicks expose 3' hydroxyl groups which are used by DNA polymerase as a primer to replace the RNA with a second strand of cDNA (Figure 6).

2.6 Linkers and Adaptors

Ligation of blunt-ended DNA fragments is not as efficient as ligation of sticky ends, therefore with cDNA molecules additional procedures are undertaken before ligation with cloning vectors.[12] One approach is to add small double-stranded molecules with one internal site for a restriction endonuclease, termed nucleic acid linkers, to the cDNA. Numerous linkers are commercially available with internal restriction sites for many

---

[12] T. A. Brown, in 'Gene Cloning, An Introduction', 3rd Edn, Chapman & Hall, London, 1995.

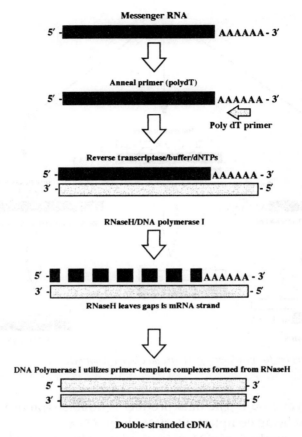

Figure 6   *Second-strand cDNA synthesis using the RNaseH method*

of the most commonly used restriction enzymes. Linkers are blunt-end ligated to the cDNA but, since they are added in great excess of the cDNA, the ligation process is reasonably successful. Subsequently the linkers are digested with the appropriate restriction enzyme which provides the sticky ends for efficient ligation to a vector digested with the same enzyme. This process may be made easier by the addition of adaptors instead of linkers. Adaptors are identical to linkers except that the sticky ends are preformed, obviating the need for restriction digestion following ligation.

2.7   Enrichment Methods for RNA

It is possible to isolate the mRNA transcribed from a desired gene from a cell or tissue that synthesizes and produces a protein in large amounts. In this case a large fraction of the total mRNA will code for that protein, *e.g.* the β-cells of the pancreas contain high levels of proinsulin mRNA. In such a case it is possible to precipitate polysomes which are actively translating the mRNA by using antibodies to the ribosomal proteins. Following precipitation mRNA can be dissociated from the precipitated ribosomes. Generally the mRNA required is only a minor component of the total cellular mRNA and, in such cases, total mRNA may be fractionated by size using sucrose density gradient centrifugation. Each fraction is then used to direct the synthesis of proteins using an *in vitro* translation system (Section 3.7).

2.8 Subtractive Hybridization

In general, genes are either transcribed in a specific cell or tissue type, or differentially activated during a particular stage of development or growth. The levels of the mRNA are usually very low, but it is possible to isolate these mRNA transcripts by subtractive hybridization.[13] First, the mRNA species common to the different cell types are removed, leaving the cell type or tissue-specific mRNAs for analysis. This may be carried out by isolating the mRNA from the so-called subtracter cells and producing a first-strand cDNA. The original mRNA from the subtracter cells is then degraded and the mRNA from the target cells isolated and mixed with the cDNA. All the complementary mRNA-cDNA molecules common to both cell types will hybridize, leaving the unbound mRNA. This may then be isolated and further analysed. A more rapid approach of analysing the differential expression of genes has been developed using the PCR. This technique, termed differential display, is explained in greater detail in Section 8.2.

2.9 Cloning PCR Products

While PCR has to some extent replaced cloning as a method for the generation of large quantities of a desired DNA fragment there is, in certain circumstances, still a requirement for the cloning of PCR amplified DNA.[14] For example, certain techniques such as *in vitro* protein synthesis are best achieved with the DNA fragment inserted into an appropriate plasmid or phage cloning vector (Section 7). Cloning methods for PCR products closely follow the cloning of DNA fragments

---

[13] C. W. Schweinfest, K. W., Henderson, J.-R. Gu, S. D. Kottaridis, S. Besbeas, E. Panatopoulou and T. S. Papas, *Genet. Anal. Techn. Appl.*, 1990, **7**, 64.
[14] B. White, in 'PCR Cloning Protocols', Humana Press, Totowa, NY, 1997.

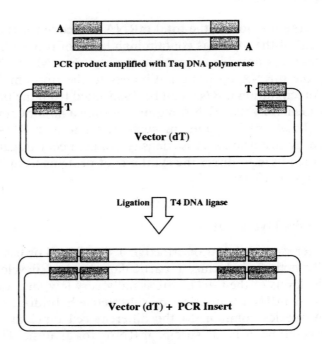

Figure 7   *Cloning of PCR products using dA:dT cloning*

derived from the conventional manipulation of DNA, and may be achieved by either blunt-ended or cohesive-ended cloning. Certain thermostable DNA polymerases such as *Taq* DNA polymerase and *Tth* DNA polymerase give rise to PCR products having a 3′ overhanging A residue. It is possible to clone the PCR product into dT vectors, termed dA:dT cloning.[15] This makes use of the fact that the terminal additions of A residues may be successfully ligated to vectors prepared with T residue overhangs to allow efficient ligation of the PCR product (Figure 7). The reaction is catalysed by DNA ligase, as in conventional ligation reactions (Section 2.2)

It is also possible to carry out cohesive-ended cloning with PCR products. In this case oligonucleotide primers are designed with a restriction endonuclease site incorporated into them. Since the complementarity of the primers needs to be absolute at the 3′ end, it is usually the 5′ end of the primer which is the region for location of the restriction site. This needs to be designed with care since the efficiency of digestion with certain restriction endonuclease decreases if extra nucleotides, not involved in recognition, are absent at the 5′ end. In this case the digestion

---

[15] J. M. Clark, *Nucleic Acids Res.*, 1988, **16**, 9677.

and ligation reactions are the same as those undertaken for conventional reactions (Section 2.1)

3 CLONING VECTORS

For the cloning of any molecule of DNA it is necessary for that DNA to be incorporated into a cloning vector. These DNA elements may be stably maintained and propagated in a host organism for which the vector has replication functions. A typical host organism is a bacterium, such as *E. coli*, which grows and divides rapidly. Thus any vector with a replication origin in *E. coli* will also replicate (together with any incorporated DNA) efficiently. It follows that any DNA cloned into a vector will enable the amplification of the inserted foreign DNA fragment and will also allow any subsequent analysis to be undertaken. In this way the cloning process resembles the PCR, although there are some major differences between the two techniques. By cloning, it is possible to not only store a copy of any particular fragment of DNA, but also to produce unlimited amounts of it.

The vectors used for cloning vary in their complexity, ease of manipulation, their selection and the amount of DNA sequence they can accommodate (the insert capacity). Vectors have, in general, been developed from naturally occurring molecules such as bacterial plasmids, bacteriophages or combinations of the elements that constitute them, such as cosmids (Section 3.4). For gene library constructions there is a choice and trade-off between various vector types; this is usually related to the ease of the manipulations needed to construct the library and the vector's maximum foreign DNA insert size capacity (Table 1). Thus, vectors with the advantage of large insert capacities are usually more difficult to manipulate, although there are many more factors to be considered, which are indicated in the following treatments of vector systems.

Table 1 *Comparison of vectors generally available for cloning DNA fragments*

Vector	Host cell	Vector structure	Insert range (kb)
M13	*E. coli*	Circular virus	1–4
Plasmid	*E. coli*	Circular plasmid	1–5
Phage λ	*E. coli*	Linear virus	2–25
Cosmids	*E. coli*	Circular plasmid	35–45
BACs	*E. coli*	Circular plasmid	50–300
YACs	*S. cerevisiae*	Linear chromosome	100–2000

BAC: bacterial artificial chromosome, YAC: yeast artificial chromosome.

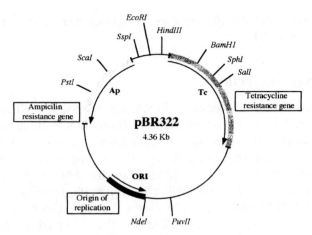

Figure 8 *Map and important features of pBR322*

3.1 Plasmid Derived Cloning Vectors

Many bacteria contain an extrachromosomal element of DNA, termed a plasmid. This is a relatively small, covalently closed circular molecule, which carries genes for antibiotic resistance, conjugation or the metabolism of 'unusual' substrates. Some plasmids are replicated at a high rate by bacteria such as *E. coli* and so are excellent potential vectors. In the early 1970s a number of natural plasmids were artificially modified and constructed as cloning vectors by a complex series of digestion and ligation reactions. One of the most notable plasmids, termed pBR322 after its developers Bolivar and Rodriguez (pBR), was widely adopted and illustrates its desirable features of a cloning vector as indicated below (Figure 8).[16]

(i) The plasmid is much smaller than a natural plasmid which makes it more resistant to damage by shearing and increases the efficiency of uptake by bacteria, a process termed transformation.

(ii) A bacterial origin of DNA replication ensures that the plasmid will be replicated by the host cell. Some replication origins display stringent regulation of replication, in which rounds of replication are initiated at the same frequency as cell division. Most plasmids, including pBR322, have a relaxed origin of replication and activity is not tightly linked to cell division, thus plasmid replication will be initiated far more frequently than chromosomal

[16] F. Bolivar, R. L. Rodriguez, P. J. Greene *et al.*, *Gene*, 1977, **2**, 95.

replication. Hence a large number of plasmid molecules will be produced per cell.

(iii) Two genes coding for resistance to antibiotics have been introduced. One of these allows the selection of cells which contain plasmid: if cells are plated on medium containing an appropriate antibiotic, only those which contain plasmid will grow to form colonies. The other resistance gene can be used, as described below, for detection of those plasmids which contain inserted DNA.

(iv) There are single recognition sites for a number of restriction enzymes at various points around the plasmid which can be used to open or linearize the circular plasmid. Linearizing a plasmid allows a fragment of DNA to be inserted and the circle closed. The variety of sites not only makes it easier to find a restriction enzyme which is suitable for both the vector and the foreign DNA to be inserted, but, since some of the sites are placed within an antibiotic resistance gene, the presence of an insert can be detected by loss of resistance to that antibiotic. This is termed insertional inactivation.

3.1.1 Plasmid Selection Systems. Insertional inactivation is a useful selection method for identifying recombinant vectors with inserts. For example, a fragment of chromosomal DNA digested with *Bam*H1 would be isolated and purified. The plasmid pBR322 would also be digested at a single site, using *Bam*H1, after which both samples would be deproteinized to inactivate the restriction enzyme. BamH1 cleaves to give sticky ends, and so it is possible to obtain ligation between the plasmid and digested DNA fragments in the presence of T4 DNA ligase. The products of this ligation will include plasmid containing a single fragment of the DNA as an insert, but there will also be unwanted products, such as plasmid which has recircularized without an insert, dimers of plasmid, fragments joined to each other, and plasmid with an insert composed of more than one fragment. Most of these unwanted molecules can be eliminated during subsequent steps. The products of such reactions are usually identified by agarose gel electrophoresis (see Chapter 2, Section 3).

The ligated DNA must now be used to transform *E. coli*. Bacteria do not normally take up DNA from their surroundings, but can be induced to do so by prior treatment with Ca^{2+} at 4°C. These cells are then termed competent, since DNA added to the suspension of cells will be taken up during a brief increase in temperature, termed a heat shock. Small, circular DNA molecules are taken up most efficiently, whereas long, linear molecules will be taken up less effectively by the bacteria. Plasmid

DNA can also be introduced into bacterial cells by electroporation. In this process the cells are subjected to pulses of a high voltage gradient, causing many of them to take up DNA from the surrounding solution. This technique has proved to be useful with not only bacterial cells, but also with animal, plant and microbial host cells.

After a brief incubation to allow expression of the antibiotic resistance genes the cells are plated onto medium containing the antibiotic, *e.g.* ampicillin. Colonies which grow on these plates must be derived from cells which contain plasmid, since this carries the gene for resistance to ampicillin. At this stage it is not possible to distinguish between those colonies containing plasmids with inserts and those which simply contain recircularized plasmids. To achieve this, the colonies are replica plated, using a sterile velvet pad, onto plates containing tetracycline in their medium. Since the *Bam*HI site lies within the tetracycline resistance gene, this gene will be inactivated by the presence of insert, but will be intact in those plasmids which have merely recircularized. Thus colonies which grow on ampicillin but not on tetracycline must contain plasmids with inserts. Since replica plating gives an identical pattern of colonies on both sets of plates, it is straightforward to recognize the colonies with inserts, and to recover them from the ampicillin plate for further growth. This illustrates the importance of a second gene for antibiotic resistance in a vector (Figure 9).

Although recircularized plasmid can be selected against, its presence decreases the yield of recombinant plasmid containing inserts. If the digested plasmid is treated with the enzyme alkaline phosphatase prior to ligation, recircularization will be prevented, since this enzyme removes the 5' phosphate groups which are essential for ligation. Links can still be made between the 5' phosphate of the insert and the 3' hydroxyl of the plasmid, so only recombinant plasmids and chains of linked DNA fragments will be formed. It does not matter that only one strand of the recombinant DNA is ligated, since the nick will be repaired by bacteria transformed with these molecules.

3.1.2 pUC Plasmid Cloning Vectors.   The valuable features of pBR322 have been enhanced by the construction of a series of plasmids termed pUC (produced at the University of California) (Figure 10). There is an antibiotic resistance gene for ampicillin and origin of replication for *E. coli*. In addition the most popular restriction sites are concentrated into a region termed the multiple cloning site (MCS). Moreover, the MCS is part of a gene in its own right and codes for a portion of a polypeptide called β-galactosidase.[17]

[17] J. Messing and J. Vieira, *Gene*, 1982, **19**, 259.

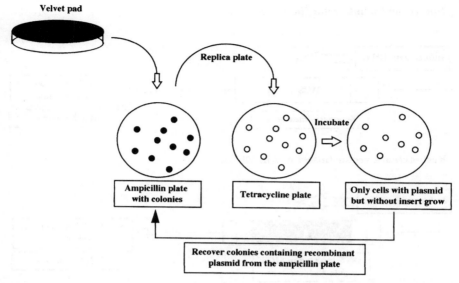

Figure 9 *Replica plating to detect recombinant plasmids. A sterile velvet pad is pressed on to the surface of an agar plate, picking up some cells from each colony growing on that plate. The pad is then pressed on to a fresh agar plate, thus inoculating it with cells in a pattern identical with that of the original colonies. Clones of cells that fail to grow on the second plate (e.g. owing to loss of antibiotic resistance) can be recovered from their corresponding colonies on the first plate*

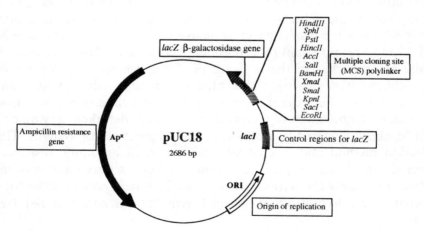

Figure 10 *Map and important features of pUC18*

When the pUC plasmid has been used to transform the host cell *E. coli* the gene may be switched on by adding the inducer IPTG (isopropyl-β-D-thiogalactopyranoside). Its presence causes the enzyme β-galactosidase to be produced. The functional enzyme is able to hydrolyse a colourless

Non-recombinant vector (no insert)

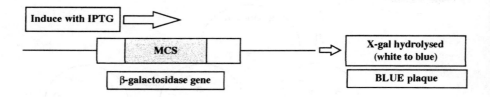

Recombinant vector (insert within MCS)

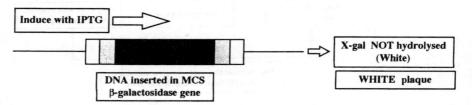

Figure 11   *Principle of blue/white selection for the detection of recombinant vectors*

substance called X-gal (5-bromo-4-chloroindol-3-yl-β-galactopyrano-side) into a blue insoluble material (Figure 11). However if the gene is disrupted by the insertion of a foreign fragment of DNA, a non-functional enzyme results which is unable to carry out hydrolysis of X-gal. Thus, a recombinant pUC plasmid may be easily detected since it is white or colourless in the presence of X-gal, whereas an intact non-recombinant pUC plasmid will be blue as its gene is fully functional and not disrupted. This elegant system, termed blue/white selection, allows the initial identification of recombinants to be undertaken very quickly and has been included in a number of subsequent vector systems. This selection method and insertional inactivation of antibiotic resistance genes do not, however, provide any information on the character of the DNA insert, only the status of the vector. To screen gene libraries for a desired insert hybridization labelled gene probes are required (see Section 5).

3.2   Virus-based Cloning Vectors

A useful feature of any cloning vector is the amount of DNA it may accept or may be inserted before it becomes unviable. Inserts greater than 5 kb in size increase plasmid size to the point at which efficient transformation of bacterial cells decreases markedly. Thus, bacterio-phages (bacterial viruses) have been adapted as vectors in order to propagate larger fragments of DNA in bacterial cells. Cloning vectors derived from one bacteriophage are commonly used since they offer approximately 16-fold advantage in cloning efficiency in comparison with the most efficient plasmid cloning vectors.[18]

Bacteriophage λ is a linear double-stranded phage approximately 49 kb in length. It infects *E. coli* with great efficiency by injecting its DNA through the cell membrane. In the wild-type phage λ the DNA follows one of two possible modes of replication. Firstly the DNA may either become stably integrated into the *E. coli* chromosome where it lies dormant until a signal triggers its excision. This is termed the lysogenic life cycle. Alternatively, it may follow a lytic life cycle where the DNA is replicated upon entry to the cell, phage head and tail proteins synthesized rapidly, and new functional phage assembled. The phage are subse-quently released from the cell by lysing the cell membrane to infect further *E. coli* cells nearby. At the extreme ends of phage λ are 12 bp sequences termed cos (cohesive) sites. Although they are asymmetrical, they are similar to restriction sites and allow the phage DNA to be circularized. Phage may be replicated very efficiently in this way, resulting in concatemers of many phage genomes which are cleaved at the cos sites and inserted into newly formed phage protein heads.

Phage λ has been used extensively in the production of gene libraries, mainly because of its efficient entry into the *E. coli* cell and the fact that larger fragments of DNA may by stably integrated. For the cloning of long DNA fragments, up to approximately 25 kb, much of the non-essential λ DNA that codes for the lysogenic life cycle is removed and replaced by the foreign DNA insert. The recombinant phage is then assembled into pre-formed viral protein particles, a process known as *in vitro* packaging. These newly formed phages are used to infect bacterial cells which have been plated out on agar.

Once inside the host cells, the recombinant viral DNA is replicated. All the genes needed for normal lytic growth are still present in the phage DNA, and so multiplication of the virus takes place by cycles of cell lysis and infection of surrounding cells, giving rise to plaques of lysed cells on a background, or lawn, of bacterial cells. The viral DNA including the cloned foreign DNA can be recovered from the viruses from these plaques and analysed further by restriction mapping and agarose gel electrophoresis (see Chapter 2, Section 3).

---

[18] R. Mutzel, λ as a cloning vector, in 'Molecular Biomethods Handbook', Humana Press, Totowa, NJ, 1998.

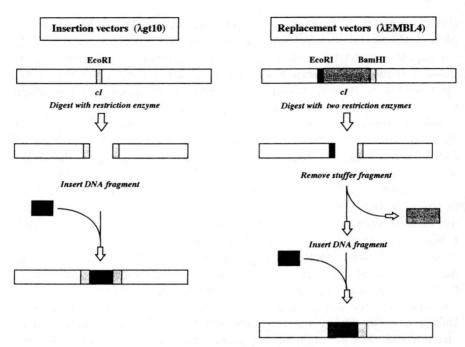

Figure 12   *General schemes used for cloning in λ insertion and λ replacement vectors*

3.2.1   Insertion and Replacement Cloning Vectors. In general, two types of λ phage vectors have been developed, λ insertion vectors and λ replacement vectors (Figure 12).[19] λ insertion vectors accept less DNA than the replacement type since the foreign DNA is simply inserted into a region of the phage genome with appropriate restriction sites; common examples are λgt10 and λcharon16A.[20] In a replacement vector a central region of DNA not essential for lytic growth is removed (a stuffer fragment) by a double digestion with, for example, *Eco*RI and *Bam*HI. This leaves two DNA fragments, called right and left arms. The central stuffer fragment is replaced by inserting foreign DNA between the arms to form a functional recombinant λ phage. The most notable examples of λ replacement vectors are λEMBL and λZAP.

λZAP is a commercially produced cloning vector which includes unique cloning sites clustered into a multiple cloning site (MCS) (Figure 13).[21] Furthermore, the MCS is located within a lacZ region providing a blue/white screening system based on insertional inactiva-

[19] A. M. Frischauf, H. Lehrach, A. Poutska and N. Murray, *J. Mol. Biol.*, 1983, **179**, 827.
[20] F. R. Blattner, B. G. Williams, A. E. Blechel *et al.*, *Science*, 1977, **196**, 161.
[21] J. M. Short, J. M. Femmde, J. A. Sorge and W. D. Huse, *Nucleic Acid Res.*, 1988, **16**, 7583.

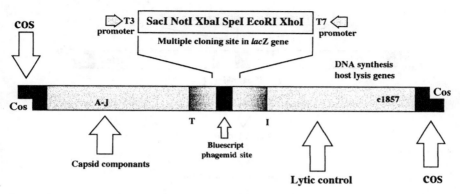

Figure 13 *General map of λZAP cloning vector indicating important areas of the vector. The multiple cloning site is based on the lacZ gene providing blue/white selection based on the β-galactosidase gene. In between the initiator (I) site and terminator (T) site sequences lie sequences encoding the phagemide Bluescript*

tion. It is also possible to express foreign cloned DNA from this vector. This is a very useful feature of some λ vectors since it is then possible to screen for protein product rather than the DNA inserted into the vector. This screening is undertaken with antibody probes directed against the protein of interest (Section 5.3). Another feature that make this a useful cloning vector is the ability to produce RNA transcripts, termed cRNA or riboprobes. This is possible because two promoters for RNA polymerase enzymes exist in the vector, a T7 and a T3 promoter, which flank the MCS (Section 4.2).

One of the most useful features of λZAP is that it has been designed to allow automatic *in vivo* excision of a small 2.9 kb colony-producing vector, the phagemid pBluescript SK (Section 3.3). This technique is sometimes called single-stranded DNA rescue and occurs as the result of a 'superinfection' process where helper phages are added to the cells which are grown for an additional period of approximately 4 hours (Figure 14).

The helper phage displaces a strand within the λZAP which contains the foreign DNA insert. This is circularized and packaged as a filamentous phage similar to M13 (Section 3.3). The packaged phagemid is secreted from the *E. coli* cell and may be recovered from the supernatant. Thus the λZAP vector allows a number of diverse manipulations to be undertaken without the necessity of re-cloning or subcloning foreign DNA fragments. The process of subcloning is sometimes necessary when the manipulation of a gene fragment cloned in a general purpose vector requires insertion into a more specialized vector for the

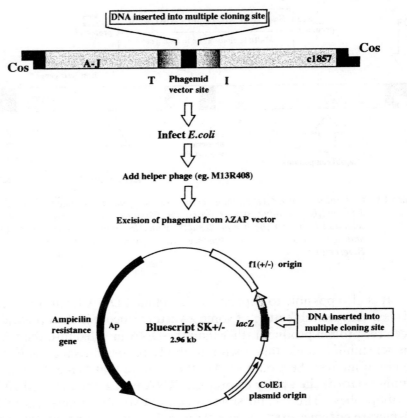

Figure 14   *Single-stranded DNA rescue of phagemid from λZAP. The single-stranded phagemid Bluescript may be excised from λZAP by addition of helper phage. This provides the necessary proteins and factors for transcription between the I and T sites in the parent phage to produce the phagemid with the DNA cloned into the parent vector*

application of techniques such as *in vitro* mutagenesis or protein production.

3.3   M13 and Phagemid-based Cloning Vectors

Single-stranded bacteriophage vectors such as M13 have been of great value in molecular biology.[22] In addition, vectors which have the combined properties of phage and plasmid, called phagemids, are also an established and frequently used reagent. M13 is a filamentous coliphage with a genome composed of single-stranded circular DNA

---

[22] J. Messing *et al.*, *Proc. Natl. Acad. Sci. USA*, 1977, **74**, 3642.

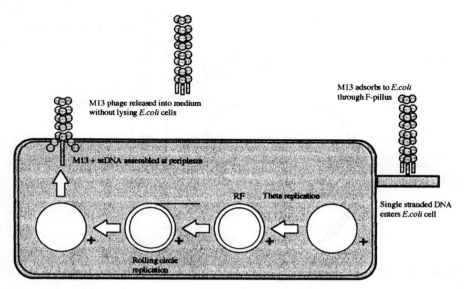

Figure 15 *Simplified life cycle of bacteriophage M13. Following infection of* E. coli, *the M13 DNA replicates initially as a double-stranded (replicative form) RF molecule but subsequently switches production to single-stranded DNA. The resulting virions are then synthesized and released into the medium for infection of further* E. coli *cells*

(Figure 15). Following infection of *E. coli*, the DNA initially replicates as a double-stranded molecule but subsequently switches production to single-stranded DNA. The resulting virions for infection of further bacterial cells are 1then synthesized. The nature of these vectors makes them ideal for techniques such as chain termination sequencing (see Chapter 2, Section 9) and *in vitro* mutagenesis (Section 6.2), since both methods require single-stranded DNA.[23]

M13 or phagemids such as pBluescript SK infect *E. coli* harbouring the male-specific structure called the F pillus. They enter the cell by adsorption to this structure and, once inside the phage DNA, are converted to a double-stranded 'replicative form' or RF DNA. Replication then proceeds rapidly until some 100 RF molecules are produced within the *E. coli* cell. DNA synthesis then switches to the production of single strands and the DNA is assembled and packaged into the capsid at the bacterial periplasm. The bacteriophage DNA is then encapsulated by the major coat protein, gene VIII protein, of which there are approximately 2800 copies with three to six copies of the gene III protein at one end of the particle. The extrusion of the bacteriophage through the

[23] F. Sanger *et al.*, *J. Mol. Biol.*, 1980, **143**, 161.

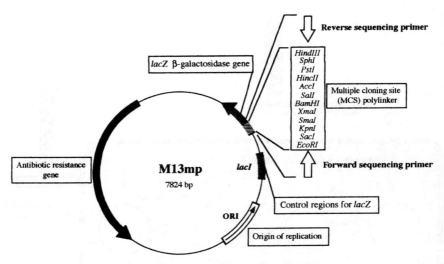

Figure 16 *Map and important features of bacteriophage vector M13*

bacterial periplasm results in a decreased growth rate of the *E. coli* cell rather than host cell lysis and is visible on a bacterial lawn as an area of clearing. Approximately 1000 packaged phage particles may be released into the medium in one cell division.

3.3.1 Cloning into Single-stranded Phage Vectors. In addition to producing single-stranded DNA the coliphage vectors have a number of other features that make them attractive as cloning vectors. Since the bacteriophage DNA is replicated as a double-stranded RF DNA intermediate, a number of regular DNA manipulations may be performed such as restriction digestion, mapping and DNA ligation. RF DNA is prepared by lysing infected *E. coli* cells and purifying the supercoiled circular phage DNA using the same methods used for plasmid isolation. Intact single-stranded DNA packaged in the phage protein coat located in the supernatant may be precipitated with reagents such as polyethylene glycol, and the DNA purified with phenol/chloroform (see Chapter 2, Section 1). Thus the bacteriophage can act as a plasmid under certain circumstances and at other times can produce DNA in the fashion of a virus. A family of vectors derived from M13 are currently widely used (M13mp8/9, mp18/19, *etc.*), all of which have a number of highly useful features (Figure 16). All contain a synthetic MCS which is located in the *lac*Z gene without disruption of the reading frame of the gene.[24] This allows efficient selection to be undertaken based on the technique of blue/

[24] C. Yanisch-Peron *et al.*, *Gene*, 1985, **33**, 103.

white screening (Section 3.1.1). As the series of vectors were developed the number of restriction sites was increased in an asymmetric fashion. Thus M13mp8, mp12 and mp18, and their sister vectors, M13mp9, mp13 and mp19, respectively, which have the same MCS but in reverse orientation, have more restriction sites in the MCS. This makes the vector more useful since a greater choice of restriction enzymes is available. However, one problem frequently encountered with M13 is the instability and spontaneous loss of inserts that are greater then 6 Kb.

Phagemids are very similar to M13 and replicate in a similar fashion.[25] One of the first phagemid vectors to be developed, pEMBL, was constructed by inserting a fragment of another phage, termed f1, containing a phage origin of replication and elements for its morphogenesis into a pUC8 plasmid. Following superinfection with helper phage, the f1 origin is activated, allowing single-stranded DNA to be produced. The phage is assembled into a phage coat extruded through the periplasm and secreted into the culture medium in a similar way to M13. Without superinfection the phagemid replicates as a pUC type plasmid and in the replicative form (RF) the DNA isolated is double stranded. This allows further manipulation such as restriction digestion, ligation and mapping analysis to be performed. The pBluescript SK vector is also a phagemid and can be used in its own right as a cloning vector and manipulated as if it were a plasmid. It can, like M13, be used in nucleotide sequencing and site-directed mutagenesis and it is also possible to produce RNA transcripts which may be used in the production of labelled cRNA probes or riboprobes.

3.4 Cosmid-based Cloning Vectors

The upper limit of the insert capacity of bacteriophage λ is approximately 21 kb. This is because of the requirement for essential genes and the fact that the maximum length between the cos sites is 52 kb.[26] Consequently cosmid vectors have been constructed which incorporate the cos sites from bacteriophage λ and also the essential features of a plasmid, such as the plasmid origin of replication, a gene for drug resistance, and several unique restriction sites for insertion of the DNA to be cloned (Figure 17). When a cosmid preparation is linearized by restriction digestion and ligated to DNA for cloning, the products will include concatamers of alternating cosmid vector and insert. Thus the

[25] R. Rapley. M13 and phagemid based cloning vectors, in 'Molecular Biomethods Handbook', ed. R. Rapley and J. M. Walker, Humana Press, Totowa, NJ, 1998.
[26] B. Honn and J. Collins, *Gene*, 1980, **11**, 291.

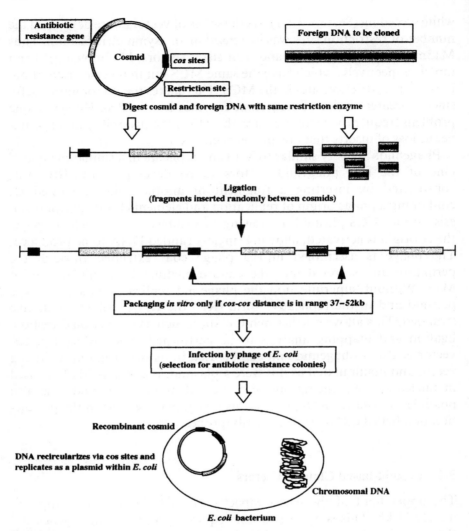

Figure 17   *Scheme of cloning foreign DNA fragments in cosmid vectors*

only requirement for a length of DNA to be packaged into viral heads is that it should contain cos sites spaced the correct distance apart; in practice this spacing can range from 37 to 52 kb. Such DNA can be packaged *in vitro* if bacteriophage head precursors, tails and packaging proteins are provided. Since the cosmid is very small, inserts of about 40 kb in length will be most readily packaged. Once inside the cell, the DNA recircularizes through its cos sites, and from then on behaves exactly like a plasmid. Cosmids are especially useful for the analysis of highly complex genomes and are an important part of various genome mapping projects (Section 10).

3.5 Large Insert Capacity Cloning Vectors

Vectors which accept larger fragments of DNA than phage λ or cosmids have the distinct advantage that fewer clones need to be screened when searching for the foreign DNA of interest. In addition they have had an enormous impact in the mapping of the genomes of organisms such as the mouse and are used extensively in the human genome mapping project. Further developments have given rise to the production of large insert capacity vectors based on bacterial and mammalian artificial chromosomes (BACs and MACs, respectively) and on the virus P1, P1 artificial chromosomes (PACs). However, perhaps most significant is the development of vectors based on yeast artificial chromosomes or YACs.[27]

3.6 Yeast Artificial Chromosome (YAC) Cloning Vectors

Yeast artificial chromosomes (YACs) are linear molecules composed of a centromere, telomere and a replication origin termed an ARS element (autonomous replicating sequence).[28] The YAC is digested with restriction enzymes at the SUP4 site (a suppressor tRNA gene marker) and *Bam*HI sites separating the telomere sequences (Figure 18). This produces two arms and the foreign genomic DNA is ligated to produce a functional YAC construct. YACs are replicated in yeast cells, however the external cell wall of the yeast needs to be removed to leave a spheorplast. These are osmotically unstable and need to be embedded in a solid matrix such as agar. Once the yeast cells are transformed, only correctly constructed YACs with associated selectable markers are replicated in the yeast strains. DNA fragments with repeated sequences which are sometimes difficult to clone in bacterially based vectors may also be cloned in YAC systems. The main advantage of YAC-based vectors, however, is the ability to clone very large fragments of DNA. Thus the stable maintenance and replication of foreign DNA fragments of up to 2000 kb have been carried out in YAC vectors and they are the main vector of choice in the various genome mapping and sequencing projects (Section 10).[29]

---

[27] A. P. Monaco and Z. Larin, *Trends Biotechnol.*, 1994, **12**, 280.
[28] A. W. Murray and J. W. Szostak, *Nature*, 1983, **305**, 189.
[29] A. Flannery and R. Anand, Yeast artificial chromosomes, in 'Molecular Biomethods Handbook', ed. R. Rapley and J. M. Walker, Humana Press, Totowa, NJ, 1998.

pYAC2 Yeast Artificial Chromosome

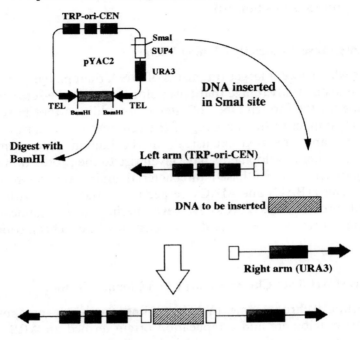

Figure 18   *Scheme for the use of yeast artificial chromosomes (YACs) in cloning large DNA fragments*

3.7   Vectors Used in Eukaryotic Cells

The application of *E. coli* as a general host for cloning and manipulation of DNA is well established. However, numerous developments have been made in cloning in eukaryotic cells such as yeasts and plants. Plasmids used for cloning DNA in eukaryotic host cells require a eukaryotic origin of replication and marker genes which will be expressed by these cells. Yeasts, like bacteria, can be grown rapidly, and are therefore well suited for use in cloning.[30] These eukaryotic cells also have a natural plasmid, the 2μ circle, however this is too large for use in cloning. Plasmids such as the the yeast episomal plasmid (YEp) have been produced by manipulation using replication origins from the 2μ circle, and by incorporating a gene which will complement a gene made defective in the host yeast cell. Thus if a strain of yeast is used which has a defective gene for the biosynthesis of an amino acid, an active copy of that gene on a yeast

[30] H. Cooke, *Trends Genet.*, 1987, **6**, 248.

plasmid can be used as a selectable marker for the presence of that plasmid. Particular use has been made of shuttle vectors which have origins of replication for both yeasts and bacteria such as *E. coli*. This allows the rapid preparation of constructs in *E. coli* which may then be delivered into yeast for expression studies.

The bacterium *Agrobacterium tumefaciens* infects plants which have been damaged near soil level, and this infection is often followed by the formation of plant tumours in the vicinity of the infected region.[31] It is well established that *A. tumefaciens* contains the 'Ti' plasmid. Part of this plasmid is transferred into the nuclei of plant cells which are infected by the bacterium. Once in the nucleus, this DNA is maintained by integrating with the chromosomal DNA. The integrated DNA carries genes for the synthesis of opines (which are metabolized by the bacteria but not by the plants) and for tumour induction (hence 'Ti'). DNA inserted into the correct region of the Ti plasmid will be transferred to infected plant cells, and in this way it has been possible to clone and express foreign genes in plants.

It is possible to deliver recombinant DNA molecules into plant cells, however there are a number of problems with this.[32] In general, the outer cell wall of the plant must be removed, usually by enzymatic digestion, to leave a protoplast. Cells are then able to take up recombinants from the supernatant. The cell wall can be regenerated by providing appropriate media. In cases where protoplasts have been generated transformation may also be achieved by electroporation. A more dramatic transformation procedure involves propelling microscopically small titanium or gold pellet microprojectiles coated with the recombinant DNA molecule into plant cells in intact tissues. This technique, termed biolistics, involves the detonation of an explosive charge which is used to propel the microprojectiles into the cells at a high velocity. The cells then appear to reseal themselves after the delivery of the recombinant molecule. This is a particularly useful technique for use with plants whose protoplasts will not regenerate whole plants.

4 GENE PROBES AND HYBRIDIZATION

4.1 Cloned DNA Probes

The availability of custom oligonucleotide synthesis and the PCR have contributed greatly to the rapid production of gene probes. These are

---

[31] K. M. A Gartland and M. R. Davey, in 'Agrobacterium Protocols', Humana Press, Totowa, NJ, 1995.
[32] H. Jones, 'Plant Gene Transfer and Expression Protocols', Humana Press, Totowa, NJ, 1995.

usually designed from gene databases or gene family related sequences. However, there are many gene probes that have been derived from cDNA or from genomic sequences and which have been cloned into plasmid and phage vectors. These require manipulation before they may be labelled and used in hybridization experiments. Gene probes may vary in length from 100 bp to a number of kilobases, although this is dependent on their origin. Many are short enough to be cloned into plasmid vectors and are useful in that they may be manipulated easily and are relatively stable both in transit and in the laboratory. The DNA sequences representing the gene probe are usually excised from the cloning vector by digestion with restriction enzymes followed by purification. In this way, vector sequences which may hybridize non-specifically and cause high background signals in hybridization experiments are removed. There are various ways of labelling DNA probes and these are described in Chapter 2, Section 6.

4.2   RNA Gene Probes

It is also possible to prepare cRNA probes or riboprobes by *in vitro* transcription of gene probes cloned into a suitable vector.[33] A good example of such a vector is the phagemid pBluescript SK, since it has promoters for T3 or T7 RNA polymerase at each end of the multiple cloning site where the cloned DNA fragment resides. Following cloning the vector is made linear with a restriction enzyme and T3 or T7 RNA polymerase is used to transcribe the cloned DNA fragment. Provided a labelled dNTP is added the reaction, a riboprobe labelled to a high specific activity will be produced (Figure 19). One advantage of riboprobes is that they are single stranded and their sensitivity is generally regarded as superior to cloned double-stranded probes. They are used extensively in *in situ* hybridization and for identifying and analysing mRNA and are described in more detail in Section 8.3.

5   SCREENING GENE LIBRARIES

5.1   Colony and Plaque Hybridization

Following the construction of a cDNA or genomic library the next step is to identify the specific DNA fragment of interest. In many cases this element may be more difficult than the production of the library itself since many hundreds of thousands of clones may be in the library.

---

[33] D. A. Melton *et al.*, *Nucleic Acids Res.*, 1984, **12**, 7035.

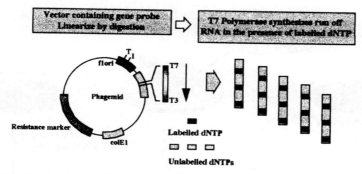

Figure 19 *Production of labelled RNA gene probes (riboprobes)*

Table 2 *Numbers of clones required for representation of DNA in genome library*

Species	Genome size (kb)	Number of clones required	
		17 kb fragments	35 kb fragments
E. coli	4000	700	340
Yeast	20 000	3500	1700
Fruit fly	165 000	29 000	14 500
Man	3 000 000	535 000	258 250
Maize	15 000 000	2 700 000	1 350 000

Clones containing the desired fragment need to be located from the library and in order to undertake this a number of techniques, mainly based on hybridization, have been developed.

Colony hybridization is a method frequently used to identify a particular DNA fragment from a plasmid gene library (Figure 20). A large number of clones are grown to form colonies on one or more plates.[34] These are then replica-plated onto nylon membranes placed on solid agar medium. Nutrients diffuse through the membranes and allow colonies to grow on them. The colonies are then lysed, and liberated DNA is denatured and bound to the membranes, so that the pattern of colonies is replaced by an identical pattern of bound DNA. The membranes are then incubated with a prehybridization mixture containing non-labelled non-specific DNA, such as salmon sperm DNA, to block non-specific sites. Following this, denatured, labelled gene probe is added. Under hybridizing conditions the probe will bind only to cloned fragments containing at least part of its corresponding gene. The

[34] M. Grunstein and D. S. Hogness, *Proc. Natl. Acad. Sci. USA*, 1975, **72**, 3961.

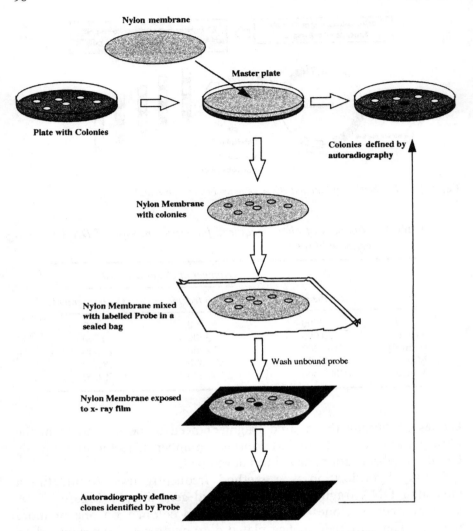

Figure 20 *Screening immobolized bacterial colonies using labelled gene probes*

membranes are then washed to remove any unbound probe and the
binding detected by autoradiography of the membranes. If non-radioactive labels have been used then alternative methods of detection must be
employed (see Chapter 2, Section 6.2). By comparing the patterns on the
autoradiograph with the original plates of colonies, those which contain
the desired gene (or part of it) can be identified and isolated for further
analysis. A similar procedure is used to identify desired genes cloned into
bacteriophage vectors. In this case the process is termed plaque hybridi-

zation. It is the DNA contained in the bacteriophage particles found in each plaque that is immobilized on to the nylon membrane. This is then probed with an appropriately labelled complementary gene probe and the detection undertaken as for colony hybridization.

5.2 Gene Library Screening by PCR

The PCR may also be applied to cDNA or genomic library screening where the libraries are constructed in vectors such as plasmids or bacteriophage.[35] This is usually undertaken with primers that are designed to anneal to the vector rather than the foreign DNA insert. The size of an amplified products may be used to characterize the cloned DNA. Subsequently restriction mapping is then usually undertaken. The main advantage of the PCR over traditional hybridization-based screening is the rapidity of the technique; PCR screening may be undertaken in 3–4 hours whereas it may be several days before detection by hybridization is achieved. The PCR screening technique gives an indication of the size of the cloned inserts rather than the sequence of the insert, however PCR primers that are specific for a foreign DNA insert may also be used. This allows a more rigorous characterization of clones from cDNA and genomic libraries.

5.3 Screening Expression cDNA Libraries

In a number of cases, the protein for which the gene sequence is required may be partially characterized and it may therefore be possible to raise antibodies to that protein. This allows immunological screening, rather than gene hybridization, to be undertaken.[36] Such antibodies are useful since they may be used as the probe if little or no nucleic acid sequence information is available. In such cases it is possible to prepare a cDNA library in a specially adapted vector, termed an expression vector. These transcribe and translate any cDNA inserted into the vector. The protein is usually synthesized as a fusion with another protein such as β-galactosidase (Figure 21). Common examples of expression vectors are those based on bacteriophage such as λgt11 and λZap or plasmids such as pEX.

The precise requirements for expression vectors are identical to vectors which are dedicated to producing proteins *in vitro* and are

[35] A. M. Griffin and H. M. Griffin, 'Molecular Biology Current Innovations and Future Trends', Horizon Scientific Press, Norfolk, UK, 1995.
[36] R. A. Young and R. W. Davis, *Proc. Natl. Acad. Sci. USA*, 1983, **80**, 1194.

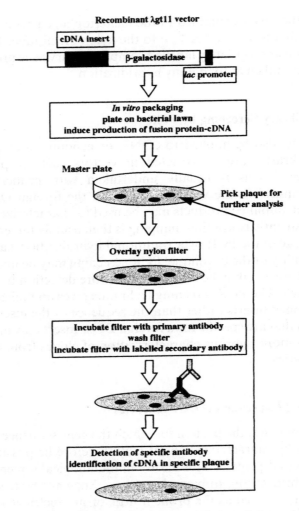

Figure 21   *Screening of cDNA libraries constructed in expression vector λgt11. The cDNA inserted upstream of the gene for β-galactosidase will give rise to a fusion protein under induction (e.g. IPTG). The plaques are then blotted on to a nylon membrane filter and probed with an antibody specific for the protein coded by the cDNA. A secondary labelled antibody directed to the specific antibody can then be used to identify the location (plaque) of the cDNA*

described in Section 7. In some cases expression vectors incorporate inducible promoters which may be activated by, for example, increasing the temperature; this allows stringent control of expression of the cloned cDNA molecules. The cDNA library is plated out and nylon membrane filters prepared as for colony/plaque hybridization. A solution containing the antibody to the desired protein is then added to the membrane. The membrane is then washed to remove any unbound protein and a further labelled antibody which is directed to the first antibody is applied. This allows visualization of the plaque or colony that contains the cloned cDNA for that protein and this may then be picked from the agar plate and pure preparations grown for further analysis.

5.4 Hybrid Select/Arrest Translation

Problems encountered in the characterization of clones and detection of desired DNA fragments from a mixed cDNA libraries may in part be overcome by two additional techniques. These are termed hybrid select (release) translation or hybrid arrest translation.[37] Following the preparation of a cDNA library in a plasmid vector the plasmid is extracted from part of each colony, and each preparation is denatured and immobilized onto a nylon membrane. The membranes are soaked in total cellular mRNA under stringent conditions, usually a temperature only a few degrees below T_m (melting temperature), in which hybridization will occur only between complementary strands of nucleic acid. Therefore each membrane will bind just one species of mRNA because it has only one type of cDNA immobilized on it. Unbound mRNA is washed off the membranes and the bound mRNA is eluted and used to direct *in vitro* translation (Section 7). By immunoprecipitation or electrophoresis of the protein, the mRNA coding for a particular protein can be detected and the clone containing its corresponding cDNA isolated. This technique is known as hybrid release translation. In a related method, termed hybrid arrest translation, a positive result is indicated by the absence of a particular translation product when total mRNA is hybridized with excess cDNA. This is a consequence of the fact that mRNA cannot be translated when it is hybridized to another molecule.

6 APPLICATIONS OF GENE CLONING

6.1 Sequencing Cloned DNA

DNA fragments cloned into plasmid vectors may be subjected to the chain termination sequencing method detailed in Chapter 2.[38] However, since plasmids are double stranded, further manipulation needs to be

[37] B. M. Paterson *et al.*, *Proc. Natl. Acad. Sci. USA*, 1977, **74**, 4370.
[38] H. G. Griffin and A. M. Griffin, 'DNA Sequencing Protocols', Humana Press, Totowa, NJ, 1993.

undertaken before this may be attempted. In these cases the plasmids are denatured, usually by alkali treatment. Although the plasmids containing the foreign DNA inserts may reanneal, the kinetics of the reaction is such that the strands are single stranded for a long enough period of time to allow the sequencing method to succeed. It is also possible to include denaturants such as formamide to the reaction to prevent further reannealing. In general, however, the superior results gained with sequencing single-stranded DNA from M13 or single-stranded phage-mids means that cloned DNA of interest are usually subcloned into such vectors.

M13 vectors are the traditional choice for chain termination sequencing because of the single-stranded nature of their DNA (Section 3.3). A further modification that makes M13 useful in chain termination sequencing is the placement of universal priming sites at -20 or -40 bases from the start of the multiple cloning site. This allows any gene to be sequenced by using one universal primer, since annealing of the primer prior to sequencing occurs outside the multiple cloning site and so is M13 specific rather than gene specific. This obviates the need to synthesize new oligonucleotide primers for each new foreign DNA insert. A further reverse-priming site is also located at the opposite end of the polylinker, allowing sequencing in the opposite orientation to be undertaken.

6.2 *In vitro* Mutagenesis

One of the most powerful developments in molecular biology has been the ability to artificially create defined mutations in a gene and analyse the resulting protein following *in vitro* expression.[39] Numerous methods are now available for producing site-directed mutations, many of which now involve the PCR. Commonly termed 'protein engineering', it is possible to undertake a logical sequence of analytical and computational techniques centred around a design cycle. This involves the biochemical preparation and analysis of proteins, the subsequent identification of the gene encoding the protein, and its modification. The production of the modified protein and its further biochemical analysis completes the concept of rational redesign to improve the structure and function of a protein. The use of design cycles and rational design systems are exemplified by the study and manipulation of engineered antibodies (see Chapter 14).

---

[39] M. Smith, *Ann. Rev. Genet.*, 1985, **19**, 423.

6.3 Oligonucleotide-directed Mutagenesis

Perhaps the most well established method for site directed mutagenesis is oligonucleotide-directed mutagenesis.[40] This demands that the gene is already cloned or subcloned into a single-stranded vector such as M13. Complete sequencing of the gene is essential to identify a potential region for mutation. Once the precise base change has been identified an oligonucleotide is designed that is complementary to most of the relevant gene sequence but has one base difference. This difference is designed to alter a particular codon which, following translation, gives rise to a different amino acid and hence may alter the properties of the protein.

The oligonucleotide and the single-stranded DNA are annealed and DNA polymerase is added together with the dNTPs. The primer for the reaction is the 3' end of the oligonucleotide. The DNA polymerase produces a new complementary DNA strand to the existing one but, importantly, incorporates the oligonucleotide with the base mutation in the new strand. The subsequent cloning of the recombinant produces multiple copies, half of which contain a sequence with the mutation and half contain the wild-type sequence. Plaque hybridization using the oligonucleotide as the probe is then used at a stringency which allows only those plaques containing a mutated sequence to be identified (Figure 22). Further methods have also been developed which simplify the process of detecting the strands with the mutations.[41]

6.4 PCR-based Mutagenesis

The PCR has been adapted to allow rapid and effective mutagenesis to be undertaken.[42] This relies on a single base being mismatched between one of the PCR primers and the target DNA which then becomes incorporated into the amplified product following thermal cycling.

The basic PCR mutagenesis system involves the use of two primary PCR reactions to produce two overlapping DNA fragments both bearing the same mutation in the overlap region, a technique called overlap extension PCR. The two separate PCR products are made single stranded and the overlap in sequence allows the products from each reaction to hybridize. Following this, one of the two hybrids bearing a free 3' hydroxyl group is extended to produce a new duplex fragment. The other hybrid with a 5' hydroxyl group cannot act as substrate in the

[40] M. J. Zoller and M. Smith, *Meth. Enzymol.*, 1987, **154**, 329.
[41] M. M. Ling and B. H. Robinson, *Anal. Biochem.*, 1997, **254**, 157.
[42] R. Higuchi, B. Krummel, and R. K. Saiki, *Nucleic Acids Res.*, 1988, **16**, 7351.

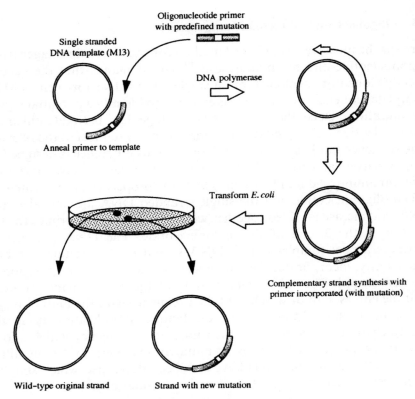

Figure 22   *Site-directed mutagenesis using an oligonucleotide primer*

reaction. Thus, the overlapped and extended product will now contain the directed mutation (Figure 23). Deletions and insertions may also be created with this method although the requirements for four primers and three PCR reactions limits the general applicability of the technique. A modification of the overlap extension PCR may also be used to construct directed mutations, termed megaprimer PCR.[43] This method utilizes three oligonucleotide primers to perform two rounds of PCR. A complete PCR product, the megaprimer is made single stranded and this is used as a large primer in a further PCR reaction with an additional primer.

The above are all methods for creating rational defined mutations as part of a design cycle system. However, it is also possible to introduce random mutations into a gene and select for enhanced or new activities of the protein or enzyme it encodes. This accelerated form of artificial

[43] G. Sarker and S. S. Sommer, *Biotechniques*, 1990, **8**, 404.

Megaprimer based PCR mutagenesis

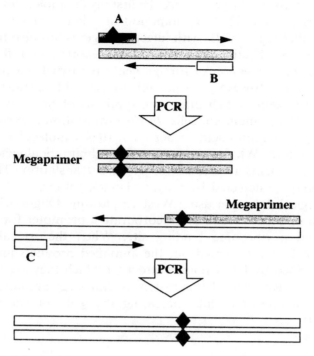

Figure 23 *Introduction of mutations using the megaprimer technique*

molecular evolution may be undertaken using error-prone PCR where deliberate and random mutations are introduced by a low fidelity PCR amplification reaction. The resulting amplified gene is then translated and its activity assayed. This has already provided novel evolved enzymes such as a *p*-nitrobenzyl esterase which exhibits an unusual and surprising affinity for organic solvents. This accelerated evolutionary approach to protein engineering has been useful in the production of novel phage-displayed antibodies and in the development of antibodies with enzymic activities (catalytic antibodies).

7 EXPRESSION OF FOREIGN GENES

In terms of biotechnology one of the most useful applications of recombinant DNA technology is the ability to artificially synthesize large quantities of natural or modified proteins.[44] This is usually carried

[44] R. S. Taun, 'Recombinant Gene Expression Protocols', Humana Press, Totowa, NJ, 1997.

out in a host cell *e.g.* bacteria or yeast. The benefits of such methods have been enjoyed for many years since the first insulin molecules were cloned and expressed in 1982. Contamination of other proteins, such as the blood product factor VIII, with infectious agents has also increased the need to develop effective vectors for *in vitro* expression of foreign genes. In general the expression of foreign genes is carried out in specialized cloning vectors. However, it is possible to use cell-free transcription and translation systems which direct the synthesis of proteins without the need to grow and maintain cells. *In vitro* translation is carried out with the appropriate amino acids, ribosomes, tRNA molecules and isolated mRNA fractions. Wheat germ extracts or rabbit reticulocyte lysates are usually the systems of choice for *in vitro* translation. The resulting proteins may be detected by polyacrylamide gel electrophoresis or by immunological detection using Western blotting. Oligonucleotide PCR primers have been designed to incorporate a promoter for RNA polymerase and a ribosome binding site. When the so called E-PCR (expression PCR) is carried out the amplified products are denatured and transcribed by RNA polymerase after which they are translated *in vitro*.[45] The advantage of this system is that large amounts of specific RNA are synthesized, thus increasing the yield of specific proteins (Figure 24).

7.1 Production of Fusion Proteins

For a foreign gene to be expressed in a bacterial cell, it must have particular promoter sequences upstream of the coding region, to which the RNA polymerase will bind prior to transcription of the gene (see Chapter 4). The choice of promoter is vital for correct and efficient transcription since the sequence and position of promoters are specific to a particular host such as *E. coli*.[46] It must also contain a ribosome binding site placed just before the coding region. If the gene has been produced *via* cDNA from a eukaryotic cell, then it will certainly not have any such sequences. Consequently, expression vectors have been developed which contain promoter and ribosome binding sites positioned just before one or more restriction sites for the insertion of foreign cDNA. These regulatory sequences are usually derived from genes which, when induced, are strongly expressed in bacteria (*e.g.* the *lac* operon).[47]

---

[45] D. E. Lanar and K. C. Kain, *PCR Methods Appl.*, 1994, **4(2)**, S92.
[46] O. Pines and M. Inouye, *Mol. Biotechnol.* 1999, **12(1)**, 25.
[47] H. Lilie, E. Schwarz and R. Rudolph, *Curr. Opin. Biotechnol.*, 1998, **9(5)**, 497.

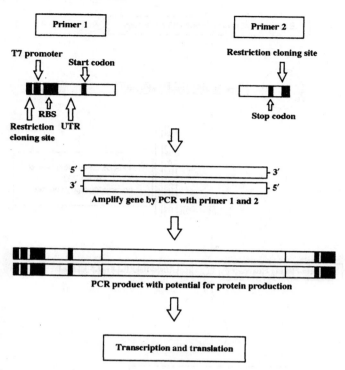

Figure 24 *Expression PCR (E-PCR). This technique amplifies a target sequence with one promoter that contains a transcriptional promoter, ribosome binding site (RBS), untranslated leader region (UTR) and start codon. The other primer contains a stop codon. The amplified PCR may be used in transcription and translation to produce a protein*

Since the mRNA produced from the gene is read as triplet codons, the inserted sequence must be placed so that its reading frame is in phase with the regulatory sequence. In some cases the protein is expressed as a fusion with a protein such as β-galactosidase or glutathione-S-transferase (GST) to facilitate its recovery.[48] It may also be tagged with a moiety such as a polyhistidine (6xHis-Tag) which binds strongly to a nickel-chelate-nitrilotriacetate (Ni–NTA) chromatography column (Figure 25). The usefulness of this method is that the binding is independent of the three-dimensional structure of the 6xHis-Tag and so recovery is efficient even under strong denaturing conditions. This may be required for membrane proteins or for those that give rise to inclusion bodies. The tags are subsequently removed by cleavage with a reagent such as

[48] N. Sheibani, *Prep. Biochem. Biotechnol.* 1999, **29(1)**, 77.

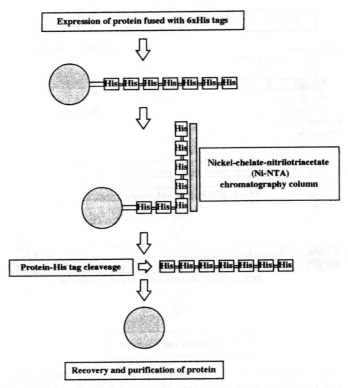

Figure 25 *Recovery of proteins using (6xHis-Tag) and (Ni-NTA) chromatography columns*

cyanogen bromide and the protein of interest purified by protein biochemical methods such as chromatography and polyacrylamide gel electrophoresis. Table 3 indicates a number of fusion expression systems that are commonly employed.

Although the use of bacterial expression systems is somewhat limited for eukaryotic proteins there are a number of eukaryotic expression systems based on plant, mammalian, insect and yeast cells. These types of cells are able to perform post-translational modifications, producing correct glycosylation pattern, a process that cannot be carried out in bacterial systems. It is also possible to include a signal or address sequence at the 5′ end of the mRNA which directs the protein to a particular cellular compartment or even into the supernatant. This makes the recovery of expressed recombinant proteins much easier as the supernatant may be drawn off while the cells are still producing protein.

Table 3   *Commonly employed fusion systems*

Fusion partner	Size	Ligand	Elution conditions
ZZ	14 kDa	IgG	Low pH
His-Tag	6–10 aa	Ni^{2+}	Imidazole
Strep-Tag	10 aa	Streptavidin	Iminobiotin
Pinpoint	13 kDa	Streptavidin	Biotin
MBP	40 kDa	Amylose	Maltose
β-lactamase	27 kDa	Phenyl-boronate	Borate
GST	25 kDa	Glutathione	Reducing agent
Flag	8 aa	Specific mAb	Low calcium

ZZ, fragment of *Staphylococcus aureus* protein A; MBP, maltose binding protein; Pinpoint, protein fragment that is biotinylated *in vivo*; Flag, peptide recognized by enterokinase; aa, amino acid.

7.2   Expression in Mammalian Cells

One useful eukaryotic expression system is based on the monkey COS cell-line. These cells each contain a region derived from a mammalian monkey virus termed simian virus 40 (SV40). A defective region of the SV40 genome has been stably integrated into the COS cell genome. This allows the expression of a protein called the large T antigen which is required for viral replication. Viral replication takes place when a recombinant vector with the SV40 origin of replication and carrying foreign DNA is inserted into the COS cells. This results in a high level expression of foreign proteins. The disadvantage of this system is the ultimate lysis of the COS cells and limited insert capacity of the vector. Much interest is also currently focused on other modified viruses, vaccinia virus and baculovirus. These have been developed for high-level expression in mammalian and insect cells, respectively. The vaccinia virus in particular has been used to correct defective ion transport by introducing a wild-type cystic fibrosis gene into cells bearing a mutated cystic fibrosis (CFTR) gene. There is no doubt that the further development of these vector systems will enhance eukaryotic protein expression in the future.

7.3   Display of Proteins on Bacteriophage

As a result of the production of phagemid vectors and as a means of overcoming the problems of screening large numbers of clones generated from genomic libraries of antibody genes a method for linking the phenotype or expressed protein with the genotype has been devised.[49]

[49] G. P. Smith, *Science*, 1985, **228**, 1315.

This is termed phage display since a functional protein is linked to a major coat protein of a coliphage whilst the single-stranded gene encoding the protein is packaged within the virion (see Chapter 14).[50] The initial steps of the method rely on the PCR to amplify gene fragments that represent functional domains or subunits of a protein such as an antibody. These are then cloned into a phage display vector which is an adapted phagemid vector (Section 3.3) and used to transform *E. coli*. A helper phage is then added to provide accessory proteins for new phage molecules to be constructed. The DNA fragments representing the protein or polypeptide of interest are also transcribed and translated, but are linked to the major coat protein gIII. Thus, when the phage is assembled, the protein or polypeptide of interest is incorporated into the coat of the phage and displayed, whilst the corresponding DNA is encapsulated (Figure 26).

There are numerous applications for the display of proteins on the surface of bacteriophage virus and commercial organizations have been quick to exploit this technology. One major application is the analysis and production of engineered antibodies from which the technology was mainly developed. In general, phage-based systems have a number of novel applications in terms of ease of selection rather than screening of antibody fragments, allowing analysis by methods such as affinity chromatography. In this way it is possible to generate large numbers of antibody heavy and light chain genes by PCR amplification and mix them in a random fashion. This recombinatorial library approach may provide new or novel partners to be formed, as well as naturally existing ones. This strategy is not restricted to antibodies and vast libraries of peptides may be used in this combinatorial chemistry approach to identify novel compounds of use to biotechnology and medicine.

Phage-based cloning methods also offer the advantage of allowing mutagenesis to be performed with relative ease. This may allow the production of antibodies with affinities approaching those derived from the human or mouse immune system. This may be brought about by using an error prone DNA polymerase in the initial steps of constructing a phage display library. It is possible that these types of libraries may provide a route to high affinity recombinant antibody fragments that are difficult to produce by more conventional hybridoma fusion techniques. Surface display libraries have also been prepared for the selection of ligands, hormones and other polypeptides in addition to allowing studies on protein–protein or protein–DNA interactions, or determining the precise binding domains in these receptor–ligand interactions.

---

[50] R. Crameri and M. Suter, *Gene*, 1993, **137**, 69.

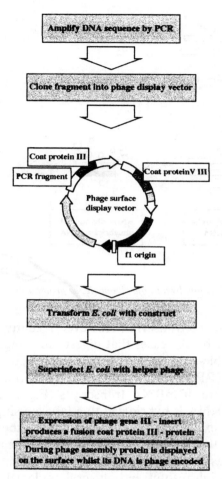

Figure 26 *General scheme for the technique of phage display*

8 ANALYSING GENES AND GENE EXPRESSION

8.1 Identifying and Analysing mRNA

A number of very useful techniques have been developed that allow the
fine structure of a particular mRNA to be analysed and the relative
amounts of an RNA quantitated. This is important not only for gene
regulation studies but may also be used as a marker for certain clinical
disorders. Traditionally the Northern blot has been used for detection of
particular RNA transcripts by blotting extracted mRNA and immobiliz-
ing it to a nylon membrane (see Chapter 2, Section 5). Subsequent
hybridization with labelled gene probes allow precise determination of

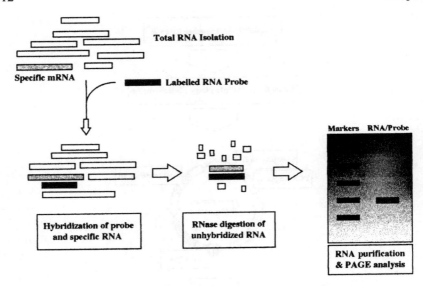

Total RNA Isolation

Specific mRNA

Labelled RNA Probe

Hybridization of probe and specific RNA

RNase digestion of unhybridized RNA

Markers   RNA/Probe

RNA purification & PAGE analysis

Figure 27   *General steps involved in the ribonuclease protection assay (RPA)*

the size and nature of a transcript. However, recently much use has been made of a number of nucleases that digest only single-stranded nucleic acids and not double-stranded molecules. In particular, the ribonuclease protection assay (RPA) has allowed much information to be gained regarding the nature of mRNA transcripts (Figure 27).[51] In the RPA single-stranded mRNA is hybridized to a labelled single-stranded RNA probe which is in excess. The hybridized part of the complex becomes protected whereas the unhybridized part of the probe made from RNA is digested with RNaseA and RNaseT1. The protected fragment may then be analysed on a high resolution polyacrylamide gel. This method may give valuable information regarding the mRNA in terms of the precise structure of the transcript (transcription start site, intron/exon junctions, *etc.*). It is also quantitative and requires less RNA than a Northern blot. A related technique, S1 nuclease mapping, is similar although the unhybridized part of a DNA probe, rather than an RNA probe, is digested, this time with the enzyme S1 nuclease.

8.2   Reverse Transcriptase PCR (RT-PCR)

The PCR has also had an impact on the analysis of RNA by the development of a technique known as reverse transcriptase-PCR (RT-PCR). Here the RNA is isolated and a first strand cDNA synthesis

---

[51] M. A. Aranda, A. Farile, F. Garcia-Arenal and J. M. Malpica, *Arch. Virol.*, 1995, **140**, 1373.

undertaken with reverse transcriptase, the cDNA is then used in a conventional PCR (Section 3.2.5). Under certain circumstances a number of thermostable DNA polymerases have reverse transcriptase activity which obviates the need to separate the two reactions and allows the RT-PCR to be carried out in one tube. One of the main benefits of RT-PCR is the ability to identify rare or low levels of mRNA transcripts with great sensitivity. This is especially useful when detecting, for example, viral gene expression and furthermore allows the means of differentiating between latent and active virus. The level of mRNA production may also be determined by a modification of the PCR termed, quantitative PCR (see Chapter 2, Section 7.7).

In many cases the analysis of tissue-specific gene expression is required and again the PCR has been adapted to provide a solution. This technique, termed differential display, is also an RT-PCR based system requiring that isolated mRNA be first converted into cDNA.[52] Following this, one of the PCR primers, designed to anneal to a general mRNA element such as the poly A tail in eukaryotic cells, is used in conjunction with a combination of arbitrary 6–7 bp primers which binds to the 5′ end of the transcripts. This consequently results in the generation of multiple PCR products with reproducible patterns (Figure 28). Comparative analysis by gel electrophoresis of PCR products generated from different cell types therefore allows the identification and isolation of those transcripts that are differentially expressed. As with many PCR based techniques, the time to identify such genes is dramatically reduced from the weeks that are required to construct and screen cDNA libraries to a few days.

8.3 Analysing Genes *in situ*

Major changes in the appearence of chromosomes (*e.g.* following rearrangements) may often be detected by light microscopic examination of the chromosomes following staining (Section 2.3). Single or restricted numbers of base substitutions, small deletions, rearrangements or insertions are far less easily detectable. However these may also induce similarly profound effects on normal cellular biochemistry. *In situ* hybridization makes it possible to determine the chromosomal location of a particular gene fragment or gene mutation.[53] This is carried out by preparing a labelled DNA or RNA probe and applying this to a tissue or chromosomal preparation fixed to a microscope slide. Any probe that

[52] P. Liang and A. B. Pardee, *Science*, 1992, **257**, 967.
[53] C. S. Herrington, *Mol. Pathol.*, 1998, **51(1)**, 8.

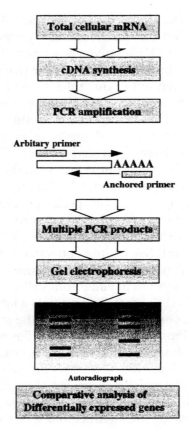

Figure 28   *Scheme showing the differential display method*

does not hybridize to complementary sequences is washed off and an image of the distribution or location of the bound probe is viewed by an appropriate detection system (Figure 29). It is also possible to carry out *in situ* PCR using tissue or cells fixed to slides. This is a highly sensitive technique where PCR is carried out directly on the tissue slide with the standard PCR reagents. Specially adapted thermal cycling machines are required to hold the slide preparations and allow the PCR to proceed. This allows the localization and identification of, for example, single copies of intracellular viruses.

An alternative labelling strategy used in karyotyping and gene localization is fluorescent *in situ* hybridization (FISH).[54] This method, sometimes termed chromosome painting, is based on *in situ* hybridization, but different gene probes are labelled with different fluorochromes, each

---

[54] P. Lichter, *Trends Genet.*, 1997, **13**, 475.

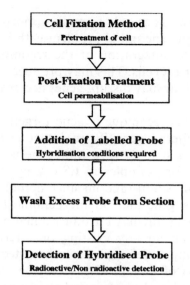

Figure 29 *Scheme for the technique of* in situ *hybridization*

specific for a particular chromosome. The advantage of this method is that separate gene regions may be identified and comparisons made within the same chromosome preparation. This technique is also likely to be highly useful in genome mapping for ordering DNA probes along a chromosomal segment (Section 3.9).

8.4 Transgenics and Gene Targeting

In many cases it is desirable to analyse the effect of certain genes and proteins in an organism (*in vivo*) rather than in the laboratory (*in vitro*). Furthermore, the production of pharmaceutical products and therapeutic proteins is also desirable in a whole organism. The introduction of foreign genes into germ-line cells and the production of an altered organism is termed transgenics. There are two broad strategies for transgenesis. The first is direct transgenesis in mammals, whereby a recombinant DNA construct is injected directly into the male pronucleus of a recently fertilized egg. This is then raised in a foster mother animal resulting in offspring that are all transgenic. The second is selective transgenesis, whereby the recombinant DNA is transferred into embryo stem cells (ES cells). The cells are then cultured in the laboratory and those found expressing the desired protein are selected and incorporated into the inner cell mass of an early embryo. The resulting transgenic animal is raised in a foster mother but, in this case, the transgenic animal is mosaic or chimaeric, since only a small proportion of the cells will be expressing the protein. The initial problem with both approaches is the random nature of the integration of the recombinant DNA into the genome of the egg or ES cells. This may produce proteins in cells where they are not required, or may disrupt genes necessary for correct growth and development.

A refinement of this technique is gene targeting which involves the production of an altered gene in an intact cell, a form of *in vivo* mutagenesis as opposed to *in vitro* mutagenesis. The gene is inserted into the genome of, for example, an ES cell by specialized viral-based vectors. The insertion is non-random since homologous sequences exist on the vector of the gene and on the genome being targeted. Thus, homologous recombination may introduce a new genetic property to the cell, or inactive in an already existing one, termed gene knockout. Perhaps the most important aspect of these techniques is that they allow animal models of human diseases to be created. This is useful since the physiological and biochemical consequences of a disease are often complex and difficult to study, impeding the development of diagnostic and therapeutic strategies.

9   MICROARRAYS AND DNA CHIPS

Perhaps one of the most exciting nucleic acid analysis techniques developed recently has been that of microarrays or DNA microchips. These provide a radically different approach to large-scale analysis and quantification of genes and gene expression patterns.[55] A microarray consists of an ordered arrangement of hundreds or thousands of DNA sequences such as oligonucleotides or cDNAs deposited onto a solid surface approximately 1.2 × 1.2 cm. The solid supports are usually wafers of glass or silicon. Arrays are synthesized on or off the glass and require complex fabrication methods similar to those used in producing microchips.[56] They may also be spotted by ultrafine robotic microarray deposition instruments which dispense volumes as low as 30 pl.

The arrays may represent mRNA produced in a particular cell type, termed cDNA expression arrays, or alternatively represent coding and regulatory regions of a particular gene or group of genes developed. A commercial example uses a 50 000 oligonucleotide array that represents known mutations in a tumour suppressor gene termed *p53*, a protein known to be mutated in many human cancers.[57] In this case, patient

[55] E. M. Southern, *Trends Genet.*, 1996, **12**, 110.
[56] K. Mir and E. M. Southern, *Nature Biotechnol.*, 1999, **17**, 788.
[57] G. Ramsey, *Nature Biotechnol.*, 1998, **16**, 40.

sample DNA is deposited and incubated on the array and any unhybridized DNA is washed off. The array is then analysed and scanned for patterns of hybridization, usually by a computer-controlled confocal fluorescence microscope. Any mutations in the *p53* gene may be rapidly analysed by computer interpretation of the resulting hybridization pattern and any mutation defined. The potential of microarrays appears to be limitless and a number of arrays have been developed for the detection of various genetic mutations including the cystic fibrosis CFTR gene (cystic fibrosis transmembrane regulator), the breast cancer gene, *BRCA1*, and for the human immunodefficiency virus (HIV).

At present microarrays require DNA to be highly purified which limits their applicability. However, as DNA purification becomes automated and microarray technology develops it is not difficult to envisage numerous laboratory tests on a single DNA microchip. This could be used not only for analysing single genes but also for large numbers of genes or DNA representing microorganisms, viruses, *etc.* Since the potential for quantitation of gene transcription exists, microarrays could also be used in defining a particular disease status. The capacity for a DNA microarray to contain numerous sequences has given rise to the possibility to perform DNA sequencing by hybridization. This technique which is currently under development may be a very significant one since it will allow large amounts of sequence information to be gathered very rapidly and assist in many fields of molecular biology, especially in large genome sequencing projects.

10 ANALYSING WHOLE GENOMES

Perhaps the most ambitious projects in the biosciences are the initiatives to map and completely sequence a number of genomes from various organisms, as indicated in Table 4. A number have already been completed, *e.g.* the bacterium *E. coli* and the yeast *S. cerevisiae*, and

Table 4 *Current selected genome sequencing projects*

Organism		Genome size
Bacteria	*Escherichia coli*	4.6 Mb
Yeast	*Saccharomyces cerevisiae*	14 Mb
Roundworm	*Caenorhabditis elegans*	100 Mb
Fruit fly	*Drosophilia melanogaster*	165 Mb
Puffer fish	*Fugu rubripes rubripes*	400 Mb
Mouse	*Mus musculus*	3000 Mb

some such as the *C. elegans*, are near completion. The mouse and the ultimate challenge, the human genome, are also both currently in progress and initially completed. The demands of such large-scale mapping and sequencing have provided the impetus for the development and refinement of even the most standard of molecular biology techniques such as DNA sequencing. It has also led to new methods of identifying the important coding sequences that represent proteins and enzymes.

10.1 Physical Genome Mapping

In terms of genome mapping a physical map is the primary goal. Genetic linkage maps have also be produced by determining the recombination frequency between two particular loci.[58] YAC-based vectors essential for large-scale cloning contain DNA inserts that are on average 300 000 bp in length, which is longer by a factor of ten than the longest inserts in the clones used in early mapping studies. The development of vectors with large insert capacity has enabled the production of contigs. These are continuous overlapping cloned fragments that have been positioned relative to one another. Using these maps any cloned fragment may be identified and aligned to an area in one of the contig maps. In order to position cloned DNA fragments resulting from the construction of a library in a YAC or cosmid it is necessary to detect overlaps between the cloned DNA fragments. Overlaps are created because of the use of partial digestion conditions with a particular restriction endonuclease when constructing the libraries. This ensures that when each DNA fragment is cloned into a vector, it has overlapping ends which theoretically may be identified and the clones positioned or ordered so that a physical map may be produced (Figure 30).

In order to position the overlapping ends it is preferable to undertake DNA sequencing, however, due to the impracticality of this approach, a fingerprint of each clone is carried out using restriction enzyme mapping. Although this is not an unambiguous method of ordering clones it is useful when also applying statistical probabilities of the overlap between clones. In order to link the contigs, techniques such as *in situ* hybridization may be used or a probe may be generated from one end of a contig in order to screen a different disconnected contig. This method of probe production and identification is termed walking, and has been used successfully in the production of physical maps of *E. coli* and some yeast genomes. This cycle of clone to fingerprint to contig is amenable to

[58] T. A. Brown, 'Genomes', Bios Scientific Publications, Oxford, 1999.

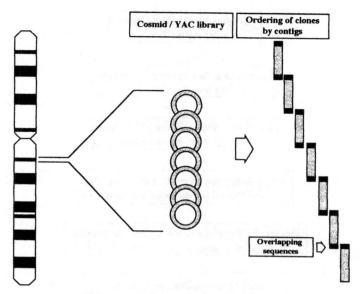

Figure 30 *Continuous overlapping cloned fragments (contigs) used in physical mapping. In order to position cloned DNA fragments resulting from the construction of a library in a YAC or cosmid, vector overlaps are detected between the cloned fragments. These are created because of the use of partial digestion conditions when constructing the libraries*

automation, however the problem of closing the gaps between contigs remains difficult.

In order to define a common way for all research laboratories to order clones and connect physical maps together an arbitrary molecular technique based on the PCR has been developed based on sequence tagged sites (STS).[59] This is a small unique sequence of 200–300 bp that is amplified by PCR (Figure 31). The uniqueness of the STS is defined by the PCR primers that flank the STS. A PCR is performed with those primers and, if the PCR results in selected amplification of target region, it may be defined as a potential STS marker. In this way, defining STS markers that lie approximately 100 000 bases apart along a contig map allows the ordering of those contigs. Thus, all groups working with clones have definable landmarks with which to order clones produced in their libraries.

An STS that occurs in two clones will overlap and thus may be used to order the clones in a contig. Clones containing the STS are usually detected by Southern blotting where the clones have been immobilized on a nylon membrane. Alternatively a library of clones may be divided

[59] T. J. Hudson, L. D. Stein, S. S. Gerety *et al.*, *Science*, 1995, **270**, 1945.

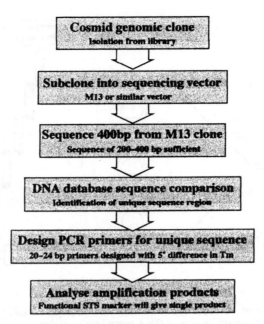

Figure 31   *Scheme for the derivation of a sequence tagged site (STS)*

into pools and each pool PCR screened. This is usually a more rapid method of identifying an STS within a clone and further refinements of the PCR-based screening method allows the identification of a particular clone within a pool (Figure 32). STS elements may also be generated from variable regions of the genome to produce a polymorphic marker that may be traced through families along with other DNA markers and located on a genetic linkage map. These polymorphic STSs are useful since they may serve as markers on both a physical map and a genetic linkage map for each chromosome and therefore provide a useful marker for aligning the two types of map.

10.2   Gene Discovery and Localization

Over the last few years there has been a convergent approach to genomics and identification resulting in a new field of functional genomics.[60] This area has been augmented by the use of *in situ* mapping techniques such as FISH (Section 8.3). In fact a number of genes have been identified and the protein determined where little was initially known about the gene except for its location. This method of gene

---

[60] S. Fields, Y. Kohara and D. J. Lockhart, *Proc. Natl. Acad. Sci. USA*, 1999, **16** 8825.

Chromosome

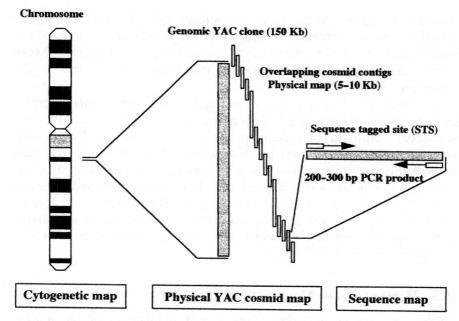

Figure 32 *Scheme for the use of STS markers in physical mapping*

discovery is known as positional cloning and was instrumental in the isolation of the CFTR gene responsible for the disorder cystic fibrosis. Further new areas based on functional genomic and termed pharmaco-genomics may also translate the information into new strategies for therapeutics and clinical management of diseases.[61]

The number of genes actively expressed in a cell at any one time is estimated to be as little as 10% of the total number. The remaining DNA is packaged and serves an as yet unknown function. Recent investigations have found that certain active genes may be identified by the presence of so called HTF (HpaII tiny fragments) islands often found at the 5' end of genes. These are CpG-rich sequences that are not methylated and form tiny fragments on digestion with the restriction enzyme HpaII. A further gene discovery method that has been used extensively in the past few years is a PCR-based technique giving rise to a product termed an expressed sequence tag (EST). This represents part of a putative gene for which a function has yet to be assigned.[62] The technique is carried out on cDNA using primers that bind to an anchor sequence such as a poly A tail and primers which bind to sequences at the

[61] W. E. Evans and M. V. Relling, *Science*, 1999, **286**, 487.
[62] M. A. Marra, L. Hiller and R. H. Waterston, *Trends Genet.*, 1997, **14**, 4.

5′ end of the gene. Such a PCR may subsequently be used to map the putative gene to a chromosomal region or the putative gene as a probe to search a genomic DNA library for the remaining parts of the gene. Much interest currently lies in ESTs since they may represent a shortcut to gene discovery.

A further gene isolation system which uses adapted vectors is termed exon trapping or exon amplification and may be used to identify exon sequences. Exon trapping requires the use of a specialized expression vector that will accept fragments of genomic DNA containing sequences for splicing reactions to take place. Following transfection of a eukaryotic cell line a transcript is produced that may be detected by using specific primers in a RT-PCR. This indicates the nature of the foreign DNA by virtue of the splicing sequences present. A list of further techniques that aid in the identification of a potential gene encoding sequence is indicated in Table 5.

10.3 Human Genome Mapping Project

There is no doubt that the mapping and sequencing of the human genome is one of the most ambitious projects in current science. It will certainly bring new insights into gene function and gene regulation, provide a means of identifying DNA mutations or lesions found in

Table 5 *Techniques used to determine putative gene encoding sequences*

Identification method	Main details
Zoo blotting (cross-hybridization)	Evolutionary conservation of DNA sequences that suggest functional significance
Homology searching	Gene database searching to gene family related sequences
Identification of CpG islands	Regions of hypomethylated CpG frequently found 5′ to genes in vertebrate animals
Identification of open reading frames (ORF) promoters/splice sites/RBS	DNA sequences scanned for consensus sequences by computer
Northern blot hybridization	mRNA detection by binding to labelled gene probes
Exon trapping technique	Artificial RNA splicing assay for exon identification
Expressed sequence tags (ESTs)	cDNAs amplified by PCR that represent part of a gene

current genetic disorders and point to new ways of potential therapy.[63] It is a multi-collaboration effort that has engaged many scientific research groups around the world and given rise to many scientific, technical, financial and ethical debates. One interesting issue is the sequencing of the whole genome in relation to the coding sequences.[64] Much of the human genome appears to be non-coding and composed of repetitive sequences. Only a small portion of the genome appears to be encode enzymes and proteins. Nevertheless, this still corresponds to approximately 90 000 genes and their mapping and sequencing is an exciting prospect. The study further aims to understand and possibly provide the eventual means of treating some of the 4000 genetic diseases in addition to other diseases whose inheritance is multifactorial. The diversity of the human genome is also an area of great interest and is currently under study. Despite some initial reservations, the complete sequence is predicted to be completed by 2003 at the earliest and, following this, the even more difficult task of decoding and interpreting the complete sequence will be required.

---

[63] J. C. Venter, M. D. Adams, G. G. Sutton, A. R. Kerlavage, H. O. Smith, and M. Hunkapiller, *Science*, 1998, **280**, 1540.
[64] E. Pennisi, *Science*, 1998, **280**, 1692.

CHAPTER 4

The Expression of Foreign DNA in Bacteria

ROBERT J. SLATER AND D. ROSS WILLIAMS

1 INTRODUCTION

The cloning of large quantities of specific DNA sequences, regardless of origin, is now straightforward. The advent of this technology has allowed the exploitation of gene products through a fuller understanding of the mechanisms of gene expression. Restriction endonuclease cleavage sites can be mapped as reference points on cloned DNA, the nucleotide sequence can be determined and coding sequences (open reading frames, or ORFs) identified. However, unless expression of cloned DNA can be obtained in a suitable host or system, many fundamental studies on the expression or function of genes could not be carried out, and most commercial applications of genetic engineering would not be possible.

The product of expression of foreign genes in host organisms is frequently referred to as 'recombinant protein' and its production is of enormous value to both fundamental studies and commercial biotechnology. Coding sequences are ligated into 'expression vectors' designed to replicate and express inserted DNA in host cells. The most popular hosts are *Escherichia coli*, *Bacillus subtilis*, yeast, and cultured cells of higher eukaryotes such as insect or mammalian cells, though the range of hosts has expanded considerably in recent years. *E. coli* has been the host of choice for much work because of the extensive knowledge of its genetics and ease of growth. Therefore, much of this article will concentrate on expression in *E. coli*, though for many purposes the bacterium is not ideal. Large (> 500 amino acids) or highly hydrophobic proteins or those requiring post-translational modification are best made in other cells, or need additional systems to ensure their correct function.

Also the reducing environment within *E. coli* is not conducive to the formation of disulfide bonds; proteins with many cysteines need to be made using a secretion system or alternative host.

To obtain expression of foreign genes in bacteria such as *E. coli*, it is first necessary to construct suitable vectors. Two principle types of vector are used: derivatives of viruses (*e.g.* bacteriophage λ) and plasmids (see Chapter 2). Specially designed bacteriophage constructs, such as the classical λ gt11 or λZAP are expression vectors for the production of polypeptides specified by DNA inserts. These cloning tools produce recombinant protein as β-galactosidase fusion products (see Section 3.4). They are used, for example, to clone genes *via* the identification of gene products.[1] Such vectors are not suitable for large-scale production because of cell lysis as a result of virus infection. They are used in small-scale production, for example, in the production of large numbers of variant recombinant proteins for functional screening. Plasmids are the vectors of choice for the large-scale production of foreign proteins in bacteria. Consequently their use will be described in this chapter after some necessary background has been discussed.

It has been known since 1944 that it is possible to transform bacteria with DNA from the same or closely related species.[2] Recombinant DNA techniques, however, have given molecular biologists the opportunity to obtain expression of foreign DNA from totally unrelated species, even mammals and higher plants, in bacteria such as *E. coli*. The expectation that DNA should be expressed in an unrelated organism was based on the observation that the genetic code is universal. That is, a DNA sequence coding for a protein in one organism should function in any organism to produce a protein with the same amino acid sequence. This was given support when it was found that not only DNA from other bacteria but also DNA from lower eukaryotes, *Saccharomyces cerevisiae*[3] and *Neurospora crassa*,[4] could be expressed in *E. coli*. Unfortunately, when similar experiments were carried out with DNA from higher eukaryotes, expression of the foreign gene was not obtained; more sophisticated techniques, discussed later, were required.

The benefits that accrue from obtaining expression of foreign DNA in bacteria are considerable. Fundamental studies are made possible concerning the relationship between the protein's primary structure (the

[1] T. V. Huynh, R. A. Young, and R. W. Davis, in 'DNA Cloning, Vol 1; A Practical Approach', ed. D. M. Glover, IRL Press, Oxford and Washington, DC, 1985, p. 45.
[2] O. T. Avery, C. M. Macleod, and M. McCarty, *J. Exp. Med.*, 1944, **79**, 137.
[3] K. Struhl, J. R. Cameron, and R. W. Davis, *Proc. Nat. Acad. Sci. USA*, 1967, **73**, 1471.
[4] D. Vapnek, J. A. Hautala, J. W. Jacobson, N. H. Giles, and S. R. Kushner, *Proc. Nat. Acad. Sci. USA*, 1977, **74**, 3508.

amino acid sequence) and its function. Large quantities of recombinant protein can be produced instead of purifying the protein from the original source, where supplies may be limited or purification difficult. In addition the DNA sequence of cloned DNA can be altered by *in vitro* mutagenesis and the effect of this mutation on subsequent protein properties studied. Amino acid sequences responsible for catalysis or membrane binding, for example, can be established. Such studies have established the discipline of protein engineering, important not only for research purposes but also in the production of polypeptides with improved properties. An example is subtilisin, a serine protease used in laundry products. Site-directed mutagenesis was used to substitute the methionine at position 222, a site on the protein that was susceptible to oxidation by bleach.[5] The mutant genes were cloned in an expression vector and the products tested. An alanine substitution was effective in making the enzyme far more stable and active in bleach. Protein engineering is of great significance to research and development within the pharmaceutical industry, for example, the production of soluble receptors, or ligand-binding, for easier screening or rational drug design.[6,7]

Antibodies are attractive targets for protein engineering.[8] They have obvious potential as therapeutics and their structure is already well understood. Strategies include replacement of parts of the molecule with a toxin. Such fusions would be highly specific in delivering the toxin to target cells. An alternative strategy is to produce hi-specificity antibodies that might, for example, bind a tumour cell with one domain and a T killer cell with the other.[9] This subject is discussed more fully in Chapter 14.

The prospect of being able to produce any protein, regardless of origin, in a fermentative organism in culture is a very attractive one, and has obvious commercial implications. Not surprisingly, proteins used in clinical diagnosis or medical treatment have received the most attention. Insulin was one of the first produced in bacteria,[10] but was only the beginning. Vaccine production is an obvious example, but a list of other products in clinical use is given in Table 1. Apart from medically

[5] L. J. Abraham, J. Tom, J. Burnier, K. A. Butcher, A. Kossiakoff, and J. A. Wells, *Biochemistry*, 1991, **30**, 4151.
[6] S. Marullo, C. Delavier-Klutchko, J. G. Guillet, A. Charbit, A. D. Strosberg, and L. J. Emorine, *Bio/Technology*, 1989, **7**, 923.
[7] B. C. Cunningham and J. A. Wells, *Proc. Natl. Acad. Sci. USA*, 1991, **88**, 3407.
[8] A. Pluckthun, *Nature (London)*, 1990, **347**, 497.
[9] J. Berg, E. Lotscher, K. S. Steiner, D. J. Capon, J. Baenziger, H. M. Jack, and M. Wabl, *Proc. Natl. Acad. Sci. USA*, 1991, **88**, 4723.
[10] D. V. Goeddel, D. C. Kleid, F. Bolivar, H. C. Heynecker, D. C. Yansura, R. Crea, T. Hirose, A. Kraszeuski, K. Itakura, and A. D. Riggs, *Proc. Natl. Acad. Sci. USA*, 1979, **76**, 106.

Table 1   *Some human proteins with therapeutic value in clinical use*

Protein	Application
Growth hormone	Pituitary dwarfism
Insulin	Diabetes
Interferon-α 2b	Hairy cell leukaemia and genital warts
Erythropoietin	Anaemia
Tissue plasminogen activator	Myocardial infarction
Interleukin-2	Cancer therapy

important proteins, enzymes such as proteases used by manufacturing and food industries, can be produced on a large scale by genetically engineered bacteria.[11] Chymosin (or renin), for example, is produced in this way for use by the cheese industry.[12,13]

Before the full potential of genetic engineering could be realized, the problems concerned with the expression of DNA from higher eukaryotes in bacteria had to be overcome. As was suggested earlier, DNA from an animal or higher plant cannot be used directly to transform *E. coli*. To overcome the problems and obtain expression of eukaryotic DNA in bacteria, it was necessary to understand as much as possible about the control of gene expression in bacteria and higher organisms. This is discussed in the next section.

2   CONTROL OF GENE EXPRESSION

The difficulties encountered in obtaining expression of genes from higher eukaryotes in bacteria can be explained if the control systems operating in the different cell types are compared. Although the basic machinery of gene action is the same, that is, protein synthesis directed by an RNA copy of a DNA template, there are a significant number of differences between the various types of organism to cause problems for genetic engineers. The mechanisms of gene expression that operate in prokaryotes and eukaryotes are described below and are summarized in Figure 1.

2.1   Prokaryotes

Probably the best understood system of gene control operating in bacteria is the operon model of gene expression originally proposed by

[11] W. Gilbert and L. Villa-Komaroff, *Sci. Am.*, 1980, **242**, 74.
[12] T. Beppu, *Trends Biochem.*, 1983, 1, 85.
[13] M. L. Green, S. Angal, P. A. Lowe, and F. A. O. Marston, *J. Dairy Res.*, 1985, **52**, 281.

A.

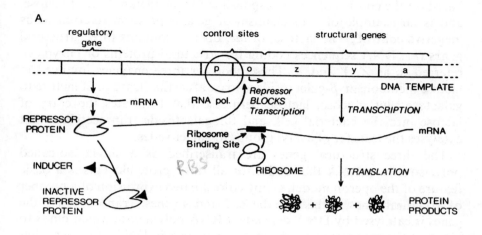

B.

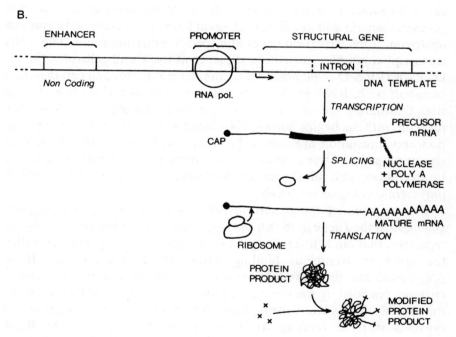

Figure 1 *The mechanism of gene expression in (A) prokaryotes and (B) eukaryotes: RNA pol, DNA dependent RNA polymerase; P, promoter; O, operator; →, transcription start site; see text for details*

Jacob and Monod in 1961.[14] The model, summarized in Figure 1A, is based on the control of genes responsible for the metabolism of lactose, and is an example of a mechanism of gene expression referred to as 'negative control'. The lactose, or *lac*, operon contains three structural genes (*i.e.* DNA sequences coding for structural proteins or enzymes), referred to as Z, Y and A. These code for three enzymes involved in lactose metabolism: β-galactosidase to catalyse the cleavage of lactose to galactose plus glucose, lactose permease that facilitates the entry of lactose into the bacterial cell, and thiogalactoside transacetylase that catalyses the transfer of acetyl groups to galactosides.

The three structural genes are transcribed as a single, so-called polycistronic, mRNA that codes for all three proteins. This is a basic feature of the operon model, giving co-ordinated expression of a number of structural genes, in this case the Z, Y and A genes. Transcripion of the genes is catalysed by DNA-dependent RNA polymerase which binds to DNA at the promoter site. Promoters are specific DNA sequences that do not code for proteins but are essential for transcription of structural genes. Promoters are an important factor in the expression of foreign genes in bacteria and are discussed again later. Transcription begins just upstream of the structural genes at a specific initiation point. The RNA transcript contains a nucleotide sequence at the 5' end called the Shine–Dalgarno (or S-D sequence) that is complementary to, and can therefore base-pair with, RNA in the ribosomes and acts as a ribosome binding site. Translation (protein synthesis) can then proceed. In prokaryotes, protein synthesis begins before RNA synthesis is terminated. Transcription and translation are said to be 'coupled'. Downstream of the final structural gene, in this case the A gene, there is a termination signal for transcription, rich in thymidine nucleotides, where the DNA–RNA polymerase complex is unstable.

Transcription of the structural genes is controlled by a regulatory gene, called the *I* gene in the *lac* operon, which codes for a protein, the *lac* repressor, that binds to a region of DNA adjacent to the promoter, called the operator. Repressor binding blocks the progress of the RNA polymerase and therefore inhibits transcription, preventing synthesis of enzymes encoded by the structural genes. An inducer acts by binding to the repressor. This alters the repressor's three-dimensional structure and prevents it from binding to the operator. In this case the RNA polymerase can continue to transcribe the structural genes coding for lactose metabolism. This system is referred to as negative control because the regulatory protein prevents gene expression.

[14] F. Jacob and J Monod, *J. Mol. Biol.*, 1961, **3**, 318.

Negative control is not the only mechanism of gene control acting in bacteria. There are additional control systems such as positive control and a process called attenuation. In positive control, an inducer binds to a protein that stimulates transcription. The best understood example is the system that operates when glucose is not available as an energy source. In this situation, the intracellular concentration of cAMP rises. The cAMP binds to a DNA-binding protein called CAP (catabolite activator protein) and the resulting complex binds to the promoters associated with operons involved with the breakdown of alternative catabolites, allowing their activation if an alternative carbon source is available. The *lac* operon is therefore under both negative (*lac* repressor) and positive (cAMP-CAP) control.

Attenuation is a control system dependent on the coupled transcription/translation mechanism that occurs in bacteria. It is a system of control that operates on genes responsible for amino acid biosynthesis. When a particular amino acid, such as tryptophan, is in short supply, the ribosomes stall during translation at tryptophan codons on the mRNA encoding a short 'leader peptide'. This allows particular secondary structures in the mRNA to form that permit transcription to proceed. If the amino acid concerned is in plentiful supply the ribosome does not stall, a different secondary structure forms in the mRNA and transcription of that gene is prematurely terminated.

It is beyond the scope of this article to describe these methods of gene control in any greater depth. Several molecular biology texts[15-17] are available that give a general overview of the subject and more specific information can be obtained from specialist texts.[18-20]

It can be seen from the model shown in Figure 1A that many separate elements are required to obtain expression of structural genes: a promoter sequence for RNA polymerase binding, a transcription initiation site, a ribosome binding site, and an inducer. The latter can be the naturally occurring inducer, such as allolactose (6-*O*-β-D-galactopyranosyl-D-glucose, an isomer of lactose in which the galactosyl residue is present on the carbon 6 rather than the carbon 4 of glucose and is produced by basal levels of β-galactosidase) or an artificial inducer such as isopropyl-β-D-thiogalactopyranoside (IPTG). Both of these molecules are active in

[15] H. Lodish, A. Berk, S. L. Zipursky, P. Matsudaira, D. Baltimore, and J. Darnell, 'Molecular Cell Biology', 4th Edn, Freeman, New York, 1999.
[16] B. Lewin, 'Genes VI', Oxford University Press, Oxford, 1997.
[17] J. D. Watson, M. Gilman, J. Witowski, and M. Zoller, 'Recombinant DNA', 2nd Edn, Scientific American Books, Freeman, New York, 1992.
[18] L. Snyder and W. Champness, 'Molecular Genetics of Bacteria', ASM Press, Washington, 1997.
[19] J. M. Fernandez and J. P. Hoeffler, 'Gene Expression Systems', Academic Press, San Diego, 1999.
[20] M. Ptashne, 'A Genetic Switch', 2nd Edn, Cell Press and Blackwell, Palo Alto, 1992.

derepressing the *lac* operon. The advantage of artificial inducers in stimulating gene expression is that they may show greater activity. In this case, IPTG, active in its native form, is taken up by cells more readily than lactose and is not degraded by β-galactosidase. The *lac* operon has received so much attention since the operon model for gene expression was first proposed that it was the obvious choice in the formation of the first generation of expression vectors, as described below.

2.2 Eukaryotes

The organization of genes and the mechanism of gene expression operating in eukaryotes is different in many respects from that in bacteria and is represented in diagrammatic form in Figure 1B. Although our knowledge has expanded at a remarkable rate, less detail is known about the control of gene expression in animals and plants than that in bacteria. The mechanism of selective gene expression during cell differentiation in multicellular organisms, in which all cells contain the same genetic material, is only partially understood. DNA in higher organisms is maintained in the cell as a complex with histones and other proteins to form a structure referred to as chromatin. Significant alterations in chromatin structure occur during transcription and these changes are associated with gene control mechanisms. Many eukaryotic transcription factors have been isolated and cloned. They interact with specific DNA sequences and each other to initiate transcription in a highly organized and controlled manner. Details are beyond the scope of this chapter but more information can be found in specialist texts.[20,21] No direct equivalent of the operon model has been found in animals and plants. Polycistronic mRNAs do not appear to exist: each structural gene is transcribed separately with its own promoter and transcription initiation and termination sites. The RNA polymerases of eukaryotes are more complex than in bacteria and exist in three forms, specific for pre-rRNA, tRNA, and 5S rRNA and mRNA synthesis. The promoter elements, while serving the same function as in bacteria, are organized in a different way and do not have the same DNA sequences. There does not appear to be a direct equivalent of the operator and once RNA polymerase has bound, RNA synthesis can begin. RNA polymerase binding is influenced by 'enhancer' elements which are non-coding DNA sequences which are *cis*-acting (*i.e.* influence the expression of genes on the same DNA molecule) and attract RNA polymerase to coding regions by binding transcription factors. The latter are referred to as *trans*-acting

[21] D. Latchman, 'Gene Regulation', 3rd Edn, Stanley Thomas, Cheltenham, 1998.

factors because the encoding genes can be on a different DNA molecule.[22,23] Following the onset of transcription, the RNA molecule is capped. That is, a 7-methyl guanosine residue is attached by a 5′–5′ phosphate linkage to the end of the RNA. Transcription is terminated and then significant RNA processing occurs prior to translation; it is generally accepted that coupled transcription–translation cannot occur in the nucleus of eukaryotes due to compartmentalization.

RNA processing involves several steps. Capping has already been mentioned, but there is also trimming of the RNA with ribonuclease and addition of between 20 and 250 adenine nucleotides (the 'poly(A) tail') to the 3′ end. Both of these processes enhance mRNA stability. Perhaps most significantly from the point of view of this article, processing involves the removal of introns. It has been known since the mid-1970s that many eukaryotic genes contain regions of DNA called intervening sequences, or introns, that do not code for an amino acid sequence. The introns considerably increase the length of DNA required to code for a protein; there are many cases known where the total length of introns greatly exceeds the length of coding regions within a gene. The number of introns is highly variable: human globin genes, for example, have two introns, whereas the chicken ovalbumin gene has seven. All introns are transcribed and thus appear in the nascent mRNA. They must be removed from the mRNA molecule before protein synthesis can occur. This is carried out by splicing enzymes, present in the nucleus, that remove the intervening sequences and precisely ligate the coding sequences, or 'exons', back together again.[15–17,23,24] Splicing has to be carried out in a very precise manner to maintain the correct reading frame of triplet codons for protein synthesis (for an explanation of reading frames see Figure 2). Splicing enzymes recognize precise sequences at the exon–intron boundaries. The following consensus sequence has emerged:

$$5'\ ^{C}/_{A}AG\ |\ GU^{A}/_{G}AGU—Y_{N}NAG\ |\ G\ 3'$$

exon | intron | exon

where Y_N represents a string of about nine pyrimidines and N denotes any base. The splicing enzymes act within large RNA–protein complexes that incorporate small nuclear RNAs that assist in sequence recognition. These complexes are called small nuclear ribonucleoproteins (snRNPs or 'snurps').

[22] T. Maniatis, S. Godbourn, and J. A. Fischer, *Science*, 1987, **236**, 1237.
[23] 'Transcription and Splicing', ed. B. D. Hames and D. M. Glover, IRL Press, Oxford, 1988.
[24] T. Maniatis and R. Reed, *Nature (London)*, 1987, **325**, 673.

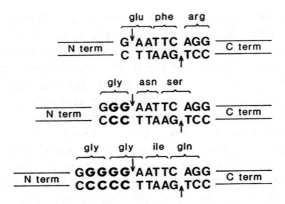

Figure 2   *The reading frame, based on triplet codons, of a fusion gene, constructed at an
EcoRI site (arrows), can be altered by the insertion of additional G-C pairs
(bold)*

The problem for genetic engineering in bacteria is that prokaryotes do not have introns and therefore possess none of the necessary machinery for their removal. It is not surprising, therefore, that an eukaryotic gene, possessing an intron, cannot be expressed in *E. coli*. Additional steps involved in the production of a mature protein in eukaryotes are post-translational modifications such as peptide cleavage, addition of prosthetic groups, glycosylation, or formation of multisubunit structures. These are specific to a cell type and are unlikely to be carried out by bacterial cells. This causes significant problems in the production of complex eukaryotic proteins by genetically engineered organisms and is a topic of considerable research interest.[25]

This section has attempted to give some background information which is necessary to understand the difficulties involved in obtaining expression of foreign genes in bacteria. The remainder of this chapter will describe how the problems have been overcome and give some examples of developments to date.

3   THE EXPRESSION OF EUKARYOTIC GENES IN BACTERIA

To obtain expression of eukaryotic genes in bacteria such as *E. coli*, the difference in mechanism of gene expression between the original organism from which the gene was obtained and the host bacterium must be overcome. The differences in gene expression mechanisms of particular

[25] R. B. Parekh and T. P. Patel, *Tibtechnology*, 1992, **10**, 276.

importance to this discussion are: the presence of introns in eukaryotic DNA; the difference in promoter sequences present in bacteria, animals, and plants; the absence of a ribosome binding site (Shine–Dalgarno sequence) on eukaryotic mRNA; preferential use of specific triplet codons in coding sequences; and, in many cases, the requirement for post-translation modification before the polypeptide is fully functional. The methods used to obviate these difficulties are discussed below.

3.1 Introns

It is apparent from the earlier discussion that a native eukaryotic gene cannot be expressed in bacteria when introns are present. There are two ways in which this problem can be overcome. Firstly, double-stranded DNA copies of mRNA molecules, referred to as complementary DNA or cDNA, can be generated by the use of an mRNA template and reverse transcriptase. This is a viral enzyme that produces a single strand of DNA, complementary in nucleotide sequence to the mRNA. The steps required to produce a double-stranded DNA molecule ready for cloning are described in Chapter 2.

The double-stranded cDNA molecule will not contain introns and can act as the coding sequence in expression vectors. There are, however, problems with the cDNA approach: if the mRNA is only present as a small constituent of a eukaryotic cell's RNA population, purification of the mRNA can be difficult. Also, the cDNA sequence synthesized by reverse transcriptase does not always include the 5′ end of the gene. Random termination of reverse transcription prior to completion of complementary strand synthesis can occur. The latter point is a serious problem but has been improved with refined cDNA synthesis approaches. Following synthesis, the cDNA is tailored to the expression vector by the addition of restriction enzyme linkers and can then be cloned.

A second approach that solves the intron problem is to synthesize the gene by chemical means without a template. If the amino acid sequence of the desired protein is known, it is possible to chemically synthesize a DNA molecule with the necessary sequence.[26-29] The advantages of this

[26] M. H. Caruthers, S. L. Beaucage, C. Becker, W. Efcavitch, F. F. Fisher, G. Galluppi, R. Coldman, P. Dettaseth, F. Martin, M. Matteucci, and Y. Stabinsley, in 'Genetic Engineering', ed. J. K. Setlow and A. Hollaender, Plenum Press, New York and London, 1982, p. 119.

[27] S. Narang, *J. Biosci.*, 1984, **6**, 739.

[28] M. D. Edge, A. R. Green, G. R. Heathcliffe, P. A. Meacock, W. Shuch, D. B. Scanlon, T. C. Atkinson, C. R. Newton, and A. F. Markham, *Nature (London)*, 1981, **191**, 756.

[29] D. G. Yanasura and D. J. Henner, in 'Gene Expression Technology', 'Methods in Enzymology', ed. D. V. Goeddel, Academic Press, San Diego, 1990, **185**, pp. 54–60.

technique over the cDNA approach are considerable: the complete sequence is obtained, the DNA can be tailored to the vector as desired, and particular codons, preferred by the organism chosen as a host for the expression vector, can be incorporated into the gene. There is no theoretical limit to the size of DNA that can be synthesized, but in practice large genes must be synthesized as fragments which are subsequently ligated.

3.2 Promoters

Promoters are sequences of DNA that are necessary for transcription. In *E. coli* the RNA polymerase recognizes the promoter as the first step in RNA synthesis. A similar system operates for the transcription of mRNA in eukaryotes but analysis of promoter sequences in bacteria and eukaryotes reveals that there are important differences. The nucleotide sequence of some characterized promoters is shown in Figure 3. There is a marked similarity between the various promoter sequences found in *E. coli* represented by the consensus sequence given in Figure 3. The promoters for mRNA synthesis in eukaryotes, however, although similar in principle are more complex, show less overall similarity, and are not recognized by bacterial RNA polymerase. The important sequences for *E. coli* RNA polymerase are the TTGACA ('-35 sequence') and TATAAT ('Pribnow box') sequences found 35 and 10 base pairs upstream of the transcription initiation point respectively.

In order to obtain expression of an eukaryotic DNA sequence in *E. coli* it is necessary to place the coding sequence downstream from a bacterial promoter. The most commonly used promoters are those from the *lac* or *trp* operons (or a combination of the two called a '*tac*' promoter), the *recA* promoter of *E. coli,* and the leftward promoter (λ P$_L$) from bacteriophage λ. Many vectors constructed using the *lac* promoter are based on the principle of using the entire *lac* control region and the first few nucleotides of the Z gene coding for β-galactosidase. The foreign DNA to be expressed is then inserted downstream of the β-galactosidase N-terminal coding sequence, giving rise to a fusion gene,

	'-35 sequence'		'Pribnow box'	
p*trp*	TTGACA	–17 bp–	TTAACTA	– transcription
p*lac* uv5	TTTACA	–18 bp–	TATAATG	– transcription
p*tac*	TTGACA	–16 bp–	TATAATG	– transcription
Prokaryote consensus	**TTGACA**		**TATAAT**	
Human β globin	CCAAT	–39 bp–	CATAAA	– transcription

Figure 3 *DNA sequence of some characterized promoters*

discussed in more detail later. One of the effects of this is to maintain the control mechanism that normally operates on the *lac* operon. Expression from the promoter is regulated by the *lac* repressor. Transcription, or derepression of the gene is then brought about by the addition of an inducer such as IPTG. Similarly, expression vectors based on the *trp* promoter incorporate control regions normally in operation for the *trp* operon. The *recA* promoter is induced by nalidixic acid. The λ P$_L$ promoter is used in specially designed host cells that produce a temperature-sensitive repressor. Raising the temperature of the culture activates expression from the promoter.

3.3 Ribosome Binding Site

The initiation of translation can be a significant limiting factor in expression of cloned genes.[30] An initiation codon, AUG, is required but other precise nucleotides, particularly in the 5' untranslated leader of the mRNA, are needed to facilitate suitable secondary and tertiary structures in mRNA and interaction between mRNA and the ribosome. Perhaps the best known of these sequences is the Shine–Dalgarno sequence (or S-D sequence or RBS: ribosome binding site). This is a stretch of 3–9 bases lying between 3 and 12 bases upstream from the AUG codon. It allows a complex to form between the mRNA and the 30S subunit of the ribosome via hydrogen bonding to the 16S rRNA. Not all *E. coli* mRNAs have an identical S-D sequence but a consensus can be identified (Figure 4). An S-D sequence is essential for translation. Other sequences that help boost translation level include translation enhancers (named after transcriptional enhancers because their sequence appears more important than their precise location) such as the Epsilon sequence in the g10-L ribosome binding site of bacteriophage T7.[31,32]

To obtain expression of foreign genes in *E. coli* it is necessary to incorporate ribosome binding motifs into the recombinant DNA molecule. Furthermore, some sequences (such as the S-D sequence) must be located at an optimal distance from the translation start codon. This is most readily achieved by construction of fusion genes where an entire untranslated leader and 5' coding sequence from a naturally occurring

[30] H. A. de Boer and A. S. Hui, in 'Gene Expression Technology', 'Methods in Enzymology', ed. D. V. Goeddel, Academic Press, San Diego, 1990, **185**, p. 103.
[31] P. O. Olins and S. H. Rangwala, in 'Gene Expression Technology', 'Methods in Enzymology', ed. D. V. Goeddel, Academic Press, San Diego, 1990, **185**, p. 115.
[32] P. O. Olins, C. S. Devine, and S. H. Rangwala, in 'Expression Systems and Processes for rDNA Products', ed. R. T. Hatch, C. Goochee, A. Moreira, and Y. Alroy, ACS, Washington, DC, 1991, p. 17.

met arg ala
5' GAUUCCU**AGGAGGU**UUGACUA UGCGAGCU --- mRNA
3' AUUCCUCCACUAG
16S rRNA

Figure 4 *Base pairing between the Shine–Dalgarno sequence (in bold) on the mRNA and a complementary region on the 3′ end of 16S rRNA*

gene is present. Nonetheless, all expression cassettes need to be tested thoroughly and sequences reorganized if necessary to optimize translation initiation.

3.4   Expression of Foreign DNA as Fusion Proteins

The problems associated with procuring a prokaryotic promoter and S-D sequence to obtain expression of eukaryotic DNA in *E. coli* can be obviated by constructing a fusion gene. The control region and *N*-terminal coding sequence of an *E. coli* gene is ligated to the coding sequence of interest. When introduced and cloned in *E. coli*, RNA polymerase recognizes the promoter as native and transcribes the gene. The 5′ end of the mRNA is also native and consequently interacts normally with a ribosome to commence protein synthesis. The protein that results will be chimaeric, the *N*- and *C*-terminals being derived from the prokaryotic and eukaryotic genes, respectively. There are a number of advantages in taking this approach. First, expression of the foreign DNA should be efficient. Second, the foreign gene can be placed under the control of the induction/repression system of the *E. coli* promoter/operator used. Third, with some constructs, the fusion peptide may be exported to the periplasmic space via the signal sequence (discussed later) and fourth, the protein should be relatively stable. This last point is important and worthy of some discussion. For reasons not fully understood, foreign proteins, particularly short peptides, in *E. coli* are recognized as such and are broken down by endogenous proteases. Foreign proteins expressed as fusion proteins, however, appear to be more stable. The *N*-terminal part of the chimera is recognized as 'self' by the cell. In this respect, the length of the *N*-terminal sequence coded for by *E. coli* DNA is important, the longer it is the more likely the fusion product is to be stable. Peptides shorter than about 80 amino acids are best synthesized as fusion proteins.

The majority of fusion genes that have been created for expression of foreign DNA in *E. coli* are based on the *lac* operon, using the *N*-terminal

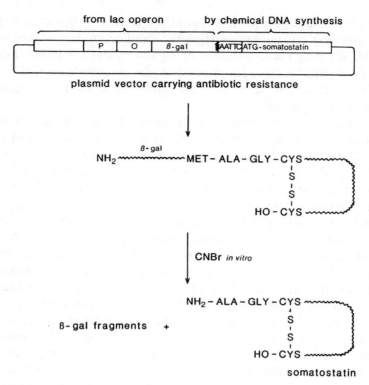

Figure 5   *The synthesis in* E. coli *of somatostatin as a fusion protein*

sequence of β-galactosidase to form the fusion peptide. The first example of this approach was to produce somatostatin in *E. coli*[33] by the approach illustrated in Figure 5. Somatostatin is a peptide hormone of 14 amino acids with the physiological role of inhibiting secretion of growth hormone, glucagon and insulin. The coding sequence for somatostatin was obtained by chemical synthesis, based on knowledge of the somatostatin amino acid sequence. The DNA was constructed in such a way as to incorporate codons which are preferentially used in *E. coli*, and to leave single-stranded projections, corresponding to the cohesive ends produced by *Eco*RI and *Bam*HI digestion, at each end. The coding sequence was preceded by a codon for methionine and terminated by a pair of nonsense codons to stop translation.

The initial hybrid genes that were constructed should have produced a hybrid protein containing the first seven amino acids of β-galactosidase,

[33] K. Itakura, T. Hirose, R. Crea, A. D. Riggs, H. L. Heyneker, F. Bolivar, and W. H. Boyer, *Science*, 1977, **198**, 1056.

but no somatostatin-like proteins could be detected. When an alternative hybrid gene was created by inserting the synthetic somatostatin gene at an *Eco*RI site near the *C*-terminus of β-galactosidase, a stable fusion of this large β-galactosidase fragment and the somatostatin was synthesized. This illustrates very well the importance of fusion genes in maintaining protein stability.

The presence of the methionine residue at the *N*-terminus of the somatostatin amino acid sequence allows for the purification of somatostatin from the fusion protein. Cyanogen bromide treatment *in vitro* cleaves peptides at the carboxyl side of methionine residues. This process, of course, is only applicable to the production of eukaryotic proteins not containing internal methionines.

A foreign gene will only be expressed as a fusion protein if it is placed in the correct translational reading frame (Figure 2). Since codons are based on triplets of nucleotides there is only a one in three chance that two randomly selected coding sequences will be ligated in phase and this assumes that they are correctly oriented (a further one in two chance). To obtain expression, the foreign DNA can be ligated into a position, known by the DNA sequence to be in the exact reading frame, as in the somatostatin experiment described above. Or else several recombinant molecules need to be constructed, all in different reading frames, to enable selection of the successful clone. There are different ways of achieving this second approach. For example, successive additions of two G-C base pairs to the *Eco*RI fragment of the *lac*Z gene has created vectors with three different reading frames.[34] Alternatively, infection of foreign DNA can employ the method of homopolymer tailing, in which differing lengths of a linker are randomly constructed.[35] In this case, each of the recombinant molecules constructed will have different lengths of the repeating G-C pair and at least some (one in six) should be in the correct reading frame and orientation. The fusion peptide approach has been used in a considerable number of cases following the somatostatin experiment, for example in the synthesis of thymosin,[36] neo-endorphin[37] and human insulin.[10] In the latter case the initial approach was to synthesize the A and B chains as separate fusion products. This was later refined to a strategy where a single β-galactosidase–insulin fusion was cleaved in one step.

[34] P. Charney, M. Perricaudet, F. Galibert, and P. Tiollais, *Nucleic Acids Res.*, 1978, **5**, 4479.

[35] L. Villa-Komaroff, A. Efstratiadas, S. Broome, P. Lomedico, R. Tizard, S. P. Naber, W. L. Chick, and W. Gilbert, *Proc. Natl. Acad. Sci. USA*, 1978, **75**, 3727.

[36] R. Wetzel, H. L. Heyneker, D. V. Goeddel, G. B. Thurman, and A. L. Goldstein, *Biochemistry*, 1980, **19**, 6096.

[37] S. Tanaka, T. Oshima, K. Ohsue, T. Ono, S. Oikawa, I. Takano, T. Noguchi, K. Kangawa, N. Minamino, and H. Matsuo, *Nucleic Acids Res.*, 1982, **10**, 1741.

In all these cases the desired peptide was prepared from the fusion peptide by CNBr cleavage which depends on the absence of methionine within the required polypeptides. The procedure, however, is not the only method that has been used for the cleavage of fusion peptides. β-Endorphin was synthesized as a β-galactosidase fusion protein in *E. coli.*[38] In this case, because of the presence of internal methionine residues in β-endorphin, the native hormone was prepared by citracony-lation and trypsin treatment. For most larger proteins, however, it is not possible to remove the β-galactosidase residues, and the fusion protein is the final product.

Not all fusion genes are based on the *lac* operon. For example, fusion genes have also been constructed using the β-lactamase gene carried in plasmid pBR322. Although expression from this promoter is not particularly efficient, its use is of interest because β-lactamase is a secreted protein, responsible for conferring ampicillin resistance, and carries a signal sequence. Signal peptides are sequences of amino acids at the *N*-terminus of proteins; they have an affinity for the cell membrane and are responsible for the export of proteins that carry them; they are cleaved by enzymes present in target organelles of eukaryotes or the periplasmic space in bacteria. Fusion genes, such as β-lactamase-proinsulin hybrid, have been constructed at the unique *Pst*I restriction site located between codons 183 and 184 of the β-lactamase coding sequence.[35] The resulting chimaeric proteins contain a substantial proportion of the β-lactamase amino acid sequence and they are, not surprisingly therefore, often found in the periplasmic space.

An alternative approach is to produce expression vectors incorporating a sequence from the gene for haemolysin.[39] This is a protein product of some pathogenic *E. coli*: it is secreted by the cells via a signal at the *C*-terminus of the protein. Ligation of this signal onto mammalian prochymosin and immunoglobulin genes has resulted in successful secretion of fusion products.

Clearly, secretion of recombinant protein has considerable advantages for production on a commercial scale: extraction and purification are simpler, unwanted *N*-terminal methionines are removed, and proteins fold better. Secretion is one of the reasons for the success of *Bacillus subtilis* as a host organism (see Section 6).

[38] J. Shine, I. Fettes, N. C. Y. Lan, J. L. Roberts, and J. D. Baxter, *Nature (London)*, 1980, **285**, 456.
[39] L. Gray, K. Baker, B. Kenny, N. Mackman, R. Haugh, and I. B. Holland, in John Innes Symposium: 'Protein Targeting', ed. K. Chater, *J. Cell Science*, Supplement, 1989.

3.5   Expression of Native Proteins

It is possible to obtain expression of native proteins using a nucleotide sequence that codes only for the peptide required and contains no codons from a prokaryotic gene. The advantage of this approach is that the amino acid sequence produced should be identical to the naturally occurring eukaryotic protein and therefore should exhibit full biological activity without the need to remove a fusion peptide; a difficult if not impossible task in most cases. Direct expression is best suited to proteins between 100 and 300 amino acids, without a great number of cysteines.

To obtain direct expression of a native protein, it is necessary to place a coding sequence, with an ATG translation initiation codon, downstream from a bacterial promoter and ribosome binding site. There will be difficulties if a cDNA is being used for the coding sequence. Depending on the nature of the mRNA template used, there may be a long leader sequence upstream of the ATG codon. As it is most unlikely that there will be a restriction site in exactly the right position to allow this leader to be removed and the coding sequences joined to the vector perceiving the correct SD-AUG spacing, a variety of methods have had to be developed. The human growth hormone gene was successfully expressed in bacteria[40] using the procedure outlined in Figure 6. The methods used are a good example of where many different techniques or approaches can be combined together in one experiment. The human growth hormone (HGH) gene was rather too long for complete synthesis by chemical means. The bulk of the coding sequence was therefore obtained by cDNA synthesis and cloning. The cDNA was cut with *Hae*III to give a defined length lacking the mammalian *N*-terminal signal sequence that would not operate in *E. coli*. This fragment was tailored to the expression vector using a chemically synthesized fragment that coded for the first 24 amino acids of HGH and included an *Eco*RI site for attachment to the plasmid vector. The cDNA and synthetic fragments were ligated together and inserted via *Eco*RI and *Sma*I sites into an expression vector containing two copies of the *lac* promoter. The resulting plasmid contained a ribosome binding sequence, AGGA, eleven base pairs upstream from the ATG initiation codon for the HGH gene. In the *lac* operon, the Shine–Dalgarno sequence lies seven base pairs upstream of the *β*-galactosidase initiation codon. A derivative of the original expression vector was therefore constructed in which four base pairs between the AGGA and ATG sequences were removed. This was achieved by

---

[40] D. V. Goeddel, H. L. Heyneker, T. Hozumi, R. Arentzen, K. Itakura, D. G. Yansura, M. J. Ross, G. Miozzari, R. Crea, and P. H. Seeburg, *Nature (London)*, 1979, **281**, 544.

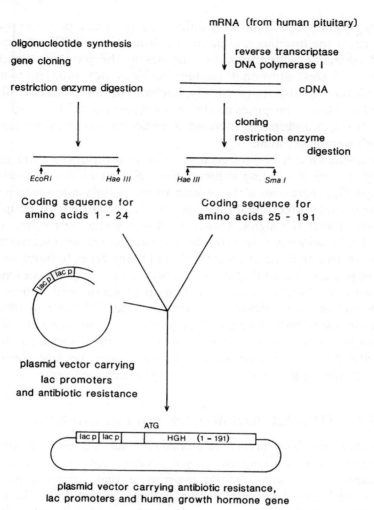

oligonucleotide synthesis

gene cloning

restriction enzyme digestion

EcoRI   Hae III

Coding sequence for
amino acids 1 - 24

mRNA (from human pituitary)

reverse transcriptase
DNA polymerase I

cDNA

cloning
restriction enzyme
                digestion

Hae III   Sma I

Coding sequence for
amino acids 25 - 191

lac p lac p

plasmid vector carrying
lac promoters
and antibiotic resistance

ATG

lac p lac p   HGH   (1 - 191)

plasmid vector carrying antibiotic resistance,
lac promoters and human growth hormone gene

Figure 6 *Construction of an expression vector that directs the synthesis of human growth hormone (HGH) in* E. coli

opening the plasmid with *Eco*RI, digesting the single-stranded tails with S1 nuclease and religating the blunt ends. Surprisingly, this new plasmid produced less HGH when introduced into *E. coli* than the original construct, containing the full 11 basepair sequence that had been deliberately shortened. This illustrates the subtleties involved in the relationship between the leader sequence and the initiator codon in protein synthesis and is discussed further in Section 5.

The HGH produced as above in *E. coli* is a soluble protein and can be readily purified. It has the same biological activity as the HGH from human pituitary and apparently differs in only one respect: the presence of an extra methionine residue at the *N*-terminus.[41] This amino acid would normally be removed by enzymes in the pituitary gland. The presence of this additional methionine does not interfere with the protein's biological activity. Ideally, bacterial products that are likely to be used for clinical purposes need to be as close as possible in structure to their natural counterparts, to avoid complications such as reaction by a patient's immune system.

The hormone can be produced in bacteria without the extra methionine by cloning the coding sequence next to a bacterial signal sequence. This specifies secretion of the protein to the periplasmic space between the inner and outer membranes of the bacterial cell. A periplasmic protease cleaves the signal sequence including the methionine leaving native HGH, which can be extracted by a hypotonic periplasmic shock.

Human growth hormone was the first of many genes to be expressed as a native protein. Those that quickly followed included human leucocyte interferon (LeIF-A), murine, and human prolactin and human interleukin, cloned downstream of a *trp* promoter,[29,40] human fibroblast interferon, using both the *trp* and *lac* promoters,[42] human perproinsulin downstream of a *lac* promoter,[27] and mouse dihydrofolate reductase, using the *β*-lactamase promoter.[43] The approach became the one of choice for the synthesis of medium-sized proteins.

4   DETECTING EXPRESSION OF FOREIGN GENES

Using the methods of classical microbial genetics, it is relatively simple to detect expression of foreign prokaryotic genes in host organisms. A host is chosen for a transformation experiment that carries a particular mutant, such as a deficiency in the synthesis of histidine. The mutant would therefore be a *his* auxotroph which would normally require a histidine supplement in its growth medium. If the foreign gene to be introduced into the cells codes for histidine synthesis (*his*⁺) successful transformants can be detected by their ability to grow without the histidine supplement, *i.e.* they can be 'selected'. This relatively simple technique, however, cannot generally be employed when expression of foreign eukaryotic DNA is desired. The reason is simple: it is unlikely

[41] D. V. Goeddel, E. Yelverton, A. Ullrich, H. L. Heyneker, G. Miozzari, W. Holmes, P. H. Seeburg, T. Dull, L. May, N. Stebbing, R. Crea, S. Maeda, N. McCandliss, A. Sloma, J. M. Tahar, M. Cross, P. C. Familletti, and S. Pestka, *Nature (London)*, 1980, **287**, 411.

[42] T. Taniguchi, L. Guarente, T. M. Roberts, D. Kimelman, J. Douhan, and M. Ptashne, *Proc. Natl. Acad. Sci. USA*, 1980, **77**, 5230.

[43] A. C. Y. Chang, H. A. Erlich, R. P. Gunsalas, J. H. Nunberg. R. J. Kaufman. R. T. Schimke, and S. M. Cohen, *Proc. Natl. Acad. Sci. USA*, 1980, **77**, 1442.

that expression of an eukaryotic gene will be able to confer an advantage to its bacterial host. Insulin synthesis, for example, is hardly likely to enhance the growth advantage of an *E. coli* cell. Alternative methods are required to detect expression of eukaryotic genes. If the function of the desired gene is known, a suitable test can be developed. This is relatively straightforward if the protein required is an enzyme; an assay can be applied to a host-cell extract although renaturation may be required. If, however, the desired protein is not an enzyme a more complex test is necessary. These tests usually employ immunodetection techniques.[1,44] For example, bacterial colonies, suspected of containing the desired protein, are grown on nitrocellulose or other filters as a replica of a master agar plate. The colonies are lysed, their contents bound to the filter and incubated with a solution of radioactively labelled antibody, specific to the protein required. Following thorough washing of the filter and autoradiography, colonies containing the required protein can be identified. The relevant colony or colonies on the master plate can be selected and cultured. Alternatively, proteins can be extracted and separated on a polyacrylamide gel. Western blotting is then used to transfer the protein to a membrane for immunodetection. Detecting expression of foreign eukaryotic genes where there is no enzyme assay, or when an antibody is not available, is more difficult. In this case, novel proteins over and above the natural background of host proteins need to be detected by protein separation techniques such as polyacrylamide gel electrophoresis.[29] If expression levels are low, as was the case in early experiments, a system is required to lower the background. Such systems include the use of mini-cells, maxi-cells, or a coupled transcription–translation system, discussed elsewhere.[45–47]

5 MAXIMIZING EXPRESSION OF FOREIGN DNA

Until now this article has been concerned with describing the principles and techniques involved in obtaining detectable levels of expression of foreign genes in bacteria. Many commercial applications of genetic engineering, however, depend on obtaining high levels of expression such that production of, for example, hormones or vaccines is a realistic economic proposition. The number of cases where genuinely high levels

[44] S. Broome and W. Gilbert, *Proc. Natl. Acad. Sci. USA*, 1978, **75**, 2746.
[45] B. Oudega and F. R. Mooi, in 'Techniques in Molecular Biology', ed. J. M. Walker and W. Gaastra, Croom Helm, London and Canberra, 1983, Chap. 13, p. 239.
[46] J. M. Pratt, in 'Transcription and Translation, A Practical Approach', ed. B. D. Hames and S. J. Higgins, IRL Press, Oxford and Washington, DC, 1984, Chap. 7, p. 179.
[47] R. J. Slater, in 'Techniques in Molecular Biology', Vol. 2, ed. J. M. Walker and W. Gaastra, Croom Helm, London and Canberra, 1987, Chap. 12, p. 203.

of expression have been obtained is relatively small. Not surprisingly, the achievement of high production levels is a result of considerable research effort and investment. Some of the best examples are the production of insulin, growth hormone and interferon with levels of expression approaching 10^5–10^6 molecules per cell.[48] In the case of insulin this is equivalent to nearly 40 mg of product per 100 g wet weight of cells.

Several factors can be identified as being involved in influencing the level of expression.[49] These include: promoter strength, codon usage, secondary structure of mRNA in relation to position of a ribosome binding site, efficiency of transcription termination, plasmid copy number and stability, and the host cell physiology. These factors are discussed below.

Optimal promoters need not necessarily be the naturally occurring ones. Sequences nearest to the consensus sequence are the most efficient, as illustrated by experiments employing the *tac* promoter,[50] a hybrid between the *lac* and *trp* promoters (Figure 2). Mutations can be made in promoters to alter their characteristics and thereby influence expression. For example, L8 and uv5 mutations in the *lac* promoter render the promoter insensitive to catabolite repression and improve RNA polymerase binding respectively.[51]

Codon usage influences levels of expression and can be accommodated if genes are synthesized chemically. The genetic code is degenerate, so for many amino acids there is more than one codon. Efficiency is related to the abundance of tRNA in the cell and the codon–anticodon interaction energy.[52,53]

The distances between the promoter, ribosome binding site and ATG initiation codon can have a profound effect on levels of expression; more than 2000-fold differences have been recorded. This is probably due to different secondary structures forming in the mRNA following transcription. It appears that to obtain high levels of expression the initiation codon (AUG) and the ribosome binding regions need to be present as single-stranded structures. Secondary structures can be predicted with computer analysis based on the maximum possible changes in free energy associated with folding of an RNA chain. The most pragmatic approach to the problem is to construct a series of vectors with different

[48] B. R. Click and G. K. Whitney, *J. Ind. Microbiol.*, 1987, **1**, 277.

[49] S. J. Coppella; G. F. Payne, and N. Dela Cruz, in 'Expression Systems and Processes for rDNA Products', ed. R. T. Hatch, C. Goochee, A. Moreira, and Y. Alroy, ACS, Washington, DC, 1991, Chapt. 1, p. 1.

[50] H. A. de Boer, L. J. Comstock, and M. Vasser, *Proc. Natl. Acad. Sci. USA*, 1983, **80**, 21.

[51] F. Fuller, *Gene*, 1982, **19**, 43.

[52] M. Gouy and C. Gautier, *Nucleic Acids Res.*, 1982, **10**, 7055.

[53] J. Grosjean and W. Fiers, *Gene*, 1982, **18**, 199.

distances, for example, between the S-D and ATG sequences. A series of clones can then be screened for levels of expression and the optimum selected. 'Cassette' vectors have now been available for several years; all the relevant signals are optimally situated with a unique restriction site for the inclusion of a coding sequence.[54] Screening can be difficult if there is no convenient assay for the desired product. A novel solution is to produce a recombinant DNA molecule containing the desired sequence upstream from the carboxy terminal sequence of β-galactosidase placed in the same reading frame. In this case production of a fusion protein can be detected by its β-galactosidase activity. The most suitable of several clones, incorporating varying distances between the S-D and ATG sequences can then be selected and the carboxy terminal of β-galactosidase removed to leave the required gene sequence as desired.[30,55]

Levels of expression are also influenced by transcription terminator (A-T rich) sequences that must be included at the end of coding regions. Read-through of the RNA polymerase may interfere with other genes downstream or may produce unnecessarily long mRNA molecules that could have reduced translation efficiency and be an undue strain on the cell's energy resources.[56]

Plasmid copy number and stability are important factors in successfully exploiting genetic engineering.[49,56] There is little to be gained from constructing expression vectors that are lost during large-scale fermentation. Losses of plasmids occur by several means: a slow rate of plasmid replication compared with cell division, insertion or deletion events in regions of a plasmid necessary for replication, inefficient partitioning at cell division, and plasmid multimerization. The stability and copy number of plasmids in bacteria is dependent on the plasmid construction, growth conditions, such as temperature, and growth rate. Plasmid stability can be maintained by antibiotic selection ('active selection') but this may be undesirable during mass production because of costs and waste disposal problems. The plasmid pBR322, one of the original cloning vectors, segregates randomly during division but naturally occurring plasmids contain partitioning functions, Par, which ensure segregation at cell division. Incorporation of features such as Par regions into expression vectors is advantageous to large-scale production.[49,57] Alternatively, genes could be cloned in expression vectors that confer an

[54] N. Panayotatos, in 'Plasmids: A Practical approach', ed. K. G. Hardy, IRL Press, Oxford and Washington, DC, 1987, p. 163.
[55] L. Guarente, G. Lauer, T. M. Roberts, and M. Ptashne, *Cell*, 1980, **20**, 543.
[56] P. Balbas and F. Bolivar, in 'Gene Expression Technology', 'Methods in Enzymology', ed. D. V. Goeddel, Academic Press, San Diego, 1990, **185**, p. 14.
[57] G. Skogman, J. Nilsson, and P. Gustafsson, *Gene*, 1983, **23**, 105.

advantage to cells carrying the plasmid; this is called passive selection. Such a gene, for example, *valS*, might code for an essential function (in this case a tRNA synthetase) deficient in the host chromosome.[58]

Once expression of foreign DNA has been successfully achieved, the host organism will need to be maintained in culture for as long as required. The growth conditions that produce optimum levels of expression at maximum stability for the least expenditure are required. Effects of batch or continuous culture, choice of growth medium, *etc.*, need to be considered. The choice of media can have significant effects on the production of recombinant protein. For example, conditions that are required to select for maintenance of the plasmid may be poor for synthesis of high levels of product. Alternatively, conditions optimal for production may limit cell growth. In short, media can affect the production rate, yield and secretion efficiency.[59]

Further problems are apparent even after the protein is synthesized. Frequently it has been found that foreign proteins expressed at high levels are deposited into insoluble inclusion bodies. Extraction then involves denaturing agents such as thiourea (hardly conducive to the extraction of biologically active products) followed by protein purification and refolding. Redox couples such as a mixture of oxidized and reduced glutathione can be used to form the disulphide bridges within the protein. Furthermore, foreign proteins can be partially degraded by protease.[60] Such degradation products are potentially immunogenic and present a problem in the production of therapeutics.

5.1 Optimizing expression in *E. coli*

When the expression of proteins toxic to the cell is required in *E. coli*, there is a requirement for tight control over their expression. The *lac* promoter can be used for expression if the *lacI* gene encoding the repressor is also expressed in the cell, usually as the constitutive *lacI*^{q} variant. This repression is rather leaky and better results can be obtained using the λ P$_L$ promoter under the control of a temperature sensitive λ cI repressor.

Even tighter control of gene expression can be obtained using the promoter from bacteriophage T7. This is recognized only by T7 RNA polymerase, and can be controlled by placing the T7 polymerase gene

[58] J. Nilsson and S. G. Skogman, *Bio/Technology*, 1986, **4**, 901.
[59] T. Kohno, D. F. Carmichael, A. Sommer, and R. C. Thompson, in 'Gene Expression Technology', 'Methods in Enzymology', ed. D. V. Goeddel, Academic Press, San Diego, 1990, **185**, p. 187.
[60] S.-O. Enfors, *Tibtechnology*, 1992, **10**, 310.

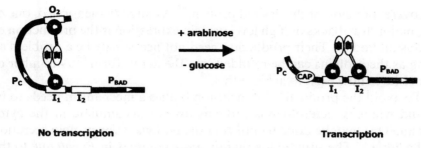

No transcription   **Transcription**

Figure 7 *Control of the* ara P$_{BAD}$ *promoter by looping caused by AraC. N and C, the N-*
and C-terminals of AraC protein monomers; O$_2$, I$_1$ and I$_2$, AraC binding sites;
P$_{BAD}$, the araBAD operon promoter; P$_C$, the araC gene promoter; CAP, the
cAMP-CAP complex. See text for description

itself under tight control, *e.g.* the *lac* uv5 promoter.[61–63] Further developments of this approach have allowed high level expression of highly
toxic proteins.[64]

Another tight control system employs the promoter of the arabinose
utilization operon, *araBAD*. This is present upstream of a multiple
cloning site on vectors which also express the regulatory *araC* gene. The
AraC regulator can act as a repressor or an activator of transcription
(Figure 7). When arabinose is absent, AraC dimers form loops in the
promoter DNA between the O$_2$ and I$_1$ binding sites, shutting down
transcription. But in the presence of arabinose the binding of AraC is
altered and it binds the I$_1$ and I$_2$ sites, releasing the loop and allowing
transcription. Maximal transcription, as with the *lac* operon, requires the
presence of the CAP–cAMP complex, which forms in the absence of
glucose. Hence, cloned gene expression can be almost completely
repressed when glucose is present and arabinose absent. Moreover, with
this approach the level of expression may be carefully controlled by the
concentration of arabinose in the medium.[65]

High level expression can be enhanced by increasing gene dosage.
Some systems have been developed that take advantage of the placement
of heterologous protein genes on plasmids. By the induction of subtle
changes in its control, runaway plasmid replication can be allowed to
levels up to 1000 per genome. The increased gene dosage is accompanied

[61] S. Tabor and C. C. Richardson, *Proc. Natl. Acad. Sci. USA*, 1985, **82**, 1074.
[62] F. W. Studier, A. H. Rosenberg, J. J. Dunn, and J. W. Dubendorff, *Methods Enzymol.*, 1990, **185**, 60.
[63] F. W. Studier, *J. Mol. Biol.*, 1991, **219**, 37.
[64] A. J. Doherty, S. J. Ashford, J. A. Brannigan, and D. B. Wigley, *Nucleic Acids Res.*, 1995, **23**, 2074.
[65] S. M. Soisson *et al.*, *Science*, 1997, **276**, 421.

by overexpression of the desired protein.[66] As already mentioned, one of the major drawbacks of high level protein expression is the production of inclusion bodies. Their production need not necessarily be a problem so long as the proteins can be refolded into the active form.[67] A number of procedures now exist to achieve this.[68]

To avoid the production of inclusion bodies a mechanism needs to be found whereby heterologous proteins do not accumulate in the cytoplasm. This may be done by utilizing the pathways for protein secretion in bacteria.[69] The situation is slightly more complex in *E. coli* due to the presence of two membranes. Secretion into the extracellular medium, though possible, is harder to achieve than in Gram-positive bacteria. Periplasmic secretion has a number of advantages. Periplasmic proteins may be selectively purified by mild osmotic shock and are contaminated with fewer proteins than in a whole cell extract. More importantly the secretion pathway involves the activity of molecular chaperones which target proteins to the periplasm.[70]

Chaperones tend to ensure the correct folding of secreted proteins increasing the likelihood that the product can be recovered in an active form. In a number of cases changes in the level of expression of chaperones in the cell have been sufficient to allow the foreign proteins to fold correctly. Even though in some cases it is lower levels of chaperones which have the desired effect.

Other host proteins important for correct folding are the thiol-disulfide oxido-reductases (Dsb proteins) responsible for the formation of disulfide bridges in proteins. Again alteration of their levels (in either direction) can promote correct disulfide bridge formation.[71] Fusions to DsbA have also been used for the expression of genes located on the chromosome.[72]

An overview of some of these expression optimization approaches has been published.[73]

6 ALTERNATIVE HOST ORGANISMS

The commercial exploitation of genetically engineered bacteria will involve, in most cases, the purification of the foreign product. This is

[66] K. Nordström and B. E. Uhlin, *Bio/Technology*, 1992, **10**, 661.
[67] H. Lilie, E. Schwarz, and R. Rudolph, *Curr. Opin. Biotechnol.*, 1998, **9**, 497.
[68] R. Rudolph, G. Böhm, H. Lilie, and R. Jaenicke, in 'Protein Function, A Practical Approach', 2nd Edn, ed. T. E. Creighton, IRL Press, Oxford and Washington, DC, 1997, p. 57.
[69] P. Braun, G. Gerritse, J.-M. van Dijl, and W. J. Quax, *Curr. Opin. Biotechnol.*, 1999, **10**, 376.
[70] S. C. Wu, R. Ye, X. C. Wu, S. C. Ng, and S. L. Wong, *J. Bacteriol.*, 1998, **180**, 2830.
[71] J. C. Joly, W. S. Leung, and J. R. Swartz, *Proc. Natl. Acad. Sci. USA*, 1998, **95**, 2773.
[72] P. Olson, Y. Zhang, D. Olsen, A. Owens, P. Cohen, K. Nguyen, J.-J. Ye, S. Bass, and D. Mascarenhas, *Protein Expression Purific.*, 1998, **14**, 160.
[73] G. Hannig and S. C. Makrides, *Tibtechnology*, 1998, **16**, 54.

best achieved using a host that is capable of secreting protein products into the growth medium. *Bacillus subtilis* is such an organism: it is a non-pathogenic, Gram-positive bacterium, and is amenable to large-scale fermentation technology.[74] It has a well developed natural transformation system (*i.e.* for taking up linear fragments of homologous DNA) which was invaluable in early genetic analysis. Importantly, if *B. subtilis* were to be a host cell for the synthesis of products relevant to the food and beverage industries, it has the official status of being 'generally regarded as safe' (GRAS) accorded by the US Food and Drug Administration. Despite the attractions the application of *B. subtilis* for the production of recombinant protein has been limited. There are some difficulties that must be overcome. Our fundamental knowledge and application of molecular biology has been primarily directed towards the use of *E. coli*. Nothing equivalent to the numerous derivatives of λ vectors exists in *B. subtilis*, for example. An important development has been the production of shuttle vectors that can be used both in *E. coli* and *B. subtilis* or other Gram-positive bacteria. Promoters and ribosome binding sites are subtly different from those of *E. coli*, in other words it is not possible to simply use expression vectors already developed for this organism in *B. subtilis*. However, constitutive and controllable promoters have been developed for *B. subtilis*, for example, from growth-phase dependent genes for *Bacillus* α-amylase and protease. An alternative approach has been to create hybrid systems such as a *Bacillus* penicillinase promoter linked to the *E. coli lac* operator. This system requires a host cell containing the *lac* repressor gene under the control of a *Bacillus* promoter. Expression of foreign proteins is controlled by IPTG as with the *E. coli* systems. A disadvantage of *B. subtilis* is that cells produce several different extracellular proteases which inevitably cause degradation of secreted foreign proteins.[60,74] Much work has gone into producing strains with reduced levels of this protease activity. Another limitation of *B. subtilis* has been limited understanding of the secretion pathway. The basic mechanism involves a signal sequence present on pre-proteins but it has been observed that some foreign proteins are exported far more efficiently than others even if they contain the same pre-sequence. It is not clear at which stage transport is blocked. Despite these difficulties, *B. subtilis* has been used successfully for the production of high levels of foreign proteins. These include α-interferon, growth hormone, epidermal growth factor and pepsinogen secreted to levels of 15, 200, 240 and 500 mg l^{-1} of medium, respectively.[74]

The ongoing development of genetic systems for use in *Bacillus* has

[74] C. R. Harwood, *Tibtechnology*, 1992, **10**, 247.

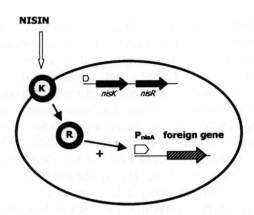

Figure 8   *The NICE (nicin controlled expression) system of promoter regulation in* Lactococcus lactis. *The* P_{nisA} *promoter is regulated through a two-component system. Nisin in the medium is detected by the sensor kinase nisK. NisK phosphorylates nisR which then activates genes transcribed from* P_{nisA}

been aided by its similarity to other low GC Gram-positive bacteria. Many of these are of considerable medical and industrial importance, such as members of the genera *Staphylococcus, Streptococcus, Lactococcus, Lactobacillus* and *Clostridium*, as well as other bacilli. Their similarity extends to their genetics and many of the vectors and gene transfer techniques developed for one low GC Gram-positive organism can be used in others.[75]

One feature of the physiology of these bacteria has been exploited to aid in the controlled expression of foreign genes. In common with many bacteria, lactococci exhibit cell-density dependent gene expression. Known as 'quorum sensing', this mechanism turns on gene expression in response to the build up of an autoinducer, in this case a peptide, secreted into the medium as the culture grows. The control of production of the lantibiotic nisin operates in this way. If a foreign gene is placed under the control of a nisin inducible promoter then it will be expressed only during the later stages of growth. Alternatively, in a strain defective in pheromone production the gene will only be expressed on addition of the inducing peptide (Figure 8).[76]

Lactic acid bacteria in general have been considered as ideal hosts for heterologous gene expression. They are non-pathogenic food-grade organisms whose use in industrial scale fermentations is highly developed. Moreover, there has been considerable impetus to develop the

[75] W. M. de Vos, M. Kleerebezem, and O. P. Kuipers, *Curr. Opin. Biotechnol.*, 1997, **8**, 547.
[76] O. P. Kuipers, P. G. G. A. de Ruyter, M. Kleerebezem, and W. M. de Vos, *Tibtechnology*, 1997, **15**, 135.

genetic engineering of these organisms because of their commercial significance. As well as using nisin, induction systems exist that rely on phage genes, temperature sensitive repressors or pH changes as triggers of expression. Like *Bacillus*, lactic acid bacteria have efficient protein secretion systems that can be used to aid the correct folding of hetero-logous proteins and ease their recovery from the surrounding medium.[77]

Streptomycetes, which are high GC Gram-positive bacteria, also have a long history of industrial use in the production of antibiotics. Genetic exchange mechanisms and vector systems have been developed for these organisms and *Streptomyces lividans* in particular has been used as a host for the production of pharmaceutical proteins.[78]

7 FUTURE PROSPECTS

Clearly there are considerable commercial and scientific opportunities in this area of molecular biology. Many of the basic techniques required to obtain expression of any DNA sequence in any host organism are now available and levels of expression are commercially viable. Future work is likely to concentrate on maximizing expression of proteins that retain biological activity; characterizing the best and most cost-effective growth conditions for mass production; the synthesis of novel, genetically engineered proteins, such as designer antibodies or receptor fragments for drug development. Further development of alternative host organisms is also continuing, particularly *B. subtilis*, lactic acid bacteria and streptomycetes. As discussed earlier, however, bacteria are not capable of carrying out many of the functions required to produce protein altered by post-translational modifications. In this respect eukaryotic hosts have the advantage. Their use is discussed in the following chapters.

8 FURTHER READING

1. R. W. Old and S. B. Primrose, 'Principles of Gene Manipulation: An Introduction to Genetic Engineering', 5th Edn, Blackwell, Oxford, 1994.
2. J. D. Watson, M. Gilman, J. Witkowski, and M. Zoller, 'Recombinant DNA', 2nd Edn, Freeman, New York, 1992.
3. 'Systems for Heterologous Gene Expression', in *Methods in Enzymology*, ed. D. V. Geoddel, Academic Press, New York, 1990, 185.

[77] W. M. de Vos, *Curr. Opin. Microbiol.*, 1999, **2**, 289.
[78] C. Binnie, J. D. Cossar, and D. I. H. Stewart, *Tibtechnology*, 1997, **15**, 315.

CHAPTER 5

Yeast Cloning and Biotechnology

BRENDAN P. G. CURRAN AND VIRGINIA C. BUGEJA

1 INTRODUCTION

The yeast *Saccharomyces cerevisiae* was developed as the first eukaryotic heterologous protein expression system because as a unicellular organism it could be genetically manipulated using many of the techniques commonly used for *Escherichia coli* and as a eukaryote it was a more suitable host for the production of authentically processed eukaryotic proteins.

The last edition of this chapter concluded with the statement '... *S. cerevisiae*'s well characterized genetic systems will ensure its central role in eukaryotic cloning and biotechnology into the foreseeable future', but even we did not realize how prophetic this statement was to be: *S. cerevisiae*'s most valuable contribution to modern biotechnology five years ago was as a host for the production of heterologous proteins; this has now been exceeded by its use as a tool in genome analysis, genome manipulation and the characterization of protein–protein interactions from a wide variety of species.

2 GENE MANIPULATION IN *S. CEREVISIAE*

2.1 Introducing DNA into Yeast

There are several procedures that achieve efficient transformation of *S. cerevisiae*. Exponentially growing cells can be enzymatically treated to remove their cell walls and the resulting sphaeroplasts exposed to DNA in the presence of calcium ions and polyethylene glycol, before being embedded in hypertonic selective agar to facilitate cell wall regeneration

155

and subsequent colony formation. Frequencies of up to 10^6 transfor-mants/μg DNA can be obtained using this technique.[1,2]

Intact yeast cells can be transformed by treating them with alkali cations (usually lithium), in a procedure analogous to *E. coli* transforma-tion,[3] or by electroporation,[4] which involves using a brief voltage pulse to facilitate entry of DNA molecules into the cells. Transformation efficiencies obtained using whole cells can be lower than those obtained using the protoplasting method, but transformed cells can be spread directly onto selective plates (rather than embedded in agar) thus facilitating procedures that require the use of colony screens.

2.2   Yeast Selectable Markers

Unlike the dominant antibiotic resistance markers used in *E. coli* transformations many yeast selectable markers are genes which comple-ment a specific auxotrophy (*e.g.* Leu-, His-, Trp-, *etc.*) and thus require the host cell to contain a recessive, non-reverting mutation. The most widely used selectable markers and their chromosomal counterparts are listed in Table 1. Further details of these and other recessive selectable markers can be found on the World Wide Web.[5] The most widely used dominant selectable marker systems are also included in Table 1.

2.3   Vector Systems

S. cerevisiae cloning plasmids are shuttle vectors containing a segment of bacterial plasmid DNA to facilitate DNA manipulation/large-scale plasmid purification in *E. coli* and a yeast selectable marker plus one or more structural elements to allow for selection, manipulation and/or expression in yeast. The vectors can be divided into a number of different types depending on their structural elements but here we restrict ourselves to those that are most commonly used for biotechnological applications. Schematic diagrams of these can be found in Figure 1.

YEp (yeast episomal plasmid) vectors are based on the ARS (auton-omously replicating sequence) sequence from the endogenous yeast 2μ plasmid which contains genetic information for its own replication and segregation (for a review see reference 6). They are capable of autono-

[1] A. Hinnen, J. B. Hicks, and G. R. Fink, *Proc. Natl. Acad. Sci. USA*, 1978, **75**, 1929.
[2] J. D. Beggs, *Nature (London)*, 1978, **275**, 104.
[3] H. Ito, Y. Fukada, K. Murata, and A. Kimura, *J. Bacteriol.*, 1983, **153**, 163.
[4] E. Meilhoc, J.-M. Masson, and J. Teissie, *Bio/Technology*, 1990, **8**, 223.
[5] http://genome-www.stanford.edu/Saccharomyces/alleletable.html.
[6] A. B. Futcher, *Yeast*, 1988, **4**, 27.

Table 1 *Selectable markers for yeast transformation*

Auxotrophic markers

Gene	Chromosomal mutation	Reference
HIS3	his3–Δ1	a
LEU2	leu2–3,leu2–112	b
TRP1	trp1–289	c
URA3	ura3–52	a

Dominant markers

Gene	Selection	Reference
CUP1	Copper resistance	d
G418ᴿ	G418 resistance (Kanomycin phosphotransferase)	e,f
TUNᴿ	Tunicamycin resistance	g

[a]K. Struhl, D. T. Stinchcomb, S. Scherer, and R. W. Davis, *Proc. Natl. Acad. Sci. USA*, 1979, **76**, 1035.
[b]J. D. Beggs, *Nature (London)*, 1978, **275**, 104.
[c]G. Tschumper and J. Carbon, *Gene*, 1980, **10**, 157.
[d]S. Fogel, and J. Welch, *Proc. Natl. Acad. Sci. USA*, 1987, **79**, 5342.
[e]T. D. Webster and R. C. Dickson, *Gene*, 1983, **26**, 243.
[f]D. D. Shoemaker, D. A. Lashkari, D. Morris, M.Mittmann, and R. W. Davis, *Nature Genetics*, 1996, **14**, 450.
[g]J. Rine, W. Hansen, E. Hardeman, and R. W. Davis, *Proc. Natl. Acad. Sci. USA*, 1983, **80**, 6750.

mous replication, are present at 20–200 copies per cell and under selective conditions are found in 60% to 95% of the cell population (Figure 1A). Some of the many applications (*e.g.* expression of heterologous proteins and the yeast two hybrid system) of these vectors will be discussed below.

The addition of a centromeric sequence converts a YEp into a yeast centromeric plasmid (YCp) (Figure 1B). YCps are normally present at one copy per cell, can replicate without integration into a chromosome and are stably maintained during cell division, even in the absence of selection.

Plasmids can be modified for use as expression vectors by the addition of suitable regulatory sequences. A typical expression vector is shown in Figure 1C. It consists of a YEp vector carrying yeast promoter and terminator sequences on either side of a unique restriction site. Promoterless heterologous genes can be expressed by inserting them into this site in the appropriate orientation (see Section 3.2). An extremely versatile family of expression vectors providing alternative selectable markers, a range of copy numbers and differing promoter strengths (see Figure 1D) has been produced by Mumberg *et al.*[7]

[7]D. Mumberg, R. Müller, and M. Funk, *Gene*, 1995, **156**, 119.

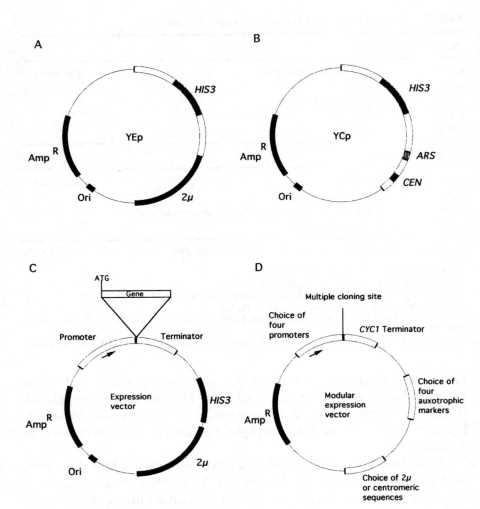

Figure 1 *Schematic diagrams of yeast cloning vectors. A yeast episomal (YEp) vector*
(A) consists of a prokaryotic gene for resistance to ampicillin (Amp^R), *a*
prokaryotic replication origin (ori), a yeast auxotrophic marker (HIS3) and a
yeast 2μ DNA sequence. A yeast centromeric (YCp) vector (B) contains a yeast
centromere (CEN) and an autonomous replication sequence (ARS) instead of
the 2μ DNA sequence. The addition of yeast promoter and terminator sequences
generates a yeast expression vector (C); transcription initiation of heterologous
genes cloned into a unique cloning site is indicated by the arrow. An extremely
versatile series of modular expression vectors (D) provides a choice of promoters
(CYC1, ADH, TEF and GPD); a choice of selectable marker genes (HIS3,
LEU2, TRP1 and URA3); a choice of copy number (centromeric or 2μ
plasmid) and contains a multiple cloning site between the promoter and
terminator sequences (see reference 7)

3 HETEROLOGOUS PROTEIN PRODUCTION

Yeast expression systems have been used for the recovery of high levels of authentic, biologically active heterologous proteins. Successful recovery depends on a variety of factors including: the source of the DNA; the overall level of mRNA present in the cell; the overall amount of protein present in the cell; and the nature of the required product (for reviews see references 8–10).

3.1 The Source of Heterologous DNA

As a general rule *S. cerevisiae* neither recognizes regulatory sequences of genes from other species nor efficiently excises heterologous introns. It is therefore necessary to obtain the gene of interest as a cDNA clone and splice it into an expression vector to obtain acceptable levels of protein production.

The level of expression may also be affected by the codon bias in the original organism. Impressive production of many heterologous genes in yeast has been achieved without optimizing codon usage patterns, but in other cases a correlation between the presence of rare codons and a drop in heterologous protein production has been noted.[11,12]

3.2 The Level of Heterologous mRNA Present in the Cell

The overall level of heterologous mRNA in the cell is a balance between the production of mRNA and its stability in the cytoplasm. The former is determined by the copy number of the expression vector and the strength of the promoter; the latter by the specific mRNA sequence.

Expression vectors based on YEp technology have a high copy number but require selective conditions to ensure their stable inheritance. Plasmid instability can be prevented by introducing a centromere into the vector, but at the cost of reducing the plasmid copy number to 1–2 copies per cell.

High level mRNA production is also dependant on the type of promoter chosen to drive expression. The major prerequisite for an

[8] J. E. Ogden, in 'Applied Molecular Genetics of Fungi', ed. J. F. Peberdy, C. E. Caten, J. E. Ogden, and J. W. Bennett, Cambridge University Press, Cambridge, 1991, p. 66.
[9] M. F. Tuite, in 'Biotechnology Handbooks, 4. Saccharomyces', ed. M. F. Tuite and S. G. Oliver, Plenum Press, New York, 1991, p. 169.
[10] D. J. King, E. F. Walton, and G. T. Yarranton, in 'Molecular and Cell Biology of Yeasts', ed. E. F. Walton and G. T. Yarranton, Blackie, London, 1989, p. 107.
[11] A. Hoekema, R. A. Kastelein, M. Vasser, and H. A. diBoer, *Mol. Cell. Biol.*, 1987, **7**, 2914.
[12] L. Kotula and P. J. Curtis, *Bio/Technology*, 1991, **9**, 1386.

expression vector promoter is that it is a strong one. Those most frequently encountered are based on promoters from genes encoding glycolytic enzymes, *e.g.* phosphoglycerate kinase (*PGK*), alcohol dehydrogenase 1 (*ADH1*) and glyceraldehyde-3-phosphate dehydrogenase (*GAPDH*), all of which facilitate high level constitutive mRNA production. Constitutive expression can be disadvantageous when the foreign protein has a toxic effect on the cells but this can be circumvented by using a regulatable promoter to induce heterologous gene expression after the cells have been grown to maximum biomass. The most commonly used one, based on the promoter of the galactokinase gene (*GAL1*), is induced when glucose is replaced by galactose in the medium, but a number of others are also available. Transcript stability is also vitally important in maintaining high levels of mRNA in the cell. Despite the fact that 'instability elements' have been identified[13] mRNA half-lives cannot be accurately predicted from primary structural information so they must be empirically determined; unstable transcripts curtail high level expression. *S. cerevisiae* often fails to recognize heterologous transcription termination signals because its own genes lack typical eukaryotic terminator elements. This can result in the production of abnormally long mRNA molecules which are often unstable.[14] As this can result in a dramatic drop in heterologous protein yield,[15] expression vectors frequently contain the 3' terminator region from a yeast gene (*e.g. CYC1, PGK* or *ADH1*) to ensure efficient mRNA termination (Figure 1D).

3.3 The Amount of Protein Produced

A high level of stable heterologous mRNA does not necessarily guarantee a high level of protein production. The protein level depends on the efficiency with which the mRNA is translated and the stability of the protein after it has been produced.

The site of translation initiation in 95% of yeast mRNA molecules corresponds to the first AUG codon at the 5' end of the message. It is advisable to eliminate regions of dyad symmetry and upstream AUG triplets in the heterologous mRNA leader sequence to ensure efficient initiation of translation. The overall context of the sequence on either side of the AUG codon (with the exception of an A nucleotide at the -3 position) and the leader length do not appear to affect the level of translation (for a review see reference 16).

[13] R. Parker and A. Jacobson, *Proc. Natl. Acad. Sci. USA*, 1990, **87**, 2780.
[14] K. S. Zaret and F. Sherman, *J. Mol. Biol.*, 1984, **177**, 107.
[15] J. Mellor, M. J. Dobson, N. A. Roberts, A. J. Kingsman, and S. M. Kingsman, *Gene*, 1985, **33**, 215.
[16] T. F. Donahue and A. M. Cigan, *Methods Enzymol.*, 1990, **185**, 366.

Achieving high level transcription and translation of a heterologous gene in any expression system does not necessarily guarantee the recovery of large amounts of heterologous gene product. Some proteins are degraded during cell breakage and subsequent purification; others are rapidly turned over in the cell. The powerful tools provided by a detailed knowledge of yeast genetics and biochemistry can be used to minimize this problem in *S. cerevisiae*. Genes for proteolytic enzymes can be inactivated to prevent degradation during cell lysis and the yeast secretory pathway exploited to smuggle heterologous proteins out of the cell into the culture medium where protease levels are low.

The use of protease deficient host strains can improve both the yield and the quality of heterologous proteins.[17,18] One mutant (*PEP4–3*) is widely used because it is responsible for the activation of vacuolar zymogen proteases and so its absence prevents the activation of a range of different proteinase activities. Nevertheless, protease inhibitors must also be present during extraction to maximize protein production.

The intracellular turnover of cytoplasmic proteins can also affect the levels of protein produced during heterologous gene expression. In eukaryotes a ubiquitin mediated proteolytic pathway[19] appears to play a major role in regulating the rate of turnover of cytoplasmic proteins. The sensitivity of some proteins to degradation via this pathway can be dramatically altered by changing the *N*-terminal amino acid of the protein;[20] such alterations could, however, cause unacceptable immuno-genicity problems if the heterologous protein is intended for therapeutic purposes.

Degradation both by vacuolar proteases and the ubiquitin degradative pathway can be avoided in many instances by exploiting the yeast secretory pathway. This not only minimizes the exposure of heterologous proteins to intracellular protease activity but also facilitates their recovery and purification due to the very low levels of native yeast proteins normally present in culture media.

Entry into the secretory pathway is determined by the presence of a short hydrophobic 'signal' sequence on the *N*-terminal end of secreted proteins. The 'signal' sequences from *S. cerevisiae*'s four major secretion products have been attached (by gene manipulation) to the *N*-terminus of heterologous proteins and used with varying degrees of success to direct their secretion. Invertase[21] and acid

[17] T. Cabezon, M. De Wilde, P. Herion, R. Loriau, and A. Bollen, *Proc. Natl. Acad. Sci. USA*, 1984, **81**, 6594.
[18] S. Rosenberg, P. J. Barr, R. C. Najarian, and R. A. Hallewell, *Nature (London)*, 1984, **312**, 77.
[19] D. Wilkinson, *Methods Enzymol.*, 1990, **185**, 387.
[20] A. Bachmair, D. Finley, and A. Varshavsky, *Science*, 1986, **234**, 179.
[21] D. T. Moir and D. R. Dumais, *Gene*, 1987, **56**, 209.

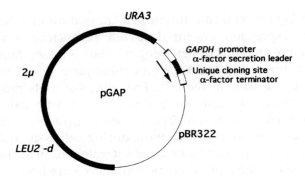

Figure 2 *Schematic representation of the yeast secretion vector pGAP (J. Travis, M. Owen, P. George, R. Carrell, S. Rosenberg, R. A. Hallewell, and P. J. Barr, J. Biol. Chem., 1985, 260, 4384.). The vector contains LEU2-d and URA3 yeast selectable marker genes, pBR322 sequences for amplification in E. coli and 2μ sequences for autonomous replication in yeast. The expression 'cassette' contains a unique cloning site flanked by GAPDH promoter, α-factor secretion leader and α-factor terminator sequences. Transcription initiation is indicated by the arrow*

phosphatase[22] signal sequences target proteins to the periplasmic space whereas α-factor[23] and killer toxin[24] signals target the proteins to the culture medium. A typical secretion vector is shown in Figure 2. Secretion can also be used to produce proteins that have an amino acid other than methionine at their *N*-terminus. If a secretory signal is spliced onto the heterologous gene at the appropriate amino acid (normally the penultimate one) then the *N*-terminal methionine which is obligatory for translation initiation will be on the secretory signal. Proteolytic cleavage of this signal from the heterologous protein in the endoplasmic reticulum (ER) will generate an authentic *N*-terminal amino acid (Figure 3).

3.4   The Nature of the Required Product

The objective of heterologous gene expression is the high level production of biologically active, authentic protein molecules. It is therefore important to consider the nature of the final product when choosing the expression system. The protein size, hydrophobicity, normal cellular location, need for post-translational modification(s) and ultimate use, must be assessed before an appropriate expression system is chosen.

---

[22] A. Hinnen, B. Meyhack, and R. Tsapis, in 'Gene Expression in Yeast', Kauppakirjapaino, Helsinki, 1983, p. 157.

[23] A. J. Brake, J. P. Merryweather, D. G. Coit, U. A. Heberlein, F. R. Masiarz, G. T. Mullenbach, M.S. Urdea, P. Valenzuela, and P. J. Barr, *Proc. Natl. Acad. Sci. USA*, 1984, **81**, 4642.

[24] N. Skiper, M. Sutherland, R. W. Davies, D. Kilburn, R. C. Miller, A. Warren, and R. Wong, *Science*, 1985, **230**, 958.

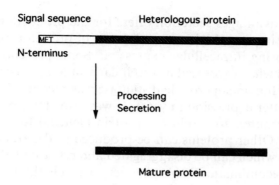

Signal sequence   Heterologous protein

MET

N-terminus

Processing
Secretion

Mature protein

Figure 3 *Schematic diagram showing the secretion of a heterologous protein using a signal sequence*

The secretory pathway is often chosen for heterologous protein production because, apart from the fact that it enhances protein stability and can generate proteins lacking an *N*-terminal methionine (see Section 3.3), secretion facilitates the accurate folding of large proteins and contains the machinery for post-translational modification. A direct comparison between the intracellular production and extracellular secretion of prochymosin and human serum albumin resulted in the recovery of small quantities of mostly insoluble, inactive protein when they were produced intracellularly but the recovery of soluble, correctly folded, fully active protein when they were secreted.[25,26]

The biological activity and/or stability of heterologous proteins can also be affected by the post-translational addition of carbohydrate molecules to specific amino acid residues. Glycosylation in yeast is of both the *N*-linked (via an asparagine amide) and *O*-linked (via a serine or threonine hydroxyl) types, occurring at the sequences Asn-X-Ser/Thr and Thr/Ser, respectively. Inner core *N*-linked glycosylation occurs in the ER and outer core glycosylation in the Golgi apparatus. However, it is important to note that the number and type of outer core carbohydrates attached to glycosylated proteins in yeast are different to those found on mammalian proteins. In many cases these differences can be tolerated but if the protein is being produced for therapeutic purposes they may cause unacceptable immunogenicity problems. One approach to overcoming this problem is to remove the glycosylation recognition site by site directed mutagenesis. This strategy was successfully used to produce urokinase type plasminogen activator.[27]

[25] R. A, Smith, M. J. Duncan, and D. T. Moir, *Science*, 1985, **229**, 1219.
[26] T. Etcheverry, W. Forrester, and R. Hitzeman, *Bio/Technology*, 1986, **4**, 726.
[27] L. M. Melnick, B. G. Turner, P. Puma, B. Price-Tillotson, K. A. Salvato, D. R. Dumais, D. T. Moir, R. J. Broeze, and G. C. Avgerinos, *J. Biol. Chem.*, 1990, **265**, 801.

Despite the advantages secretion offers for the production of heterologous proteins in yeast, higher overall levels of protein production are
often possible using intracellular expression. Some proteins form insoluble complexes when expressed intracellularly in *S. cerevisiae* but many
others do not. Human superoxide dismutase was recovered as a soluble
active protein after expression in yeast. It was also efficiently acetylated
at the amino terminus to produce a protein identical to that found in
human tissue.[28] Other proteins can be produced as denatured, intracellular complexes which can be disaggregated and renatured after harvesting. The first recombinant DNA product to reach the market was a
hepatitis B vaccine produced in this way.[29]

4 USING YEAST TO ANALYSE GENOMES, GENES AND PROTEIN–PROTEIN INTERACTIONS

Initially exploited as a system for heterologous protein production, *S.
cerevisiae* is now increasingly used as a tool in the analysis of genomes,
genes and protein function because its entire genome has been
sequenced,[30] its homologous recombination system facilitates *in vivo*
gene manipulation, and its well characterized genetics and biochemistry
make it ideally suited for developing novel reporter systems.

4.1 YAC Technology

Yeast artificial chromosomes (YACs) are specialized vectors capable of
accommodating extremely large fragments of DNA (100–1000 kb).[31]
Schematic diagrams of a YAC and its use as a cloning system are shown
in Figure 4. YACs contain a centromere, an autonomously replicating
sequence, two telomeres and two yeast selectable markers separated by a
unique restriction site. They also contain sequences for replication and
selection in *E. coli*. YACs are linear molecules when propagated in yeast
but must be circularized by a short DNA sequence between the tips of the
telomeres for propagation in bacteria. When used as a cloning vehicle,
the YAC is cleaved with restriction enzymes to generate two telomeric
arms carrying different yeast selectable markers. These arms are then
ligated to suitably digested DNA fragments, transformed into a yeast
host and maintained as a mini chromosome.

---

[28] R. A. Hallewell, R. Mills, P. Tekamp-Olsen, R. Blacker, S. Rosenberg, F. Otting, F. R. Masiarz, and C. J. Scandella, *Bio/Technology*, 1987, **5**, 363.
[29] D. E. Wampler, E. D. Lehman, J. Boger, W. J. McAleer, and E. M. Scolnick, *Proc. Natl. Acad. Sci. USA*, 1985, **82**, 6830.
[30] http://genome-www.stanford.edu/Saccharomyces/
[31] D. T. Burke, G. F. Carle, and M. V. Olsen, *Science*, 1987, **236**, 806.

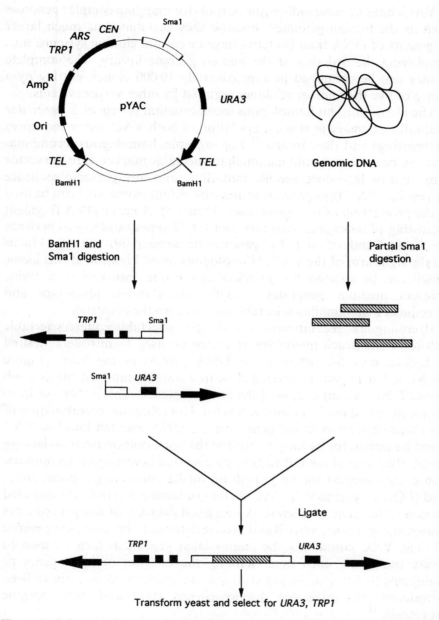

Figure 4 *Schematic diagram of a YAC cloning vector, indicating prokaryotic gene for resistance to ampicillin (Amp^R), prokaryotic replication origin (ori), yeast auxotrophic markers (URA3, TRP1), autonomous replication sequence (ARS), yeast centromere (CEN) and telomeres (TEL)*

YACs have become indispensable tools for mapping complex genomes such as the human genome[32] because they accommodate much larger fragments of DNA than bacteriophage or cosmid cloning systems, thus simplifying the ordering of the human genome library. The complete library can be contained in approximately 10 000 clones, cutting by a factor of five the number of clones required by other vector systems.

The highly efficient homologous recombination system of *S. cerevisiae* facilitates extensive *in vivo* manipulation of both YAC vector sequences (retrofitting) and their inserts.[33] For example, homologous recombination can be used to retrofit mammalian selectable markers into the vector arms and/or introduce specific mutations into any genomic sequence carried in a YAC thus generating artificial chromosomes that can be used in the production of transgenic mice (Figure 5). A linear DNA fragment consisting of neomycin resistance and *LYS2* genes sandwiched between the 5' and 3' ends of the *URA3* gene can be targeted into the *URA3* locus on the right arm of the YAC. Homologous recombination at this locus, which can be selected for by selecting for transformants on a lysine deficient medium, generates a useful uracil minus phenotype and introduces a mammalian selectable marker into the construct.

Homologous recombination and the selectable/counterselectable nature of the uracil phenotype can then be used to introduce desired mutations into the heterologous DNA carried in the YAC (Figure 6A,B). A *URA3* gene is inserted close to a suitably mutated site in a sub clone of the relevant section of the gene of interest. This is then cut from the plasmid and used to transform yeast. Homologous recombination of the mutated version of the gene into the DNA inserted into the YAC insert is selected for by simply growing the yeast cells on medium lacking uracil. Removal of the *URA3* gene by a second homologous recombination event, selected for by growth on media containing 5-fluoroorotic acid (FOA),[34] generates a YAC vector containing a specifically mutated version of the gene of interest. A simplified diagram of this procedure is shown in Figures 6A and 6B and precise details can be found in reference 33. The YAC containing the manipulated genes can then be used to create transgenic mice thus enabling the analysis of large genes or multigenic loci *in vivo*. Many such murine and human genes have been introduced into mice and display correct stage and tissue specific expression.[33]

[32] P. Sudbery, in 'Human Molecular Genetics', Addison Wesley Longman Ltd, England, 1998, p. 209.
[33] K. R. Peterson, C. H. Clegg, Q. Li, and G. Stamatoyannopoulos, *TIGS*, 1997, **13**, 61.
[34] J. D. Boeke, F. LaCroute, and G. R. Fink, *Mol. Gen. Genet.*, 1984, **197**, 345.

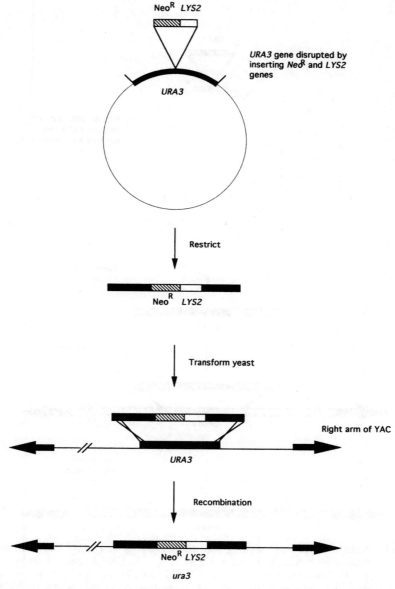

Figure 5 *Schematic diagram showing the integration of linear yeast DNA into a homologous region of chromosomal DNA carried in a YAC. A linear fragment of DNA carrying the neomycin resistance gene (Neo^R) and the LYS2 gene flanked by URA3 gene sequences is isolated from plasmid DNA and transformed into the YAC carrying yeast strain. Homologous recombination events which result in the replacement of the wild-type (URA3) gene on the right arm of the YAC with this linear fragment are selected for by growth on lysine deficient medium*

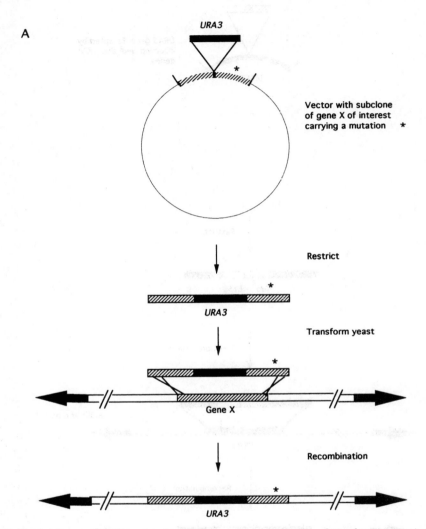

Figure 6   *Schematic diagram showing the introduction of a desired mutation into a heterologous gene X carried in a YAC. (A) The* URA3 *gene is inserted into a suitably mutated subclone of gene X (the asterisk indicates a mutation in the gene DNA sequence). Homologous recombination and co-selection with* URA3 *is then used to introduce this mutation into the YAC insert.*

B

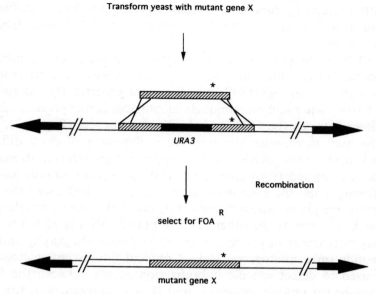

Figure 6 *(continued) (B) Homologous recombination using the mutated subclone without the URA3 gene and counterselecting on FOA medium is then used to generate a specifically mutated version of gene X*

4.2 Gene Knockouts

As the first eukaryotic organism to have its entire genome sequenced, *S. cerevisiae*[30] is currently in the process of providing a unique insight into eukaryotic cell metabolic function. PCR-based gene disrupted procedures[35] can be used to disrupt open reading frames (ORFs) affording analysis of any resultant physiological effects. Long PCR primers, homologous to a section of the DNA sequence under investigation at the 5′ end and homologous to a selectable marker (frequently the *Kan*[R] gene) at their 3′ terminus end, are used to generate a PCR fragment which consists of two short target gene segments on either side of a selectable marker. The target gene is then disrupted by homologous recombination and the event selected for by growth on appropriate selective medium. The genes carried on yeast chromosome III have already been examined in this way[36] and of the 55 ORFs disrupted or

[35] A. Baudin, O. Ozier-Kalogeropoulos, A. Denouel, F. Lacroute, and C. Cullin, *NAR*, 1993, **21**, 3329.
[36] S. G. Oliver, Q. J. van der Aart, M. L. Agostoni-Carbone, M. Aigle, L. Alberghina, D. Alexandraki, G. Antoine, R. Anwar, J. P. Ballesta, P. Benit *et al.*, *Nature*, 1992, **357**, 38.

deleted only three were essential with a further 21 revealing a specific phenotype. The phenotypic effect of deleting individual genes is frequently masked by functionally homologous genes but a multideletion approach such as that outlined by Güldener *et al.*[37] should circumvent this problem.

A PCR-based gene deletion strategy is currently being used by a consortium of 16 American and European laboratories to systematically disrupt each of the 6000 ORFs in the yeast genome. By incorporating one shared 18mer and one uniquely identifiable 20mer sequence into the 5′ primer between the ORF and the *Kan*^{R} sequences (Figure 7),[38] PCR can be used to monitor the population dynamics of these differently tagged yeast strains growing under competitive growth conditions. Total DNA is extracted from such a population and the unique sequence, identifying a specific gene disruption event, amplified using the shared 18mer as one primer and a fluorescently labelled primer from the middle of the *Kan*^{R} gene as the other. The amplified DNA is added to a high density microarray of probes generated by photolithography and oligonucleotide on a microscope slide. Confocal fluorescent microscopy is then used to detect specific hybridization events between the fluorescently labelled unique sequences and their counterparts in the microarray. Computer analyses of fluorescent signals reveals the relative numbers of each strain present in that particular population.

At the time of writing, deletion strains for 1700 ORFs were available[39] – a resource that is now matched by the commercial availability of constitutive and inducible versions of all the yeast genes (Genestorm^{TM} expression-ready yeast clones from invitrogen).

Not surprisingly, many *S. cerevisiae* genes show sequence homology to human genes. Comparative sequence analysis between *S. cerevisiae* and human genes has already provided valuable insight into human cellular metabolism. A variety of examples is given in Table 2, of which perhaps the most dramatic to date is the analysis of an autosomal recessively inherited disease rhizomelic chondrodysplasia punctata which presents with symptoms of severe growth and mental retardation. Patients with this condition were found to carry mutations leading to defects in peroxisome biogenesis which are functionally equivalent to the yeast peroxisome targeting mutants (*pex5* and *pex7*).[40,41] Thus, quite apart

[37] U. Güldener, S. Heck, T. Fiedler, J. Beinhauer, and J. H. Hegemann, *Nucleic Acids Res.*, 1996, **24**, 2519.
[38] D. D. Shoemaker, D. A. Lashkari, D. Morris, M. Mittmann, and R. W. Davis, *Nature Genetics*, 1996, **14**, 450.
[39] http://www-sequence.stanford.edu/group/yeast_deletion_project/images/EnterDB_frame.html.
[40] P. E. Purdue, J. W. Zhang, M. Skoneczny, and P. B. Lazarow, *Nature Genetics*, 1997, **15**, 381.
[41] A. M. Motley *et al.*, *Nature Genetics*, 1997, **15**, 377.

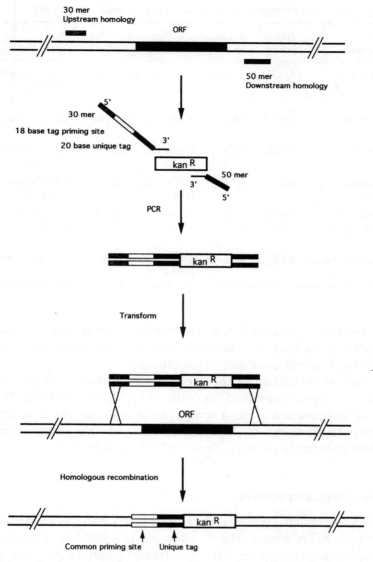

Figure 7 *PCR gene deletion strategy. The Kanamycin resistance gene (Kan^R) is amplified using an 86mer forward primer that contains 30 bases of upstream homology to the yeast gene of interest, an 18 base tag priming site, a uniquely identifiable 20 base sequence tag and 18 bases of homology to the Kan^R gene. The reverse primer is a 68mer that contains 50 bases of downstream homology to the yeast gene of interest and 18 bases of homology to the Kan^R gene. The PCR products are transformed into a haploid yeast strain and selected for on G-418 containing medium. Homologous recombination replaces the targeted yeast ORF with a common 18mer priming site, a unique 20 base tag and the Kan^R gene*

Table 2 *Yeast genes homologous to positionally cloned human genes*

Human disease	Human gene	Yeast gene	Yeast gene function	Yeast phenotype
Hereditary non-polyposis colon cancer	MSH2	MSH2	DNA repair protein	Increased mutation frequency
Cystic fibrosis protein	CFTR	YCF1	Membrane transport	Cadmium sensitivity
Wilson's disease	WND	CCC2	Copper transport atpase	Iron uptake deficiency
Glycerol kinase deficiency	GK	GUT1	Defective glycerol kinase	Defective glycerol utilization
Rhizomelic chondrodysplasia punctata	PEX5, PEX7	PEX5, PEX7	Peroxisome targeting mutants	Peroxisome dysfunction
Ataxia telangiectasia	ATM	TEL1	Phosphoinositol 3-kinase	Short telomeres

from providing a model of how a eukaryotic cell carries out its metabolic functions, yeast is likely to contribute to a basic understanding of human cell growth, differentiation and development.

The ultimate extension of this technology is the use of known yeast mutants in complementation analysis. Heterologous cDNAs from a variety of organisms including fungi, plants and humans have already been used to complement yeast mutations and as a direct screening procedure to identify genes present in cDNA libraries.[42,43]

4.3 Novel Reporter Systems

Our extensive understanding of the biochemistry and molecular biology of yeast cells has resulted in the development of a number of imaginative manipulations of promoter elements, transcription factors and signal cascade proteins to transduce heterologous molecular interactions into easily scoreable yeast phenotypes.

The yeast cell is insensitive to oestrogen, yet when the human oestrogen receptor was expressed in yeast cells and the oestrogen receptor element (ERE) was cloned into a disabled *CYC1* promoter fused to a β-galactosidase gene, oestrogen induced β-galactosidase

[42] P. Ullmann, L. Godet, S. Potier, and T. J. Bach, *Eur. J. Biochem.*, 1996, **236**, 662.
[43] S. Tanaka, J. Nikawa, H. Imai, S. Yamashita, and K. Hosaka, *FEBS Let.*, 1996, **393**, 89.

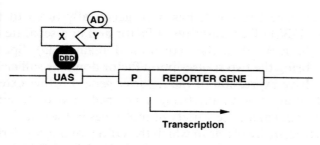

Figure 8 *The two-hybrid system. Protein X is fused to a DNA binding domain protein (DBD) and protein Y is fused to an activation domain (AD) of a transcription factor. Specific interaction of proteins X and Y brings the DBD and AD together, thus driving reporter gene transcription*

enzyme production.[44] This reporter system was sufficiently sensitive to analyse the effect of site specific mutations on hormone binding efficiency and to measure the effectiveness of agonists and antagonists on hormone action.

Likewise yeast cells have been manipulated to provide a 'readout' of heterologous G-protein-coupled-receptor (GPCR) activity. The well characterized mating–signal–transduction pathway has been extensively re-engineered to replace a cell arrest phenotype with a HIS + one.[45] The membrane bound α-factor receptor can be deleted and replaced with a heterologous one. The regulatory subunit of the yeast G-protein can then be manipulated to ensure activation of the mating pathway when the agonist binds to the foreign receptor. Deletion of the gene responsible for initiating cell cycle arrest and the fusion of a *HIS3* gene to a mating pathway activated promoter effectively transduces the receptor–agonist interaction to a scorable HIS + phenotype. As GPCRs represent the targets for the majority of presently prescribed pharmaceutical drugs this system has exciting potential for the development of high throughput screening technology.[46]

In a totally separate technique the interaction of two proteins X and Y can be transduced into a scorable phenotype by subjecting them to analysis by the extensively used yeast two-hybrid system (Figure 8).[47] The interaction of X and Y depicted in Figure 8 results in the expression

[44] C. K. Wrenn and B. S. Katzenellenbogen, *J. Biol. Chem.*, 1993, **268**, 24089.
[45] M. H. Pausch, *Tibtech*, 1997, **15**, 487.
[46] J. R. Broach and J. Thorner, *Nature*, 1996, **384**, 14.
[47] S. Fields and O. Song, *Nature*, 1989, **340**, 245.

of a reporter gene because X has been genetically fused to the DNA binding site (DBS) of a transcription factor and in a separate construction Y has been fused to the activation domain (AD). Specific X–Y interactions bring the two transcription factor domains sufficiently close together to drive expression of the reporter gene. Further extensions of this procedure are continually being developed in yeast to address the complex interplay of a wide variety of molecules in the cell.[48]

Yeast vectors are available in which the DBS (commonly derived from *GAL4* or *Lex A*) and the AD (commonly derived from *GAL4* or the viral activator VP16) are on separate 2μ-based expression plasmids carrying different selectable markers (for a review see reference 49). The genes for the proteins under analysis are inserted into these specific vectors and the plasmids are co-transformed into a suitable yeast host. Expression from a reporter gene (*e.g. Lac Z, HIS3, LEU2* or *URA3*) is used to test for a proposed protein interaction; to establish the effect of site-directed mutagenesis on previously characterized interacting proteins, X may be used as a bait to trap unknown interacting Ys from an expression library fused to the AD. The two-hybrid system has been successfully applied to study a broad spectrum of protein–protein interactions from a wide variety of different species.[50]

5 FUTURE PROSPECTS

A paradigm shift is taking place in biology as we enter the new millennium. A synthesis of molecular biology, microarrays and information technology is facilitating movement from reductionism to holism in molecular experimentation. At the leading edge of this revolution from the purely empirical to the increasingly rational, from manipulating molecules to manipulating their information content, sits a humble yeast cell.

S. cerevisiae's genome has been sequenced, its homologous recombination system facilitates gene and genome analysis and its elegant reporter systems reveal much about the dynamic nature of cellular macromolecules. As we accumulate an ever-increasing number of gene sequences from a variety of organisms, this yeast will be invaluable in establishing the biological importance of many of them. Likewise, as we accumulate an ever increasing amount of information about individual

[48] R. K. Brachmann and J. D. Boeke, *Curr. Opin. Biotechnol.*, 1997, **8**, 561.
[49] R. D. Gietz, B. Triggsraine, A. Robbins, K. C. Grahame, and R. A Woods, *Mol. Cellular Biochem.*, 1997, **172**, 67.
[50] K. H. Young, *Biol. Reprod.*, 1998, **58**, 302.

cellular building blocks, *S. cerevisiae* will play a central role in determining their dynamic interactions.

Suffice to say that having already made a significant contribution to heterologous protein production, the major contribution of the yeast *S. cerevisiae* to molecular biology in general and to biotechnology in particular has yet to come.

CHAPTER 6

Cloning Genes in Mammalian Cell-lines

EDWARD J. MURRAY

1 INTRODUCTION

DNA cloning in mammalian cells originated from observations that naked, uncoated protein-free viral DNA, when presented to cells, resulted in the initiation of the viral life-cycle within a small number of cells in the population.[1,2] These cells are said to be transfected, and the transfection efficiency of such experiments could be increased by the use of facilitators. Indeed, effective transfection was only seen when the viral DNA was mixed with DEAE-dextran,[3] or presented as a calcium phosphate co-precipitate.[4]

Of course, such techniques may be used to transfect cells with non-viral DNA and, since those early times, other transfection techniques have been used, such as microinjection,[5] liposome or protoplast fusion,[6] electroporation,[7] polycation-assisted transfection[8,9] and scrapefection[10] (see Sections 2.1–2.6).

DNA transfections in mammalian cells have been invaluable to molecular biologists who wish to study the mechanisms underlying gene expression. The biotechnologist will want to use this knowledge to

[1] F. Graham and A. Van der Eb, *Virology*, 1973, **52**, 456.
[2] J. M. McCutchan and J. S. Pagano, *J. Natl. Cancer Inst.*, 1968, **41**, 351.
[3] L. M. Sompayrac and K. J. Danna, *Proc. Natl. Acad. Sci. USA*, 1981, **78**, 7575.
[4] M. Wigler, S. Silverstein, L. Lih-Syng, A. Pellicer, C. Yung-Chi, and R. Axel, *Cell*, 1977, **11**, 223.
[5] M. Capecci, *Cell*, 1980, **22**, 479.
[6] M. Rassoulzadegan, B. Binetruy, and F. Cuzin, *Nature*, 1982, **295**, 257.
[7] E. Neuman, M. Schafer-Ridder, Y. Wong, and P. Hofschneider, *EMBO J.*, 1982, **1**, 841.
[8] P. Felgner, T. Gadek, M. Holm, R. Roman, H-W. Chan, C. Wenz J. Northrop, G. Ringold, and M. Danielson, *Proc. Natl. Acad. Sci. USA*, 1987, **84**, 7413.
[9] R. Aubin, M. Weinfeld, and M. Paterson, *Somatic Cell Molec. Genet.*, 1988, **14**, 155.
[10] M. Fechheimer, J. F. Boylan, S. Parker, J. E. Sisken, G. L. Patel, and S. G. Zimmer, *Proc. Natl. Acad. Sci. USA*, 1987, **84**, 8463.

generate altered cell lines which produce high yield and maximum activity of a commercially viable product.

E. coli offers many attractive features when used as a host for the production of recombinant proteins, including the ability to generate a high biomass in a short time and the relative low cost of maintaining a microbial fermentation.[11] In addition, a wide variety of prokaryotic expression vectors are commercially available which enable high inducible levels of recombinant protein fusions which bear either maltose binding protein, glutathione synthetase or hexahistidine peptides. The specific binding properties of the fusions can be exploited by using specific-affinity chromatography to facilitate purification procedures. Other prokaryotic expression vectors can exploit the secretion pathway to the periplasmic space to enable recombinant proteins to refold in a relatively non-reducing environment. Nevertheless, it is clear that the activity of some eukaryotic proteins is defective when expressed in *E. coli* and although some experimental modifications may be employed to increase the activity,[12] the problems may be insurmountable. Thus, the use of mammalian cell-lines offer the only option as host for the production of recombinant proteins with high specific activities.

There are two key areas covered in this chapter. First, the variety of transfection techniques will be described and compared and, second, the parameters governing DNA transfection will be evaluated to permit an understanding of how the expression of a transfected gene can be optimized and modulated.

Before discussing the above aims in depth, let us now consider the fate of DNA during a transfection experiment. During calcium phosphate co-precipitation transfection, up to 100% of the cells adsorb the DNA upon their cell membrane and it then enters the cell, probably by a phagocytic mechanism.[13] The function of most facilitators is to increase the ease of one or both of these processes. The DNA is able to migrate to the nucleus in a proportion of these cells and become complexed with a variety of histonal and non-histonal nuclear proteins.

The physical fate of the DNA from this point is unclear, but in general circular DNA templates become nicked and subsequently linearized within 48 hours.[14] Some laboratories note that the use of the calcium phosphate co-precipitation procedure leads to concatenerization,[15]

[11] R. E. Spier, in 'Molecular Biology and Biotechnology' ed. J. M. Walker and E. B. Gingold, Royal Society of Chemistry, Cambridge, 1985, pp. 119-134.
[12] C. S. Schein, *Curr. Opin. Biotech.*, 1991, **2**, 746.
[13] A. Loyter, G. E. Scangos, and F. Ruddle, *Proc. Natl. Acad. Sci. USA*, 1982, **79**, 422.
[14] H. Weintraub, P.-F. Cheng, and K. Conrad, *Cell*, 1986, **46**, 115.
[15] F. Weber and W. Schaffner, *Nature*, 1985, **315**, 75.

resulting in the formation of high molecular weight ligates called pekeliosomes.[16]

Within the 48 hour time span, many copies of the DNA are present and exist in an episomal (extrachromosomal) state. At this stage, expression of some DNA templates is apparent. This is called transient expression, and can be modulated by a variety of experimental factors and DNA sequences/topology which will be discussed in Section 6. After a few days, the observed expression decreases, presumably due to the loss of extrachromosomal copies which are unable to replicate and segregate in a dividing population of cells. However, a small proportion of these copies are able to integrate into the host cell genome, upon which they become a stably inherited genetic unit. Usually, the integration occurs at random sites (*i.e.* non-homologous) in the genome and multimeric copies of the DNA are present.[16] However, transfection techniques like microinjection and electroporation tend to favour single-copy integrations at a frequency which enables integration into homologous sites to be detected.[17,18] This may be important to potential clinical uses of DNA transfections in gene therapy. The process of stable integration is inefficient in that only 10^{-4} cells (or much fewer) of the initial transfected population eventually produce stable clones containing the foreign donor DNA. Thus, the success of such experiments depends upon the design of a stringent selection for the integration event. A variety of selection techniques are available which are either dominant or recessive (Section 6). Some selections cause the integrated foreign DNA to become amplified by several orders of magnitude, an observation which may suit the biotechnologists' purpose to achieve high yields of gene products.

2 METHODS OF DNA TRANSFECTION

A variety of methods are currently available for introducing DNA into mammalian cells, namely calcium phosphate co-precipitation,[4] DEAE-dextran,[3] and microinjection,[5] protoplast fusion,[6] electroporation,[7] lipofection,[8] polybrene–DMSO treatment,[9] and scrapefection.[10] Most cell-lines respond better to one or other particular technique, which has to be determined empirically. Figure 1 shows the four stages involved in a generalized transfection, and which are utilized by the different techniques.

[16] M. Perucho, D. Hanahan, and M. Wigler, *Cell*, 1980, **22**, 309.
[17] O. Smithies, R. Gregg, S. Boggs, M. Koralewski, and R. Kucherlapati, *Nature*, 1985, **317**, 230.
[18] K. Thomas, K. Folger, and M. Capecchi, *Cell*, 1986, **44**, 419.

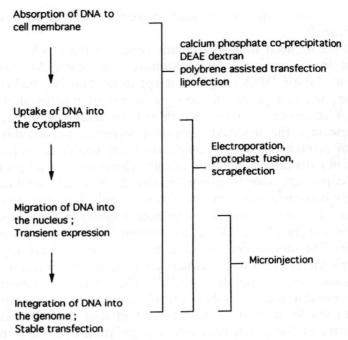

Figure 1 *The stages of DNA transfection which are involved in a variety of techniques*

2.1 Calcium Phosphate Co-precipitation

This technique has been studied by Wigler *et al.* in mouse L-cells, and has been optimized to the extent that 10^{-1}–10^{-4} cells of the population may be transfected, depending upon cell type and laboratory.[19-22] As a general rule, there is an upper limit of approximately 20 μg of DNA to be transfected. This amount is near the optimum for co-precipitate formation,[13] thus, if necessary, carrier DNA (salmon sperm DNA) is added to the test gene to achieve this total. In essence, a calcium phosphate precipitate is slowly induced to form in the presence of DNA. It is also possible to incorporate bacteriophage particles in the resulting precipitate.[23] This is then added directly to the cells and

[19] M. Wigler, S. Silverstein, L.-S. Lee, A. Pellicer, Y.-C. Cheng, and R. Axel, *Cell*, 1977, **11**, 223.

[20] M. Wigler, A. Pellicer, S. Silverstein, and R. Axel, *Cell*, 1978, **14**, 725.

[21] M. Wigler, R. Sweet, G.-K. Sim, B. Wold, A. Pellicer, E. Lacy, T. Maniatis, S. Silverstein, and R. Axel, *Cell*, 1979, **16**, 777.

[22] C. Gorman, in 'DNA Cloning – A Practical Approach' Vol. 2, eds D. M. Glover, IRL Press, pp. 143–190.

[23] M. Ishiura, S. Hirose, T. Uchida, Y. Hamada, Y. Suzuki, and Y. Okada, *Mol. Cell. Biol.*, 1982, **2**, 607.

incubated for 4–16 hours. Alternatively, the co-precipitate may be formed directly in the tissue culture dish.[24] The cells are then rinsed free of the precipitate and harvested 30–48 hours later for transient expression, or passaged in selective media for 10–14 days to isolate stable clones.

The function of the calcium component is not known, but it possibly concentrates the DNA on the cell membrane to further facilitate uptake by phagocytosis and protects the DNA from hydrolytic nucleases.[13] The transfection efficiency, or level of gene expression, depends upon both the efficiency of uptake by the cells and the transcriptional capability of each DNA template introduced into the successfully transfected cells. The efficiency of DNA uptake is critical upon the pH of the calcium phosphate DNA co-precipitate formation, and the amount of DNA in that complex.[13,20] A higher proportion of the cells can take up DNA after the application of a glycerol shock.[22] The inherent transcriptional capability of DNA templates can also be modulated by a number of factors. Sodium butyrate treatment increases expression via hyperacetylation of associated histonal proteins,[25] and enhancer sequences may be used as part of the vector (see Section 3). Commercially available facilitators such as CalPhos Maximizer are now available which both reduce transfection time and increase transfection efficiency. The purity and isomeric form of the DNA preparation is also important. Transient expression is favoured by Form I DNA (covalently closed, circular), whereas Form III DNA (linear) is successfully used for stable expression.

2.2 DEAE-Dextran

This technique has previously been used to increase the efficiency of viral infection of cell-lines. It is as simple and cheap as the calcium phosphate method, but can be used to transfect cells which will not survive even short exposures to calcium phosphate.[3]

Because the presence of serum may inhibit this method of transfection, the cells have to be washed extensively in phosphate-buffered saline prior to the addition of the DEAE-dextran–DNA solutions. The optimum DEAE-dextran concentration and length of incubation have to be determined experimentally to obviate the toxic effects of this treatment. Incubation times are in the range of 1–8 hours, after which excess DEAE-dextran–DNA is rinsed off the cell monolayer. As shown for calcium phosphate transfections, the transfection efficiency can be enhanced by glycerol shock. Also, chloroquinine treatment has been used to increase

[24] C. Chen and H. Okayama, *Mol. Cell. Biol.*, 1987, **7**, 2745.
[25] C. Gorman, B. Howard, and R. Reeves, *Nucl. Acids Res.*, 1983, **11**, 7631.

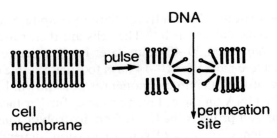

Figure 2   *The principle of electroporation. The lipid bilayer (cell membrane) interacts with an electrical pulse to generate a permeation site*

the transient expression of the DNA template.[26] However, stable expression has been difficult to demonstrate using this technique.

2.3   Electroporation

This method is now commonly used and requires the purchase of specialist equipment. It is based upon the induction and stabilization of permeation sites within the cell membrane *via* an interaction of lipid dipoles in an electric field (see Figure 2). The nature and type of electrical pulse required to optimize such structures has been evaluated by commercial companies.[27] The DNA and cell concentration needs to be fairly high (*i.e.* 50 μg DNA/5 $\times$ 10^7 cells), but of course, this ratio has to be optimized for different cell-lines. After mixing, the DNA and cell suspension is subjected to three electric pulses at 8 kV cm^{-1} with a pulse decay of 5 μs. However, these spike potentials may be now superseded in favour of a lower potential square-wave pulse which may increase the viability of the cells.[27]

Electroporation does not require carrier DNA and is very quick compared to the above techniques (*i.e.* one may select for stable transformants 10 min after pulsing). Also, it has been shown that this technique gives a higher proportion of stable transformants which harbour a single copy of exogenous DNA at single integration sites. The frequency of these events allows the detection of homologous integration sites.[17]

2.4   Protoplast Fusion

This is a general method for introducing macromolecules into cells and is therefore not restricted to DNA. The method is outlined in Figure 3. To

[26] H. Luthman and G. Magusson, *Nucl. Acids Res.*, 1983, **11**, 1295.
[27] G. Chu, H. Hayakawa, and P. Berg, *Nucl. Acids Res.*, 1987, **15**, 1311.

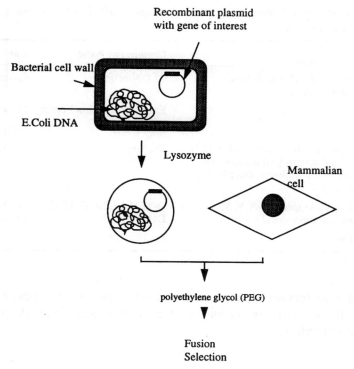

Figure 3 *A scheme illustrating cell or protoplast fusion*

obtain efficient transfection, care must be taken to ensure no lysis of sphaeroplasts takes place, and that the optimal protoplast cell ratio is achieved (usually 10^4:1). One obvious advantage to this technique is that the cloned DNA to be transfected need not be purified from *E. coli*.

2.5 Lipofection

Synthetic cationic lipids can be used to generate liposomes in the presence of DNA.[8] The resulting unilamellar structures entrap 100% of the DNA present. These vesicles may be fused to cell membranes and thus deliver the DNA directly to the cell interior. Serum appears to inhibit this process. The rationale behind this procedure is that a single DNA plasmid is surrounded by sufficient cationic lipid to completely neutralize the negative charge of DNA and provide a net positive charge which facilitates an association with the negatively charged surface of the cell. Depending upon the cell-line, lipofection is reported to be 5–100-fold more effective than either the calcium phosphate or DEAE-dextran

Table 1 *Commercial sources of cationic amphiphilic lipids for liposome mediated transfection*

Lipid	Commercial name	Company
N-t-butyl-N-tetradecyl-3-tetradecyl-aminopropion-amidine	CLONfectin	Clontech
	EasyFector, PrimeFector	FlowGen
	LipoTAXI	Stratagene
N,N[bis(2-hydroxyethyl)-N-methyl-N-[2,3-di(tetradecanoyloxy)propyl] ammonium iodide mixed with L-dioleoyl phosphatidyl ethanolamine	TransFast	Promega
N-[1-(2.3-dioleyloxy)propyl]-N,N,N,-trimethylammonium chloride	DOTMA (also DOTAP, DOSPER,)	Boehringer Mannnheim
Mixture of lipids	FuGENE-6	

mediated transfection techniques.[8] The synthetic cationic lipids necessary for lipofection are available from commercial biotechnological companies (Table 1).

2.6 Polybrene-DMSO Treatment

Polybrene is a synthetic polycation which can be used to facilitate DNA transfection by serving as an electrostatic bridge between the negatively charged DNA and positively charged components on the host cell membrane.[9] As such, this transfection technique exploits the same strategy as described above for DEAE-dextran and lipofection. Thus, similar to DEAE-dextran, the uptake of the adsorbed DNA on the cell surface is enhanced by treatment with DMSO which, in the manner of the glycerol shock, serves to permeabilize the DNA-coated cells. In contrast to DEAE-dextran, the use of polybrene as a facilitator enables the investigator to generate stably transfected clones.

2.7 Microinjection

The term transfection is a misnomer when applied to this technique. The DNA is introduced directly into the nucleus using a glass micropipette with the aid of a micromanipulator. This is obviously very time-consuming with respect to the techniques previously described. All the viable microinfected cells transiently express the DNA and between 1%

and 30% will become stable clones. No carrier DNA is required and the form and number of DNA molecules present in the host cell may be strictly controlled. Injection in the cytoplasm gives 10^{-3} fewer stable transformants compared to nuclear injections.[5] This technique is commonly used to transfect embryonal stem (ES) cells as a preliminary for generating transgenic mice.

2.8 Scrapefection

This technique was initially described as a general method for introducing functional macromolecules into adherent cell-lines.[28] It has since been demonstrated to function well in a DNA transfection experiment.[10] The protocol is very simple and involves incubating a washed monolayer of cells with a buffered DNA solution (1–50 μg ml^{-1}) prior to scraping with a rubber policeman. The scraped cells are then distributed to fresh culture plates and analysed for transient expression of the transfected DNA 1 to 5 days later.[10] Up to 80% of scrapeloaded cells were shown to express the DNA and cell viability of 70% was obtained. Stable expression of co-transfected DNAs is also possible. The method is rapid, efficient and is low in cost compared to other transfection procedures. It does not require the formation of a DNA complex or co-precipitate and therefore is a relatively facile operation. As such, this is probably the first choice in determining which transfection technique is adequate for a cell-line of choice.

3 REQUIREMENTS FOR GENE EXPRESSION

Although describing the nature and variety of DNA sequences which modulate eukaryotic transcription is beyond the scope of this chapter, a brief mention should be made of some salient points.

The DNA sequences required for efficient initiation of gene expression can be broadly classified into three categories, (i) simple promoters, (ii) enhancers and (iii) regulatory regions. Figure 4 illustrates some well characterized sequences required for expression for a number of genes.

In the simplest terms, promoters may be regarded as being a patchwork of common concensus sequence motifs which are found across a wide variety of different genes, with some sequences being more prevalent than others.[29] The TATA motif is amongst the most common

[28] P. L. McNeil, R. F. Murphy, F. Lanni, and D. L. Taylor, *J. Cell. Biol.*, 1984, **98**, 1556.
[29] P. Bucher and E. N. Trifonov, *Nucl. Acids Res.*, 1986, 10009.

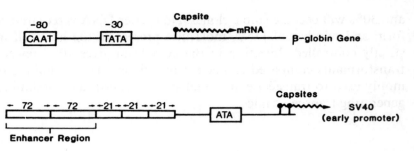

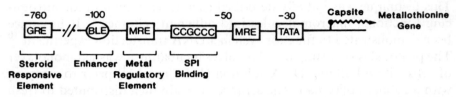

Figure 4   *Some well characterized 5 sequences required for efficient transcription. The β-globin gene may be regarded as having a fairly simple promoter. (It should be noted that the more complex sequence requirements for tissue and stage-specific erythroid expression are not shown.) The enhancer region of SV40 early promoter is characterized by the 72 basepair repeats. Each one of the three 21 basepair repeats contain two binding sites for Sp1. The multiplicity of control elements within the well studied metallothionine promoter are also shown*

and is usually found at a fixed characteristic position at ∼ 30 basepairs (bp) upstream (*i.e.* 5′) of transcriptional start-sites (cap-sites).

Other promoter sequence motifs include CAAT and CCGCCC, which are usually found upstream from the TATA box and are less fixed in their relative position or orientation to the cap-site. These three sequence motifs are operated upon by fairly ubiquitious transcription factors, namely, TF11D, CTF (also called NF-1) and Sp1, respectively.[30]

Enhancer sequences were first identified in viral genomes and may also be regarded as a patchwork of sequence motifs, some of which are also found in promoters.[30,31] They are operationally defined as DNA sequences which augment transient gene expression in both *cis* and in an orientation-independent manner.[32] Other characteristics of enhancers include the ability to exert the effect over large stretches of DNA, and the ability to activate different (heterologous) promoters other than those by which they are naturally found. Recently, host cellular genes have been

[30] N. Jones, E. Ziff, and P. Rigby, *Genes Devel.*, 1988, **2**,
[31] H. Singh, R. Sen, D. Baltimore, and P. A. Sharp, *Nature*, 1986, **319**, 154.
[32] J. Banerji, S. Rusconi, and W. Schaffner, *Cell*, 1981, **27**, 299.

found to contain enhancer sequences which appear to function in a tissue-specific manner.[33,34] Viral enhancers have a wider, more varied host range, although some retroviral enhancers do have a preference for a specific cell type under certain conditions.[25,35] The most widely used enhancer/promoter for expression vectors is from the immediate early gene from human cytomegalovirus. Studies have revealed up to nine binding sites for transcription factors over a 400 bp enhancer region which accounts for why this has been reported as the most powerful effector of heterologous gene expression in both *in vivo* and *in vitro* experiments.[36,37]

Regulatory sequences may be loosely defined as DNA sequences which interact with tissue-specific *trans*-acting proteins to ensure correct tissue- or temporal-specific expression. Included in this category are sequences which are able to induce gene expression under an appropriate stimulus, *i.e.* heat-shock,[38] presence of heavy metals,[39] or steroid hormones.[40] This ability to induce, and therefore control, gene expression is obviously very useful to the biotechnologist, especially if the protein products are toxic to the host cell. Thus, populations of stable clones harbouring such a gene need only be induced prior to harvest. An outstanding example of the precision with which investigators can increase or attentuate the expression levels of cloned genes is provided by the Tet-Off™ and Tet-On™ systems supplied by ClonTech. This enables up to 1000-fold induction of gene expression which is quantitatively regulated by addition of either tetracycline (Tc) or doxycycline (dox) to the transfected host cells.[41] In the Tet-Off™ system, gene expression is turned on when Tc or Dox is removed from the culture medium. In contrast, gene expression is activated in the Tet-On™ system by the addition of Dox to the culture medium. More importantly, the levels of expression can be tightly and quantitatively controlled by varying the concentrations of either Tc or Dox (Figure 5).

A novel type of enhancer/regulatory sequence has recently been identified in the human β-globin gene cluster.[34,42] These sequences,

[33] M. Walker, T. Edland, M. Boulet, and W. Rutter, *Nature*, 1983, **306**, 557.
[34] F. Grosveld, G. Blom van Assendeltf, D. Greaves, and G. Kollias, *Cell*, 1987, **51**, 975.
[35] C. Gorman, G. T. Merlino, M. C. Willingham, I. Pastan, and B. Howard, *Proc. Natl. Acad. Sci. USA*, 1982, **79**, 6777.
[36] P. Ghazal, H. Lubon, B. Fleckenstein, and L. Hennighausen, *Proc. Natl. Acad. Sci. USA*, 1987, **84**, 3658.
[37] J. Sinclair, *Nucl. Acids Res.*, 1987, **15**, 2392.
[38] H. Pelham and M. Bienz, *EMBO J.*, 1982, **1**, 1473.
[39] F. Lee, R. Mulligan, P. Berg, and G. Ringold, *Nature*, 1982, **294**, 228.
[40] D. H. Hamer and M. J. Walling, *J. Molec. Appl. Genet.*, 1982, **1**, 273.
[41] M. Gossen and H. Bujard, *Proc. Natl. Acad. Sci. USA*, 1992, **89**, 5547.
[42] D. Talbot, P. Collis, M. Antoniou, M. Vidal, F. Grosveld, and D. Greaves, *Nature*, 1989, **336**, 352.

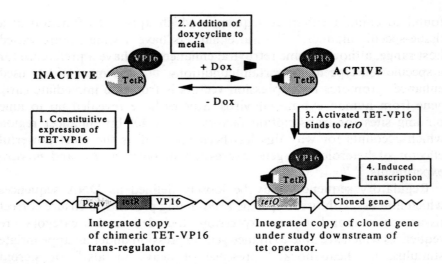

Figure 5 *The doxycycline controlled transcriptional activator is a chimeric protein of the activation domain of VP16 (from Herpes simplex virus) and a variant of the tetracycline repressor (tetR). The variant tetR contains four amino acid changes which reverse the wild-type repressor property to an activator phenotype as a result of increased binding of the liganded tetR to the tetO DNA sequence in the presence of its ligand, doxycycline (dox)*

called dominant control regions (DCRs), are required to commit adjacent genes to specific expression in erythroid tissue and have been identified by studies of the sequence requirements of human β-globin gene expression in transgenic mice.[34] Phenotypically they appear to function in a similar fashion to tissue-specific enhancers, but exhibit the novel property of isolating adjacent chromatin from the effect of the host integration site. Thus, they have the unusual ability of manifesting a degree of exogenous gene expression which is dependent upon the exogenous gene copy number. The DCR has now been identified to within a few kilobases[42] which makes it possible to construct retroviral vectors (see Sections 4.3 and 4.5) containing these important regulatory regions. Such tools may prove invaluable to correct gene defects by replacing defective globin genes in, for example, thalassaemic host bone marrow.

There are many more stages which affect expression of the final gene product such as post-transcriptional processing, translation efficiency and post-translational modifications (which this chapter will not deal with), and which will be of greater or lesser importance depending upon the gene product under study. Note that *E. coli* will not necessarily perform these functions in the same manner as in eukaryotic hosts.

4 THE DNA COMPONENT

So far we have discussed the DNA component in a transfection without defining its nature. Basically, it is possible to transfect cells with chromosomal DNA, naked genomic DNA, cloned DNA or using a bacteriophage. Many elegant experiments have been performed by transfecting genomic DNA into cell-lines. For instance, if one isolates total genomic DNA from a tissue-expressing gene X to transfect an X-negative cell-line, then it is possible to characterize gene X by selecting for a stable transfectant now exhibiting the phenotype associated with X expression. Secondary transfections will permit the eventual isolation and cloning of the gene. This type of approach has been used to isolate activated cellular oncogenes[43] and is amenable to isolate genes which express their products on the transfected cell surface, providing antibodies are available as probes for their presence. Similar strategies have been employed in the cloning of T cell specific genes in fibroblasts.[44]

4.1 Use of Vectors

The use of vectors allows considerable flexibility in regulating the expression of cloned genes. Although, in theory, any simple pBR322 plasmid vector will permit a degree of transient and subsequent stable expression of the passenger gene, many different types of vector have been constructed and imbued with characteristics which render them particularly amenable for gene cloning. Vectors can be classified into either viral- or plasmid-based, and each class may be sub-divided into general-purpose cloning vehicles or expression vectors (for cDNA inserts). Expression vectors are those which contain an efficient promoter upstream of a cloning site such that transcription of the inserted cloned cDNA sequences can be driven by the promoter in the vector. A comprehensive listing of cloning vectors can be found in reference 45.

4.2 Plasmid-based Vectors

A common feature of all plasmid-based vectors is the presence of a prokaryotic replication origin (replicon) and a selectable marker gene to permit the recombinant DNA molecule (*i.e.* vector and insert) to be amplified in *E. coli*.

[43] C. Tabin, S. Bradley, C. Burgmann, and R. Weinburg, *Nature*, 1982, **300**, 143.
[44] D. R. Littman, Y. Thomas, P. Maddon, L. Chess, and R. Axel, *Cell*, 1985, **40**, 237.
[45] P. H. Pouwels, B. E. Enger-Valk, and W. J. Brammar, 'Cloning Vectors – A Laboratory Manual', Elsevier Science Publishers B.V., 1985.

In addition, most vectors have a eukaryotic replicon and/or a eukaryotic selectable marker. This latter feature is required for stable expression and will be discussed in depth in Section 6. The selectable marker need not be present on the vector, but may be provided in *trans*, on a separate vector, to be co-transfected at an appropriate ratio.[21]

The eukaryotic replicon is usually derived from viruses such as bovine papilloma virus (BPV), Simian virus 40 (SV40) or Epstein–Barr virus (EBV). It should be noted that viral replicons require additional viral gene products to initiate DNA replication of the recombinant. Such proteins may be supplied in *trans* either by co-transfecting with the viral gene or, more uniformly, by choosing a cell-line which constituitively expresses the required viral functions. For instance, a monkey cell-line CV-1 has been stably transformed with the gene coding for the appropriate SV40 replicative functions to generate COS cell-lines, which are able to replicate plasmid vectors containing the SV40 replicon up to 4 × 10^5 copies per transfected cell. However, the viability of these transfected cells appears to decrease after a few days, probably due to this excessive DNA synthesis.[46] It has been observed that certain pBR322 sequences act in *cis* to inhibit extra chromosomal replication (so called 'poison' sequences).[47] Derivatives of pBR322 that lack these sequences are available and should therefore provide the basis for any replicative vector.

An interesting feature of chimeric plasmid-BPV vectors is their ability to induce stable transformation of recipient mouse cells, thereby reducing the serum requirement for cell-propagation and also providing an intrinsic selection for the transfection event without applying any external selection, although one is restricted in the choice of a semi-permissive cell-line.

The recombinant BPV vector is stably maintained *without* integration and exists as a stable extrachromosomal element at 10–100 copies per cell.[48]

Chimeric vectors have been constructed which contain both SV40 and BPV origins, and also BPV genes required in *trans* for the regulated replication from the BPV replicon.[46] Upon transfection of COS cells with the chimera, it appears that the regulated controlled replication from the BPV origin is dominant over the runaway SV40 replication. Thus, whereas plasmid vectors containing the SV40 replicon alone undergo excessive replication leading to cell death, the dual replicon

[46] J. M. Roberts and H. Weintraub, *Cell*, 1986, **46**, 741.
[47] M. Lusky and M. Botchan, *Nature*, 1981, **293**, 79.
[48] P. D. Mathias, H. U. Bernard, A. Scott, G. Brady, T. Hashimto-Gotoh and G. Schutz, *EMBO J.*, 1983, **2**, 1487.

chimera is maintained extrachromosomally at 500–1000 copies per cell leading to the generation of stable cell-lines. Deletion of the BPV transacting function results in runaway replication. It is evident that if conditional mutations in these transacting proteins are found, this sytem offers considerable control over recombinant copy number and cell-line viability.

Some plasmid-based vectors also contain other eukaryotic regulatory sequences such as a splicing site to promote mRNA stability and a poly-adenylation sequence to ensure appropriate termination of transcription.[49]

Vectors which contain secretory signals for the production and generation extracellular secreted proteins would have the advantage of enabling easy harvest of the cloned gene products and these are now available from commercial biotechnological companies.

4.3 Virus-based Vectors

Animal virus vectors offer an alternative to DNA transfection as a means of enabling a foreign gene to be expressed in different cell types.[50] Recombinant viral genomes carrying the cloned gene may be packaged into infectious virions and used to infect cell-lines or tissues. Factors influencing choice of viral vector include: types of cells or animals to be infected, whether cell transformation or cell lysis is required, the size of DNA to be cloned, and whether virus infectivity is to be retained.

Small DNA viruses such as SV40 have a restricted host range and limited capacity for cloning due to the constraints of genome size able to be packaged. In most cases, viral genes are deleted to permit incorporation of foreign DNA and, as a consequence, these viral replacement vectors are replication-defective, requiring helper virus or special cell-lines for their propagation and replication.

Large viruses like human adenovirus and poxvirus have a greater capacity for foreign DNA without necessarily reducing their infectivity. Murine retroviruses have also been successfully employed as replacement vectors.

4.4 Adenovirus Vectors

Infection by wild-type adenovirus induces cytopathic changes in human cell-lines and tissues (in particular adenoidal and other lymphoid

[49] P. Gruss and G. Khoury, *Nature*, 1980, **286**, 634.
[50] P. W. J. Rigby, *J. Gen. Virol.*, 1983, **64**, 255.

organs). A variety of viral strains have been isolated and it has been shown that some strains are more oncogenic than other in their ability to induce tumours in newborn rodents and rat primary cells. However, no strain which exhibits oncogenic potential in any human tissue has been reported. The virus may be propagated in Hela cells to a high titre and evince characteristic cytopathology (*i.e.* nuclear lesions). Adenovirus has a 35 kb linear genome which makes its manipulation as a potential vector difficult, although variant strains which contain single cloning sites have been developed to overcome these difficulties.[51] Their upper DNA packaging limit is 5% greater than the wild-type genome and they can therefore be used as insertion vectors for small inserts and, as such, can be propagated without a helper virus. Extra sequences can be cloned by deleting some early viral genes and replacing these with larger inserts. These replacement recombinant vectors have to be propagated in helper cell-lines (such as the adenovirus transformed human 293 cell-line) in which the deleted viral functions are provided by chromosomally integrated sequences. Thus, naked recombinant viral DNA is transfected into these cells and a productive infection ensues releasing infectious recombinant virions that may be harvested for future infections.

4.5 Retrovirus Vectors

To understand the potential advantages of using retrovirus as cloning vectors, a brief mention must be made of their life-cycle. Infectious wild-type replication competent retroviruses contain a single-stranded 7–8 kb RNA genome which is packaged by viral-coded envelope protein. The type of envelope protein determines the host range infectivity of the virus. After infection, the RNA genome is translated and the resulting viral proteins are able to convert the RNA genome into double-stranded DNA (reverse transcription). During this process, the ends of the viral genome are duplicated to generate long terminal repeats (LTRs). LTRs represent powerful promoter/enhancer elements. The DNA intermediate is able to integrate into the host genome by a specific mechanism not available to DNA viruses. The genome is now described as a provirus. The integration is precise in that the resulting proviral genes are colinear and unrearranged, being flanking by 5′ and 3′ LTRs. The integration site in the host genome is random in most cell-lines. The proviral state is stable, and the cell remains viable.

If the LTRs remain active, a genomic RNA molecule is produced which contains all the coding information represented between the two

[51] N. Jones and T. Shenk, *Proc. Natl. Acad. Sci. USA*, 1979, **76**, 3665.

LTRs. This polycistronic message undergoes a variety of different fates. It may be packaged into an infectious particle, translated to produce proteins, or spliced into sub-genomic mRNAs and subsequently translated. The packaging of the RNA is dependent upon the presence of a signal sequence at the 5' end. This is called the ψ-sequence and mRNAs which lack this signal cannot be packaged. This observation has led to the production of special helper cells called ψ-2 and ψ-AM.[52] These cell-lines have been infected with a retrovirus which lacks the ψ-sequence. Thus, the resulting provirus produces all the necessary functions for packaging but is unable to package its own genomic RNA. Therefore, transfection of these cell lines with *in vitro* constructed recombinant proviral DNA, which contains the gene to be cloned between two LTRs, results in the sole production of infectious recombinant viral particles.[52] However, in some rare instances, recombination does occur during the propagation cycle and results in wild-type infectious particles. The ψ-AM cell-line permits the recombinant virus to exhibit a very broad host range.

From the above discussion, it should be evident that a recombinant proviral DNA, into which a gene which contains introns has been inserted, will be generated as an infectious particle which now contains the cloned gene as a cDNA copy.[53]

A useful vector (pLXIN) has now been developed which contains an IRES (internal ribosome entry site) between the MCS (multiple cloning site) and a selectable marker gene which encodes resistance to the antibiotic neomycin (neo^r). Thus, expression from the 5' LTR generates a functional bicistronic mRNA encoding the gene of interest upstream of neo^r. Therefore, when ψ-AM cells are infected and selected on neomycin, only recombinant viruses will be packaged and this results in high titre virus-producing cell-lines. Additionally, later infection of target cells with the recombinant virus allows neomycin selection of infected cells which will result in 100% infected target-cell population. Note also that use of the IRES obviates potential promoter competition problems that may occur if the selectable marker was being expressed from an independent LTR.

4.6 Poxviral Vectors

Poxviruses, such as vaccinia, have a very large DNA genome (187 kb) and considerable flexibility is observed in the amount of genomic DNA

---

[52] R. D. Cone and R. C. Mulligan, *Proc. Natl. Acad. Sci. USA*, 1984, **81**, 6349.
[53] K. Shimotohno and H. M. Temin, *Nature*, 1982, **299**, 265.

to be packaged; therefore it has the capacity to accommodate very large inserts (up to 35 kb in some strains). No helper virus is required for propagation and vaccinia strain exhibit a wide host range for infectivity. The genetic make-up of vaccinia differs considerably from the viruses previously mentioned, such that a vaccinia gene cannot be recognized by host RNA polymerases. Therefore naked vaccinia DNA is non-infectious. Additionally, vaccinia-coded RNA polymerases do not transcribe promoters recognized by host RNA polymerases. Thus vaccinia promoters must be used to obtain efficient expression of foreign cDNAs. Indeed only cDNAs may be cloned in vaccinia vectors because the virus replicates in the cytoplasm of infected cells, whereas splicing is a nuclear event.

In order to construct an infectious vaccinia recombinant, a two-step strategy is employed (Figure 6). First, an intermediate plasmid is

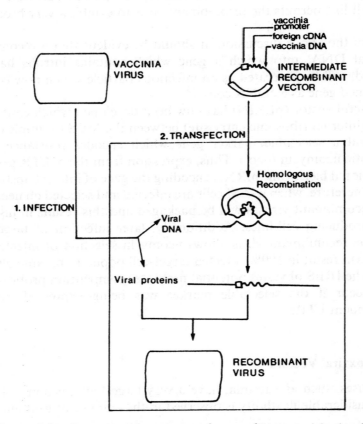

Figure 6   *A scheme illustrating the generation of a recombinant vaccinia virus is shown. See text for details*

constructed *in vitro* which contains the foreign cDNA linked to a vaccinia promoter. This chimeric transcription unit is also flanked by vaccinia DNA from a non-essential region of the viral genome. General intermediate vectors of this sort are available which contain either one or two vaccinia promoters.[54] Additionally, these intermediate vectors may contain a selectable marker driven by a viral promoter. Second, the recombinant intermediate vector is transfected into eukaryotic cells which have been previously infected with vaccinia. In a cell which contains an infecting virus, the flanking homologous sequences in the intermediate vector and virus genome enable recombination to occur with the resulting transfer of the chimeric foreign cDNA from the plasmid to the virus genome. The recombinant viral genomes are then packaged into infectious particles. Because the flanking DNA in the plasmid was from non-essential regions, the recombinant virus produced does not require any helper functions from this point on.

At best, only 0.1% of total viral progeny are recombinant, but selective techniques are available to ensure relatively easy isolation of the required virus.[54]

A variety of foreign genes have been cloned in vaccinia including rabies virus glycoprotein, HTLV III envelope protein, hepatitus B virus surface antigen and influenza virus haemagglutinin (see reference 54 and references therein).

4.7 Baculovirus Vectors

Baculovirus is an insect-specific agent which is unable to infect any other species. Viral vectors have been derived from either *Bombyx mori*[55] or, more commonly, *Autographa californica*.[56] Baculoviruses are classified according to their visible characteristics, namely, nuclear polyhedrosis viruses (NPVs), granulosis viruses (GVs) and non-occluded viruses (NOVs). *Autographa californica* (AcNPV) is composed of a nucleocapsid core which encompasses a circular DNA genome of ~130 kb.

During part of the natural viral particle maturation process, the viral nucleocapsid is enveloped by host cellular membrane before being further packaged into large polyhedral structures (occlusion bodies) via the expression of a single viral gene encoding the polyhedrin protein (~30 kDa). To propagate the virus *in vitro*, the polyhedrin protein is solubilized to release the infectious enveloped nucleocapsids. Therefore,

[54] M. Mackett and G. L. Smith, *J. Gen. Virol.*, 1986, **67**, 2067.
[55] S. Maeda, T. Kawai, M. Obinata, H. Fujiwara, T. Horiuchi, Y. Saeki, Y. Sato, and M. Furusawa, *Nature*, 1985, **315**, 592.
[56] V. Luckow and M. Summers, *Bio/Technology*, 1988, **6**, 47.

further infection of neighbouring cells is acheived independently of the polyhedral protective structure, although the polyhedrin gene continues to be extensively expressed in concert with other late viral genes, most notably the p10 gene, which encode other components of the protective polyhedral structure. It is clear that these two genes are rendered non-essential for viral propagation and therefore may be replaced by other exogenous coding sequences. Due to the complexity of the baculoviral genome, the preferred method of gene replacement is similar to the strategy described above for poxviruses. A transfer vector is intially generated by ligating a deleted version of the polyhedrin gene into a bacterial plasmid. The viral features essential for polyhedrin gene expression are retained, as in the vector pAcYM1 (see reference 56 for a comprehensive listing of available vectors). The gene sequences coding for the desired protein are appropiately ligated into the transfer vector, and subsequently co-transfected into a host insect cell-line (*Spodotera frugiperda*) with infectious AcNPV DNA. Ensuing recombination within the transfected cell generates a helper virus independent polyhedrin-negative infectious virus which is indentified by the presence of a clear polyhedrin-negative plaque upon the host cells. The ability of insect cells to correctly process exogenous gene products is reviewed in reference 57.

It is clear that other non-essential genes of the recombinant virus obtained may be similarily manipulated. Additionally, if essential viral gene functions are provided in *trans* by the host *Spodoptera frugiperda* cell-line, then these may also be replaced giving rise to a multiply-recombined infectious expression system which should be able to subvert the biochemistry of the host cell to the investigator's purpose. Since baculovirus specifically infects insects and is non-pathogenic to animals and plants, such vector systems may be used in field studies to combat agricultural pests.

Finally, the option of growing the baculoviral recombinant vectors in insect larvae offers the biotechnologist a relatively cheap but effective host to provide large and continuing amounts of exogenous protein.[55]

5 SOME CONSIDERATIONS IN CHOICE OF CELL-LINE

Established cell-lines vary considerably in the ability to take up DNA *via* transfection. Amongst the more susceptible are human Hela cells and mouse L cells, whereas human K562 and mouse MEL cell-lines are relatively refractory. It has been noted that transfection efficency will

[57] L. K. Miller, *Ann. Rev. Microbiol.*, 1988, **42**, 177.

decline upon prolonged passage of some cell lines *in vitro* (for example > 15 passages).

It is obviously desirable to have a quick and easy assay for transfection efficiency, especially with respect to determining the experimental conditions which permit optimal expression with a specific promoter. The CAT assay is a simple transient expression system which can be used to evaluate these parameters. CAT vectors contain a prokaryotic gene coding for chloramphenicol acetyl transferase (CAT) which is absent in all mammalian cell-lines. The variety of CAT vectors available allows easy cloning of either enhancer or promoter sequences.[55] After transfection the cells are lysed by successive freeze–thaw cycles and aliquots of the resulting cellular extract can be rapidly assayed for CAT activity. Alternatively, staining the transfected cell population with anti-CAT antibodies enables the fraction of successfully transfected cells to be calculated and therefore maximized.

A number of other reporter gene systems may be used including tissue plasminogen activator (tPA), β-galactosidase or the luciferase gene.[58]

Obviously the chosen cell-line has to be compatible with the gene system being used by the investigator. Thus if a regulatable promoter (such as metal ion or steroid hormone inducible) is being employed to direct gene expression, then the host cell must have the appropiate receptors to allow entry of the inducing agent. Also if the gene product has to undergo specific modifications or proteolytic cleavage to achieve full activity, then it is neccessary to choose a cell-line able to perform these processes.

Finally if the objective of the DNA transfection is to generate stable clones, one has to pick a cell-line appropriate for the selection system to be used (see Section 6.1).

6 TRANSIENT *VERSUS* STABLE EXPRESSION

It is apparent from the preceding sections that the advantages of transient expression are rapidity and the obviation of any selection. However much these features are appreciated by the molecular biologist, the disadvantages to the biotechnologist include the requirement of very clean, pure Form 1 DNA and the high background of untransfected cells. Selection for stable clones overcomes both these problems. However, the integration site within the host cell chromatin has a dominant effect over the expression levels of the integrated cloned gene. Thus,

[58] N. Wrighton, G. McGregor, and V. Giguere, Methods in Molecular Biology, Vol. 7, in 'Gene Transfer and Expression Protocols', ed. E. J. Murray, Humana Press Inc., New York, 1991.

integration in inactive chromatin (heterochromatin) will invariably repress the expression levels. This effect can be overcome by the use of DCR as discussed in preceding sections. Alternatively, random selection of high expressor integration sites can be achieved by the use of pIRES-EGFP marker vectors. Such a vector contains an attenuated IRES upstream of the enhanced green fluorescent protein gene (egfp). The gene of interest is cloned upstream of the IRES-EGFP and the resulting recombinant construct is used to generate stable integrated clones. Each clone is then analysed for GFP expression which can be measured by rapid non-invasive assays. This highlights those high GFP expressers which by necessity are also concomitant high expressers for the gene of interest since they share the same integration site.

6.1 Selection by Host Cell Defect Complementation

A number of selective techniques are based upon alleviation of inhibited synthesis of host cell nucleotides. For instance, aminopterin is a potent antagonist of dihydrofolate reductase (dhfr) which is required for both purine and pyrimidine *de novo* biosynthesis, and is therefore extremely toxic to all cells. The lethal effect of aminopterin can be overcome by supplementing the growth medium with the appropiate nucleotides percursors which are utilized *via* the cell by the salvage pathway enzymes namely; thymidine kinase (tk) for thymidine, adenine phosphoribosyl transferase (aprt) for adenine, and hypoxanthineguanine phosphoribosyl transferase (hgprt) for guanine (Figure 7). Thus *tk*⁻ cell lines will only survive in HAT medium (containing hypoxanthine–aminopterin–thymidine) if they have been transfected with the *tk* gene. Similar selective criteria can be defined for the *aprt* and *hgprt* genes for the corresponding negative cell-lines (for a full discussion, see reference 58). These selectable marker genes have been introduced in a variety of cloning vectors.[45]

6.2 Dominant Selective Techniques

A major disadvantage of HAT selection is the restriction of being limited to the relatively few *tk*⁻, *aprt*⁻ or *hgprt*⁻ cell-lines. Thus, a number of dominant selection systems have been designed which can be applied more universally.

The *E. coli xgprt* gene is functionally analogous to the mammalian *hgprt* gene, but differs from its mammalian counterpart in the ability to use xanthine as a GTP precursor via a XMP intermediate (Figure 7). Mycophenolic acid (MA) inhibits the *de novo* biosynthesis of the XMP intermediate. This inhibition is more pronounced by the addition of

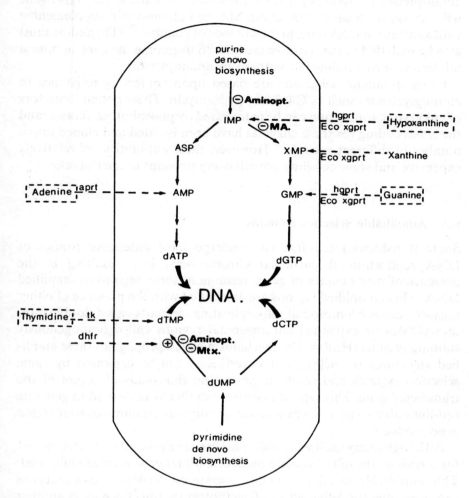

Figure 7 *A scheme illustrating* de novo *nucleotide synthesis and the inhibition of various intermediates by aminopterin (aminopt.), methotrexate (mtx) and mycophenolic acid (MA). The inhibitory effects can be alleviated by the presence of nucleotide precursors and the appropiate salvage pathway enzyme genes as indicated by dotted lines. The salvage pathway selections in which the precursor is enclosed by a dotted line are dependent upon the transfected cell being negative for the corresponding selectable marker gene. The other two selections are dominant. Eco = E. coli. See text for details*

aminopterin. Therefore only cells transfected with the *E. coli xgprt* gene will survive in medium containing MA and aminopterin supplemented with xanthine and adenine, in the absence of guanine.[59] (Thymidine must also be included in the selective medium to overcome the concommitant inhibition of pyrimidine biosynthesis by aminopterin.)

Other dominant selections are based upon conferring resistance to aminoglycosides such as G418 and hygromycin. These potent inhibitors of eukaryotic translation can be inactivated by phosphotransferases, and the genes coding for these enzymes have been isolated and cloned into a number of different vectors.[60] However, these antibiotics are relatively expensive and some cell-lines are relatively resistant to their uptake.

6.3 Amplifiable Selection Systems

Some transformed cell-lines can undergo local reiterative rounds of DNA replication at particular chromosomal loci, resulting in the presence of many copies of genes residing at those regions of amplified DNA. These amplified regions are associated with the presence of either unstable extrachromosomal self-replicating elements called double minutes (DMs) or expanded chromosomal regions called homogenously staining regions (HSRs). The mechanism which propagates these amplified structures is unclear, but nevertheless can be exploited by some selection systems and result in generating thousands of copies of the transfected gene. This type of selection can therefore be used to generate cell-lines able to produce potentially prodigious amounts of transfected gene products.

Although many such selectable markers are available (see reference 61 for a review) the dihydrofolate reductase (dhfr) gene is commonly used. This amplifiable selection system is based upon the observation that cells can overcome the lethal effect of methotrexate (mtx), which is another analogue of folate, by either amplifying the endogenous *dhfr* gene or mutating it to generate a more resistant form of the protein.[62] A mutant mtx-resistant *dhfr* gene has been cloned and may be used as a dominant selectable marker in *dhfr*$^+$ cell-lines under stringent conditions which use high levels of mtx. However, amplification is low in this system (see reference 63) and the use of *dhfr*$^-$ cell-lines (such as the Chinese hamster

[59] R. Mulligan and P. Berg, *Proc. Natl. Acad. Sci. USA*, 1981, **78**, 2072.
[60] F. Grosveld, T. Lund, E. J. Murray, A. Mellor, and R. A. Flavell, *Nucl. Acids Res.*, 1982, **11**, 6715.
[61] R. J. Kaufman, in 'Genetic Engineering' Vol. 9, ed. J. Setlow, Plenum Press, New York, 1987.
[62] M. Wigler, M. Perucho, D. Kurtz, S. Dana, A. Pellicer, R. Axel, and S. Silverstein, *Proc. Natl. Acad. Sci. USA*, 1980, **77** 3567.
[63] M. I. Cockett, C. R. Bebbington, and G. T. Yarranton, *Bio/Technology*, 1990, **8**, 662.

ovary sub-line DUKX-B11) and native wild-type *dhfr* selectable marker, for example, pSV2.dhfr,[64] represent suitable hosts and vectors for high levels of amplification. The transfected cells are treated with increasing doses of mtx, and the surviving cells amplify the *dhfr* marker gene. During this process, the flanking host DNA had also been amplified[65] and copy numbers of greater than 2000 copies per cell have been obtained.[66]

An alternative dominant system has been generated at Celltech in which high levels of amplification and gene expression can be obtained without the requirement of a mutant cell-line. The selectable marker in this system is the glutamine synthetase (gs) gene which is amplified in response to methionine sulfoximine (msx), even in cell-lines containing an active *gs* gene.[63] Yields of up to 180 mg 1^{-1} of recombinant protein have been acheived with this system.

[64] S. Subramani, R. Mulligan, and P. Berg, *Mol. Cell. Biol.*, 1981, **1**, 854.
[65] C. Gorman, L. Moffat, and B. H. Howard, *Mol. Cell. Biol.*, 1982, **2**, 1044.
[66] G. F. Crouse, R. N. McKewan, and M. L. Pearson, *Mol. Cell. Biol.*, 1983, **3**, 257.

CHAPTER 7

Plant Biotechnology

MICHAEL G. K. JONES

1 INTRODUCTION

The last decade has been a period of remarkable change which has taken plant biotechnology from study of the science itself to large scale commercial applications. This is true for almost every aspect of plant biotechnology, both applying basic knowledge of molecular biology and gene organization such as development of molecular markers to speed up plant breeding practices, and using knowledge of genes and how to control expression of those genes to generate and commercialize transgenic crops. In general, the application of plant biotechnology can be divided into two categories: those directed towards the same goals as conventional plant breeding (*e.g.* improved yield, quality, resistance to pests and diseases, tolerance to abiotic stresses), and completely novel applications (such as the use of plants as bioreactors to generate pharmaceuticals, vaccines or biodegradable plastics). The emphasis of this chapter therefore reflects these changes, and is focused more on the application of plant biotechnology rather than the detailed molecular biology which underlies those applications.

2 APPLICATIONS OF MOLECULAR BIOLOGY TO SPEED UP THE PROCESSES OF CROP IMPROVEMENT

Plant breeding is based on the principles of Mendelian genetics. In the past, plant breeding was something of an art, and selection of superior genotypes of a particular crop depended to a great extent on subjective decisions made by the breeder. With increasing knowledge of the genes underlying useful traits, plant breeding has become a more directed and

203

scientific activity. This is in part a result of the generation of molecular maps of crop genomes, extensive sequencing of expressed sequences (expressed sequenced tags or 'ESTs') and of genomic sequences, and of study of genome organization, repetitive and non-coding sequences, and the ability to identify polymorphisms at particular loci which can be exploited as molecular markers if they are closely linked to a useful trait.

2.1 Molecular Maps of Crop Plants

Arabidopsis thaliana had been used as a model plant for mutagenic and genetic studies in the 1970s. The advantages of this species as a model for molecular studies became apparent in the 1980s because of its small nuclear genome size, low repetitive DNA content, short life-cycle, large seed production and later its amenability to transformation. As a result, through international collaborations, a genomic sequencing programme was established, and the complete DNA sequence of the *Arabidopsis* genome should be completed in the year 2000.[1]

The rice genome is only about four times the size of the *Arabidopsis* genome, and as a model cereal and important food crop, sequencing of this genome is also well advanced. Massive sequencing efforts of ESTs of wheat, barley, soybean, rice, *Medicago truncatula*, and other crops are also in progress, mainly driven by major life sciences companies.

Much interesting information on genome organization has resulted from this work, including knowledge of location of genes, gene clustering, and repetitive and non-coding sequences.[2] For cereals, gene arrangements show 'synteny' in that major blocks of genes are arranged in similar sequences in rice, maize, barley, wheat, *etc.* The major differences in genome size being the result of different amounts of repetitive/non-coding sequences, and for wheat, the fact that it is hexaploid and contains three sets of progenitor genomes.

For *Arabidopsis*, it emerges that there are about 22 000 genes required to contain all the information for this organism. For other plants, we might expect the number of genes present to be between this figure and about 50 000 genes.

Research on mapping genome organization, sequences and synteny has led to many practical applications, some of which are discussed below.

[1] E. M. Meyerowitz and C. R. Somerville, ed., 'Arabidopsis', Cold Spring Harbor Laboratory Press, USA, 1994.
[2] C. Dean and R. Schmidt, *Ann. Rev. Plant Physiol. Plant Mol. Biol.*, 1995, **46**, 395.

2.2 Molecular Markers

A genetic marker is any character that can be measured in an organism which provides information on the genotype (*i.e.* genetic make-up) of that organism. A genetic marker may be a recognizable phenotypic trait (*e.g.* height, colour, response to pathogens), a biochemical trait (*e.g.* an isozyme) or a molecular trait (*i.e.* DNA-based). Whereas phenotypic markers depend on expression of genes and are limited to those genes expressed at a particular time or under particular developmental or environmental conditions, DNA-based markers provide an almost unlimited supply of markers that identify specific sequences across the genome.[3] Their advantages are:

   (i) single base changes in DNA can be identified, providing many potential marker sites across a genome.

  (ii) they are independent of developmental stage, environment or expression.

 (iii) markers can be found in non-coding or repetitive sequences.

 (vi) most DNA marker sequences are selectively neutral.

Thus, for example, because about 80% of the wheat genome is non-coding DNA, only molecular markers can be used to identify polymorphisms and to map 'loci' in these regions of the genome.

2.3 Types of Molecular Markers

There are many potential approaches to identify molecular markers. Most are based on using the polymerase chain reaction (PCR) to amplify specific DNA sequences.[3-5] They include:

   (i) RFLPs (restriction fragment length polymorphisms).

  (ii) RAPD-PCR (random amplified polymorphic DNA).

 (iii) microsatellites or simple sequence repeats (SSRs).

 (iv) AFLP (amplified fragment length polymorphisms).

RFLPs rely on the combination of a probe and restriction enzymes to identify polymorphic DNA sequences using Southern blotting. This

[3] M. Mohan, S. Nair, A. Bhagwat, T. G. Krishna, M. Yano, C.R. Bhatia, and T. Sasaki, *Molecular Breeding*, 1997, **3**, 87.
[4] W. Powell, G. C. Machray, and J. Provan, *Trend. Plant Sci.*, 1996a, **1**, 215.
[5] W. Powell, M. Morgante, C. Andre, M. Hanafey, J. Vogel, S. Tingey, and A. Rafalski, *Molecular Breeding*, 1996b, **2**, 225.

approach requires either radioactive or non-radioactive detection methods to identify polymorphic DNA bands, and is therefore more time-consuming than PCR-based methods.

RAPD-PCR does not require sequence information, and involves amplifying random pieces of DNA in which PCR is primed by a single 10 base primer at low stringency, such that random sequences of DNA are amplified based on homologous sequences to the primer being present in the target DNA. It is a useful initial approach to identify polymorphisms, but is not regarded as reproducible enough between laboratories.

Microsatellites or SSRs are groups of repetitive DNA sequences that are present in a significant proportion of plant genomes. They consist of tandemly repeated mono-, di-, tri-, tetra- or pentanucleotide units. The number of repeats varies in different individuals, and so the different repeats can be regarded as 'polymorphic' alleles at that 'locus'. To reveal polymorphic microsatellite sequences it is necessary to sequence the conserved flanking DNA, and to design PCR primers that will amplify the repeat sequences. (Because of the repetitive nature of the amplified sequences typically the main amplified PCR band and additional 'stutter' bands are generated.) For example, at microsatellite locus *Hsp176* of soybean, there is an AT repeat with 13 different numbers of bases in the repeated units in different soybean accessions.[4,5] Microsatellites provide reliable, reproducible molecular markers.

AFLP is also a PCR-based technique, in which selective pre-amplification and amplification steps are carried out to amplify a subset of fragments of the genome, depending on the linkers added and primers used. Many potentially polymorphic fragments are generated by this approach. Polymorphic bands between parents can be identified and linked to useful traits.

Both microsatellite and AFLP markers can be analysed using autoradiography or a DNA sequencer, using fluorescent tags. The latter allows multiplexing such that three different coloured tags plus a size marker can be run in each lane. A single multiplexed AFLP gel can generate 100 polymorphic bands.

2.4   Marker-assisted Selection

Using one of the above approaches to identify molecular markers, in combination with an appropriate mapping population of plants plus or minus the trait of interest, many markers have been identified which are closely linked to genes for agronomic traits of interest.[3-7] These include markers for genes for:

(i) pest and disease resistance (against viruses, fungi, bacteria, nematodes, insects).
(ii) quality traits (*e.g.* malting quality barley, bread or noodle wheats, alkaloid levels, *etc.*).
(iii) abiotic stresses (*e.g.* tolerance to salinity or toxic elements such as boron or aluminium).
(iv) developmental traits (*e.g.* flowering time, vegetative period).

If the molecular marker is in the target gene itself, it has been called a 'perfect' marker. Clearly, the closer the molecular marker is linked to the target gene the better. The overall process of developing a marker thus involves setting up appropriate mapping populations, looking for poly-morphic DNA sequences closely linked to the trait of interest, conver-sion of the polymorphism to a routine marker (usually PCR-based), validation and implementation.

Quantitative trait loci (QTLs) are the genes which control quantitative traits such as yield for which the final character is controlled by several genes.[6] To identify and map QTLs a defined mapping population is required which is screened for polymorphisms by RFLPs, AFLPs, SSRs which can be mapped. Statistical approaches are then used to identify associations between the traits of interest and specific markers. Although the location of QTLs is usually not known exactly, the association of a genotype at a marker/locus and a contribution to the trait indicates that there is a QTL near that marker.

The promise of molecular marker-assisted selection for crop improve-ment is in: increased speed and accuracy of selection; stacking genes, including minor genes; following genes in backcross populations; and reduced costs of field-based selection. Thus, rather than growing breed-ing lines in the field and challenging or testing for important traits over the growing season, it is possible to extract DNA from 50 mg of a seedling leaflet, and test for presence or absence of a range of traits in that DNA sample in one day. Plants lacking the required traits can then be removed early in the breeding programme.[7] With the availability of more validated molecular markers, marker-assisted selection therefore becomes a highly cost-effective and efficient process.

[6] J.-M. Ribaut and D. Hoisington, *Trend. Plant Sci.*, 1998, **3**, 236.
[7] A. Briney, R. Wilson, R. H. Potter, I. Barclay, R. Appels, and M. G. K. Jones, *Molecular Breeding*, 1998, **4**, 427.

2.5 Examples of Marker-assisted Selection

There are now many examples of the use of molecular markers for selection in plant breeding.[3] Examples include (i) the microsatellite locus *HSP176* of soybean (Section 2.3) which is closely linked to a gene (*Rsv*) conferring resistance to soybean mosaic virus, (ii) a perfect marker for noodle quality starch in wheat[7] and (iii) a marker for early flowering in lupins.[8]

Western Australia exports specialty wheat to the Asian market to make white alkaline salted noodles. This segment of the export trade is worth $250 million per annum. White noodles require specific swelling properties of starch. Noodle quality wheats all have two rather than three copies of granule bound starch synthase (GBSS), the enzyme which synthesizes amylose, the linear polymer of starch. This reduces the ratio of amylose to amylopectin by 1.5–2%, increasing the flour swelling volume. A 'perfect' PCR molecular marker was developed which identified presence or absence of the GBSS gene on chromosone 4A, *i.e.* 'bad' or 'good' noodle starch. This molecular marker test is now used as a primary screen for all noodle wheat breeding lines in Western Australia, and has resulted in the accelerated production of a series of new noodle quality wheat varieties.[7]

Narrow-leafed lupin is the major grain legume grown in Australia. Early flowering is required for the crop to complete its life-cycle before the rain limits growth in areas of Mediterranean climate where it is grown. Using fluorescent AFLPs, a marker linked to early flowering of lupins was identified (using a DNA sequencer). The AFLP was then run as a radioactive version, the polymorphic band isolated, cloned and sequenced, and a co-dominant PCR-based marker developed for routine implementation.[8]

2.6 Molecular Diagnostics

The same principles used in developing molecular markers can be applied for a range of molecular diagnostic purposes in plants, including:

    (i) identification of plant pathogens (viruses, fungi, nematodes, bacteria, insects).

   (ii) studying population structure/variations in pathogens.

  (iii) identifying presence and quantifying presence of transgenes in transgenic foods.

  (iv) following possible pollen transfer of transgenes.

[8] S. J. Brien, R. H. Potter, and M. G. K. Jones (in press).

All that is needed is to identify specific nucleic acid (RNA or DNA) sequences unique to the target organism, and then to develop a reliable extraction/PCR analysis system such that a DNA fragment is only amplified if the target organism or target sequence is present in a sample. The methods and scale by which such analyses (and also marker-assisted selection) can be carried out, are advancing rapidly. Analysis can be by:

(i) PCR and gel electrophoresis.
(ii) real time fluorescent PCR (*e.g.* using an ABI Taqman 7700) to quantify the original amount of target sequence without gel electrophoresis.
(iii) matrix-assisted laser desorption ionization time-of-flight mass spectrometry (MALDI-TOF ms).
(iv) use of DNA chips and hybridization of labelled samples to bound DNA sequences.

The latter two methods (MALDI-TOF ms) and chip technology, including microarrays (Section 2.8) promise to speed up all DNA fragments analysis applications by an order of magnitude over current gel-based DNA separation technologies.

Typical examples of applications of molecular diagnostics are (1) routine analysis of farmers' seed samples (such as lupin) for presence of seed-borne cucumber mosaic virus (CMV).[9] This can done by RT-PCR (to detect viral RNA, sensitivity < 1 infected seed per 1000 seeds), or by real-time fluorescent PCR. Farmers must buy clean seed if infection levels of CMV are above 0.5%; (2) routine analysis of farmers' lupin seed for the fungal disease anthracnose, caused by *Colletotrichum acutatum*. A PCR test, based on repeated ribosomal DNA sequences specific to the fungal pathogen, allows infection levels of 1 seed in 10 000 to be detected. In this case, only clean anthracnose-free seed can be sown.

2.7 DNA Fingerprinting, Variety Identification

The same processes of DNA fragment production and analysis can be applied to DNA fingerprint plants. The main use is in variety identification and quality control, but the techniques can equally be applied to study plant populations, taxonomy, conservation biology and rehabilitation of mine-sites or cleared forests. For rehabilitation studies, DNA fingerprinting data can be used to ensure that an appropriate range of

[9] S. J. Wylie, C. R. Wilson, R. A. C. Jones, and M. G. K. Jones, *Aust. J. Agric. Res.*, 1993, **44**, 41.

genotypes of species removed is used to rehabilitate cleared land.[10] With
the advent of end-point levies on delivered bulk grains rather than on
royalties from seed sales (in some countries) there is a need for rapid and
accurate identification of crop varieties at the receival depots. This can
be achieved by rapid analysis of DNA fingerprints of specific crop
varieties.

2.8 DNA Microarrays

DNA microarrays can be set up robotically by depositing specific
fragments of DNA at indexed locations on microscope slides.[11] With
current technology, cDNAs, EST clones or open reading frames (ORFs)
from sequenced genomes can be set up in microarrays of 10 000 spots per
$3.24 \, cm^2$, thus the whole genome of *Arabidopsis* could be displayed
on one microscope slide. Fluorescently labelled mRNA probes are
hybridized onto the array, and specific hybridizing sequences are
identified by their fluorescent signals. Microarray technology can be
used to study gene expression patterns, expression fingerprints, DNA
polymorphisms and, in theory, as a breeding tool to evaluate new genetic
materials, for a specific trait (*e.g.* drought tolerance), together with
phenotypic tests.[12]

2.9 Bioinformatics

The generation of massive amounts of molecular information on plant
genomes and their products from large-scale sequencing programmes,
DNA fingerprinting and fragment analyses, mapping, molecular diagnostics, marker-assisted selection and DNA microarrays requires a
concomitant increase in the ability to handle and analyse such data.
The handling and interpretation of molecular data is generally referred
to as bioinformatics, and marries requirements of computing power,
data handling and appropriate software. Depending on the level of
analysis, the area can be divided into 'genomics' (DNA level), 'proteomics' (protein expression level) and 'metabolomics' (metabolic level).
In many cases it is useful to 'mine' DNA or other databases to look for
new or useful genes or sequences, using specific software programs and
algorithms. Bioinformatics is discussed in more detail in Chapter 15.

[10] A. Karp, K. J. Edwards, M. Bruford, S. Funk, B. Vosman, M. Morgante, O. Seberg, A. Kremer,
    P. Boursot, P. Arctander, D. Tantz, and G. M. Hewitt, *Nature Biotechnology*, 1997, **15**, 625.
[11] T. Desprez, J. Amselem, M. Caboche, and H. Hofte, *Plant J.*, 1998, **14**, 643.
[12] D. M. Kehoe, P. Villand, and S. Somerville, *Trend. Plant Sci.*, 1999, **4**, 38.

3 TRANSGENIC TECHNOLOGIES

It is now possible to generate transgenic plants from every major crop plant species. Some are easier to transform than others, but if there are sufficient economic reasons to fund research and development, then, for almost any crop, transgenic plants can be produced. Detailed procedures to generate transgenic plants have been described in the literature, with many incremental improvements in efficiencies of transformation, or applications to new genotypes. Two approaches have survived the test of time: transformation using *Agrobacterium tumefaciens* as a vector, and particle bombardment.

3.1 *Agrobacterium*-mediated Transformation

The use of *A. tumefaciens* (or *A. rhizogenes*) as a gene vector is the method of choice for plant transformation. The development of dis-armed binary vectors, in which the virulence (*vir*) genes are separated from the genes to be transferred ('T-DNA'), with the latter on a small easily manipulated plasmid, made the processes of recombinant DNA manipulations more routine. The process of transfer of gene(s) of interest and selectable marker genes have been extensively reviewed,[13] and will not be covered further here. However, one topic that deserves further discussion is that of the selectable marker genes used to identify and extract transgenic cells from the non-transgenic cells, and new reporter genes used to monitor gene expression.

3.2 Selectable Marker and Reporter Genes

The neomycin phosphotransferase (*nptII*) gene is still the most widely used selectable marker gene, and has been used extensively to generate many transgenic dicotyledonous plants, and also some monocotyledonous plants. It confers tolerance to aminoglycoside antibiotics (kanamycin, geneticin, puromomycin). For various reasons (including intellectual property considerations, fear of transfer of kanamycin tolerance to gut bacteria, *etc.*) it would be useful to have alternative selectable marker genes. Some alternatives include: *hpt* (hygromycin tolerance), *bar* or *pat* (phosphinothricin/Basta tolerance),[14] *aroA* (EPSP synthase/glyphosate

[13] R. G. Birch, *Ann. Rev. Plant Physiol. Plant Mol. Biol.*, 1997, **48**, 297.
[14] C. J. Thompson, N. R. Movva, R. Tichard, R. Crameris, J. E. Davies, and M. Lauwereys, *EMBO J.*, 1987, **6**, 2519.

tolerance),[15] modified *als* (chlorsulfuron tolerance),[16] bromoxynil tolerance, and carbon utilizing genes mannose phosphate isomerase (mannose utilization) and xylose isomerase (xylose utilization).[17]

To study specificity of expression, widely used reporter genes include
those encoding β-glucuronidase, (*gus*), luciferase (*luc*) and more recently
green fluorescent protein (*gfp*) from the jellyfish *Aequoria victoriae*.
There are now engineered variants of the latter for improved plant
expression (cryptic intron removed) or altered colours of fluorescence
(*e.g.* blue, red, yellow). An advantage of *gfp* is that no substrate is
required, and so *in vitro* expression can be followed in the same cells or
tissues over long periods (weeks/months), and high-resolution studies in
live cells can be undertaken using confocal scanning laser microscopy.[18]

3.3 Particle Bombardment

Many different strategies have been tried to introduce naked DNA into
plant cells (including microinjection, DNA pollination, silicon whiskers,
electroporation of cells or tissues, electroporation or chemically induced
introduction of DNA into protoplasts), but the major alternative to
Agrobacterium transformation is particle bombardment. This approach
now normally involves coating 1 μm diameter particles of tungsten or
gold with DNA, putting them on a support, accelerating them to high
speed using a pulse of high pressure helium into an evacuated chamber
containing the target tissues. Particles can penetrate up to about six cell
layers, and the DNA released from the particles in surviving cells may be
expressed transiently and, in a small proportion of cases, it becomes
integrated into the nuclear genome of that cell. With appropriate tissue
culture and selection, transgenic plants can be regenerated.

Particle bombardment is usually limited to generation of transgenic
cereals, and its use has decreased since *Agrobacterium* transformation
now works for cereals such as rice and barley.[19] It is still used widely for
maize and wheat transformation, although there is one report of wheat
transformation using *Agrobacterium* as a vector.[20]

[15] D. M. Shah, R. B. Horsch, H. J. Klee, G. M. Kishore, J. A. Winter, N. E. Tumer, C. M. Hironaka,
P. R. Sanders, C. S. Gasser, S. Aykent, N. R. Siegel, S. R. Rogers, and R. T. Fraley, *Science*, 1986,
233, 478.
[16] G. W. Haughn, J. Smith, B. Mazur, and C. Somerville, *Mol. Gen. Genet.*, 1998, **211**, 266.
[17] A. Haldrup, S. G. Petersen, and F. T. Okkels, *Plant Cell Rep.*, 1998, **18**, 76.
[18] J. Haseloff, K. R. Siemering, D. C. Prasher, and S. Hodge, *Proc. Nat. Acad. Sci. USA*, 1997, **94**,
2122.
[19] S. Tingay, D. McElroy, R. Kalla, S. Fieg, M. Wang, S. Thornton, and R. Brettel, *Plant J.*, 1997,
11, 1369.
[20] M. Cheng, J. E. Fry, S. Pang, H. Zhou, C. M. Hironaka, D. R. Duncan, T. W. Connor, and Y.
Wan, *Plant Physiol.*, 1997, **115**, 971.

4 APPLICATIONS OF TRANSGENIC TECHNOLOGIES

The application of transgenic technologies in plant biotechnology has the potential to exceed the technical advances that have taken place in previous 'revolutions' in production agriculture. It has only taken four years of commercial growth of transgenic crops in North America for 52% of the soybean, 30% of the maize and 9% of both cotton and canola grown in 1999 to be transgenic, with increasing production of a wide range of other transgenic crops such as rice, wheat, barley, sorghum, sugar cane, sugar beet, tomato, potato, sunflower, peanut, papaya, tree species and horticultural crops such as carnations. Major companies involved in commercialization of transgenic crops and their alliances are given in Table 1.[21]

Of the 69.5 million acres of transgenic crops grown (27.8 million hectares) in 1998, 74% were in the USA, 15% in Argentina, 10% in Canada and 1% in Australia. The transgenic traits grown commercially were dominated by herbicide tolerance (71%) and insect resistance (28%), with only 1% for the other traits.[21]

However, the transgenic plants currently being commercialized are the first generation of transgenic crops, and three generations of transgenic crops can be envisaged:

First generation   – production traits (*e.g.* herbicide tolerance, insect/ disease resistance)
Second generation – stacked genes for multiple traits (*e.g.* combinations of disease resistance genes plus quality traits)
Third generation   – varieties tailored for specific end uses (*e.g.* food, fibre, fuel, lubricants, plastics, pharmaceuticals and raw materials for industrial processes)

Already, the production of second generation transgenic crops is in progress, and some specific examples of applications are given in the following sections. However, to achieve the potential benefits of transgenic plant biotechnology, there are many additional factors to consider, which include regulation of biotechnology, intellectual property, food safety, public acceptance, allergenicity, labelling, choice, the environment, segregation of transgenic products and international trade.

The emphasis here is on application and potential benefits of transgenic technologies. As in any developing technologies, the aim is to maximize the benefits and minimize the risks.

[21] A. M. Thayer, *Chemical & Engineering News*, 1999, **77**, 21.

Table 1 *Major companies involved in commercialization of transgenic crops and their alliances* (modified from Thayer 1999)[21]

Company	Partner	Technology basis of collaboration
AgroEvo	Gene Logic	Genomics for crops/crop protection
	Kimeragen	Gene modification technology
	Lynx Therapeutics	Genetics for crops/crop protection
American Cyanamid	Acacia Biosciences	Compounds for agrochemicals
	AgriPro Seeds	Herbicide-tolerant wheat
	Zeneca Seeds	Transgenic canola
BASF	Metanomics	Functional plant genomics
	SunGene	Testing genes in crops
Bayer	Exelixis Pharmaceuticals	Screening targets for agrochemicals
	Lion Bioscience	Genomics for crop protection products
	Oxford Asymmetry	Compounds for agrochemicals
	Paradigm Genetics	Screening targets for herbicides
Dow Chemical	Biosource Technologies	Functional genomics for crop traits
	Demegen	Technology to increase protein content
	Oxford Asymmetry	Compounds for agrochemicals
	Performance Plants	Gene technology to increase yield/content
	Proteome Systems	Protein production in plants
	Ribozyme Pharmaceuticals	Technology to modify oil/starch content
	SemBiosys Genetics	Commercialize proteins produced in plants
DuPont	CuraGen	Genomics for crop protection products
	3-D Pharmaceuticals	Compounds for agrochemical target
	Lynx Therapeutics	Genetics for crops/crop protection
	Pioneer Hybrid	Crop breeding/varieties
FMC	Xenova	Compounds for agrochemicals
Monsanto	ArQule	Compound libraries for agrochemicals
	GeneTrace	Genomics for crops
	Incyte Pharmaceuticals	Plant, bacterial, fungal genomics
	Mendel Biotechnology	Functional genomics in plants
	Cereon Genomics	Plant genomics
Novartis	Chiron	Compounds for agrochemicals
	CombiChem	Compounds for agrochemicals
	Diversa	Plant genetics for transgenic crops
Rhone-Poulenc	Agritope	Genomics joint venture for plant traits
	Celera AgGen	Corn genomics
	Mycogen/Dow AgroSciences	Genetic traits in crops, marketing
	RhoBio	Genetics for disease resistance
Zeneca	Alanex	Compound libraries for agrochemicals
	Incyte Pharmaceuticals	Plant genomics
	Rosetta Inpharmatics	Compounds for crop protection products

5 ENGINEERING CROP RESISTANCE TO HERBICIDES

Herbicides are the method of choice to control weeds in most broadscale agricultural systems. They play a major role in maximizing crop yields by reducing competition from weeds for space, light, water and nutrients, and help control erosion by enabling weed control where crops are drilled directly without ploughing. Weeds can also act as a reservoir for crop pathogens.

Because herbicide resistance genes are also effective selectable marker genes in culture, herbicide tolerant crop varieties were the first major transgenic trait to be produced and commercialized, and herbicide tolerant varieties are still the most widely grown transgenic crops.

Based either on expression of a herbicide insensitive gene, degradation of the herbicide or overexpression of the herbicide target gene product, engineered resistance is now available to a range of herbicides[21,22] including: glyphosate ('Roundup Ready', 'Touchdown'), glufosinate ('Liberty Link'), imidazalonones (IMI), protoporphyrinogen oxidase inhibitors ('Acuron'), bromoxynil, triazines, 2,4-D, chlorsulfuron/sulfonylureas, and, isoxazoles.

There is now good evidence that far from increasing the application of herbicides, the control that transgenic herbicide resistant crops provide to the farmer has resulted in a reduction in glyphosate usage of 33% on Roundup Ready soybeans, and a reduction in glufosinate usage of about 20% for Liberty Link canola.[21]

6 ENGINEERING RESISTANCE TO PESTS AND DISEASES

6.1 Insect Resistance

Chemical control of insect pests is both expensive and environmentally unfriendly. Worldwide expenditure on insecticides and the value of crop losses from insect predations[22,23] have been estimated in Table 2.

Transgenic cotton, maize and potato crops are being grown commercially which express *Bacillus thuringiensis* (*Bt*) toxins to confer resistance to chewing insects. On sporulation, *B. thuringiensis* synthesizes δ-endotoxin crystalline proteins encoded by *Cry* genes. On ingestion by an insect, prototoxins are cleaved in the alkaline midgut to the active toxin. This binds to specific receptors in the gut epithelial cells which

[22] P. J. Dale, *TIBTech*, 1995, **13**, 398.
[23] R. A. de Maagd, D. Bosch, and W. Stiekma, *Trend. Plant Sci.*, 1999, **4**, 9.

Table 2   *Worldwide costs of insecticides and losses caused*
              by insects

Cost of insecticides (millions of US$)		Losses caused by insects (millions of US$)	
Cotton	1870	Fruit	20 000
Fruit and vegetables	2465	Vegetables	25 000
Rice	1190	Rice	45 000
Maize	620	Maize	8 000
Other crops	1965		

results in the formation of pores and eventually to the death of the insect. Some advantages of *Bt* toxins include:

  (i) specificity – each *Cry* protein is active against only one or a few insect species.
 (ii) diversity – many different *Cry* proteins have been identified.
(iii) reduced or no detrimental effects identified on non-target insects or natural enemies of insects.
 (iv) very low mammalian toxicity.
  (v) easily degradable.

The production of transgenic *Bt* expressing insect resistant crops has been a high commercial priority, but effective production of *Bt* toxins required re-engineering of *Cry* genes for plant codon usage and removal of cryptic signals (*e.g.* splice sites, polyadenylation signals). These changes enable efficient expression of *Bt* toxins in plants, with both full-length and truncated versions of *Cry* genes used successfully for insect resistance. Now, more than 40 different genes containing insect resistance have been incorporated into transgenic crops with several commercialized in different countries such as the USA and Australia.[24]

Given the usefulness of *Bt* toxins for insect control, various management strategies must be adopted to delay development of insect resistance to *Bt*. These include:

  (i) setting aside areas of non-*Bt* cotton as refuges to reduce the selection pressure towards insect resistance.
 (ii) deploying different insect resistance genes (*e.g.* protease inhibitors).
(iii) using multiple *Bt* toxins which target different receptors.

[24] T. H. Schuler, G. M. Poppy, B. R. Kerry, and I. Denholm, *TIBTech*, 1998, **16**, 168.

(iv) use of spray inducible promoters to control expression of *Bt* genes.

(v) use of tissue-specific promoters such that insects can feed unharmed on economically less important parts of the plant.

It is mandatory to set aside non-transgenic refuges when growing *Bt* cotton, maize and potatoes. Even with additional costs associated with this and other agronomic management practices, the growth of *Bt* cotton gives higher returns to farmers, environmental benefits (50–80% less chemicals used, spraying reduced from 10–12 per year to 4–5 per year) and less occupational exposure of farm workers to sprays.[25]

There are other approaches under development for transgenic insect resistance,[24] including those based on: protease inhibitors, α amylase inhibitors, lectins, chitinases, cholesterol oxidase, cloned insect viruses, tryptophan decarboxylase, anti-chymotrypsin, anti-elastace, bovine pancreatic trypsin inhibitor, and spleen inhibitor.

6.2 Engineered Resistance to Plant Viruses

Viruses cause significant losses in most major food and fibre crops worldwide. A range of strategies can be used to control virus infection, including chemical treatments to kill virus vectors, identification and introduction of natural resistance genes from related species, and use of diagnostics and indexing to ensure propagation of virus-free starting material (seeds, tubers, *etc.*). However, the major development has been exploitation of pathogen-derived resistance, *i.e.* use of virus-derived sequences expressed in transgenic plants to confer resistance to plant viruses.[26] This approach is based on earlier observations that inoculation or infection of a plant initially with a mild strain of a virus confers protection against subsequent inoculation with a virulent strain of the same or a closely related virus.[27] Pathogen-derived resistance thus involves transformation of plants with virus-derived sequences; host resistance appears to result from two different mechanisms: (i) protection thought to be mediated by expression of native or modified viral proteins (*e.g.* coat protein, replicase, defective replicase), and (ii) protection mediated at the transcriptional level ('RNA-mediated resistance') which requires transcription of RNA either from full or partial sequences

[25] N. W. Forrester, in 'Commercialization of Transgenic Crops: Risk, Benefit and Trade Consideration', eds G. D. McLean, P. M. Waterhouse, G. Evans, and M. J. Gibbs, Bureau of Resource Science, Canberra, Australia, 1997, p. 239.
[26] M. Prins and R. Goldback, *Archiv. Virol.*, 1996, **141**, 2259.
[27] J. C. Sanford and S. A. Johnstone, *J. Theor. Biol.*, 1985, **113**, 395.

derived from the target virus (including genes for coat protein, replicase, defective replicase, protease, movement protein, *etc.*).

The molecular events which underlie pathogen-derived resistance are the subject of intensive research.[28] The basis of RNA-mediated virus resistance and post-transcriptional gene silencing are probably similar and reflect fundamental activities in plant cells to detect, inactivate, and eliminate foreign DNA or RNA.[29] For example, endogenous plant genes inserted into viruses such as PVX can silence expression of the endogenous plant gene.[30] It is probable that low molecular weight double-stranded RNA sequences homologous to the gene message to be silenced or degraded travel systemically through the plant from the site of induction, to ensure that viruses with homologous sequences are degraded when they arrive elsewhere in the plant.[31] Understanding the basis of resistance is needed to ensure practical applications of transgenic virus resistance are stable and have the least environmental risks when deployed on a wide scale.

Most of the major crops have been transformed with genes from major viral pathogens based on the concept of pathogen-derived resistance. For example, by expression of viral replicase derived sequences, host resistance has to be obtained to 13 genera of viruses representing 11 plant virus taxa.[32]

Major crops transformed for virus resistance include potato, tomato, canola, soybean, sugar beet, rice, barley, sugar cane, papaya, melons/cucurbits, peanut, horticultural and tree species. Effective resistance against a wide range of viruses has been achieved,[22] including PVX, PVY, PLRV, CMV, BYMV, PRSV, ACMV, CPMV, TYLCV, PPV, PMMV, TMV, PEBV, CymRSV, BYDV, RTV, BBTV.

A good practical application of this technology is that of effective protection of transgenic papaya (*Carica papaya*) against papaya ringspot virus (PRSV).[33] In Hawaii, PRSV has devastated papaya production. Resistance to PRSV (mediated *via* viral coat protein constructs) has held up under field conditions in Hawaii, and these results suggest that long-term protection of perennial crops, such as papaya, will be possible using pathogen-derived resistance.

[28] A. van Kammen, *Trend. Plant Sci.*, 1997, **2**, 409.
[29] S. P. Kumpatla, M. B. Chandrasekharan, L. M. Iyer, G. Li, and T. C. Hall, *Trend. Plant Sci.*, 1998, **3**, 97.
[30] J. J. English and D. C. Baulcombe, *Plant J.*, 1997, **12**, 1311.
[31] H. Vaucheret, C. Beclin, T. Elmayan, F. Feuerbach, C. Godon, J.-B. Morel, P. Mourrain, J.-C. Palauqui, and S. Vernheltes, *Plant J.*, 1998, **16**, 651.
[32] R. Yang 1999, PhD Thesis, Murdoch University, Perth, Western Australia.
[33] S. Luis, R. M. Manshardt, M. M. M. Fitch, J. L. Slightom, J. C. Sanford, and D. Gonsalves, *Molec. Breeding*, 1997, **3**, 161.

6.3 Resistance to Fungal Pathogens

Plants react to attack by fungal and other pathogens by activating a series of defence mechanisms, both locally and throughout the plant.[34] The responses may be non-specific induction of defence reactions to pathogens, or specific responses based on the race of the pathogen and the genotype of the host plant. Local resistance may appear as a hypersensitive response in which a local necrotic lesion restricts the growth and spread of a pathogen. Systemic resistance, which may take several hours or days to develop, provides resistance to pathogens in parts of the plant remote form the initial site of infection, and longer term resistance to secondary challenge by the initial pathogen and also to unrelated pathogens.

The hypersensitive response is characterized by rapid reactions to invasion by a potential pathogen, through recognition of pathogen or cell wall derived elicitors.[35] It involves:

   (i) opening of specific ion channels.
   (ii) membrane potential changes.
   (iii) oxidative burst (generation of reactive oxygen species).
   (iv) synthesis of peroxidase.
   (v) production of secondary metabolites (phenylpropanoids and phytoalexins).
   (vi) synthesis of pathogenesis-related (PR) proteins (*e.g.* β-1,3-glucanases, chitinases).(vii)
   (vii) cell wall changes (*e.g.* suberin, lignin).

It results in death of host cells, formation of a necrotic lesion and restriction or death of the pathogen.

Transduction of a signal following pathogen recognition can be both local and systemic, and involves a number of different pathways.[36,37] The synthesis and accumulation of salicylic acid appears to be necessary both for local and systemic induction of defence responses, and salicylic acid (or methyl salicylate) is a major signalling molecule. However, other compounds can activate plant defence genes[34] (*e.g.* 2,6-dichlororisonicotinic acid, benzothiadiazole, ethylene, abscisic acid, jasmonic acid and systemin). Systemic signals may lead to induction of systemic acquired resistance.

[34] J. Durner, J. Shah, and D. F. Klessig, *Trend. Plant Sci.*, 1997, **2**, 266.
[35] N. Benhamou, *Trend. Plant Sci.*, 1996, **1**, 233.
[36] C. Wasternack and B. Parthier, *Trend. Plant Sci.*, 1997, **2**, 302.
[37] E. Blumwald, G. S. Aharon, and B. C.-H. Lam, *Trend. Plant Sci.*, 1998, **3**, 342.

For salicylic acid signalling, salicylic acid moves in the phloem and its presence may be required to establish and maintain systemic acquired resistance. Its arrival in tissues leads to expression of plant defence-related genes in sites distant from the initial challenge, such as PR proteins and production of hydrogen peroxide and reactive oxygen species, cross-linking of cell wall proteins and lignin synthesis.

There are a series of other defence systems that plants use to combat pathogens, and these include natural resistance genes and antifungal proteins.

6.4 Natural Resistance Genes

The hypersensitive response often results from a specific interaction between a biotrophic pathogen and its host plant. This is known as a 'gene-for-gene' interaction between pathogen and host, in which an avirulence gene product from the pathogen is recognized by a resistance gene in the host plant. This recognition leads to induction of hypersensitive defence responses.[38]

Avirulence (*Avr*) genes encode a variety of polypeptides, some of which may be required for pathogenicity but have then become avirulence factors once they have been detected by the plant. The best characterized *Avr* genes are *Avr4* and *Avr9* of the fungal pathogen *Cladosporium fulvum*.[39] These encode pre-proteins which are processed to mature, extracellular cysteine-rich peptides of 86 and 28 amino acids, respectively. These peptides induce a hypersensitive response in plants which contain the matching resistance genes *Cf-4* and *Cf-9*, respectively.

A series of plant resistance (R) genes active against a range of pathogens have been cloned and characterized. The R genes *Cf-4* and *Cf-9* encode transmembrane glycoproteins, in which the extracellular portion has characteristic leucine rich repeats (LRR), a transmembrane domain and a *C*-terminal cytoplasmic domain. The *Avr* gene products are recognized by the LRR receptor regions, which results in signal transduction, gene activation and the hypersensitive response.

With the characterization of R-genes against fungal, bacterial, viral and nematode pathogens, common patterns have emerged.[38] Depending on the site of recognition of the elicitor, the R gene products may either span the plasma membrane and detect the elicitor extracellularly, or be located in the cytoplasm for intracellular elicitor recognition. Intra-

---

[38] P. J. G. M. de Witt, *Trend. Plant Sci.*, 1997, **2**, 452.
[39] M. Kooman-Gersmann, R. Vogelslang, E. C. M. Hoogendijk, and P. J. G. M. de Witt, *Molec. Plant Microbe Interactions*, 1997, **10**, 821.

cellular recognition would be expected for virus infection, but pathogens growing extracellularly appear to be recognized by the presence of extracellular elicitors or signal molecules from the pathogens that cross the host plasma membrane.

Five classes of R-genes have been proposed (Table 3 and Figure 1), in

Table 3 *Natural resistance genes characterized*

Class	R-gene	Feature(s)	Location	Pathogen
I	*Pto*	Kinase site	Intracellular	*P. syringae* pv. *tomato*
II	*RPS2*	LRR, NBS, LZ	Intracellular	*P. syringae* pv. *tomato*
	RPM1	LRR, NBS, LZ	Intracellular	*P. syringae* pv. *maculicula*
	I2	LRR, NBS, LZ	Intracellular	*F. loxsporium* f.sp. *lycopersici*
	Mi	LRR, NBS, LZ	Intracellular	*Meloidogyne* spp. (root-knot nematodes)
III	*N*	LRR, NBS, TIR	Intracellular	TMV
	L6	LRR, NBS, TIR	Intracellular	*Melamspora lini*
	RPP5	LRR, NBS, TIR	Intracellular	*Peronospora parasitica*
IV	*Cf2, Cf4, Cf5, Cf9, HS1 pro-1*	LRR	Transmembrane	*C. fulvum* *Heterodera schachtii*
V	*Xa21*	LRR, kinase site	Transmembrane	*Xanthomonas campestris* pv. *vesicatoria*

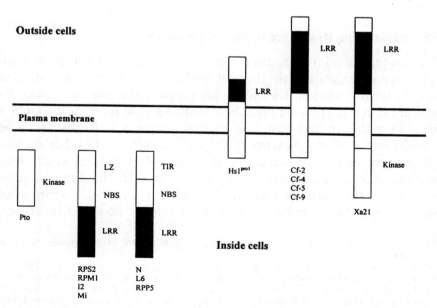

Figure 1 *Diagnostic representation of different classes of resistance (R)-genes*

which common features occur – leucine-rich repeats (LRR), nucleotide binding site (NBS), leucine zippers (LZ), toll and interleukin-like receptors (TIR), and kinase sites.[38,40] Some of these are shown in Table 3.

Natural R-genes of this type can be transferred to other plants and, depending on presence of elicitors/races of pathogen, may confer resistance in other plants. Modification of R-genes to alter the specificity of recognition of elicitors could also be achieved, and various schemes have been developed to convert the specific recognition (*e.g.* of the *avr9/Cf9* system) to a general defence response switched on by damage caused by non-specific pathogens.[41]

As indicated above, most of the R-genes identified so far have common sequence and structural motifs. It is possible to align published sequences and synthesize PCR primers complementary to the conserved sequences, and to amplify classes of 'resistance gene analogues' (RGAs).[42] RGAs can be mapped, and exist in local clusters of related sequences in the genome. This approach can aid identification of natural resistance genes, and mapping of RGAs can also provide molecular markers closely linked to known resistance genes.

However, not all R-genes operate via a gene-for-gene mechanism. The *HM1* R-gene encodes a reductase which inactivates toxins produced by the fungal pathogen *Cochliobulus carbonum*, and the *mlo* gene for powdery mildew resistance encodes a negative regulator of cell death.[43]

6.5 Engineering Resistance to Fungal Pathogens

The strategy used to engineer resistance to fungal pathogens often depends on the nature of the the host–pathogen interaction. As indicated in the previous section, for biotrophic fungal pathogens, a specific R-gene approach can be used since there is often a gene-for-gene interaction between pathogen and host, and natural or modified R-genes may be transferred to other genotypes of the same species, or to other species, which may confer resistance to the race of pathogen which they recognized in the host plant. However, for necrotrophic fungal pathogens, which kill tissues in advance of hyphal invasion, other approaches are required. These include induction of sytemic acquired resistance,

[40] S. B. Milligan, J. Bodeau, J. Yaghoobi, I. Kaloshian, P. Zabel, and V. M. Williamson, *Plant Cell*, 1998, **10**, 1307.
[41] K. E. Hammond-Kosack, S. Tang, K. Harrison, and J. G. Jones, *Plant Cell*, 1998, **10**, 1251.
[42] V. Kanazin, L. F. Marek, and R. C. Shoemaker, *Proc. Natl. Acad. Sci. USA*, 1996, **93**, 11746.
[43] R. Buschges, K. Hollricher, R. Panstruga, G. Simons, M. Wolter, A. Frijters, R. van Daelen, T. van der Lee, P. Diergaarde, J. Groenendijk, P. Topsch, P. Vos, F. Salamini, and P. Schulze-Lefert, *Cell*, 1997, **88**, 695.

production of a range of antifungal proteins,[44] or introduction of genes which can degrade fungal toxins. Examples include:

(i) genes for toxin inactivation (*e.g. HM1*).
(ii) genes encoding anti-fungal proteins (*e.g.* plant defensins such as radish anti-fungal protein, thionins such as macadamia nut anti-fungal protein).
(iii) genes encoding PR proteins (*e.g.* chitinases, β1,3-glucanases).
(iv) genes that will activate the systemic acquired resistance response.
(v) artifically induced hypersensitive reaction.

In general, the approaches which involve transformation of plants with genes for anti-fungal proteins do not give complete resistance to fungal pathogens. As a result, it is envisaged that stacking of such resistance genes will be required to provide effective fungal resistance.[44] This may be achieved by multiple transformations, or by joining the coding sequences of different anti-fungal protein genes with linkers for peptides recognized by proteases, such that the anti-fungal proteins are translated as one polyprotein and subsequently cleaved to their separate active constituents by protease digestion.

6.6 Resistance to Bacterial Pathogens

Pathogenic bacteria have developed pathogenicity or virulence factors which allow the pathogen to multiply in infected tissues. As indicated above for fungal pathogens, in which R-genes resistant to bacterial pathogens are also given, host R-genes recognize some of these factors, encoded by *avr* genes, which trigger a similar range of defence responses.[45,46] Approaches to engineer resistance to bacterial pathogens include:

(i) production of anti-bacterial proteins of non-plant origin (*e.g.* antibacterial proteins *Shiva1, MB39, AttacinE* from the giant silk moth; lysozymes of various origins, *lactoferrin* and *tachyplesin* from the horseshoe crab).
(ii) inhibition of bacterial pathogenicity or virulence factors (*e.g.* tab toxin-resistance protein, phaseolotoxin-insensitive ornithine carbamoyltyransferase from *P. syringae*).

---

[44] W. Broekaert, B. P. A. Cammue, M. F. C. De Bolle, K. Thevissen, G. W. De Samblanx, and R. W. Osborn, *Crit. Rev. Plant Sci.*, 1997, **16**, 297.
[45] U. Bonas and G. Van den Ackerveken, *Plant J.*, 1997, **12**, 1.
[46] F. Mourgues, M. Brisset, and E. Chevreau, *TIBTech*, 1998, **16**, 203.

(iii) enhanced plant defences (*e.g.* pectate lyase, R-genes *Xa21* and *Pto*, glucose oxidase, thionins).
(iv) artificial induction of the hypersensitive response.

The insect lytic peptides (*e.g.* cecropins from the giant silk moth, and synthetic analogues) form pores in bacterial membranes. Transgenic potato, tobacco and apples with genes for these lytic peptides have been generated, which confer some protection against bacterial wilt and fireblight (*Erwinia amylovora*). Lysozymes digest bacterial cell wall peptidoglycans, and expression of different lysozyme genes can give partial resistance to *E. caratova atroseptica* and *P. syringae* pv. *tabaci* in potato and tobacco, respectively.

The R-gene *Xa21* confers resistance to bacterial leaf blight (*Xanthomonas oryzae* pv. *oryzae*), and has been transferred into susceptible rice genotypes to confer resistance to *X. oryzae*, and overexpression of the resistance conferred by the R-gene *Pto* to *Pseudomonas syringae* pv. in tomato confers resistance to other pathogens.[47]

6.7 Resistance to Nematode Pathogens

The major nematode pathogens of crop plants are the sedentary endoparasites root-knot (*Meloidogyne* spp.) and cyst-nematodes (*Heterodera/Globodera* spp.). Root-knot nematodes are more prevalent in subtropical and tropical regions, whereas cyst-nematodes are more of a problem in temperate regions. There are also semi-endoparasites and ecto-parasites.

The endoparasites develop an intimate association with hosts, with the formation of specific feeding cells (giant cells for *Meloidogyne* spp. and syncytia for *Heterodera/Globodera* spp.), which take the form of multinucleate transfer cells.[48] The nematodes are dependent on these cells for nutrients, and will not complete their life-cycles if the feeding cells are damaged. The feeding cells therefore provide a major target for engineered resistance.[49] This can take the form of genes that prevent feeding cell induction, genes that prevent or inhibit feeding cell function, or toxic compounds delivered to the nematode *via* the feeding cells. For endo-, semi endo- and ectoparasitic nematodes, it is also possible to attack the body wall of the nematode itself, by expression of specific cuticle degrading enzymes such as collagenase.

[47] X. Tang, M. Xie, Y. J. Kim, J. Zhou, D. F. Klessig, and G. B. Martin, *Plant Cell*, 1999, **11**, 15.
[48] M. G. K. Jones, *Annals Appl. Biol.*, 1981, **97**, 353.
[49] V. M. Williamson and R. S. Hussey, *Plant Cell*, 1996, **8**, 1735.

Some resistance to nematode attack has been achieved by expression of protease inhibitors in feeding cells,[50] and by damaging feeding cells by feeding cell specific expression of a ribonuclease (*barnase*). Because of the clear requirement for feeding cell formation and function, and the differences between the nematode and its host plant, it is probable that effective broadly applicable synthetic resistance genes to nematodes will soon be available.

It is also of interest that cloned nematode R-genes (to sugar beet and cereal cyst-nematodes and to root-knot nematodes) have been characterized. The *Mi* R-gene[40] also confers resistance to aphids,[51] and this is the first example of an R-gene which gives resistance to two different classes of pests (*i.e.* nematodes and aphids).

7 MANIPULATING MALE STERILITY

For many crops, it was either impossible or very difficult to generate commercial hybrid seed by conventional means. Hybrid seed is attractive to seed companies because farmers must purchase new seed from them each year, since hybrid varieties do not breed true. It is now possible to engineer male sterility by expression of a ribonuclease gene (*barnase*) specifically during development of the tapetal layer that nourishes developing pollen grains. Developmental regulation of the ribonuclease (by the TA29 tapetum specific promoter) kills the tapetal cells leading to male sterility. Male sterile plants can be used as the female parent to produce hybrid seed. Fertility can be restored by expression of the *barstar* gene, which inactivates *barnase*. This technology can be used to produce hybrids of crops such as maize or sugar beet, or canola/rapeseed. It is not possible to produce hybrid canola conventially, and hybrid canola can exhibit hybrid vigour and increased yields. The same basic technology can also be used to induce sterility in transgenic trees, to prevent gene flow *via* cross-pollination.

8 TOLERANCE TO ABIOTIC STRESSES

The full potential yield of crops is rarely met because of environmental stresses. These include drought, cold and salt stresses, and mineral toxicities. In most cases, advances to generate stress tolerant plants by

[50] P. E. Urwin, C.-J. Lilley, M. J. McPherson, and H. J. Atkinson, *Plant J.*, 1997, **12**, 455.
[51] M. Rossi, F. L. Goggin, S. B. Milligan, I. Kaloshian, D. E. Ullman, and V. M. Williamson, *Proc. Natl. Acad. Sci., USA*, 1998, **95**, 9750.

traditional breeding have been slow. This is partly because tolerance to these stresses usually involves many genes and physiological processes.

For cold and drought tolerance, recent research has shown that a series of functionally different cold (COR) and drought response genes show common promoter regulatory sequences (CRT/DRE).[52,53] Transcription factors which bind to these regulatory sequences have been identified which switch on the stress response genes in concert. Overexpression of the transcription factor (CBFI or DREB) in *Arabidopsis* by two separate groups conferred tolerance to freezing.[54,55] In one case, the plants also withstood high salt stress.[56] These results are significant because they show that introduction of a single regulatory gene can confer tolerance to stresses.

It may be that freezing injury in plants is mainly caused by dehydration as ice crystals form, so that drought and freezing tolerance may share common protective mechanisms.

Other approaches to stress tolerance include expression of compatible solutes[57] (*e.g.* trehalose or glycine betaine or the amino acid proline[58]) or antifreeze proteins, which order the water in cell cytoplasm and so reduce the freezing point or size of ice crystals which may form. They may also help in maintaining membrane integrity.

The application of microarray technology has major applications in helping to identify genes that are up-regulated when plants encounter environmental stresses, including genes involved in stress perception, signalling and tolerance. For example, this approach is being used to identify genes involved in response to salt tolerance.

9 MANIPULATING QUALITY

Quality may be defined simply as the nutritional or technological properties of a product. However, the required quality of a harvested product depends on its intended use, whether it be fresh produce for human or animal consumption, for processing as food or as a raw

[52] E. J. Stockinger, S. J. Gilmour, and M. F. Thomashow, *Proc. Natl. Sci. USA*, 1997, **94**, 1035.
[53] S. J. Gilmour, D. G. Zarka, E. F. Stockinger, M. P. Salazar, J. M. Houghton, and M. F. Thomashow, *Plant J.*, 1998, **16**, 433.
[54] K. R. Jaglo-Ottosen, S. J. Gilmour, D. G. Zarka, O. Schabenberger, and M. F. Thomashow, *Science*, 1998, **280**, 104.
[55] Q. Liu, M. Kasuga, Y. Sakuma, H. Abe, S. Miura, K. Yamaguch-Shinozaki, and K. Shinozaki, *Plant Cell*, 1998, **10**, 1391.
[56] M. Kasuga, Q. Liu, K. Yamaguchi-Shinozaki, and K. Shinozaki, *Nature Biotechnology*, 1999, **17**, 287.
[57] H. J. Bohnert and R. G. Jensen, *Trends Biotechnol.*, **14**, 89.
[58] P. B. K. Kishor, Z. Hong, G.-H. Miao, C.-A. A. Hu, and D. P. S. Verma, *Plant Physiol.*, **108**, 1387.

material for an industrial or other commercial process. It is therefore a complex subject. Improved taste, storage life and quality for current uses, novel uses and aspects such as partitioning of metabolites when products are altered are considered here.

9.1 Prolonging Shelf Life

Much produce is lost between harvesting and the point of sale to consumers, and before final consumption. A delay in ripening of climacteric fruit (*e.g.* tomato, melon, in which ripening is accompanied by a burst of ethylene production and respiration) and longer storage life is therefore useful, because it allows produce to be left on the plant longer to ripen (providing better flavour) and reduces losses in transport and storage.

Approaches in commercial use to prolong shelf-life include: switching off polygalacturonase genes (*e.g.* 'Flavr-Savr' tomato); switching off genes in the ethylene biosynthesis pathway, or degrading intermediates in that pathway; and expression of cytokinin genes.

Initial commercialization was done with anti-sense polygalacturonase tomatoes, with both fresh fruit ('Flavr-Savr' sold under the 'Macgregor' label) in the USA and engineered canned tomato puree in the UK. The latter was carefully marketed in two major supermarkets in 1996 by Zeneca, and all production was sold.[60] This was the first food product from a transgenic plant sold in the European Union.

In Australia, transgenic long vase-life carnations have been commercialized, with ethylene production inhibited by down-regulation of the ACC synthase gene. Similar technology is now being applied to a wide range of other crop and horticulture plants to enhance storage life.

9.2 Nutritional and Technological Properties

9.2.1 Proteins. Animals, including humans, cannot synthesize 10 of the 20 essential amino acids, which must therefore be obtained from the diet.[61] It is a paradox that in the past the nutritional balance of foods for human consumption was of less concern to breeders than nutritional properties of products for animal feed. Thus, limiting factors for animal

---

[59] D. Worrall, L. Elias, D. Ashford, M. Smallwood, C. Sidebottom, P. Lillford, J. Telford, C. Holt, and D. Bowles, *Science*, **282**, 115.

[60] N. J. Poole, in 'Commercialization of Transgenic Crops: Risk, Benefit and Trade Considerations', Bureau of Resource Sciences, Canberra, eds G. D. McLean, P. M. Waterhouse, G. Evans, and M. J. Gibbs, Canberra, Australia, 1997, p. 17.

[61] L. Tabe and T. J. V. Higgins, *Trend. Plant Sci.*, 1998, **3**, 282.

feed for grains such as barley, maize and wheat were the levels of the amino acids lysine, threonine and tryptophan, whereas sulfur-containing amino acids (methionine and cysteine) were limiting in grain legumes and pulses. It is now possible to overexpress genes encoding proteins with high proportions of limiting amino acids, either from natural sources or as synthetic proteins.

One example is synthesis of enhanced levels of the sulfur-rich amino acid methionine in grain legumes (*e.g.* lupin), by transfer of a sulfur-rich sunflower seed albumin into lupins. Feeding trials of sulfur-enriched lupins gave increased live weight gain, showing that effective improvements in nutritive value of grain crops is achievable by genetic engineering.[62]

The use of plants to produce speciality proteins is considered in Section 10.

The breadmaking quality of wheat results from the visco-elastic properties of the storage proteins. Wheat gluten is a complex mixture of more than 50 proteins, mainly prolamins. The most important of these are high molecular weight subunits of glutenin, because these determine much of the elastic properties of gluten for wheat doughs to make bread, pasta and other foods. High molecular weight glutenin genes have now been transferred to a number of wheat varieties, with modification to dough elasticity.[63,64] This work shows that the technological properties of wheat storage proteins can be modified usefully by genetic engineering.

9.2.2 Oils. Plant oils are normally stored as triacylglycerols, with fatty acids and glycerol separated in downstream processing. Oil crops are second in importance to cereals as food sources for humans, and provide many industrial products. The major oil products are derived from: soybean (18 Mt), oilpalm (15 Mt), canola/rapeseed (10 Mt), sunflower (8 Mt) and other sources (20 Mt), with a total value of US$45 billion per annum.[65] This value is expected to increase to US$70billion by about 2010.

Rapeseed (or canola) has been a model oil crop, in great part because it is closely related to *Arabidopsis*, and information from study of *Arabidopsis* can be applied directly to rapeseed. Identification and

[62] L. Molvig, L. M. Tabe, B. O. Eggum, A. E. Moore, S. Craig, D. Spencer, and T. J. V. Higgins, *Proc. Natl. Acad. Sci. USA*, 1997, **94**, 8393.
[63] F. Barro, L. Rooke, F. Bekes, R. Gras, A. S. Tatham, R. Fido, P. A. Lazzeri, P. R. Shewry, and P. Barcelo, *Nature Biotechnology*, 1997, **15**, 1295.
[64] I. K. Vasil and O. D. Anderson, *Trend. Plant Sci.*, 1997, **2**, 292.
[65] D. J. Murphy, *TIBTech*, 1996, **14**, 206.

isolation of genes involved in pathways of oil synthesis has led to a range of transgenic rapeseed varieties with modified oils. The term 'designer oils' has been coined to indicate that chain length, degree of saturation and position of double bands can be manipulated. Field trials and commercial release of many new oil quality lines are in progress and are shown in Table 4.[65]

The percentage content of a specific oil can probably be raised up to at least 90% of the oil product. With depletion of world hydrocarbon reserves, in the future it is probable that plant oils (*e.g.* biodiesel) will compete in terms of price and quality with oil, coal and gas, and will become the major large-scale source of renewable industrial hydrocarbons.

Oilseed meal, which is high in protein, is also a valuable commodity used to supplement animal and fish feeds and its protein can be used for human consumption in products such as soy milk and textured vegetable meat substitutes.

Although most advances in biotechnology of oils have come from rapeseed/canola improvement, the same principles can be applied to other oil crops, although this is technically more difficult to achieve for example, with oilpalms.

9.2.3 Manipulation of Starch. Starch constitutes 50–80% of the dry weight of starch-storing organs (*e.g.* potato tubers, cereal endosperm). It occurs in granules as amylose, a linear α1–4 glucan polymer (15–35%) and amylopectin, a linear α1–4 glucan chains connected with α1–6 branches (65–85%). Transgenic expression of enzymes in sucrose meta-

Table 4 *Transgenic rapeseed varieties under development*[65]

Seed product	Industrial product
40% stearic (18:0)	Margerine, cocoa butter
40% lauric (12:0)	Detergents
60% lauric (12:0)	Detergents
80% oleic (18:1 Δ_9)	Food, lubricants, inks
Petroselinic (18:1 Δ_6)	Polymers, detergents
Jojoba wax (C_{20}, C_{22})	Cosmetics, lubricants
40% myristate (14:0)	Detergents, soaps, personal care
90% erucic (22:1)	Polymers, cosmetics, inks, pharmaceuticals
Riconoleic (18:1 –OH)	Lubricants, plasticisers, cosmetics, pharmaceuticals
Polyhydroxybutyrate	Biodegradable plastics
Phytase	Animal feeds
Industrial enzymes	Fermentation, paper manufacture, food processing
Novel peptides	Pharmaceuticals

bolism have shown that it is possible, for example in potato, to modify tuber size, number, yield and starch content.[66]

There has been increased commercial interest in amylose-free (waxy) or high amylopectin starches, because they make processing easier, and because of their gel stability and clarity.[67] Amylose-free 'waxy' mutants have been found for many species (*e.g.* maize, rice, barley, wheat, potato, pea), and this correlates with loss of granule bound starch synthase (GBSS) enzyme activity. Waxy starches can also be generated using anti-sense technology by switching off GBSS gene expression, and its relative accumulation and form can be manipulated (*e.g.* in potato tubers). Applications include: generating noodle quality wheat starch (lacking GBSS on chromosone 4A); reduced amylose or waxy wheat starch (*e.g.* two or all three GBSS loci—4A, 7A, 7D—inactive); and generating starches with novel structures.

Modified starches have novel properties that can be used to develop new applications in the food processing industry.

9.2.4   Fructans.   Sugar beet normally stores sucrose, but by expression of a 1-sucrose–sucrose fructosyl transferase (1-ISST) gene in sugar beet transgenic plants now store fructans. The significance of this work is that humans cannot digest fructans, so that they can be used as low calorie food ingredients. Low molecular weight fructans (up to five monosaccaride units) taste like sugar, and can be used as low calorie sweeteners, whereas long chain fructans form emulsions which can be used to replace fats in foods such as creams and spreads.[68,69]

9.3   Manipulation of Metabolic Partitioning

The flow of metabolites through metabolic and biosynthetic pathways is regulated to respond to developmental and environmental conditions. The distribution of metabolites (partitioning) controls the flux of carbon compounds towards the synthesis of sugars, starch and oils in storage organs, and may change to meet developmental states, environmental constraints or in the activation of the diverse array of defence responses when a plant is attacked by pathogens. Metabolic fluxes can be manipulated, and an understanding of the control of partitioning between metabolic pathways is required for metabolic engineering – this

[66] U. Sonnewald, M.-R. Hajirezaei, J. Kossman, A. Heyer, R. M. Trethewey, and L. Willmitzer, *Nature Biotechnology*, 1997, **15**, 794.
[67] S. G. Ball, M. H. B. J. van de Wal, and R. G. F. Visser, *Trend. Plant Sci.*, 1998, **3**, 462.
[68] S. Smeekens, *Nature Biotechnology*, 1998, **16**, 822.
[69] R. Sevenier, R. D. Hall, I. M. van der Meer, H. J. C. Hakkert, A. J. van Tunen, and A. J. Koops, *Nature Biotechnology*, 1998, **16**, 843.

knowledge can be used to increase the production or yield of commercially important metabolites and products.[70,71] Strategies to modify partitioning include:

(i) changing levels of signalling metabolites and hormones.
(ii) removal of end products to change reaction equilibria.
(iii) expression of heterologous enzymes to bypass endogenous regulatory mechanisms.
(iv) expression of transcriptional regulators that control pathways.
(v) switching off gene expression by sense/antisense suppression.

Most commercial applications of metabolic partitioning involve altered starch and oil biosynthesis, reducing lignin content or synthesis of biodegradable plastic (polyhydroxybutyrate, PHB), but this aspect is also relevant to all uses of plants as bioreactors to produce more of current or novel compounds efficiently.

10 PRODUCTION OF PLANT POLYMERS AND BIODEGRADABLE PLASTICS

Plastics are important materials in everyday use. They are polymers that can be molded using heat and pressure. However, they present a serious problem of disposal, persistence and environmental pollution, and it would be preferable if biodegradable plastics could be generated from agricultural products, which could be degraded biologically to carbon dioxide and water.[72] Potential sources of biological polymers include starch, cellulose, pectins, proteins and PHB. Various native starches from wheat, rice, maize and potato have been used to make plastics, and a range of shaped products have been manufactured.[73] However, production of PHB or related polyhydroxy alkanoates in plants provide the most attractive sources of biodegradable plastics at low cost. These crops will enhance the value of production and benefit land use and waste disposal.

A combination of microbiology, polymer chemistry and plant biotechnology is likely to provide new crops providing valuable non-food products. For example, there is the prospect of growing 'easy-care' polyester cotton, in which PHB is deposited in the hollow central lumen

[70] M. Stitt and U. Sonnewald, *Ann. Rev. Plant Physiol. Plant Mol. Biol.*, 1995, **46**, 341.
[71] K. Herbers and U. Sonnewald, *TIBTech*, 1996, **14**, 198.
[72] C. Nawrath, Y. Poirier, and C. Somerville, *Molecular Breeding*, 1995, **1**, 105.
[73] J. J. G. Van Soest and J. F. G. Vliegenthart, *TIBTech*, 1997, **15**, 208.

of the cotton fibre. Less than 1% PHBs can improve the insulating property of the cotton fibre by 8–9%.[74]

11 PLANTS AS BIOREACTORS: BIOPHARMING AND NEUTRACEUTICALS

Plant products have long been used in medicine, now plant biotechnology is entering the medical field in a spectacular way, with the creation of plants which produce proteins of pharmaceutical value or subunit vaccines.

11.1 Edible Vaccines

Traditionally, attenuated strains of pathogenic organisms have been injected or delivered orally to invoke an immune response. Now it is possible to clone genes encoding immunogenic subunits of pathogen proteins, and to express these either in transgenic plants or in plant viruses. Transgenic plants have a permanent capacity to express the vaccine, whereas engineered viruses enable transient production of large quantities of immunogenic protein.[75] The aim of this work is that by eating fresh fruit (*e.g.* banana) containing an antigen, individuals can develop immunity to the pathogenic organism *via* the gut immune system. In some cases human trials are in progress to test the efficiency of this approach for the following diseases and applications: hepatitis B; cholera; *E. coli* heat labile enterotoxin; Norwalk virus; rhinovirus; HIV; rabies; antimalarial parasites; immunocontraception (*ZP3* zona pellucida protein); and inhibition of late onset diabetes.[76]

The same approach is being used to engineer plant vaccines for domestic animals and livestock (*e.g.* parvoviruses such as feline panleukopenia virus, canine virus, mink enteritis virus);[77] swine fever, foot and mouth disease and rabies.[78]

[74] C. Bryne, *Chemistry & Industry*, May 1999, 343.
[75] C. J. Arntzen, *Nature Biotechnology*, 1997, 15, 221.
[76] S. W. Ma, D.-L. Zhao, Z.-Q. Yin, R. Mukherjee, B. Singh, H.-Y. Qin, C. R. Stiller, and A. M. Jevnikar, *Nature Medicine*, 1997, 3, 793.
[77] K. Dalsgaard, A. Uttenthal, T. D. Jones, F. Xu, A. Merryweather, W. D. O. Hamilton, J. P. M. Langeveld, R. S. Boshuizen, S. Kamstrup, G. P. Lomonossoff, C. Porta, C. Vela, J. I. Casal, R. H. Meloen, and P. B. Rodgers, 1997, *Nature Biotechnology*, 1997, 15, 248.
[78] Modelska, B. Dietzschold, N. Sleysh, Z. F. Fu, K. Steplewski, D. C. Hooper, H. Koprowski, and V. Yusibov, *Proc. Natl. Acad. Sci. USA*, 1998, 95, 2481.

11.2 Production of Antibodies in Plants

The 'biopharming' approach indicated above can also be used to produce either large quantities of antibodies in plants as bioreactors, or to deliver pharmacologically active antibodies in food. A good example of the latter is the synthesis in plants of antibodies against the bacterium (*Streptococcus mutans*) that causes dental caries in teeth. Extracted antibodies confer protection in human trials.[79] Expressed in an apple, this transgenic apple a day would also keep the dentist away!

11.3 Plant Neutraceuticals

The concept of using plants as bio-factories to produce pharmaceuticals also extends to the potential production of a wide range of other products. These may be in the health care area (*e.g.* dermatology, cardiology, human metabolism, endocrinology, respiration, transplantation and oncology),[80] or production of compounds that improve human diet or health (the formation of a new 'wellness complex' by food). For example, Prodi Gene and Stauffer seeds are growing maize that produces avidin for use in health diagnostic kits,[81] and Brazzein, a low calorie natural sweetener.[80] Other neutraceuticals include overproduction of vitamin A in rice, or phytosterols to reduce cholesterol in humans. Clearly, identity preservation and separate storage/transport and handling will be required for such products, as for most high value transgenic crops.

12 PLANT BIOTECHNOLOGY IN FORESTRY

Forests are both economically and environmentally important. The demand for wood products, such as timber for construction, paper, pulp and energy is increasing, and expanded plantation forestry will be required to meet the global demand. The time lines for tree breeding have made progress in improving forest productivity slow. Now, the full range of biotechnological technologies are being applied to tree improvement. These include clonal micropropagation of superior trees, hybridization, molecular markers and marker-assisted selection, and transformation.

[79] J. K.-C. Ma, B. Y. Hikmal, K. Wycoff, N.-D. Vine, D. Chargeleque, L. Yu, M. B. Hein, and T. Lehmer, *Nature*, 1998, **4**, 601.
[80] J. Olson, *Farm Industry News*, 1999, **32**, 1.
[81] E. E. Hood, D. Witcher, S. Maddock, T. Meyer, C. Baszczynski, M. Bailey, P. Flynn, J. Register, L. Marshall, D. Bond, E. Kulisek, A. Kusnadi, R. Evangelista, Z. Nikolov, C. Wooge, R. J. Mehigh, R. Hernan, W. K. Kappel, D. Ritland, C. P. Li, and J. A. Howard, *Molecular Breeding*, 1997, **3**, 291.

A range of tree species has been transformed using both *Agrobacterium* and particle bombardment approaches:[82] poplar, European larch, hybrid larch, Norway spruce, Scots pine, white spruce, black spruce, eastern larch, radiata pine, Tasmanian bluegum, and a range of Eucalyptus species. The targets for tree improvement involve improved growth rate, wood characteristics, pulp quality, pest and disease resistance and tolerance to abiotic stresses.

A major target is to engineer reduced lignin content, since lignin represents about 25% of the wood biomass, and lignins reduce the efficiency of pulp and paper production. High energy usage and production of chemical pollutants also results from de-lignification processes. This has been achieved by down-regulation of key enzymes in the lignin biosynthetic pathway (*e.g.* cinnamyl alcohol dehydrogenase, CAD; *O*-methyl transferase, OMT).

Transgenic trees have also been generated with modified form, quality and performance using auxin biosynthetic genes, *rol* c genes and peroxidase,[82] and also with herbicide resistance and Bt insect resistance. Environmental considerations are particularly important for growth of transgenic trees. In order to reduce gene flow *via* pollen to native species, engineered sterility of transgenics is desirable.

13 INTELLECTUAL PROPERTY

Because of patenting of novel genes, promoters, generic technology and applications, the commercialization of plant biotechnology can be a complex issue. The costs associated with establishing, contesting, defending and monitoring patents are a major reason why large life sciences companies rather than public organizations (Universities and Government funded research institutes) are exploiting the new technologies. This trend will undoubtedly continue, and the major companies listed in Table 1, such as Monsanto, Dupont-Pioneer and Novartis, will increase their hold on commercialization. Plant breeding is already essentially in the private domain in Europe, and is rapidly moving that way in North America and Australia.

Although intellectual property and patents must now be considered from the start of any scientific work in molecular biology that is intended to be commercialized, and freedom-to-operate from gene to the sale of seeds established, there is another level to consider. That is, the ownership of germplasm. Breeding materials from many crops often originate in centres of origin which are frequently in developing countries, in

---

[82] T. Tzfira, A. Zuker, and A. Altman, *TIBTech*, 1998, **16**, 439.

which germplasm may be held by international research centres. The countries where the germplasm originated may well require some equity or payment for use of their germplasm, and accusations of 'biopiracy' have been made against breeders in developed countries. Greater sensitivity will be needed in this area if access to new or wild germplasm is not to be blocked.

14 PUBLIC ACCEPTANCE

The speed at which genetically modified ('GM') crops and food has arrived has been unexpectedly rapid. In North America, where labelling of GM food or products is not undertaken unless there is a substantial difference from non-GM food, there are more than 300 million people routinely eating GM produce. In Europe, different forces prevail on public opinion, many of which are more connected with other events such as the 'mad cow disease' (BSE) outbreak in the UK, trust of scientists, politics and trade. These will slow acceptance and development of GM foods and biotechnology, and put Europe at a competitive disadvantage compared with North America. Australia lies between Europe and North America in terms of public acceptance of GM crops and food, with transgenic cotton and carnations currently commercialized.

Major questions in acceptance of GM technology arise:

Is GM food safe?
Are there any long-term affects from eating GM food?
Should all GM food be labelled?
Should consumers have the right to choose between GM and non-GM food?
Will they contain new allergens?
What are the environmental consequences?

There is, of course, no logical reason why any organization would wish to produce unsafe food. At present, GM food is subjected to much more stringent testing than conventional foods, and is probably safer as a result. The labelling issue is not as simple as it may seem. It is straightforward to label a transgenic tomato as such but, for example, soy products are used in differing amounts in thousands of processed foods and it is difficult to label these in a meaningful and informative way. Similarly, cotton oil is relatively pure and free of DNA or protein derived from transgenes conferring herbicide tolerance. Does it need to be labelled as GM? Is it substantially equivalent to the non-GM product or not – where should the line be drawn? There are also concerns about 'ethnically' sensitive genes (*e.g.* a gene derived from a pig inserted in a crop plant), and access of developing countries to GM technologies.

The large-scale consumption of GM food in North America by 300 million people since its introduction, without any reported problems, supports the general view based on the science that GM food is safe. Environmental issues should perhaps be of more concern. Pollination of organic crops by transgenic maize is an issue,[83] as is possible injury to Monarch butterfly caterpillars eating food plants dusted with Bt maize pollen. In contrast, there are reports of increased numbers of birds of prey after three years growth of Bt insect-resistant crops, as fewer non-target insects are killed and this effect is transmitted through the food chain.

At present, public debate often has little to do with the science and actual risks, and more to do with sensationalism and disinformation. Banner headlines occur regularly in the press, on 'Demon Seeds', 'Mutant Foods from the Gene Giants', 'Frankenstein foods', *etc.* However, this phase is expected to pass fairly rapidly, as understanding of the benefits of the technologies in efficient food production, improved food quality, protection of the environment and of biodiversity are realized.

15 FUTURE PROSPECTS

Although transgenic technologies will not provide all the answers to generating sustainable food production in the next millenium, the 'gene revolution' will undoubtedly have more impact than the 'green revolution' on all aspects of agriculture and biotechnology. The potential is that 80% of the food eaten in developed countries will have transgenic content by the year 2010, and that up to 10% of the US maize crop will be devoted to bioreactors, biopharming and production of neutraceuticals. Whether or not these percentages will be achieved in practice will depend very much on public attitudes and acceptance of transgenic technologies over the next few years.

[83] R. B. Jorgensen, T. Hanser, T. R. Mikkelsen, and H. Ostergaard, *Trend. Plant Sci.*, 1996, **1**, 356.

CHAPTER 8

Molecular, Structural and Chemical Biology in Pharmaceutical Research

TOMI K. SAWYER

1 INTRODUCTION

Pharmaceutical research has been significantly influenced by the applications of recombinant DNA technology, genomic target identification and validation, protein structure determination (X-ray crystallography and NMR spectroscopy), computational chemistry, synthetic combinatorial libraries, high-throughput screening, informatics, and other drug discovery technologies that integrate molecular, structural and chemical biology. The emergence of several hundred biotechnology companies throughout the world illustrates the scope and diversity of efforts to accelerate and expand the discovery of new drugs, including recombinant proteins, designed peptidomimetics and nonpeptides, and gene therapies. Collaborative networks of biotechnology and large pharmaceutical companies have led to a so-called 'paradigm shift' with respect to the implementation of various 'gene to drug' campaigns to discover and develop new therapeutics.

The human genome has nearly been sequenced, and the predicted number of new targets (estimated at 3000–10 000) provides an exciting opportunity for pharmaceutical research as it makes a transition into the new millennium. Relative to the precedence of recombinant proteins such as human growth hormone, insulin, tissue plasminogen activator, erythropoietin, myeloid colony stimulating factors, and interferons (α, β and γ), the principles of molecular biology have been further exploited to advance new generations of genetically engineered proteins, antibodies, vaccines, antisense oligonucleotides and regulated transcription therapies. Peptidomimetic and nonpeptide drug discovery has also significantly evolved to utilize sophisticated structural biology, drug design and combinatorial chemistry methodologies to facilitate the discovery and optimization of lead compounds. A plethora of examples exist, including HIV protease inhibitors, *Ras* farnesyl transferase inhibitors, fibronectin receptor antagonists, somatostatin receptor agonists, angiotensin receptor antagonists, stromelysin inhibitors, tyrosine kinase inhibitors, and numerous other receptor-, peptidase- and signal transduction-targeted therapeutic agents. Knowledge of chemical biology is manifest in such pharmaceutical research, with particular significance to molecular recognition, biochemical mechanisms, pro-drug design and other concepts that are critical to the development of potent, specific, metabolically stable, bioavailable and *in vivo* efficacious drugs.

2 MOLECULAR BIOLOGY OF DISEASE AND *IN VIVO* TRANSGENIC MODELS

The development of molecular biology methodologies has revolutionized the drug discovery process over the past two decades. The applications of molecular biology to identify drug targets, decipher complex disease mechanisms, and support biochemical screening and structural determination studies are particularly important. Target validation using *in vivo* transgenics involves deletion, addition or mutation of specific genes in animals to produce phenotypes that manifest a particular disease (Table 1). Gene deletions may be used to mimic the effects of a single gene defect or the physiological function of a specific gene product (protein). For example, gene deletion of c-Src, a tyrosine kinase, in mice results in osteopetrosis.[1] Similarly, it has been shown that gene deletion of cathepsin-K, a cysteine protease, in mice results in osteopetrosis.[2] On the other hand, gene deletion of osteoprotegerin in mice results in osteoporosis.[3] As expected from the successful development of inhibitors of angiotensin-converting enzyme (ACE), gene deletion of ACE or the renin substrate, angiotensinogen, in mice results in hypotension.[4,5] Gene deletion of α-inhibin, a TGF-β-related growth factor, in mice results in gonadal stromal tumours, therefore identifying α-inhibin as a critical

[1] P. Soriano, C. Montgomery, R. Geske, and A. Bradley, *Cell*, 1991, **64**, 693.
[2] P. Saftig, E. Hunziker, O. Wehmeyer, S. Jones, A. Boyde, W. Rommerskirch, J.D. Moritz, P.Schu, and K. VonFigura, *Proc. Natl. Acad. Sci. USA*, 1998, **95**, 13453.
[3] N. Bucay, I. Sarosi, C. R. Dunstan, S. Morony, J. Tarpley, C. Capparelli, S. Scully, H. L. Tan, W. Xu, D. L. Lacey, W. J. Boyle, and W. S. Simonet, *Genes Dev.*, 1998, **12**, 1260.
[4] J. H. Krege, S. W. M. John, L. L. Langenbach, J. B. Hodgin, J. R. Hagaman, E. S. Bachman, J. C. Jennette, D. A. O'Brien, and O. Smithies, *Nature (London)*, 1995, **375**, 146.
[5] H. S. Kim, J. H. Krege, K. D. Kluckman, J. R. Hagaman, J. B. Hodgin, C. F. Best, J. C. Jennette, T. M. Coffman, N. Maeda, and O. Smithies, *Proc. Natl. Acad. Sci. USA*, 1995, **92**, 2735.

Table 1   *Transgenic animal models to study the relationship of disease and gene expression*

Specific gene	Gene modification	Disease phenotype
Src tyrosine kinase	Gene deletion	Osteopetrosis
Cathepsin-K	Gene deletion	Osteopetrosis
Nf-κB	Gene deletion	Osteopetrosis; impaired B-cell function
Osteoprotegerin	Gene deletion	Osteoporosis
α-inhibin	Gene deletion	Gonadal stromal tumours
Angiotensin-converting enzyme	Gene deletion	Hypotension
Angiotensinogen	Gene deletion	Hypotension
PTP1B tyrosine phosphatase	Gene deletion	Diabetes; obesity
Zap70 tyrosine kinase	Gene deletion	Severe combined immunodeficiency
Recombinant *bcr-abl*	Gene addition	Leukaemia
Transforming growth factor-α	Gene addition	Psoriasis
HIV-TAT (from HIV-LTR)	Gene addition	Karposi's sarcoma
Interleukin-5	Gene addition	Eosinophilia
Melanocortin-4 receptor	Gene mutation	Obesity syndrome
Leptin	Gene mutation	Obesity syndrome
Leptin receptor	Gene mutation	Obesity syndrome

tumour-suppressing protein.[6] Gene deletion of Zap70, a tyrosine kinase, in mice results in severe combined immunodeficiency due to impaired development of thymocytes.[7] Gene additions may be used to analyse the pathophysiological consequences of inappropriate protein expression. Such studies may utilize constructs carrying proto-oncogenes, growth factors and cell-surface antigens that are under transcriptional control of heterologous promoters.[8,9] For example, recombinant methods used to prepare a *bcr-abl* gene fusion in mice has been shown to mimic the Philadelphia chromosome translocation that results in leukaemia.[10] Finally, gene mutations may give rise to loss of function or, in some cases, constituitive activity. Several independent gene mutations (each effecting loss of function) in mice result in obesity syndrome. These include the melanocortin-4 receptor,[11] leptin,[12] and

[6] M. M. Matsuk, M. J. Finegold, J.-G. Su, A. J. W. Hsueh, and A. Bradley, *Nature (London)*, 1992, **360**, 313.

[7] Q. Gong, L. White, R. Johnson, M. White, I. Negishi, M. Thomas, and A. C. Chan, *Immunity*, 1997, **7**, 369.

[8] R. Jaenisch, *Science*, 1988, **240**, 1468.

[9] G. T. Merlino, *FASEB J.*, 1991, **5**, 2996.

[10] N. Heisterkamp, G. Jenster, J. ten Hoeve, D. Zovich, P. K. Pattengale, and J. Groffen, *Nature (London)*, 1990, **344**, 251.

[11] D. Huszar, C. A. Lynch, V. Fairchild-Huntress, J. H. Dunmore, Q. Fang, L. R. Berkemeier, W. Gu., R. A. Kesterson, B. A. Boston, R. D. Cone, F. J. Smith, L. A. Campfield, P. Burn, and F. Lee, *Cell*, 1997, **88**, 131.

[12] Y. Zhang, R. Proenca, M. Maffei, M. Barone, L. Leopold, and J. M. Friedman, *Nature (London)*, 1994, **372**, 425.

the leptin receptor.[13] A comprehensive compilation of mouse gene knockout and mutation data is accessible on the internet (http://www.biomednet.com/db/mkmd). Relative to the approximately 5000 inherited diseases known in man,[14] including extremely rare cases, such genetic mutations and pathogenesis provide yet significant challenges to drug discovery.

The emerging 'gene to drug' philosophy for pharmaceutical research initially involves two critical ingredients: gene identification and functional genomics.[15] Some of the major molecular biology technologies for gene discovery include: expressed sequence tag (EST) sequencing, secreted protein analysis, differential display, expression profiling and positional cloning.[16] These gene discovery approaches vary significantly in terms of throughput (genes/year) and relevance to disease. Overall, the EST and secreted protein approaches may rank amongst the most likely to identify therapeutic targets. In particular, EST methods facilitate the ability to sift through relatively large numbers of novel gene sequences for subsequent selection of targets on basis of homology to known classes of drug (*e.g.* G-protein coupled receptors, ion channels, secreted protein hormones, kinases, proteases). With respect to functional genomics, the integration of several drug discovery technologies (*e.g.* knockout mice, ribozyme, antisense oligonucleotides and DNA microarray screening) provide powerful approaches to elucidate the gene function and support target validation. As exemplified above, transgenic knockouts refer to germ-line genomic deletion of a target gene,[17] whereas conditional transgenics that use inducible, gene-specific knockout provide an opportunity to eliminate a gene in a tissue rather than the entire organism.[18] Relative to drug discovery, transgenic animals are useful tools for the identification of therapeutic targets, distinction of target isoforms, differentiation of signalling pathways, generation of disease animals and toxicological testing.[19] Ribozymes are RNAs that can catalytically cleave specific target sequences and may be used to functionally knockout genes in both cells and *in vivo*.[20] Antisense oligonucleotides provide the opportunity to reduce message levels for any

---

[13] M.-Y. Wang, K. Koyama, M. Shimabukuro, C. B. Newgard, R. H. Unger, *Proc. Natl. Acad. Sci. USA*, 1998, **95**, 714.

[14] A. Hamosh, A. F. Scott, J. Amberger, D. Valle, and V. A. McKusick, *Human Mutation*, 2000, **15**, 57.

[15] G. C. Kennedy, *Drug Dev. Res.*, 1997, **41**, 112.

[16] S. R. Wiley, *Curr. Pharm. Design*, 1998, **4**, 417.

[17] B. P. Zambrowicz, G. A. Friedrich, E. C. Buxton, S. L. Lilleberg, C. Person, and A. T. Sands, *Nature (London)*, 1998, **392**, 608.

[18] R. Kuhn, F. Schwenk, M. Aguet, and K. Rajewsky, *Science*, 1995, **269**, 1427.

[19] U. Rudolph and H. Mohler, *Curr. Opin. Drug Disc. Dev.*, 1999, **2**, 134.

[20] T. Cech, *Curr. Opin. Struct. Biol.*, 1992, **2**, 605.

gene.[21] Antisense oligonucleotide technologies have also led to potential therapeutics[22] (see above). DNA microarray screening can be used to measure gene expression levels in a high-throughput manner for target identification as well as to monitor changes in gene expression as affected by drug treatment.[23]

3 GENOMIC PROTEIN TARGETS AND RECOMBINANT THERAPEUTICS

The mapping and sequence of the human genome continues to be a major scientific undertaking over the past several years[24] as first hallmarked by the 1988 inauguration of the Human Genome Project (http://nhgri.nih.gov/HGP/). In retrospect, a key objective of the Human Genome Project was to sequence each of the 3 billion nucleotide base pairs in the human genome, and then identify the structure of the approximately 80 000–140 000 genes by 2003. A physical map of > 30 000 human genes, including most genes that encode proteins of known function, is currently available (http://www.ncbi.nlm.nih.gov/ Genbank) as a resource tool for analysis of complex genetic traits, positional cloning of disease genes, cross-referencing of mammalian genomes and validated human transcribed sequences for large-scale studies of gene expression.[25] In parallel with these efforts, a number of academic and industrial groups are advancing high-throughput sequencing of expressed genes and EST databases to explore comparisons between species, discover new gene families involved in human disease, and provide detailed analysis of specific tissues.[26,27] Relative to the fact that past drug discovery has been focused on approximately 400 human therapeutic targets, the predicted forthcoming 3000–10 000 genomic targets will significantly impact pharmaceutical research in the new millennium. Nevertheless, the ultimate task to identify the causative

[21] R. W. Wagner, M. D. Matteucci, J. G. Lewis, A. J. Gutierrez, C. Moulds, and B. C. Froehler, *Science*, 1993, **260**, 1510.

[22] C. Wahlestedt and L. Good, *Curr. Opin. Drug Disc. Dev.*, 1999, **2**, 142.

[23] C. Debouck and P. N. Goodfellow, *Nature (Genetics)*, 1999, **21**, 48.

[24] F. S. Collins, A. Patrinos, E. Jordan, A. Chakravarti, G. Aravinda, R. Gesteland, and L. Walters, *Science*, 1998, **282**, 682.

[25] P. Deloukas, G. D. Schuler, G. Gyapay, E. M. Beasley, C. Soderlund, H. L. Rodriguez-Tome, T. C. Matise, K. B. McKusick, J. S. Beckmann, S. Bentolila, M.-T. Bihoreau, B. B. Birren, J. Browne, A. Butler, A. B. Castle, N. Chiannilkulchai, C. Clee, P. J. R., Day, A. Dehejia, T. Dibling, N. Drouot, S. Duprat, C. Fizames, S. Fox, S. Gelling, L. Green, P. Harrison, R. Hocking, E. Holloway, S. Hunt, S. Kell, P. Lijnzaad, C. Louis-Dit-Sully, J. Ma, A. Mendis, J. Miller, J. Morissette, D. Muselet, H. C. Nusbaum, A. Peck, S. Rozen, D. Simon, D. K. Slonim, R. Staples, L. D. Stein, E. A. Stewart, M. A. Suchard, T. Thangarajah, N. Vega-Czarny, C. Webber, X. Wu, J. Hudson, C. Auffray, N. Nomura, *et al.*, *Science*, 1998, **282**, 744.

[26] P. Spence, *Drug Dis. Today*, 1998, **3**, 179.

[27] J. Drews, *Science*, 2000, **287**, 1960.

genes for complex polygenic diseases (*e.g.* diabetes, asthma, athero-sclerosis, Alzheimer's disease) still poses an extraordinary challenge to genomics-based drug discovery. The so-called 'genomics revolution' has emerged in the industrial arena with numerous strategic alliances between biotechnology and pharmaceutical companies, including some academic institutions (Table 2). The spectrum of such genomics-based collaborations varies from large-scale gene sequencing and analysis to focussed genomic research in one or more specific disease areas, including varying applications of high-throughput screening, combinatorial chemistry, structure-based design, and other drug discovery technologies.[28,29]

Beyond the human genome, the complete sequences of several bacterial, yeast and nematode genomes have been determined (Table 3). These include major milestones such as the first complete genome of a free-living organism, *Haemophilus influenza* Rd,[30] and that of the first eukaryote, the yeast *Saccharomyces cerevisiae*.[31] Overall, the complete genome sequences of many other microorganisms have now been determined, including *Escherichia coli*,[32] *Bacillus subtilis*,[33] *Myobacterium tuberculosis*,[34] *Helicobacter pylori*,[35] *Mycoplasma pneumoniae*,[36] *Mycoplasma genitialium*,[37] *Borrelia burgdorferi*,[38] *Thermotoga maritima*,[39]

[28] L. J. Beeley, D. M. Duckworth, and D. Malcolm, *Drug Disc. Today*, 1996, **1**, 474.
[29] D. F. Veber, F. H. Drake, and M. Gowen, *Curr. Opin. Chem. Biol.*, 1997, **1**, 151.
[30] R. Fleischmann, M. Adams, O. White, R. Clayton, E. Kirkness, A. Kerlavage, C. Bult, J. Tomb, B. Dougherty, J. Merrick, K. McKenney, G. Sutton, W. FitzHugh, C. Fields, J. Gocayne, J. Scott, R. Shirley, L. Liu, A. Glodek, J. Kelley, J. Weidman, C. Phillips, T. Sprigs, E. Hedblom, M. Cotton, T. Utterback, M. Hanna, D. Nguyen, D. Saudek, R. Brandon, L. Fine, J. Fritchman, J. Furhmann, N. Geoghagen, C. Gnehm, L. McDonald, K. Small, C. Fraser, H. O. Smith, and J. C. Venter, *Science*, 1995, **269**, 496.
[31] H. W. Mewes, K. Albermann, M. Bahr, D. Frishman, A. Gleissner, J. Hani, K. Heumann, K. Kleine, A. Maierl, S. G. Oliver, F. Pfeiffer, and A. Zollner, *Nature (London)*, 1997, **387**, 7.
[32] F. R. Blattner, G. Plunkett III, C. A. Bloch, N. T. Perna, V. Burland, M. Riley, J. Collado-Vides, J. D. Glasner, C. K. Rode, G. F. Mayhew, *et al.*, *Science*, 1997, **277**, 1453.
[33] F. Kuntz, N. Ogansawara, I. Mosner, A. M. Albertini, G. Alloni, V. Azevedo, M. G. Bertero, P. Bessieres, A. Bolotin, S. Borchert, *et al.*, *Nature (London)*, 1997, **390**, 249.
[34] S. T. Cole, R. Brosch, J. Parkhill, T. Garnier, C. Churcher, D. Harris, S. V. Gordon, K. Eiglmeier, S. Gas, C. E. Barry III, T. Tekaia, K. Badcock, D. Basham, D. Brown, T. Chillingworth, R. Connor, R. Davies, K. Devlin, T. Feltwell, S. Gentles, N. Hamlin, S. Holroyd, T. Hornsby, K. Jagels, A. Krogh, J. McLean, J. Moule, L. Murphy, K. Oliver, J. Osborn, M. A. Quail, M.-A. Rajandream, J. Rogers, S. Rutter, K. Seeger, J. Skelton, R. Squares, S. Squares, J. E. Sulston, K. Taylor, S. Whitehead, and B. G. Barrell, *Nature (London)*, 1998, **393**, 537.
[35] J.-F. Tomb, O. White, A. R. Kerlavage, R. A. Clayton, G. G. Sutton, R. D. Fleischmann, K. A. Ketchum, H. P. Klenk, S. Gill, B. A. Dougherty, K. Nelson, J. Quackenbush, L. Zhou, E. F. Kirkness, S. Peterson, L. Scott, B. Loftus, D. Richardson, R. Dodson, H. G. Khalak, A. Glodek, K. McKenney, L. M. Fitzgerald, N. Lee, M. D. Adams, E. K. Hickey, D. E. Berg, J. D. Cocayne, T. R. Utterback, J. D. Peterson, J. M. Kelley, M. D. Cotton, J. M. Weldman, C. Fujii, C. Bowman, L. Watthey, E. Wallin, W. S. Hayes, M. Borodovsky, P. D. Karp, H. O. Smith, C. M. Fraser, and J. C. Venter, *Nature (London)*, 1997, **389**, 412.
[36] R. Himmelreich, H. Hilbert, H. Plagens, E. Pirkl, B.-C. Li, R. Herrmann, *Nucleic Acids Res.*, 1996, **24**, 4420.

Table 2   *Some examples of strategic alliances in genomics-based drug discovery*

Genomics partner	Pharmaceutical company	Major focus of research
Human Genome Sciences	SmithKline Beecham	Human genome sequencing
Human Genome Sciences	Schering-Plough	Human genome sequencing
Human Genome Sciences	Takeda Chemical Industries	Human genome sequencing
Human Genome Sciences	Synthelabo	Human genome sequencing
Human Genome Sciences	Merck KGaA	Human genome sequencing
Incyte Pharmaceuticals	Pfizer	Human genome sequencing
Incyte Pharmaceuticals	Pharmacia & Upjohn	Human genome sequencing
Incyte Pharmaceuticals	Hoechst Marion Roussel	Human genome sequencing
Incyte Pharmaceuticals	Abbott	Human genome sequencing
Incyte Pharmaceuticals	Johnson & Johnson	Human genome sequencing
Incyte Pharmaceuticals	Roche	Human genome sequencing
Incyte Pharmaceuticals	Zeneca	Human genome sequencing
Incyte Pharmaceuticals	BASF	Human genome sequencing
ARIAD Pharmaceuticals	Hoechst Marion Roussel	Genomics and specific diseases (osteoporosis, cardiovascular, cancer)
Darwin Molecular Corporation	Rhone-Poulenc Rorer	Genomics and specific diseases (cancer)
Millennium Pharmaceuticals	Bayer	Genomics and specific diseases (cardiovascular diseases, cancer, osteoporosis, pain, viral infections, liver fibrosis, haematology)
Millennium Pharmaceuticals	Eli Lilly	Genomics and specific diseases (atherosclerosis, oncology)
Millennium Pharmaceuticals	Roche	Genomics and specific diseases (obesity, diabetes)
Millennium Pharmaceuticals	Astra-Zeneca	Genomics and specific diseases (inflammatory respiratory disorders)
Millennium Pharmaceuticals	American Home Products	Genomics and specific diseases (CNS diseases)
Myriad Genetics	Novartis	Genomics and specific diseases (CNS diseases)
Sequana Therapeutics	GlaxoWellcome	Genomics and specific diseases (diabetes)
Sequana Therapeutics	Boehringer Ingelheim	Genomics and specific diseases (asthma)

Table 3 *Some examples of microbial and animal genomic sequence determinations*

Organism	Class	Number of genes	Genome size (basepairs)
Mycoplasma genitalium	Bacteria	470	580 070
Mycoplasma pneumoniae	Bacteria	706	816 394
Rickettsia prowazekii	Bacteria	834	1 111 523
Thermotoga maritima	Bacteria	1014	1 860 725
Helicobacter pylori	Bacteria	1590	1 667 867
Haemophilus influenza Rd	Bacteria	1746	1 830 140
Mycobacterium tuberculosis	Bacteria	3974	4 411 529
Bacillus subtilis	Bacteria	4221	4 214 814
Escherichia coli K-12	Bacteria	4668	4 639 221
Saccharomyces cerevisiae	Yeast	6526	12 147 823
Caenorhabditis elegans	Nematode	19 099	~97 000 000

and *Rickettsia prowazekii*.[40] Such microbial genomic information is hoped to be translated into the discovery of new targets, especially those that may not give rise to drug resistance in contrast to that which has emerged amongst many existing antimicrobial agents.[41] Also, microbial genomics may provide a means to predict the spectrum and selectivity of yet novel drugs.[42] The recent achievement in sequencing the complete genome of an animal, namely the nematode, *Caenorhabditis elegans*, has revealed more than 19 000 protein-coding genes, including a significant number of seven transmembrane receptors, protein tyrosine and serine/threonine kinases, zinc fingers, RNA recognition motifs, protein tyrosine phosphatases, ion channels, and other intracellular signal transduction and gene-related proteins.[43] Finally, the National Institute of Health has recently launched a

[37] C. M. Fraser, J. D., Gocayne, O. White, M. D. Adams, R. A. Clayton, R. D. Fleishmann, C. J. Bult, A. R. Kerlavage, G. Sutton, J. M. Kelley, *et al.*, *Science*, 1995, **270**, 397.

[38] C. M. Fraser, S. Casjens, W. M. Huang, G. G. Stutton, R. Clayton, R. Lathigra, O. White, K. A. Ketchum, R. Dodson, E. K. Hickey, *et al.*, *Nature (London)*, 1997, **390**, 580.

[39] K. E. Nelson, R. Clayton, S. R. Gill, M. L. Gwinn, R. J. Dodson, D. H. Haft, E. K. Hickey, J. D. Peterson, W. C. Nelson, K. A. Keechum, L. McDonald, T. R. Utterback, J. A. Malek, K. D. Linher, M. M. Garrett, A. M. Stewart, M. D. Cotton, M. S. Pratt, C. A. Phillips, D. Richardson, J. Heidelberg, G. G. Sutton, R. D. Fleishmann, J. A. Eisen, O. White, S. L. Salzberg, H. O. Smith, J. C. Venter, and C. M. Fraser, *Nature (London)*, 1999, **399**, 323.

[40] S. G. E. Andersson, A. Zomorodipour, J. O. Andersson, T. Sicheritz-Ponten, U. C. M. Alsmark, R. M. Podowski, A. K. Naslund, A.-S. Eriksson, H. H. Winkler, and C. G. Kurland, *Nature (London)*, 1998, **396**, 133.

[41] S. B. Levy, *Sci. Amer.*, 1996, **278**, 46.

[42] M. B. Schmid, *Curr. Opin. Chem. Biol.*, 1998, **2**, 529.

[43] *C. Elegans* Sequencing Consortium (for list of authors, see: genome.wustl.edu/gsc/C_elegans/ and www.sanger.ac.uk/Projects/C_elegans/), *Science*, 1999, **282**, 2012.

major research effort to decipher the mouse genome (*i.e.* Mouse Genome Sequencing Network) that is anticipated to be completed by 2005.

Several human genomic targets and recombinant protein therapeutics have become the focus of major drug discovery efforts. Examples include cathepsin-K, leptin, osteoprotegerin, MPIF-1, KGF-2, and BLyS (Table 4).[44] The discovery of cathepsin-K as an apparent osteoclast-specific protease was facilitated by the analysis of a human osteoclast cDNA library using EST methodologies.[45] Inhibitors of cathepsin-K may provide novel drugs for the treatment of osteoporosis. The discovery of leptin, a secreted protein, was achieved by positional cloning of a mutated gene that causes severe obesity syndrome in the *ob/ob* mouse.[46] Administration of leptin to obese mice results in weight loss. Recombinant leptin or designed leptin mimics may provide novel drugs for the treatment of obesity. A combination of the EST approach and secreted protein analysis has recently led to the discovery of the novel secreted protein osteoprotegerin.[47] Recombinant osteoprotegerin increases bone density *in vivo* in animal models of osteoporosis. Recombinant osteoprotegerin or designed osteoprotegerin mimics may provide novel drugs for the treatment of osteoporosis. The use of EST methods have also led to the discoveries of three secreted proteins, myeloid progenitor inhibitory factor (MPIF-1), keratinocyte growth factor 2 (KGF-2), and B lymphocyte stimulator (BLyS). Specifically, MPIF-1 is a chemokine that regulates the proliferation of bone marrow stem cells, and recombinant MPIF-1 may provide a novel therapeutic to protect the bone marrow of cancer patients from the toxic effects of chemotherapy.[48] The cellular selectivity properties of KGF-2 to promote the growth of keratinocytes versus fibroblasts provides promise for the application of recombinant KGF-2 for wound healing.[49] In the case of Blys, a member of the tumour necrosis factor family, this secreted protein is a specific B-cell stimulant

[44] S. J. Rhodes and R. C. Smith, *Drug Disc. Today*, 1998, **3**, 361.
[45] F. H. Drake, R. A. Dodds, I. E. James, C. Debouck, S. Richardson, E. Lee-Rykaeczewski, L. Coleman, D. Rieman, R. Barthlow, G. Hastings, and M. Gowen, *J. Biol. Chem.*, 1996, **271**, 12511.
[46] Y. Zhang, R. Proenca, M. Maffei, M. Barone, L. Leopold, and J. Friedman, *Nature (London)*, 1994, **372**, 425.
[47] W. S. Simonet, D. L. Lacey, C. R. Dunstan, M. Kelley, M.-S.Chang, R. Luthy, H. C. Nguyen, S. Wooden, L. Bennet, T. Boone, G. Shimamoto, M. DeRose, R. Elliott, A. Colombero, H.-L. Tan, G. Trail, J. Sullivan, E. Davy, N. Bucay, L. Benshaw-Gegg, T. M. Hughes, D. Hill, W. Pattison, P. Campbell, S. Sander, G. Van, J. Tarpley, P. Derby, R. Lee, and W. J. Boyle, *Cell*, 1997, **89**, 309.
[48] B. Nardelli, H. L. Tiffany, G. W. Bong, P. A. Yourey, D. K. Morahan, Y. Li, P. M. Murphy, and R. F. Anderson, *J. Immunol.*, 1999, **162**, 435.
[49] P. M. Soler, T. E. Wright, P. D. Smith, S. P. Maggi, D. P. Hill, P. A. Jimenez, and M. C. Robson, *Wound Repair Regen.*, 1999, **7**, 172.

Table 4   *Some examples of genomic targets or recombinant protein therapeutics*

Protein target or therapeutic	Biotechnology and/or pharmaceutical company	Disease application
Cathepsin-K	SmithKline Beecham	Osteoporosis
Leptin	Amgen	Obesity
Osteoprotegerin (OPG)	Amgen	Osteoporosis
Myeloid progenitor inhibitory factor-1 (MPIF-1)	Human Genome Sciences	Cancer chemotherapy
Keratinocye growth factor-2 (KGF-2)	Human Genome Sciences	Wound healing; mucositis
B lymphocyte simulator (BLyS)	Human Genome Sciences	Vaccine adjuvant; leukaemia, lymphoma
Osteogenic protein-1 (OP-1/BMP-7)	Creative Biomolecules/ Stryker	Bone fractions
Nerve growth factor (NGF)	Genentech/ CytoTherapeutics	Peripheral neuropathies
Brain-derived neurotrophic factor (BDNF)	Amgen/Regeneron	Amyotrophic lateral sclerosis
Glial cell line-derived neutrophic factor (GDNF)	Amgen	Parkinson's disease
Erythropoietin (EPO)	Amgen	Chronic renal failure
Granulocyte colony stimulating factor (GCSF)	Amgen	Chemotherapy (bone marrow transplant-induced neutropenia)
Interferon-α	Biogen	Hepatitis B, hepatitis C
Interferon-$\beta 1\alpha$	Biogen	Multiple sclerosis
Tissue-plasminogen activator (TPA)	Genentech	Clot lysis (heart attack, ischaemic stroke, pulmonary embolism)
Growth hormone	Genentech	Hypopituitary dwarfism
Insulin	Lilly	Diabetes
CD20 monoclonal antibody (chimeric IgG)	IDEC Pharmaceuticals/ Genentech	Non-Hodgkin's lymphoma
p185^{HERB2} monoclonal antibody (humanized IgG)	Genentech/Roche	Breast cancer
CD33 monoclonal antibody (humanized IgG conjugate)	Celltech/American Home Products	Acute myeloid leukemia
IgE monoclonal antibody (humanized IgG)	Genentech	Allergic asthma, allergic rhinitis
CD4 monoclonal antibody (primatized IgG)	SmithKline Beecham/ IDEC Pharmaceuticals	Rheumatoid arthritis
CMV monoclonal antibody (human IgG)	Novartis/Protein Design Labs	CMV infection

in terms of growth and antibody production.[50] Used as a vaccine adjuvant, BLyS may augment the effectiveness of such agents by strengthening the immune response *via* stimulated B cell production, whereas inhibition of BLyS as a therapeutic target may provide a novel drug for leukemia and lymphomas that arise from abnormal proliferation of B cells. In retrospect, it is noted that the development of such new recombinant protein therapeutics has significant precedence by several marketed drugs including growth hormone, insulin, tissue-plasminogen activator, erythropoietin, granulocyte colony stimulating factor, and interferons-α and β (Table 4). Finally, recombinant antibodies have also been successfully developed for the diagnosis and treatment of human disease.[51] Noteworthy for this field of pharmaceutical research, numerous examples of monoclonal antibodies have been developed that exhibit specificity to a target antigen and relatively low immunogenecity.

4 STRUCTURAL BIOLOGY AND RATIONAL DRUG DESIGN

The contribution of molecular biology to the determination of three-dimensional structure of a therapeutic target, including key catalytic or non-catalytic domains, is also well recognized with respect to structure-based drug design. And, in cases where both X-ray crystallography and NMR spectroscopy have not been successful, the application of molecular biology to refine computer-designed models of ligand-target complexes, by site-directed mutagensis of predicted binding or catalytic residues has supported drug discovery efforts. Within the scope of known therapeutic targets (for a non-comprehensive listing, see Table 5), significant progress has been made to integrate structural biology and drug design technologies.

A significant impact of structure-based drug design has emerged over recent years in the discovery of novel peptidomimetics and nonpeptides.[52] Such research has been catalyzed by X-ray crystallography and NMR spectroscopy[53–56] as well as computational chemistry methodolo-

[50] P. A. Moore, O. Belvedere, A. Orr, K. Piere, D. W. LaFleur, P. Feng, D. Soppert, M. Charters, R. L. Gentz, D. Parmelee, Y. Li, O. Galperina, J. G. Giri, V. Roschke, B. Nardelli, J. Carrell, S. Sosnovtseva, W. Greenfield, S. M. Ruben, H. S. Olsen, J. Fikes, and D. M. Hilbert, *Science*, 1999, **285**, 260.

[51] D. J. King and J. R. Adair, *Curr. Opin. Drug Disc. Dev.*, 1999, **2**, 110.

[52] T. K. Sawyer, in 'Structure-Based Drug Design: Diseases, Targets, Techniques and Developments', ed. V. Veerapandian, Marcel Dekker, New York, 1997, p. 559.

[53] P. Veerapandian, 'Structure-Based Drug Design: Diseases, Targets, Techniques and Developments', Marcel Dekker, New York, 1997.

[54] J. Greer, J. W. Erickson, J. J. Baldwin, and M. D. Varney, *J. Med. Chem.*, 1994, **37**, 1035.

[55] P. M. Colman, *Curr. Opin. Struct. Biol.*, 1994, **4**, 868.

[56] C. L. M. J. Verlinde and W. G. J. Hol, *Structure*, 1994, **2**, 577.

Table 5   *Some examples of receptor, signal transduction, and protease targets for drug discovery*

G protein-coupled receptors	*Receptor/non-receptor kinases*	*Transcription factors*
Angiotensin (AT$_1$, AT$_2$)	*Receptor tyrosine kinases*	NF-κB
Bradykinin B$_1$, B$_2$)	Epidermal growth factor	STAT
Cholecystokinin (CCK$_A$)	Fibroblast growth factor	NFAT
Gastrin (CCK$_B$)	Insulin	SMAD
Endothelin (ET$_A$, ET$_B$,	Nerve growth factor	CREB
α-Melanotropin (MCR1)	Platelet-derived growth factor	
Adrenocorticotropin (MCR2)	*Non-receptor tyrosine kinases*	*Transferases*
Substance P (NK1)	Src and Src-family (Lck, Hck)	Farnesyl transferase
Neurokinin-A (NK2)	Abl	Geranyl-geranyl transferase
Neurokinin-B (NK3)	Syk	
δ-opioid (Enkephalin)	Zap70	*Proline cis-trans isomerases*
μ-opioid (Endorphin)	*Receptor serine/threonine kinases*	Cyclophilin
κ-opioid (Dynorphin)	Transforming growth factor	FKBP-12
Oxytocin	*Non-receptor serine/threonine*	
Somatostatin (sst$_1$–sst$_5$)	*kinases*	*Lipases*
Vasopressin (V$_{1A_{1A}}$, V$_{1B}$)	cAMP-dependent protein kinase	Phospholipase-C
Neuropeptide-Y (Y$_1$–Y$_5$)	Phosphoinositol-3-kinase (P13K)	Phospholipase-A$_2$
Calcitonin	Cyclin-dependent kinases (CDKs)	
Adenosine (A$_1$–A$_3$)	Mitogen-activated protein kinase	*Aspartic proteases*
Cathecholamine (α_1, α_2, β_1–β_3)	Protein kinase C (PKC)	Pepsin
Histamine (H$_1$, H$_2$)	Janus family kinases (JAKs)	Renin
Muscarinic acetylcholine	IκB family kinases (IKKs)	Protein kinase C (PKC)
Seratonin (5HT$_1$–5HT$_7$)		Cathepsins (D,E)
Melatonin (ML$_{1A}$, ML$_{1B}$)	*Receptor/non-receptor phosphatases*	HIV-1 protease
Dopamine (D$_1$, D$_2$, D$_4$, D$_5$)	*Receptor tyrosine phosphatases*	
γ-Amino butyric acid (GABA$_B$)	CD45	*Serinyl proteases*
Leukotrienes (LTB$_4$, LTC$_4$, LTD$_4$)	LAR	Trypsin
	Non-receptor tyrosine	Thrombin
Cytokine/lymphokine receptors	*phosphatases*	Chymotrypsin-A
Interleukins (IL-1α, IL-1β, IL-2)	PTP1B	Kallikrein
Erythropoietin	Syp	Elastase
Growth hormone	*Non-receptor serine/threonine*	Tissue plasminogen activator
Prolactin	*phosphatases*	Factor Xa
Interferons (INF-α, β, and γ)	PP-1	
Tumour necrosis factor (many	Calcineurin	*Cysteinyl proteases*
subtypes)	VH1	Cathepsins (B,H,K,M,S,T)
Granulocyte colony stimulating		Proline endopeptidase
factor	*Adapter proteins*	Interleukin-converting enzyme
	Grb2	Apopain (CPP-32)
Cell adhesion integrin receptors	Crk	Picornavirus C3 protease
αvβ3 (Fibrinogen)	IRS-1	Calpains
αIIbβ3 or gpIIaIIIb (Fibrinogen)	Shc	
α5β1 (Fibronectin)		*Metallo proteases*
α4β1 (VCAM-1)[Oestrogen	*Nuclear hormone receptors*	*Exopeptidase group*
	Oestrogen	Peptidyl dipeptidase-A (ACE)
Ion channel receptors	Aldosterone	Aminopeptidase-M
Glutamate (NMDA, AMPA)	Cortisol	Carboxypeptidase-A
γ-Amino butyric acid (GABA$_A$)	Cortisone	*Endopeptidase group*
Nicotinic acetylcholine	Retinoic acid	Endopeptidases (24.11, 24.15)
	Vitamin D	Stromelysin
		Gelatinases (A,B)
		Collagenase

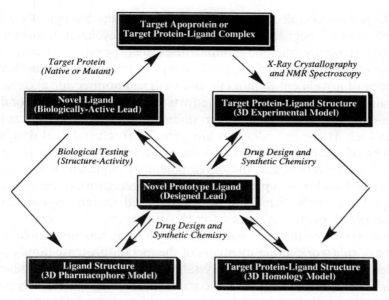

Figure 1   *Typical iterative structure–design–testing cycle used in drug discovery*

gies[56–58] (*e.g.* QSAR, homology modelling, ligand docking, molecular dynamics and mechanics, solvent-accessible surface visualization and analysis, two- and three-dimensional database searching, and *de novo* drug design programmes). Overall, the iterative cycle of structure–design–testing (Figure 1) has evolved to provide an 'engine of invention' for the discovery of many examples of peptidomimetic and *de novo* designed nonpeptide lead compounds (see above).

Relative to the superfamily of more than 1000 cloned G protein-coupled receptors, including genomics-derived orphan receptors, there exists substantial sequence, mutation and ligand binding information[59] (see http://www.gpcr.org/7tm and http://www/gcdb.uthscsa.edu). A majority of G protein-coupled receptors share sequence homology with rhodopsin, and a 6 Å resolution structure of rhodopsin has been determined for it using electron cryomicroscopy.[60] Homology modelling of a number of G protein-coupled receptors have been developed using rhodopsin or bacteriorhodopsin structures as templates.[61] In the case of

[57] D. C. Roe and I. D. Kuntz, *Pharm. News*, 1995, **2**, 13.
[58] Ajay and M. A. Murko, *J. Med. Chem.*, 1995, **38**, 4973.
[59] F. D. King and S. Wilson, *Curr. Opin. Drug Disc. Dev.*, 1999, **2**, 83.
[60] V. M. Unger, P. A. Hargrave, J. M. Baldwin, and G. F. X. Schertier, *Nature (London)*, 1997, **389**, 203.
[61] J. A. Bikker, S. Trump-Kallmeyer, and C. Humblet, *J. Med. Chem.*, 1998, **41**, 2911.

many G protein-coupled receptors, knowledge of the biologically active conformation of cognate peptide ligands using cyclization strategies, secondary structure mimetic modifications, incorporation of unnatural amino acids or functionalization of nonpeptide templates has resulted in a plethora of novel peptidomimetic and designed nonpeptide agonists or antagonists (Figure 2).[62,63] Such efforts are transcending beyond G protein-coupled receptors to other therapeutic target receptor super-families such as the cell adhesion integrins.[64] Structure-based design of inhibitors of each of the major classes of proteases has been achieved (Figure 2).[52] In retrospect, the development of angiotensin-converting enzyme (ACE) inhibitors provided significant precedence to the potential of using both mechanism- and substrate-based design approaches to advance novel peptidomimetics for the treatment of hypertension.[65] Although the three-dimensional structure of ACE was not available at the time of this work, other members of the metalloprotease family have provided high resolution X-ray crystallographic data of both the apo-protein (free) and inhibitor-bound complexes thereof. Particularly note-worthy has been the structure-based design of both peptidomimetic and nonpeptide inhibitors of HIV-1 protease,[66-69] including several promis-ing drugs for the treatment of AIDS. Finally, an increasing number of intracellular proteins important to signal transduction pathways are providing new therapeutic targets for structure-based design of novel drugs (Figure 2).[52] These include numerous kinases,[70-73] phosphatases,[74]

[62] V. J. Hruby, F. Al-Obeidi, and W. Kazmierski, *Biochem. J.*, 1990, **268**, 249.

[63] T. K. Sawyer, in 'Peptide and Protein Drug Analysis' ed. R. Reid, Marcel Dekker, Inc., New York, 1999, p. 81.

[64] J. A. Zablocki, S. N. Rao, D. A. Baron, D. L. Flynn, N. S. Nicholson, and L. P. Feigen, *Curr. Pharm. Des.*, 1996, **1**, 533.

[65] M. J. Wyvratt and A. A. Patchet, *Med. Res. Rev.*, 1985, **5**, 483.

[66] D. J. Kempf, and H. L. Sham, *Curr. Pharm. Des.*, 1996, **2**, 225.

[67] N. A. Roberts, J. A. Martin, D. Kinchington, A. V. Broadhukrst, J. C. Craig, I. B. Duncan, S. A. Galpin, B. K. Handa, J. Kay, A. Krohn, R. W. Lambert, J. H. Merret, J. S. Mills, K. E. B. Parkes, S. Redshaw, A. J. Ritchie, D. L. Taylor, G. J. Thomas, and P. J. Machin, *Science*, 1990, **248**, 358.

[68] P. Y. S. Lam, P. K. Jadhav, C. J. Eyermann, C. N. Hodge, Y. Ru, L. T. Bacheler, J. L. Meek, M. J. Otto, M. M. Rayner, Y. N. Wong, C.-H. Chang, P. C. Weber, D. A. Jackson, T. R. Sharpe, and S. Erickson-Viitanen, *Science*, 1994, **263**, 380.

[69] J. V. N. Vara Prasad, K. S. Para, E. A. Lunney, D. F. Ortwine, J. B. Dunbar, Jr., D. Ferguson, P. J. Tummino, D. Hupe, B. D. Tait, J. M. Domagala, C. Humblet, T. N. Bhat, T. N. Liu, D. M. A. Buerin, E. T. Baldwin, J. W. Erickson, and T. K. Sawyer, *J. Am. Chem. Soc.*, 1994, **116**, 6989.

[70] G. McMahon, L. Sun, C. Liang, and C. Tang, *Curr. Opin. Drug Disc. Dev.*, 1998, **1**, 131.

[71] J. L. Adams, and D. Lee, *Curr. Opin. Drug Disc. Dev.*, 1999, **2**, 96.

[72] M. Mohammadi, G. McMahon, L. Sun, C. Tang, P. Hirth, B. K. Yeh, S. R. Hubbord, and J. Schlessinger, *Science*, 1997, **276**, 955.

[73] D. C. Dalgarno, C. A. Metcalf, W. C. Shakespeare, and T. K. Sawyer, *Curr. Opin. Drug Disc. Dev.*, 2000, **3**, 549.

[74] T. R. Burke and Z.-Y. Zhang, *Biopolymers (Peptide Science)*, 1998, **47**, 225.

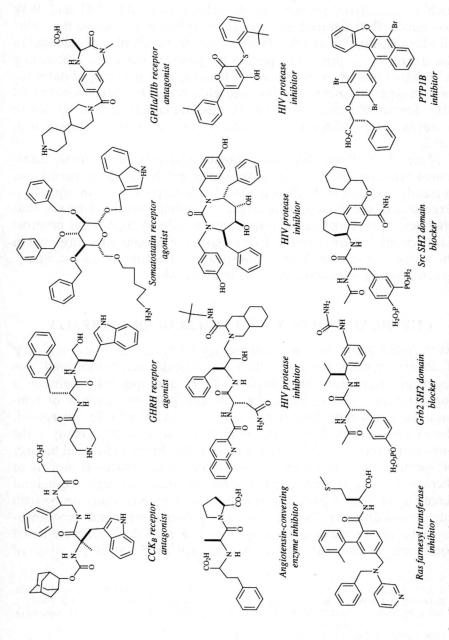

Figure 2   *Some examples of peptidomimetic and nonpeptide drug discovery using structure-based design*

transferases,[75,76] and non-catalytic proteins or, in many cases, binding domains of the catalytic signal transduction target proteins. Examples of such non-catalytic, protein motifs include SH3, SH2, PH and WW domains.[77,78] Structure-based design of tyrosine and serine/threonine kinase inhibitors[70,71] and Ras farnesyl transferase inhibitors[76] have led to the discovery of promising peptidomimetic or nonpeptide (including ATP-based inhibitors in the case of kinases) therapeutic candidates for various types of cancers. Similar research exists for the discovery of novel SH2 domain blockers[79] aimed at therapeutic targets such as related to osteoporosis (*e.g.* Src), immune diseases (*e.g.* Zap70) and cancer (*e.g.* Grb2).

More than 10 000 three-dimensional structures of proteins, determined primarily by X-ray crystallography and NMR spectroscopy, are currently available at the Protein Data Bank (http://www.rcsb.org/pdb/). Examples of some therapeutic targets, including ligand complexes, that are relevant to drug discovery are given in Table 6. Significant progress has been made relative to the determination of enzyme structures within the scope of protease, kinase, phosphatase, phosphodiesterase, lipase, gyrase, cyclase, isomerase and transferase families.

5 CHEMICAL BIOLOGY AND MOLECULAR DIVERSITY

At the interface of chemistry and biology exists the very essence of drug discovery, namely, the unravelling of molecular recognition and biochemical mechanisms that impart affinity and specificity between a cognate ligand and its target (*e.g.* receptor, protease, catalytic or non-catalytic signal transduction protein, DNA and RNA). In this regard, there co-exists exquisite fidelity and phenomenal diversity in many of the biological processes which are essential to life. From a chemical biology perspective, the integration of three-dimensional structural models of therapeutic targets with combinatorial libraries and high-throughput screening of ligands provides an approach to systematically probe such fidelity and diversity. Even in cases where three-dimensional experimental data of the therapeutic targets is not available, structure-based combinatorial chemistry approaches are still quite feasible by virtue of

[75] S. M. Sebti and A. D. Hamilton, *Drug Disc. Today*, 1998, **3**, 26.
[76] S. J. O'Connor, K. J. Barr, L. Wang, B. K. Sorensen, A. S. Tasker, H. Sham, S.-C. Ng, J. Cohen, E. Devine, S. Cherian, B. Saeed, H. Zhang, J. Y. Lee, R. Warner, S. Tahir, P. Kovar, P. Ewing, J. Alder, M. Mitten, J. Leal, K. Marsh, J. Bauch, D. J. Hoffman, S. M. Sebti, and S. H. Rosenberg, *J. Med. Chem.*, 1999, **42**, 3701.
[77] G. B. Cohen, R. Ren, and D. Baltimore, *Cell*, 1995, **80**, 237.
[78] T. Pawson, *Nature*, 1994, **373**, 573.
[79] T. K. Sawyer, *Biopolymers (Peptide Science)*, 1998, **47**, 243.

Table 6 *Some examples of structural determinations of therapeutic targets or related proteins*[52]

Protein or domain	Structure (resolution)	Method
Rhodopsin receptor	Apoprotein (6 Å)	Electron cryomicroscopy
Bacterial rhodopsin receptor	Apoprotein (2.5 Å)	Electron cryomicroscopy
Src tyrosine kinase (SH3-SH2-kinase)	Apoprotein (1.7 Å)	X-ray crystallography
Src SH3 domain	SH3–peptide complex	NMR spectroscopy
Src SH2 domain	SH2–phosphopeptide complex (2.7 Å)	X-ray crystallography
Insulin kinase domain	Kinase–ATP analogue–peptide complex (1.9 Å)	X-ray crystallography
cAMP-dependent protein kinase	Kinase–peptide inhibitor complex (2.9 Å)	X-ray crystallography
PTP1B phosphatase domain (catalytically inactive C215S mutant)	Phosphatase–peptide substrate complex (2.6 Å)	X-ray crystallography
FKBP-12 (proline *cis-trans* isomerase)	FKBP-12–FK-506 complex (1.7 Å)	X-ray crystallography
Endothiapepsin	Protease–peptidomimetic inhibitor complex (1.6 Å)	X-ray crystallography
Human renin	Protease–peptidomimetic inhibitor complex (2.4 Å)	X-ray crystallography
HIV protease	Protease–peptidomimetic inhibitor complex (2.3 Å)	X-ray crystallography
Trypsin	Protease–peptidomimetic inhibitor complex (1.8 Å)	X-ray crystallography
Thrombin	Protease–peptidomimetic inhibitor complex (1.9 Å)	X-ray crystallography
Elastase	Protease–peptidomimetic inhibitor complex (1.8 Å	X-ray crystallography
Interleukin-converting enzyme	Protease–peptidomimetic inhibitor complex (2.5 Å)	X-ray crystallography
Cathepsin-K	Protease–peptidomimetic inhibitor complex (2.2 Å)	X-ray crystallography
Thermolysin	Protease–peptidomimetic inhibitor complex (1.7 Å)	X-ray crystallography
Carboxypeptidase-A	Protease–peptidomimetic inhibitor complex (2.0 Å)	X-ray crystallography
Collagenase	Protease–peptidomimetic inhibitor complex (2.1 Å)	X-ray crystallography
Stromelysin	Protease–peptidomimetic inhibitor complex (1.9 Å)	X-ray crystallography

having adequate information on the ligand (*e.g.* three-dimensional pharmacophore model) to identify possible 'privileged' scaffolds or templates for drug discovery. Recently, noteworthy progress has been achieved for such three-dimensional structure-exploiting design approaches including, to a varying extent, certain molecular dimension-

alities (*e.g.* charge, H-bonding donor or acceptor groups, size, lipophili-
city and related physicochemical properties) that are believed to be
important for 'drug-like' compounds.[80–84] Some examples include novel
combinatorial library-derived peptidomimetic or nonpeptidic lead compounds for receptors[85–87] and proteases[88–90] (Figure 3).

Exploitation of chemical biology principles has impacted the investigation of tyrosine kinase signal transduction pathways[91] as well as the
development of regulated gene therapy systems based on the dimerization properties of the natural product rapamycin.[92] Sophisticated structure-based design strategies involving both mutated protein targets and
complementary modified ligands (Figure 4) have provided new insight
towards harnessing molecular recognition and biochemical specificity in
drug discovery.[93] In the first case, the structural degeneracy of kinase
active sites provided an opportunity to design a highly specific and
potent inhibitor of Src kinase (top panel, Figure 4) that was also effective
as predicted only in cells expressing the rationally engineered, mutant
I338G *versus* wild-type Src kinase.[91] Of further impact to such research
has been the use of combinatorial chemistry to generate and optimize
chemically novel ATP-related inhibitors of kinases.[94] In the second case,

[80] J. Antel, *Curr. Opin. Drug Disc. Dev.*, 1999, **2**, 224.
[81] F. R. Salemme, J. Spurlino, and R. Bone, *Structure*, 1997, **5**, 319.
[82] E. J. Martin, D. C. Spellmeyer, R. E. Critchlow, and J. M. Blaney, *Rev. Comp. Chem.*, 1997, **10**, 75.
[83] A. Ajay, W. P. Walters, and M. A. Murko, *J. Med. Chem.*, 1998, **41**, 3314.
[84] B. A. Bunin, J. M. Dener, and D. A. Livingson, *Ann. Rep. Med. Chem.*, 1999, **34**, 267.
[85] R. N. Zuckermann, E. J. Martin, D. C. Spellmeyer, G. B. Stauber, K. R. Shoemaker, J. M. Kerr, G. M. Figliozzi, D. A. Goff, M. A. Siani, R. J. Simon, S. C. Banville, E. G. Brown, L. Wang, L. S. Richter, and W. H. Moos, *J. Med. Chem.*, 1994, **37**, 2678.
[86] S. P. Roher, E. T. Birzin, R. T. Mosley, S. C. Berk, S. M. Hutchins, D.-M. Shen, Y. Xiong, E. C. Hayes, R. M. Parmar, F. Foor, S. W. Mitra, S. J. Degrado, M. Shu, J. M. Klopp, S.-J. Cai, A. Blake, W. W. S. Chan, A. Pasternak, L. Yang, A. A. Patchett, R. G. Smith, K. T. Chapman, and J.M . Schaeffer, *Science*, 1998, **282**, 737.
[87] W. J. Hoekstra, B. E. Marynoff, P. Andrade-Gordon, J. H. Cohen, M. J. Costanzo, B. P. Damiano, B. J. Haerlein, B. D. Harris, J. A. Kauffman, P. M. Keane, D. M. McComsey, F. J. Villani, Jr., and S. C. Yabut, *Bioorg. Med. Chem. Lett.*, 1996, **6**, 2371.
[88] E. K. Kick, D. C. Roe, A. G. Skillman, G. Liu, T. J. A. Ewing, Y. Sun, I. D. Kuntz, and J. A. Ellman, *Chem. Biol.*, 1997, **4**, 297.
[89] C. Illig, S. Eisennagel, R. Bone, A. Radzicka, L. Murphy, T. Randle, J. Spurline, E. Jaeger, F. R. Salemme, and R. M. Solle, *Med. Chem. Res.*, 1998, **8**, 244.
[90] A. K. Szardenings, D. Harris, S. Lam, L. Shi, D. Tien, Y. Wang, D. V. Patel, M. Narve, and D. A. Campbell, *J. Med. Chem.*, 1998, **41**, 2194.
[91] A. C. Bishop, C. Kung, K. Shah, L. Wituchi, K. M. Shokat, and Y. Liu, *J. Am. Chem. Soc.*, 1999, **721**, 624.
[92] T. Clackson, W. Yang, L. W. Rozamus, M. Hatada, J. F. Amara, C. T. Rollins, L. F. Stevenson, S. R. Magari, S. A. Wood, N. L. Courage, X. Lu, F. Cerasoli, Jr., M. Gilman, and D. A. Holt, *Proc. Natl. Acad. Sci. USA*, 1998, **95**, 10437.
[93] T. Clackson, *Curr. Opin. Struct. Biol.*, 1998, **8**, 451.
[94] N. S. Gray, L. Wodicka, A.-M. W. H. Thunnissen, T. C. Norman, S. Kwon, F. H. Espinoza, D. O. Morgan, G. Barnes, S. LeClerc, L. Meijer, S.-H. Kim, D. J. Lockart, and P. G. Schultz, *Science*, 1998, **281**, 533.

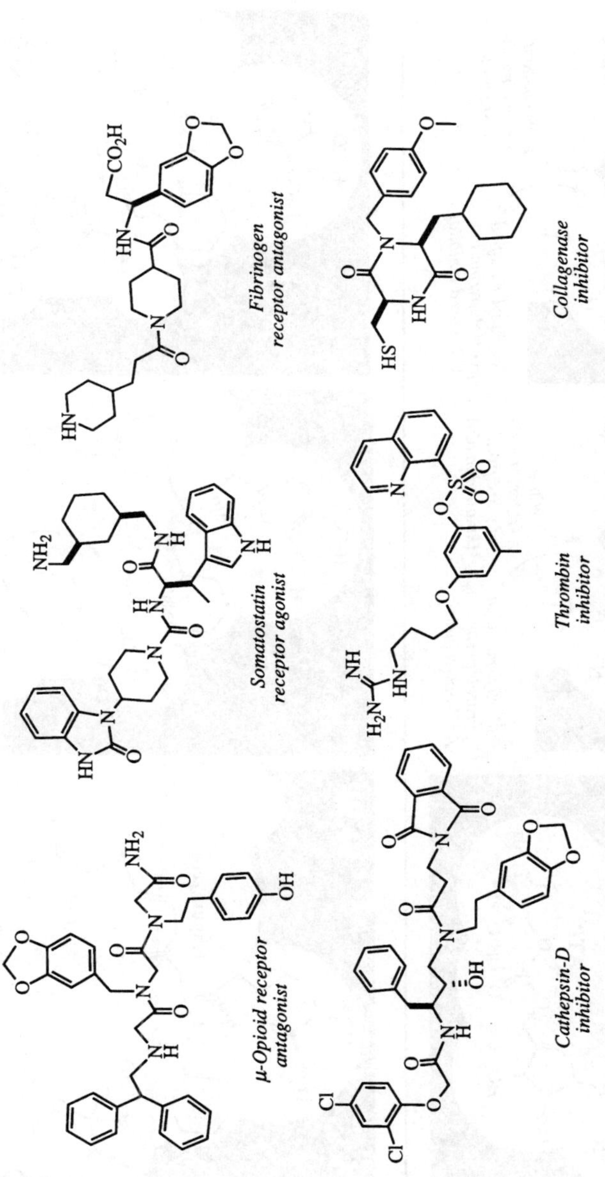

Figure 3   *Some examples of peptidomimetic and nonpeptide drug discovery using molecular diversity strategies*

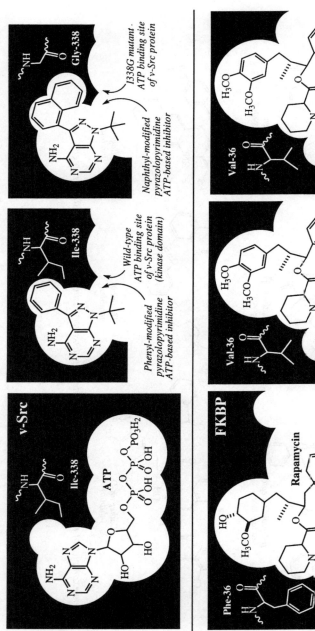

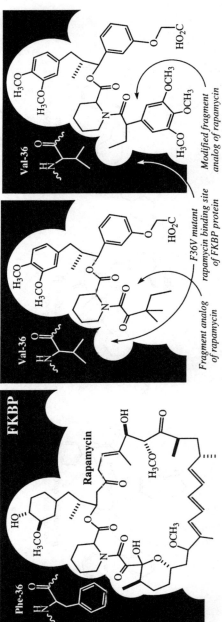

Figure 4 *Examples of designed ligand–protein complex interfaces exploiting molecular recognition and biochemical specificity*

a mutated FKBP was tested with a series of modified fragment analogues of the natural product rapamycin (bottom panel, Figure 4) to discover a highly specific and potent FKBP ligand. This compound was further synthetically altered to provide a dimerizer capable of complexing two of the complementary mutated FKBP proteins.[92] Such work also exemplifies the use of immunophilin ligands (*e.g.* rapamycin, FK506 and cyclosporin) which naturally effect heterdimerization of their respective immunophilins (*e.g.* rapamycin-induced, FKBP-FRAP dimerization and cyclosporin-induced, cyclophilin–cacineurin dimerization) to develop novel methods for the pharmacological regulation of cellular receptor signalling[95] and gene transcription (see above).[96,97] Yet other examples of designing cognate ligand–target interfaces exist, including studies on proteases[98,99] and nuclear hormone receptors.[100]

6   GENE THERAPY AND DNA/RNA-TARGETED THERAPEUTICS

The potential of gene therapy to utilize engineered genes to treat various diseases (*e.g.* cancer, cardivascular and metabolic diseases, and infectious diseases) provides tremendous incentive for drug discovery.[101–103] Also, DNA/RNA-targeted antisense oligonucleotide therapeutics[104] have advanced significantly in synchrony with gene therapies in terms of drug development (Table 7).

Most recently, *in vivo* effective and pharmacologically regulated transgenic expression systems have been developed (*e.g.* rapamycin[105,106] and the progesterone antagonist, RU486[107,108]). In the case of rapamycin-regulated gene therapies, long-term regulated *in vivo*

[95] D. M. Spencer, T. J. Wandless, S. L. Schreiber, and G. R. Crabtree, *Science*, 1993, **262**, 1019.

[96] S. N. Ho, S. R. Biggar, D. M. Spencer, S. L. Schreiber, and G. R. Crabtree, *Nature*, 1996, **382**, 822.

[97] P. J. Belshaw, S. N. Ho, G. R. Crabtree, and S. L. Schreiber, *Proc. Natl. Acad. Sci. USA*, 1996, **93**, 4604.

[98] P. Carter and J. A. Wells, *Science*, 1987, **237**, 394.

[99] G. K. Smith, S. Banks, T. A. Blumenkopf, M. Cory, J. Humphreys, R. M. Laethem, J. Miller, C. P. Moxham, R. Mullin, P. H. Ray, *et al.*, *J. Biol. Chem.*, 1997, **272**, 15804.

[100] D. J. Peet, D. F. Doyle, D. R Corey, and D. J. Mangelsdorf, *Chem. Biol.*, 1998, **5**, 13.

[101] R. C. Mulligan, *Science*, 1993, **260**, 926.

[102] T. Clackson, *Curr. Opin. Chem. Biol.*, 1997, **1**, 210.

[103] M. A. Barry, *Curr. Opin. Drug Disc. Dev.*, 1999, **2**, 118.

[104] J. K. Taylor and N. M. Dean, *Curr. Opin. Drug Disc. Dev.*, 1999, **2**, 147.

[105] V. M. Rivera, X. Ye, N. L. Courage, J. Sachar, F. Cerasoli, Jr., J. M. Wilson, and M. Gilman, *Proc. Natl. Acad. Sci. USA*, 1999, **96**, 8657.

[106] X. Ye, V. M. Rivera, P. Zoltick, F. Cerasoli, Jr., M. A. Schnell, G. Gao, J. V. Hughes, M. Gilman, and J. M. Watson, *Science*, 1999, **283**, 88.

[107] Y. Wang, B. W. O'Mally, Jr. S. Y. Tsai, and B. W. O'Mally, *Proc. Natl. Acad. Sci. USA*, 1994, **91**, 8180.

[108] Y. Wang, F. J. DeMayo, S. Y. Tsai, and B. W. O'Mally, *Nature (Struct. Biol.)*, 1997, **15**, 239.

Table 7 *Some examples of gene therapy or DNA/RNA-targeted therapeutics*

Gene therapy orDNA/RNA-targeted therapeutic	Biotechnology and/or pharmaceutical company	Disease application
Regulated therapeutic protein expression gene therapy	ARIAD Pharmaceuticals	Multiple (*e.g.* growth hormone, erythropoietin or insulin)
Regulated cellular proliferation gene therapy	ARIAD Pharmaceuticals	Multiple (*e.g.* T-cell for graft-versus-host disease)
Vascular endothelial growth factor (VEGF) gene therapy	GenVec/Warner-Lambert	Coronary artery disease
Tumour necrosis factor-α (TNF-α) gene therapy	GenVec	Cancer radiotherapy
Cystic fibrosis conductance regulator gene therapy	Genzyme	Cystic fibrosis
Cytomegalovirus gene therapy	Targeted Genetics	CMV infection; AIDS
DNA vaccine (HIV)	Apollon	HIV infection
DNA vaccine (B-cell lymphoma)	Vical	Non-Hodgkin lymphoma
DNA vaccine (influenza)	Vical/Merck	Influenza virus infection
DNA vaccine (plasmodium falciparum)	Virogenetics	Parasitic infection
Antisense oligonucleotide	ISIS Pharmaceuticals	Cytomegalovirus infection
Antisense oligonucleotide	ISIS Pharmaceuticals	Breast, colon, lung, ovary, pancrease and prostrate tumours
Antisense oligonucleotide	Hybridon	Cytomegalovirus infection, retinitis, viral infections
Antisense oligonucleotide	Lynx Therapeutics	Leukaemia

expression of growth hormone and erythropoietin in animal models was achieved using rapamycin-induced reconstitution of chimeric transcription factors (*i.e.* one fusion construct of FKBP-DNA binding domain which binds rapamycin and the other being a DNA activation domain fusion constuct with FRB, the rapamycin-FKBP binding domain of FRAP). Activation of this heterologous transcription factor system by rapamycin then drives the transgene (*e.g.* growth hormone or erythropoietin) via an inducible promoter in host cells that had been previously transfected with both chimeric transcription factor and therapeutic gene vectors using adeno-associated virus (AAV). From a perspective of regulated gene therapy,[102] the rapamycin system described here provides a unique type of gene switch as compared to allosteric 'off-switch' or allosteric 'on-switch' systems based on tetracyclin[109-111] or

[109] M. Grossen and H. Bujard, *Proc. Natl. Acad. Sci. USA*, 1992, **89**, 5547.
[110] A. Kistner, M. Gossen, F. Zimmerman, J. Jerecic, C. Ullmer, H. Lubbert, and H. Bujard, *Proc. Natl. Acad. Sci. USA*, 1996, **93**, 10933.
[111] D. Bohl, N. Naffakh, and J. M. Heard, *Nature (Medicine)*, 1997, **3**, 299.

RU486[107,108] gene therapy approaches. Again, such work illustrates the application of regulated gene therapy to the expression of therapeutic proteins (Table 7). As aforementioned, exploitation of the molecular recogniton properties of rapamycin for FKBP led to the design of analogues capable of binding two FKBP molecules and the basis to develop a novel gene therapy approach for regulated cellular proliferation. Specifically, addition of the designed FKBP homodimerizer to transfected T-cells expressing Fas-FKBP chimeras led to the expected Fas-dependent apoptosis.[92]

Other noteworthy drug discovery efforts involving control of gene transcription or translation processes includes that of DNA vaccines[103] and antisense oligonucleotides.[104] In the case of DNA vaccines, genetic immunization is guided by the principle that a gene for an antigen is introduced into the host to stimulate the production of vaccine antigens *in vivo*. An advantage of such DNA vaccines is the production of protective cellular immune responses as well as antibody responses to combat viral pathogens (*e.g.* HIV, cytomegalovirus, influenza virus and parasitic infections), and some cancers[112] (Table 7). In the case of antisense oligonucleotide therapeutics, design strategies have been focused on the 'sense' (or coding) strand of nucleic acids *via* base-pairing complementarity and inactivation of transcription through steric hinderance or invoking nucleic acid processing activities.[113] A majority of antisense oligonucleotides are DNA analogues having backbone modifications[114] (*e.g.* peptide nucleic acids or PNAs[115]). Current development of antisense oligonucleotide therapies include those for viral infections, transplant rejection, cancer and multi-drug resistance (Table 7).

7 FUTURE PROSPECTS IN PHARMACEUTICAL RESEARCH

The genomic era is providing new momentum to pharmaceutical research. The future prospects are both compelling and challenging as the human genome will be translated into viable therapeutic targets for drug discovery and development. It is anticipated that several major drug discovery technologies will contribute to accelerate this 'gene to drug' campaign for future pharmaceutical research endeavours. They include proteomics, functional genomics, high-throughput screening, combinatorial chemistry, structural biology, drug design, informatics and pharmacogenomics.

[112] S. W. Barnett and M. A. Liu, *Ann. Rep. Med. Chem.*, 1999, **34**, 149.
[113] S. T. Crooke, *Biotechnol. Genet. Eng. Rev.*, 1998, **15**, 121.
[114] N. M. Dean and R. H. Griffey, *Antisense Nucleic Acid Drug Dev.*, 1997, **7**, 229.
[115] P. E. Nielsen, *Curr. Opin. Struct. Biol.*, 1999, **9**, 353.

With genome sequencing from different species, including pathogenic organisms, it will become possible to more completely define essential biological pathways such as cellular metabolism, signal transduction, gene expression and replication, and intracellular transport. Such comparative genomic information is expected to impact the identification of potential new drug discovery targets. Proteomics will be a very important technology to translate genetic information into the structural and functional characterization of validated therapeutic targets (*e.g.* the Human Proteome Project).[116] In addition to mass spectrometry for protein characterization and analysis,[117] this will require that robust structural biology methods are implemented to provide both rapid prediction and determination of three-dimensional structures using computational modelling,[118] X-ray crystallography[119] and NMR spectroscopy.[120] Exploitation of sophisticated drug design strategies (see above), ranging from virtual screening of combinatorial libraries using both computational chemistry and cheminformatics tools to *de novo* structure-based design methods, will provide powerful applications to forthcoming three-dimensional structures of therapeutic targets. Finally, both bioinformatics[121] and pharmacogenomics[122] are expected to further contribute to future drug discovery and development efforts in terms of addressing genetic variations at the molecular level of therapeutic targets as well as other *in vivo* pharmacological interactions affecting the overall efficacy and safety of drugs in the patient population.

8 CONCLUSIONS

The impact of molecular, structural and chemical biology in pharmaceutical research has been extraordinary over the past decade. The integration of drug discovery technologies which will play significant roles in the transition of pharmaceutical research from a 'gene to drug' concept to a 'gene to drug' reality in the new millennium has been established. Such efforts will exploit genomics and proteomics, provide validated new therapeutic targets, and integrate the use of structure-based and molecular diversity approaches to identify novel lead compounds.

[116] G. T. Montelione and S. Anderson, *Nature (Struct. Biol.)*, 1999, **6**, 11.
[117] P. Roepstorff, *Curr. Opin. Biotech.*, 1997, **8**, 6.
[118] R. Sanchez and A. Sali, *Proc. Natl. Acad. Sci. USA*, 1998, **95**, 13597.
[119] A. Sali, *Nature (Struct. Biol.)*, 1998, **5**, 1029.
[120] P. J. Hajduk, T. Gerfin, J.-M. Boehlen, M. Haberli, D. Marek, and S. W. Fesik, *J. Med. Chem.*, 1999, **42**, 2315.
[121] D. T. Kingsbury, *Drug Dev. Res.*, 1997, **41**, 120.
[122] D. S. Bailey and P. M. Dean, *Ann. Rep. Med. Chem.*, 1999, **34**, 339.

CHAPTER 9

Genetically Modified Foods

ROSA K. PAWSEY

1 INTRODUCTION

Today, biotechnology offers the prospect of changes in the raw materials supplying the food industry; the production of new seed varieties of staple monocotyledenous food grains and of dicotyledenous plants such as tomatoes through genetic manipulation. Changes which were foreseen and are now impacting at the point of consumption are in the resistance of plants to adverse growth conditions, in increased yields, in food storage, and in food-processing qualities. Biotechnology can now offer changes in food animals – in their productivity, in the composition of their meat, and in their milk or egg yields. The research into the genetic modification of organisms to produce new starter strains, new enzymes and new products has reached the point where the dairying and brewing industries in particular are beginning to make regular use of these developments. There are prospects for the development of novel foods, *e.g.* single cell protein, and for both new and well established food ingredients from new, often microbial sources. The development of microbial products, particularly of enzymes and the use of immobilized organisms including those produced by genetic modification offer prospects of processing or monitoring food quality in faster or cheaper ways.

Some genetically modified (GM) products are already available to the food industry and to the consumer. The legal framework which protects the health and safety of consumers and the environment is in place, although it is continuously being developed. Although the timescale for product development and safety testing is considerable, some inventions have been launched, with launches of many more imminent. However, although some products have already been on the market for several

261

years, future widespread acceptance of food products containing GM constituents is not guaranteed.

2 LEGAL REQUIREMENTS IN THE PRODUCTION OF NOVEL FOODS AND PROCESSES

In the UK, mycoprotein ('Quorn'®), the novel food produced from the mould *Fusarium graminareum*, was launched in 1986 after approval from the Ministry of Agriculture, Fisheries and Food (MAFF) that it was safe for human consumption. It was first released only in premanufactured foods such as stews and curries but in 1992 became available through retail food outlets for home use. In a sense this was a project which, while not involving GM organisms, laid the ground rules for approval procedures for all other novel products and processes.

UK law has aimed to cover three aspects of the development of genetically modified organisms (GMOs): the experimentation and development in the laboratory; the experimental, accidental and full-scale release into the environment; and the health and safety of consumers of food products which may be or may contain GM material. Until May 1997 the requirement for companies to submit their novel products to the approval process was voluntary. However, since that date the law governing the use of genetically modified foods and food materials changed to comply with Regulation (EC) 258/97[1] of the European Parliament. Thus the voluntary scheme for the assessment of novel foods was replaced in the UK by the compulsory one required by the Novel Foods and Novel Food Ingredients Regulations 1997.[2] Other member states of the European Union (EU) similarly implemented regulations compliant with the above EC Regulation.

The new UK Regulations define the meaning of a novel food as 'food which has not been used for human consumption to a significant degree within the Community' and details of the categories, which include GM materials, are shown in Table 1. Applications must be made by companies to the Advisory Committee on Novel Foods and Processes (ACNFP) which makes recommendations regarding the safety for human consumption of the material to the Government through the MAFF, which is currently designated the competent authority. However, this arrangement may change if the proposed new Food Standards Agency is set up. When safety assessment of release into the environment

[1] Regulation(EC) 258/97 European Commission concerning novel foods and novel food ingredients. *Official Journal of the European Communities*, No L43 14th February 1997.
[2] SI 1997 No 1335 The Novel Foods and Novel Food Ingredients Regulations 1997; and SI 1997 No 1336. The Novel Foods and Novel Foods Ingredients (Fees) Regulations 1997.

Table 1 *Novel Foods: definitions under the Novel Foods and Novel Food Ingredients Regulations 1997.*
(Adapted from the Annual Report, 1997, of the Advisory Committee on Novel Foods and Processes. Department of Health and Ministry of Agriculture, Fisheries and Food, 1997. MAFF Publications, 1998.)

(i) Foods and food ingredients containing or consisting of GMOs within the meaning of Directive 90/220/EEC[a].
(ii) Foods and food ingredients produced from, but not containing GMOs.
(iii) Foods and food ingredients with a new or intentionally modified primary molecular structure.
(iv) Foods and food ingredients consisting of or isolated from microorganisms, fungi or algae.
(v) Foods and food ingredients consisting of, or isolated from plants and food ingredients isolated from animals, except for foods and food ingredients obtained by traditional propagation or breeding practices and having a history of safe food use.
(vi) Foods and food ingredients to which has been applied a production process not currently used, where that process gives rise to significant changes in the composition or structure of the foods or food ingredients which affect their nutritional value, metabolism or level of undesirable substances.

[a]Directive 90/220/EEC on the deliberate release into the environment of genetically modified organisms. Official Journal L117 8 May 1990. Implemented in the UK by the Genetically Modified Organisms (Deliberate Release) Regulations 1992. SI 1992 No 3280. HMSO, London, and SI 1995 No. 304, HMSO, London.

is involved, the Advisory Committee on Releases to the Environment (ACRE) may be referred to by the ACNFP. Some of the submissions concerning GM foods addressed by the ACNFP since 1989 are summarized in Table 2.

The system of positive approvals for each and every novel food as carried out in the UK does not apply similarly in the USA. Rather, under a policy published in 1992,[3] GM foods are required to meet the same standards as all other foods, namely that producers and sellers must offer only safe products to consumers. The policy[3] regards 'substances deliberately added to food through GM as additives if they are significantly different in structure, function or amount than substances currently found in food'. Since many GM food crops currently being developed do not contain substances significantly different from those already in the diet, the Food and Drugs Administration (FDA) policy implies such products do not need pre-market approval, although the producers go through a detailed consultative process with the scientists of the FDA. For example, in the case of the FLAVR SAVR[TM] tomato,

[3] *Federal Register*, 'Statement of policy: foods derived from new plant varieties'. Volume 57, No. 104/ May 29, 1992.

Table 2 *Some submissions considered by ACNFP since 1989 involving GM foods or GM technology*

Legal	EC Regulations on novel foods	ACNFP annual reports (1992–1997)[a]	Clearance by ACNSF as safe food
Food crop plants	Chicory (GM) salad leaves	1996	
	Cotton (GM) as a source of cotton seed oil	1996, 1997	1997
	Maize (GM) herbicide resistant	1996, 1997	1997
	Maize (GM) insect resistant	1996, 1997	1997
	Oilseed rape (GM) oil	1995, 1996	1997
	Pollen from GM plants in honey	1990	1995, 1996
	Potatoes (GM) insect resistant	1997	
	Soya beans (GM) herbicide resistant[b]	1994	1995
	Tomatoes (GM) to be eaten fresh	1995	1996
Food animals	Transgenic animals	1989, 1990, 1991, 1992, 1994	
Manufactured foods	Microorganisms and their products		
	Amylolytic yeast for use in beer making	1992, 1993	1994
	Bakers yeast (GM)	1989, 1990	1990
	Chymosin from *Kluyveromyces lactis* (GM)[b]	1989, 1990	1991
	Chymosin from *Aspergillus niger* var. *awamori* (GM)	1990	1991
	Chymosin from *E. coli* K-12 (GM)	1991, 1992	1992
	Lipase from *Aspergillus oryzae* (GM)	1992, 1994, 1995	
	Vitamin (riboflavin) from *Bacillus subtilis* (GM)	1996	1997
	Fats and fatty acids		
	Oils from GM rape seed, GM cotton seed		
	Carbohydrates and sugars		
	Hemicellulase enzymes from GM sources	1995, 1996, 1997	
	Whole products		
	Processed products from GM tomatoes[b]	1994, 1995, 1997	1995, 1996
		1990, 1991	
Consumer concerns			

[a] All annual reports published by HMSO the following year.
[b] Known to be on the UK market (1997) but since at that time labelling was not required this probably represents an underestimate (ACNFP, Annual Report 1996).

which remains firm when ripe, the pectin of the tomatoes is identical to that in non-GM tomatoes and is therefore 'GRAS' (generally accepted as safe). However, the marker gene for kanamycin resistance that encodes the enzyme aminoglycoside-3'-phosphotransferase II (APH(3')II), and which was used to select plant cells carrying the antisense gene, was regarded as a new substance, an additive, and evaluated as such. In the submissions from the company (Calgene) to the FDA, the company showed it not to have the characteristics of an allergen, not to contain a sugar molecule attached to a protein, and to be degraded in gastric juice (*i.e.* when consumed). It was also shown not to interfere with the medicinal/therapeutic effects of kanamycin and neomycin when they were administered orally. Additionally, since there is no known mechanism by which plant genes can be transferred to microbes (albeit a concept not allowing for future knowledge), the possibility of transfer of the antibiotic resistance marker to gut microbes was considered to be an extremely low risk. At the end of extensive consultations the FDA concluded the new tomato had not been significantly altered when compared to varieties of tomatoes with a safe history of use, and the consultative process in the context of safety for consumers was complete.

Thus, the European and American systems differ, with Europe operating a positive approval system and the USA requiring consultation with the FDA for GM crops and other foods in which a new or changed component, an 'additive', is deemed to be present. Table 3 lists a number of changes to food plants which have been through and completed a similar consultative process with respect to their safety for consumption.

3 FOOD CROPS

Genetic modification of food crops is a rapidly developing field and is discussed more fully in Chapter 7. Theoretically, strains of food plants specifically designed to possess almost any combination of desirable qualities are possible because the advantage that genetic modification has over conventional plant breeding is that genes can be transferred between any pair of organisms, one of which does not have to be a plant. Some current possibilities are summarized in Table 4.

The first stage of the development of the transgenic strain relies on the insertion of specific genes derived from donor sources and the second stage requires the subsequent growth of whole plants from these modified undifferentiated cells. Gene insertion (see Chapter 5) has largely been achieved by using, for example, modified strains of the bacterium

Table 3 Examples of genes modifying food crops[47]

Date	Company[a,b]	Food plant	Character change(s)	Gene(s)	Gene source
1998	Agrevo	Canola	Glufosinate tolerant/male sterility	Phosphinothricin acetyltransferase Barnase	Streptomyces viridochromogenes Bacillus amyloliquifaciens
1998	Agrevo	Canola	Glufosinate tolerant/sterility restorer	Phosphinothricin acetyltransferase Barstar	Streptomyces viridochromogenes Bacillus amyloliquifaciens
1998	Agrevo	Corn	Glufosinate tolerant/insect resistant	Bar cry9C	Streptomyces hygroscopicus Bacillus thuringensis subsp. tolworthi
1996	Dekalb	Corn	Glufosinate tolerant	Phosphinothricin acetyltransferase	Streptomyces hygroscopicus
1997	Dekalb	Corn	Insect protected	cryIA(c)	Bacillus thuringensis (Bt) sp. tenebrionis
1996	Monsanto	Corn	Insect protected	cryIA(b)	Bacillus thuringensis subsp. kurstaki
1998	Monsanto	Corn	Glyphosate tolerant	Modified enolpyruvylshikimate-3-phosphate synthase	Corn
1996	Northrup	Corn	Insect protected	cryIA(b)	Bacillus thuringensis subsp. kurstaki
1996	Monsanto	Corn	Insect protected and glyphosate tolerant	Enolpyruvylshikimate-3-phosphate synthase Glyphosate oxidoreductase cryIA(b)	Agrobacterium sp. strain CP4 Ochrobactrum anthropi Bacillus thuringensis subsp. kurstaki
1998	Calgene	Cotton	Bromoxynil tolerant Insect protected	Nitrilase cryIA(c)	Klebsiella pneumoniae Bacillus thuringensis subsp. kurstaki
1996	Dupont	Cotton	Sulfonylurea tolerant	Acetolactate synthase	Tobacco, Nicotiniana tabacum cv. Xanthi
1998	Univ/Sask	Flax	Sulfonylurea tolerant	Acetolactate synthase	Arabidopsis
1996	Pl Genetic	Oilseed rape	Male sterile	Barnase	Bacillus amyloliquifaciens

Year	Company	Crop	Trait	Gene/Protein	Source organism
1996	PI Genetic	Oilseed rape	Fertility restorer	Barstar	*Bacillus amyloliquifaciens*
1997	Univ H/C	Papaya	Virus resistant	Coat protein	Papaya ringspot virus
1998	Monsanto	Potato	Insect protected	cryIIIA	*Bacillus thuringiensis* (Bt) sp. *tenebrionis*
			Virus protected	Potato Leafroll Virus replicase	
1998	Agrevo	Soya bean	Glufosinate tolerant	Phosphinothricin acetyltransferase	*Streptomyces viridochromogenes*
1997	Dupont	Soya bean	High oleic acid	Sense suppression of the *GmFad2*–1 gene which encodes a delta-12 desaturase enzyme	
1997	Seminis	Squash	Virus resistant	Coat protein	Cucumber mosaic virus / Zucchini yellow mosaic virus / Watermelon mosaic virus
1998	Agrevo	Sugar beet	Glufosinate tolerant	Phosphinothricin acetyltransferase	*Streptomyces viridochromogenes*
1998	Mon/Nov	Sugar beet	Glyphosate tolerant	Enolpyruvylshikimate-3-phosphate synthase / Truncated glyphosate oxidoreductase	*Agrobacterium* sp. strain CP4 / *Ochrobactrum anthropi*
1996	Agrotope	Tomato	Modified fruit ripening	S-adenosylmethionine hydrolase	*E. coli* bacteriophage T3
1998	Calgene	Tomato	Insect protected	cryIA(c)	*Bacillus thuringiensis* subsp. *kurstaki*
1994	DNA PI Tech	Tomato	Improved ripening	Aminocyclopropane carboxylic acid synthase fragment	Tomato
1994	Zeneca	Tomato	Delayed softening	Polygalacturonase gene fragment	Tomato
1994	Calgene	Tomato (Flavr savr)	Firm when ripe	Antisense polygalacturonase	Tomato

[a] Full company names are given apart from: Dekalb, Bekalb Genetics Corporation; DNA PI Tech, DNA Plant Technology; Mon/Nov, Monsanto/Novartis; Northrup, Northrup King; PI Genetic, Plant Genetic Systems; Seminis, Seminis Vegetable Seeds; Univ/Sask, University of Saskatchewan; Univ H/C, University of Hawaii and Cornell University; Zeneca, Zeneca Plant Science.

[b] The developers listed had completed the consultations with the US FDA prior to commercial distribution consistent with the 1992 policy of the FDA.[46]

Source: FDA website: dated November 6 1998 (http://vm.efsan.fda.gov/~lrd/biotechm.htm).

Table 4 *Goals for biotechnological improvements in crop plants*

Goals	Comments
Resistance to disease (insect pests, microbial pathogens, competing weeds)	Several companies are already trading plants with these characters
Resistance to herbicides	Maize, soya bean
Resistance to drought and soil salinity	Potential to offer improved prospects for third world agriculture
Nitrogen fixing ability	Long-term goal
Cheaper to grow (requiring less pesticide)	Less environmental damage
Higher yielding (more tonnes per acre) the traditional goal; higher yields of essential components	Traditional methods of plant breeding have given higher yields of rice in India, wheat in Europe and India, cloned palm oil, soybean, higher lauric acid in oil seed rape
Easier handling, rapid and synchronous ripening	Cloned sesame seeds, tomatoes
Better nutritional properties	Higher protein content; increased amounts of essential amino acids, *e.g.* methionine, lysine; different lipid composition of oil seeds
Better storage properties	Freezing resistance, *e.g.* tomatoes, potatoes; reduction of post harvest losses through delay in ripening, *e.g.* tomatoes
Better processing qualities	Retention of texture; improvement of colour; changes in viscosity; changes in elasticity; changes in emulsifying properties
Absence of allergens	Gluten-free grains
Novel products	Unique tasty processed vegetable snacks, *e.g.* 'vegisnax'
Novel sources of proteins and vitamins	From seeds, plants and single-celled organisms
Modified sensory attributes	Increased sweetness due to genetic modification, *e.g.* thaumatin expression in temperate climate plants, *e.g.* monellin

Adapted from R. K. Pawsey in 'Molecular Biology and Biotechnology', 3rd Edn, ed. J. M. Walker and E. B. Gingold, Royal Society of Chemistry, London, 1993, with additional materials from ACNSF Annual Report 1997, HMSO, London, 1998.

Agrobacterium tumiefaciens as the vector. The successfully modified cells are subsequently cultured onto platelets, and develop then into mature plants.

In 1992, approximately 30–40 crops had been genetically modified,[4,5] but in the intervening years the number has expanded to more than 80,[6]

[4] J. L. Jones, *Trends Food Sci. Technol.*, 1992, **3**, 55.
[5] R. Fraley, *Bio/Technology*, 1992, **10**, 40.
[6] Institute of Biology: Genetically modified crops: the social and ethical issues, 1998.

Table 5   *Some crops so far successfully genetically modified*

GM crops	Marketing consents in Europe as of December 1997*	Marketing consents in Europe pending as of December 1997*
Alfalfa		
Apple	Experimental field trials UK*	
Asparagus		
Banana		
Barley		
Cabbage		
Capsicum		
Carrot		
Celery		
Chicory	Non food use (Beijo Zaden)	
Chicory		Food use (Beijo Zaden)
Clover		
Coffee		
Cotton		
Cowpea		
Cucumber		
Frenchbean		
Grapevine		
Lettuce		
Liquorice		
Maize	Insect resistant (Ciba Geigy)	Insect resistant (AgroEvo, Monsanto, Novartis)
Melon		
Oilseed rape	Herbicide tolerant (Plant Genetic Systems)	Herbicide tolerant (AgroEvo, Plant Genetic Systems)
Pear	Experimental field trials UK*	
Potato		Modified starch content (Avebe)
Rice		
Rye		
Soya bean	Herbicide tolerant	
Sprouts		
Strawberry	Experimental field trials UK*	
Sugarbeet		
Sunflower		
Tobacco		
Tomato		
Turnip		
Wheat		

Source: J. L. Jones, *Trends Food Sci. Technol.*, 1992, **3**, 55; R. Fraley, *Bio/Technology*, 1992, **10**, 40; MAFF – Consumer Panel paper CP(98) 33/2 (*).

with over 10 000 trials currently occurring.[7] Some of these are listed in Table 5. The number of plant species and varieties at the commercially usable stage is increasing, with applications to use or grow them in the UK increasing in parallel (see Tables 2 and 5). The monocotyledons are

[7] MAFF Consumer Panel CP(98) 33/2.

very important economically, encompassing many of the worlds staple crops, *e.g.* rice, wheat and barley, and the successful development right to the market place of mature genetically modified monocotyledenous plants is beginning to be seen. In 1997 the ACNFP cleared the safety of food products made from maize that had been genetically modified for resistance to the European corn borer insect and to the herbicide glufosinate (Table 2).

However, a greater number of GM dicotyledenous plants containing stable genetically engineered changes have been achieved (Tables 3 and 5). Several of these, which have been developed and have successfully completed the consultative process in the USA, are shown in Table 3. As indicated earlier, the tendency of stored tomatoes to soften during storage has been modified. The activity of the gene encoding for the enzyme polygalacturonase has been substantially reduced by the insertion of the antisense RNA to polygalacturonase[8] resulting in red, but firm tomatoes.[9,10] These tomatoes are now being grown commercially in the USA and their products are available in retail outlets in the UK. ACNSF examined the submission in 1994–7 and safety clearance was granted.

Initially it was not known whether the tomatoes with genetically modified storage properties had retained the good flavour properties of the donor strains from which they derived because taste trials had not been permitted. Concern centred around the dissemination of GM seeds in human faeces which could potentially lead to the uncontrolled release of genetically modified seeds carrying their modified genes and selectable marker genes which in some cases encoded for resistance to antibiotics such as kanamycin, neomycin or ampicillin. Guidelines for the conduct of taste trials using human volunteers were set up[11] which ensure that the volunteers understand the potential risks, and that their medical records are retained. However, the issue of the control of dissemination of viable GM seeds through human faeces in the country of product growth and elsewhere through those of air travellers has not yet been resolved through international laws. In the case of the GM tomato, the applicants were able to demonstrate that the kanamycin resistance marker gene product in the fruit was degraded rapidly in the digestive fluids in the normal gut and in patients where gut contents were neutralized, efficacy of neomycin would not be compromised.[12]

[8] J. E. Blalock, *Trends Biotechnol.*, 1990, **8**, 140.
[9] G. Hobson, *Acta Horticulturae*, 1989, **258**, 593.
[10] A. Flaherty, *Grower*, 1988, **11**, 6.
[11] Advisory Committee on Novel Foods and Processes, Annual Report 1991, HMSO, 1992.
[12] Advisory Committee on Novel Foods and Processes, Annual Report 1995, HMSO, 1996.

4 FOOD ANIMALS

The large-scale production of the anabolic steroid hormones such as bovine somatotrophin (BST) and porcine somatotrophin (PST) in microorganisms through recombinant DNA techniques is a developed technology. Pigs have been shown to produce leaner pork when reared with the use of pork somatotrophin (PST). When injected into cows, it is claimed BST, another growth hormone, increases a cow's milk production by 10–15%. Yet in spite of huge investment by the manufacturers into the development of BST, the commercial success of the product remains to be seen. In the USA, the FDA concluded that milk from such cows presented 'no increased health risk to consumers',[13] and it is now widely used in milk production. In the UK, milk from cows treated with BST during a trial was permitted to mingle with the milk supply; some sections of the public reacted to this with alarm. In Europe the view is different from that in the USA and there is currently a moratorium (which finishes at the end of 1999) on its use. Surveys show that the public do not accept such products, partly because they are legitimately concerned as to the long-term effect on human health[14] through consumption of either milk or meat from treated animals, and partly because of the social and ethical issues involved; generally consumers are extremely wary of genetic engineering of animals.[15]

The development of transgenic animals remains contentious. While much of the research work is related to the elucidation of the development of embryos, some is more directly concerned with the development of food species. For example, transgenic catfish carrying growth hormones from rainbow trout have been developed in the USA, the idea being to farm catfish whose growth rate is the same as that of the rainbow trout. However, the environmental implications of the accidental release of such farmed, transgenic fish could be serious. The ABRAC (Agricultural Biotechnology Research Advisory Committee) of the US Department of Agriculture gave limited approval to the continued research by only permitting study of the catfish in specially constructed containment ponds.[16]

The question of the food use of animals resulting from experimental work in the production of transgenic animals has arisen several times in recent years. Much of the work with transgenic animals has been directed towards the incorporation of human genes for the generation of products

[13] J. C. Juskevich and C. G. Guyer, *Science*, 1990, **249**, 875.
[14] J. H. Hulse, *IFST Proceedings*, 1986, **19**, 11.
[15] A. H. Scholten, M. H. Feenstra, and A. M. Hamstra, *Food Biotechnol.*, 1991, **5**, 331.
[16] J. Fox, *Bio/Technology*, 1992, **10**, 492.

for medical use, *e.g.* human insulin, organs for human transplant purposes. In these programmes the gene incorporation success rate in producing transgenic animals has been low, about 0.1%. The issue then arose as to whether the other offspring, not showing the transgenic characteristics, could be released into the food chain. In the UK, the ACNFP safety committee concluded[17] that the next generation of animals where the gene(s) could not be detected by sensitive techniques (such as PCR) could be released into the food chain, there being no safety issues in this case. However, when consulted about labelling, the Food Advisory Committee recommended that such meat should *not* enter the food chain.[18] Additionally, the Government, sensitive to the ethical dimension, set up a committee to examine ethical issues surrounding genetic modification of life forms.

5 CURRENT TRENDS IN MANUFACTURED FOODS

It can be seen from Table 2 that, at that time (1996), the foods involving genetic manipulation and known to be reaching the consumer in the UK,[19] were products containing GM soya, cheeses made using GM chymosin and processed products from GM tomatoes. The probability is that there were actually more GM foods in circulation for more clearances on safety grounds had certainly been granted. Attitudes to labelling have hardened since then and at issue today is whether all foods made either wholly from or containing GM materials have to be labelled, and if so how.

Yeasts have been used in food fermentations for centuries, in breadmaking and in the production of fermented beverages. Recent developments in understanding the genetics and biochemistry of yeasts are leading to the development of new yeast strains through both the exploitation of classical yeast genetics and through the development of recombinant DNA techniques.[20]

In yeast production, new rapidly growing, high cell yielding strains have potential value; in breadmaking rapid growth and good carbon dioxide yield in different dough systems, including those of high osmotic pressure, are goals already achieved in the laboratory. In the UK, one strain of genetically manipulated baker's yeast has progressed beyond development to achieve Government approval (Table 2) and, with this approval, no requirement for specific labelling. The genes responsible for

---

[17] Advisory Committee on Novel Foods and Processes, Annual Report 1994, HMSO, 1995.
[18] Food Advisory Committee, Annual Report 1994, HMSO, 1995.
[19] Advisory Committee on Novel Foods and Processes, Annual Report 1996, HMSO, 1997.
[20] J. W. Chapman, *Trends Food Sci. Technol.*, 1991, **2**, 176.

maltase and maltose permease in the recipient yeast have been modified by sequences from the donor strain resulting in a recombinant strain which expresses the enzymes independent of the presence of maltose and not inhibited by glucose, thus permitting more efficient utilization of maltose.[21] Short synthetic sequences have been used to link the implant into the recipient genome but, apart from this, the manipulated yeast contains no genetic material other than that from strains of *Saccharomyces cerevisiae*. Because of this, when considering the first genetically manipulated food organisms evaluated for sale, the ACNFP was satisfied that the risk that the manipulated strain would produce toxic metabolites was no greater than that of unmodified strains used previously. The risk of genetic transfer from the modified yeast to human consumers or their gut flora was also considered no greater than might be anticipated from any other strain of bakers yeast.[22] They were using the 'substantial equivalence' concept to safety that food produced using a GM organism should be compared with closely related products already consumed. This approach has informed the processes by which safety assessment is achieved, not only in the UK but also elsewhere in the world.[23]

In the brewing industry a number of changed attributes in yeasts, such as the production of strains for the production of light beers through more efficient amylase activity, control of the development of off-flavours produced by the production of diacetyl and phenolics, improvement of filterability through reduction in residual β-glucans and improvement in storage quality by the control of haze development, have been foreseen as being valuable to the industry.

Genes encoding for a number of enzymes (amylases, acetolactate dehydrogenase and β-glucanases), which would contribute to the solution of these problems, derived from a range of both microbial and non-microbial sources, have been successfully cloned into yeasts.[24] Approval was granted in 1994 for the use of a strain of amylolytic GM yeast in beer making,[17] with no labelling requirement.

Equally, a number of targets have been foreseen in dairying for dairy starter cultures: in bacteriophage resistance, in resistance to antibiotics and herbicides, in the ability to produce acid steadily and facilitate cheese making, in the production of enhanced flavours and cheese texture, and in the acceleration of ripening. A number of new strains

---

[21] Gist Brocades NV, European patent A2 0 306 107, 1989.
[22] Advisory Committee on Novel Foods and Processes, Annual Report 1989, HMSO, 1990.
[23] OECD, Safety evaluation of foods derived by modern technology. Concepts and Principles OECD, 1993.
[24] W. J. Donnelly, *J. Soc. Dairy Technol.*, 1991, **112**.

produced through non-recombinant techniques such as conjugation and cell fusion are currently on commercial trial. One new patented strain, *Lactobacillus casei* spp. *rhamnosum* GG, has the following characteristics: ability to attach to human intestinal mucosa cells, stability to acid and bile, *in vitro* production of an anti-microbial substance, and hardy *in vitro* growth. It is to be used to produce a fermented whey drink and a yoghurt-type product, both of which are aimed at promoting health benefits in humans being both helpful in the control of diarrhoea and constipation.[25] This organism is illustrative of new opportunities perceived by the dairy industry. Being a human strain, and also not produced by GM techniques, its launch onto the market is, in spite of extensive clinical trials, simpler than for GM organisms of similar potential. Although application for the use another organism, *Enterococcus faecium strain K77D*, for production of cultured milk products was received and cleared in 1995,[12] no application for the use of a GM strains of organisms in dairy foods had been received in the UK by 1998.

The most significant change in the food industry in the UK has been the approval of the enzyme chymosin, the principal component of calf rennet, for food use (Table 2). Chymosin is produced by three separate transgenic organisms, *Fluyveromyces lactis*, *Aspergillus niger* var. *awamori*, and *Escherichia coli K-12*, performing satisfactorily in the production of cheeses from cows'[26] and ewes'[27] milk. The organisms have been achieved by the cloning of the appropriate gene sequences from calf cells. In each case the chymosin, which is responsible for coagulation of milk in the manufacture of cheese, has been shown to be identical to that produced from calves and because of this, specific labelling of such products was not recommended by the advisory committees.[28] Plant and fungal rennets (used for vegetarian cheese) have their own properties and confer their own characteristic flavours on the cheeses in which they are used. Chymosin produced from transgenic microorganisms, in addition to being identical to the natural product, has other advantages: it is more readily harvested than rennet from the traditional source, the calves' stomach; it may contain a higher concentration of active enzyme, and yet it still produces cheese with texture and flavour properties comparable to traditionally produced cheese. The incentive for the production of chymosin from these new sources is in the economics; the current

[25] S. Salminen, S. Gorbach, and K. Salminen, *Food Technol.*, June 1991, 112.
[26] G. Van den Berg and P. J. de Koning, *Neth. Milk Dairy J.*, 1990, **44**, 189.
[27] M. Nunez, M. Medina, P. Gaya, A. M. Guillen and M. A. Rodriguez-Marin, *J. Dairy Res.*, 1992, **59**, 81.
[28] Advisory Committee on Novel Foods and Processes, Annual Report 1990, HMSO, 1991.

estimated chymosin market value is £100 million for 18 tonnes of chymosin,[24] so clearly consumer acceptance is vital in order to realize this market.

More generally there are many processes in the food industry where enzymes can be used. The sources of the enzymes are varied, but many are microbially derived. The potential exists for the genetic modification of microbial species to improve expression, stability, specificity, activity and yield. Currently, a number of food enzymes (β-glucanase, glucan α-1,4-glucosidase, glucoamylase, amyase, pullulanase, acetolactate decarboxylase, β-galactosidase, chymosin, α-galactosidase, phospholipase, lipoxygenase) have been cloned in yeasts. In use the enzymes may be applied as additives, immobilized or encapsulated. As far as the dairy industry is concerned the applications of enzyme technology are very varied. In his review of the applications of biotechnology in the dairy industry, Donnelly[24] discusses the use of enzymes in the hydrolysis of milk proteins tailored to products in health care and baby foods, a market related to that referred to earlier in respect of patented *Lactobacillus casei* spp. *rhamnosum* GG. There is also potential for the use of enzymes to accelerate cheese ripening. However, in both situations the production of bitter flavours, due to high contents of hydrophobic amino acids such as leucine, proline and phenylalanine in casein, tend to partly counteract the advantages offered.

The wide range of other products, food additives such as polysaccharides, low calorie sweeteners, flavour modifiers, vitamins, colourings, water binding agents and nutritional supplements, are still awaiting their entry onto the UK market.

In food quality monitoring, recombinant DNA technology offers the prospect for more rapid tests for detection and sometimes enumeration or quantification of pathogens, for spoilage organisms, for chemical indications of spoilage, for contaminants, and for adulterants. Many systems which exploit these techniques are already on the market which, because they are exclusively used in the laboratory analysis of food samples, do not have to go through the same approval procedures as GM affected foods for consumption. An increasing number of DNA probes (some produced by cloning) are available for the detection of food borne pathogens such as *Salmonella*, *Listeria*, *Escherichia coli*, *Yersinia*, *Campylobacter*, *Staphylococcus*, *Pseudomonas*, *Shigella*, *Vibrio* and viruses.[29] The use of such probes, being quicker than conventional techniques, will help in the evaluation of the quality of raw materials and that of the finished products.

[29] M. J. Woolcott, *J. Food Protect.*, 1991, **54**(5), 387.

Another technique offering a rapid and sensitive method for the detection of antimicrobial substances in food is through the production of GM bioluminescent sensor strains. These organisms fail to emit light because they die in the presence of the inhibitor, thus, by monitoring the reduction in luminosity the presence of antimicrobial material can be determined. This clearly has value in the cheese industry where, for example, traces of antibiotic inhibit starter organisms. Equally, in defined conditions the same detector can be used to evaluate preservative systems in foods, or the efficacy of biocides.[30,31]

The potential for biotechnological techniques in monitoring food quality is wide and offers many possibilities for detecting and quantifying analytes at trace levels not possible by other techniques. The wider use of these techniques should lead to a raising of standards of food quality.[32]

6   CONSUMER ACCEPTANCE AND MARKET FORCES

In the late 1980s to early 1990s period in the development of GM foods and food products, the submission for their safety assessment was voluntary and also largely confidential. Press releases were careful and limited in information while a progression of safety clearances of products were granted for which no special labelling was required. However, the UK Government decided to deposit data supporting submissions for novel foods (including those involving GM organisms) in the British Library, available for public perusal,[22] and has also since striven to increase transparency of the process.[33,34] However, with the application for the release into the food chain of non-transgenic animals originating from experimental GM programmes using human genes it became clear that the ethical dimension of GM needed consideration. Hence a Committee under the chairmanship of the Reverend Polkingthorne was convened, which reported in 1993.[35]

The structure of the market in the European Union meant that products produced by genetic modification and approved elsewhere in the EU could enter the UK market without further formality of any sort. Furthermore, the power of world trade impinged on the UK during 1997–1998 with the approval and large-scale production of GM soya in

[30] G. S. A. B. Stewart, S. P. Denyer, and J. Lewington, *Trends Food Sci. Technol.*, 1991, **2**, 7.
[31] J. M. Baker, M. W. Griffiths, and D. L. Collins-Thompson, *J. Food Protection*, 1992, **55**(1), 62.
[32] 'Biotechnology and Food Quality', ed. S. Kung, D. B. Bills, and R. Quaytrana, Butterworths, 1989.
[33] Advisory Committee on Novel Foods and Processes, Annual Report 1997, HMSO, 1998.
[34] MAFF Consumer Panel Paper CP(98)33/2.
[35] MAFF Report of the committee into the ethics of genetically modified foods and food use. London, HMSO, 1993.

the USA. The consumer learnt both that many processed food products contained soya, 60% was stated,[36] and that the American soya bean producers claimed it was impossible to separate GM soya from non-GM soya.

These developments led to higher public awareness of the fact that GM food products were now a reality and much public debate followed. Professional groups published their position statements, usually carefully neutral,[6,37] while pressure groups publicised their view on environmental, socio-economic and ethical considerations.[38] The need for public information was propounded by consumer groups[39] and debated among scientists.[40] Food retailers had to decide whether they would[41] or would not[36] sell GM foods, while producers resorted to advertising, from their perspective, the benefits of GM.[42,43]

From the food, rather than the agricultural, ethical or socio-economic perspective, one issue stood out – that of labelling. Governments in Europe responded to consumer demand to be able to choose whether to consume GM foods. In 1993 a framework proposal was put forward[44] which suggested only labelling foods which contained ethically sensitive copy genes, such as those where animal genes had been inserted into plants. However, possibly in response to greater public awareness, the EC Regulation 1139/98,[45] which came into force on 1 September 1998, supercedes this with a much more open approach and requires the labelling of all foods containing GM soya and GM maize. The UK Government's view[46] of this is that it also opens the way for the labelling of all GM foods and those containing GM ingredients. Rather than focusing on the presence of copy genes, and construing them to be absent if destroyed by processing[44] (Table 6), it requires food labelling, except when detectable protein or DNA resulting from GM is absent. This opens the question of surveillance of compliance, and the European

---

[36] The Iceland Food Company, 'Important Information for our Customers. Genetic modification and how it affects you', 18 March 1998.

[37] Institute of Food Science and Technology, Position Statement on Genetically Modified Foods, Keynote, p. 3–5, 1998.

[38] Greenpeace, 'European opposition to genetically engineered foods hardens'. *Greenpeace Business*, Oct/Nov 1998, p. 6.

[39] Consumer Association leaflet: Genetically modified maize, *Which?*, February 1997.

[40] C. Reilly. Biotechnological Science and the Consumer. Conference Report, *Trends Food Sci. Technol.*, 1996, 7, 336.

[41] Tesco customer information. Genetically modified maize. The facts.

[42] Monsanto advertising campaign, *e.g. Guardian Weekend*: 6 and 24 June 1998.

[43] J. Ramsay 'Monsanto fights back', *Chem. Ind.*, 20 July 1998.

[44] Advisory Committee on Novel Foods and Processes, Annual Report 1993, HMSO, 1994.

[45] Regulation(EC) 1139/98 European Commission on the labelling of GM soya and maize. *Official Journal of the European Communities*. No L159 3 June 1998.

[46] MAFF, News Press release 21 May 1998.

Table 6   *Labelling of products from GM sources – European requirements*

Proposals 1993[44]
Food should be labelled appropriately if:
It contains a copy gene originally derived from a human.
It contains a copy gene originally derived from an animal which is the subject of religious dietary restrictions.
Is plant or microbial material and contains a copy gene originally derived from an animal.
These rules would NOT apply if the inserted copy gene had been destroyed by processing and was not therefore present in food.

EC Regulation 1139/98[45]
EC Regulation 1139/98 will require all foods containing ingredients from GM soya and GM maize to be labelled except when neither protein nor DNA resulting from GM is present.

Commission is to look into the question of a *de minimis* threshold for the presence of DNA or protein resulting from genetic modification.[46]

Currently in the USA, unless a novel component (additive) is present, specific labelling of GM products is not required to identify a product as being or containing materials derived from a GM source, although a producer may voluntarily choose to do so.[47]

It has become clear to the public that many GM food materials are now available and ready for launch onto the market. The problem still remains that the speed of technological change together with great commercial power are pitted against the reluctance of the public to be assured of the safety of GM foods and the time required for the necessary evaluation of the environmental and socio-economic change they imply. The October 1998 moratorium on field trials or commercial growing in the UK may represent a breathing space in which the public may have a chance to assimilate the implications of GM foods and to react to them. The proposed labelling regimes will inform the public debate and help the public to decide whether the widespread change in methods of food supply through the availability of GM foods is acceptable or not.

[47] FDA website (dated 1998, November 06): http://vm.efsan.fda.gov/~lrd/biotechm.htm 'Future Foods' exhibition in the Science Museum, London to March 1988, sponsored by MAFF.

CHAPTER 10

Molecular Diagnosis of Inherited Disease

ELIZABETH GREEN

1 INTRODUCTION

Molecular biology has advanced dramatically during the 1990s. The major international effort into the Human Genome Project has provided much of the driving force behind this progress. This huge and ambitious project's aim is to map and sequence the entire human genome. The early part of the project focused on defining collections of partially over-lapping cloned human DNA segments to represent each of the 24 different types of human chromosome (22 autosomes plus the X and Y sex chromosomes). Each set of overlapping DNA clones is known as a clone contig map. Several clone contig maps have been published[1] which form the framework for the second part of the project. More recently the focus has moved towards characterizing the DNA sequence of each chromosome and the mapping and identification of all 65 000–80 000 human genes. The work of the Human Genome Project and the resources it provides has led to an exponential increase in the rate at which single gene disorders are mapped, cloned and characterized. As a consequence the demand for genetic testing has also increased dramati-cally. In the UK there are over 40 laboratories that are funded by the NHS to provide genetic testing. The number of disorders covered is over 100. No single laboratory offers testing for all these disorders but samples can be exchanged between laboratories to ensure that every patient in the UK has access to a comprehensive genetics testing service. The Clinical Molecular Genetics Society produce a register of UK DNA

[1] P. Little, *Nature*, 1995, **377**, 286.

Diagnostic laboratories which includes listings of all services available and can be accessed *via* the internet.[2]

To support The Human Genome project there have been many technological advances such as robotics and automated DNA analysers. The latter now also play a vital role in many DNA diagnostic laboratories. Automated DNA analysers were originally developed for high throughput DNA sequencing. Traditionally sequencing was performed using radioisotope labelling and autoradiography for the detection of products. Now automated fluorescent DNA sequencing is possible using fluorescent labelled primers or fluorescent dideoxynucleotides. A different fluorophore is used for each of the four (A, C, G, T) base-specific reactions which permits the loading of all four reactions on one lane of a denaturing polyacrylamide gel. During electrophoresis, the machine detects and records the fluorescence as the DNA migrates through a fixed point in the gel. The data is collated electronically and can be displayed as strings of coloured peaks (see Section 2.3.5). The advances in fluorescent chemistry and software for use with automated DNA analysers has meant that other forms of DNA analysis other than DNA sequencing are now possible using the same equipment, such as sizing and quantitation of PCR products.

This chapter will describe how modern methods of DNA analysis have been applied to both direct and indirect testing for mutations in a variety of important inherited disorders. Nearly all of the assays discussed rely upon the polymerase chain reaction (PCR; see Chapter 3) followed by some further manipulation. The most significant improvement has been the application of automated DNA analysis and fluorescence technology to this area of science. This technology has increased not only throughput but also accuracy of mutation detection as highlighted in the following examples.

2 DIRECT DETECTION OF GENE MUTATIONS

The structural alteration of a gene which causes a loss of, or altered, function can be anything from a single nucleotide change, *i.e.* a point mutation, to a large deletion of millions of nucleotides. The detection of gross DNA rearrangements is straightforward, whereas a variety of ingenious techniques have been developed to detect more subtle changes in gene structure.

---

[2] http://www.leeds.ac.uk/cmgs/lablist/labs.htm

2.1 Detection of Deletions, Duplications and Insertions

The causative mutation in a variety of inherited disorders can be the loss of all or part of the DNA coding for a particular gene product. Similarly duplications of all or part of a gene are quite common.

Duchenne muscular dystrophy (DMD) and Becker muscular dystrophy (BMD) are both caused by mutations in the dystrophin gene located on the X chromosome. Both disorders cause muscle weakness in affected males. In DMD the symptoms are severe and most patients are wheelchair bound by the age of 11 years and rarely survive into their twenties. BMD is a much milder disorder and overall life expectancy is only slightly reduced. In approximately 72% of males with DMD or BMD the causative mutations are deletions (65%) or duplications (7%) of one or more exons of the dystrophin gene. The remaining 28% of cases are thought to be caused by point mutations.[3]

The dystrophin gene consists of 79 exons of which some are more prone to deletion than others. This knowledge led to the development of two multiplex PCR assays for the detection of dystrophin deletions.[4,5] Key exons are simultaneously amplified using oligonucleotide primers designed such that the PCR product from each exon is a particular size. Males only have one X chromosome, therefore a deletion of an exon can be detected by the absence of a product of the appropriate size when the PCR is analysed by gel electrophoresis. Products were detected by ethidium bromide staining. More recently the method has been improved and adapted for electrophoresis on an automated DNA analyser[6] by using fluorescently labelled amplification primers. This allows accurate quantitative analysis of the data. Thus, it is now possible to detect duplications as well as deletions in male patients with DMD. A male with a duplication of an exon will have twice the quantity of product for that exon compared to that of the other exons analysed. A further advantage of the fluorescent dosage system is that female carriers of deletions or duplications can accurately be identified. All females have two X-chromosomes, so when an exon is deleted it is present at half the quantity of other exons (Figure 1). Conversely, when an exon is duplicated it is present at three times the quantity of the other exons. Fewer prenatal diagnoses for DMD and BMD are performed because it has been possible to offer women from families with a known deletion or

[3] R. G. Roberts, R. J. Gardner, and M. Bobrow, *Hum. Mutat.*, 1994, **4**, 1.
[4] J. S. Chamberlain, R. A. Gibbs, J. E. Ranier, P. N. Nguyen, and C. T. Caskey, *Nucleic Acids Res.*, 1988, **16**, 11141.
[5] S. Abbs, S. C. Yau, S. Clark, C. G. Mathew, and M. Bobrow, *J. Med. Genet.*, 1991, **28**, 304.
[6] S. C. Yau, M. Bobrow, C. G. Mathew, and S. J. Abbs, *J. Med. Genet.*, 1996, **33**, 550.

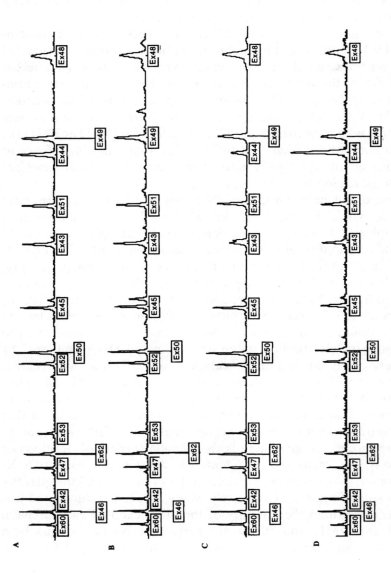

Figure 1 *Multiplex PCR amplification for the detection of deletions and duplications. Products obtained from simultaneous amplification of 13 deletion-prone exons in the dystrophin gene are separated on a size basis. Each peak corresponding to an exon is labelled. (A) Normal profile; (B) Exon 44 deletion (male); (C) Female heterozygous for exon 44 deletion; (D) Exon 44 duplication*

duplication mutation a definitive carrier test. An unexpected advantage has been the ability to detect some point mutations. The PCR products can be sized very accurately by comparison with a fluorescently labelled molecular weight standard, so it is possible to identify single base deletions and insertions.

The method of quantitative fluorescent dosage PCR can be applied to any disorder where deletion or duplications mutations are common. In the DNA Diagnostic Laboratory at Guy's Hospital the method has been used to screen for deletions and duplications of the peripheral myelin protein 22 (PMP22) gene located on chromosome 17. Hereditary motor sensory neuropathy type 1A is a duplication of the PMP22 whilst deletions of the PMP22 gene cause a disorder called hereditary neuropathy with liability to pressure palsies. Both disorders are dominant so the defect occurs in only one copy of the gene. A variety of methods have been used to diagnose these disorders but probably the simplest method is that of quantitative fluorescent dosage.

2.2 Expansion Mutations

There are now many genes identified where the disease causing mutation is an expansion of a trinucleotide repeat sequence such as CTG or CGG. Such diseases are often called triplet repeat disorders. Some examples are described below.

Huntington's disease (HD) is an autosomal dominant disorder. The symptoms include dementia, abnormal movements of the limbs called chorea and psychiatric problems. The disease is caused by the expansion of a CAG trinucleotide located in the Huntington gene located on chromosome 4.[7] Normal individuals have up to 35 copies of the CAG repeat whereas affected individuals have above 36 repeats. There is a correlation between the size of the repeat and the age of onset of symptoms. Late onset cases tend to have repeat sizes between 36–40 repeats whereas juvenile onset cases have repeat sizes of approximately 60–120. The disorder is diagnosed by sizing the HD repeat allele by PCR. Primers are designed to amplify across the repeat. The products of amplification are electrophoresed on denaturing polyacrylamide gels and the products are detected by autoradiography (if radiolabelled primers are used), silver staining, or (again) an automated DNA analyser (when fluorescently labelled primers are used). Figure 2 shows an example electropherogram. To date it has been possible to detect all expanded HD alleles by PCR but this is not the case in some

[7] Huntington's Disease Collaborative Research Group, *Cell*, 1993, **72**, 971.

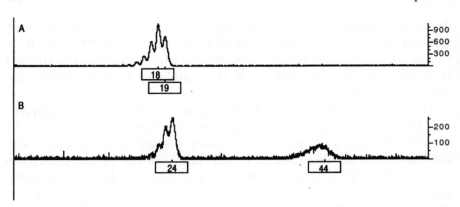

Figure 2 *Detection of triplet repeat expansion in Huntington's disease. Fluorescent PCR products of triplet repeat region in Huntington's disease: (A) Normal profile showing two alleles of 18 and 19 repeats; (B) HD profile showing normal allele of 24 repeats and mutant allele of 44 repeats*

other triplet repeat for example fragile X syndrome and myotonic dystrophy.

Fragile X syndrome, a major cause of mental retardation particularly in males, is caused by excessive amplification of a CGG repeat located in a gene called FMR1 located on the X chromosome.[8] Similarly myotonic dystrophy, a muscle disease is caused by an expansion of a CAG repeat within the myotonin gene located on chromosome 19.[9] Unlike the expansion mutation in Huntington's disease, the majority of expansion mutations for these two diseases are much larger and can only be detected on Southern blots as restriction fragments of increased size. However, it is normal to use PCR as the first step in the screening process, particularly for fragile X syndrome. The number of patient samples sent to a DNA diagnostic laboratory to be tested for fragile X syndrome is high (approximately 1000 per annum referred to the Guy's DNA Diagnostic Laboratory) but the proportion found to have the disorder is low (< 1%). Samples are subjected to a fluorescently labelled PCR that amplifies across the CGG repeat of FMR1 known as FRAXA. Products are sized by automated DNA analysis. Males with a normal FRAXA allele and females with two normal FRAXA alleles do not have fragile X syndrome. Any male with no detectable FRAXA PCR product and females who have a single FRAXA allele are tested further by Southern blot analysis (Figure 3). This 'pre-screening' by PCR allows

[8] I. Oberle, F. Rousseau, D. Heitz, C. Kretz, D. Devys, A. Hanauer, J. Boue, M. F. Bertheas, and J. L. Mandel, *Science*, 1991, **252**, 1097.
[9] C. T. Caskey, A. Pizzuti, Y. Fu, G. Fenwick, Jr, and D. L. Nelson, *Science*, 1992, **256**, 784.

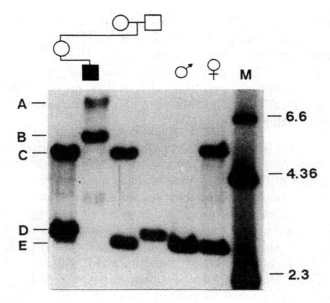

Figure 3 *Detection of insertional mutations by Southern blot analysis. Autoradiograph showing hybridization of the probe StB12.3 to DNA digested with both EcoRI and the methylation-sensitive enzyme EagI, from a family with the fragile X syndrome. Control samples from a normal male and female are on the right, showing the normal 2.8 kb fragment (E) from the normal, unmethylated X chromosomes, and in the female a 5.2 kb fragment (C) from the methylated inactive X. The affected male (filled square) has an insertional mutation detected by hybridization to a larger methylated fragment (B), which has mutated from the smaller insertional fragment (D) transmitted from his normal grandfather, through his mother. The affected male has an additional larger mutation in some of his cells, as shown by the presence of an additional higher molecular weight band (A)*

most samples to be ascertained rapidly without recourse to the more time consuming method of Southern blotting. A similar two step approach is used for myotonic dystrophy screening.

2.3   Point Mutations

2.3.1   Allele-specific Oligonucleotides.   An alteration of a single nucleotide in the sequence of a gene can be detected by synthesizing oligonuceotide probes that are complementary to either the normal or the mutated sequence. Under carefully controlled conditions, a probe of 18–20 nucleotides will hybridize to a homologous sequence but not to a sequence with a single mismatch within it.[10] The section of the gene that

[10] R. Saiki, T. L. Bugawan, G. T. Horn, K. B. Mullis, and H. A. Erich, *Nature*, 1986, **324**, 163.

contains the mutation is first amplified by PCR, and the products dotted on to a nylon membrane. This is then hybridized to a radioactively labelled allele-specific probe complementary to the normal or the mutated sequence. Unbound probe is removed by washing at a carefully controlled temperature, and then the bound probe is detected by autoradiography (Figure 4). Probes can also be labelled by nonradioactive methods, such as biotinylation.

2.3.2   Restriction Enzyme Site Analysis.   Bacterial restriction endonucleases that recognize and cleave at specific nucleotide sequences are an integral part of the cloning process. They have also been widely used in molecular diagnosis, since they can identify mutations that occur within their target recognition sequence. For example, sickle cell anaemia is caused by a mutation in the gene that codes for the β chain of haemoglobin. The DNA sequence is mutated from CCTGAGGAG to CCTG*T*GGAG. The restriction enzyme MSTII will cut DNA at the sequence CCTGAGG but will not cut the mutated sequence. Therefore, if we amplify this region of the gene by PCR and digest the products with

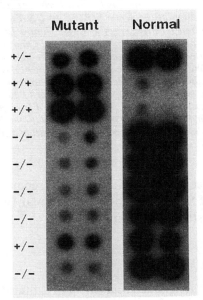

Figure 4   *Mutation analysis by ASO probe hybridization. Autoradiographs obtained after the products of amplification from exon 11 of the cystic fibrosis gene have been dot-blotted (in duplicate) onto a nylon membrane and hybridized with ASO probes which match the normal and mutant sequence of the G542X mutation. Individuals who are heterozygous for this mutation (i.e. they have one normal copy and one mutated copy) show strong hybridization with both ASOs (+ / −), whereas mutant (+ / +) or normal (− / −) homozygotes hybridize strongly only with mutant or normal ASO*

*Mst*II, a gene bearing the sickle-cell mutation can be detected as a larger (undigested) band upon gel electrophoresis.

The advantage of this method is that it is very simple to perform. No radioactive or other probe labelling is required, and multiple samples can be tested within about 6 hours. Its limitation is that many mutations do not occur within the recognition sites of any known restriction enzymes. However, this problem can be often overcome by the use of carefully designed amplification primers.[11] If no restriction site is present, we can create one in the amplified product by altering the sequence of one of the PCR primers by one or two bases. We have used this technique to devise an assay for a mutation identified in the dystrophin gene of a Duchenne muscular dystrophy patient. In this example, the sequence of one of the amplification primers was altered so as to engineer a *Pvu*II site in the sequence of the PCR product obtained from the normal gene, but not the mutant gene, as shown in Figure 5a.

The mutation is a substitution of a cystosine (C) with a thymine (T), as shown in the open box in Figure 5a. The sequence of the primer (in shaded box) is complementary to the normal sequence apart from the incorporation of an adenine (A) in place of a T at the asterisked position. Since the recognition site for PvuII is CAGCTG, the PCR products from a normal copy of the gene will be cleaved, but the product from a mutant copy of the gene will not contain a *Pvu*II site and will not be digested. The outcome of the experiment is shown in Figure 5b.

2.3.3 'ARMS'. The rather grand and cumbersome name of this technique is amplification refractory mutation systemm, or 'ARMS' for short.[12] It is based on the fact that the identity of the most 3′ nucleotide of a PCR primer is critical for the success of amplification. If this nucleotide is not complementary to the target sequence, amplification is likely to fail. Thus, if PCR primers are designed so that the last 3′ base of one of them falls right at the site of the gene where a mutation occurs, the presence of the mutation causes failure of the PCR and no product is obtained. The disadvantage of the ARMS method is that a second PCR must be set up using primers specific for the mutated sequence. Thus a product is only obtained in the second PCR if the mutation is present. However, once the assay has been devised it is simple to perform, and no digestion with a restriction enzyme is required. Furthermore, multiple mutations can be tested for simultaneously by positioning the second non-critical primer at a different distance from the mutation specific

[11] E. J. Sorscher and Z. Huang, *Lancet*, 1991, **337**, 1115.
[12] C. R. Newton, A. Graham, L. E. Heptinstall, S. J. Powell, C. Summers, N. Kalsheker, J. C. Smith, and A. F. Markham, *Nucleic Acids Res.*, 1989, **17**, 2503.

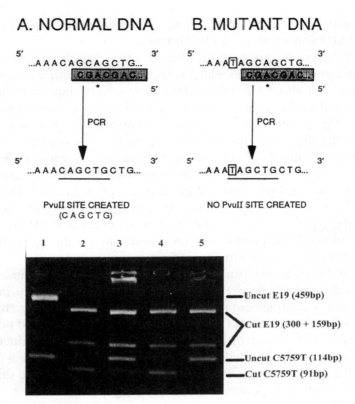

Figure 5 (a) *Diagram illustrating how a restriction site can be engineered into the products of PCR amplification from a normal sequence (A), but not from the mutant sequence (B), in cases where the mutation does not naturally create nor destroy a known restriction site. See the text for a detailed explanation of the method. The results of typing a family with this assay for a point mutation causing Duchenne muscular dystrophy are shown in Figure 5b.* (b) *Mutation analysis by introduction of a restriction enzyme site. Assay for a C to T substitution (C5759T) associated with a Duchenne muscular dystrophy family using a primer to introduce a restriction enzyme (PvuII) recognition sequence in the absence of the mutation. Exon 19 (E19) of the dystrophin gene is used as an internal control for the completeness of digestion, and undigested PCR products are shown in the first lane of the gel. The C5759T PCR product (113 bp) obtained from normal DNA digests to 91 bp and 23 bp (23 bp band not shown). Failure to digest this product, while obtaining complete digestion of the internal control indicates the presence of the mutation. Thus the affected male (lane 5) shows the mutation in his X chromosome; his mother (lane 3) is a carrier of the mutation, having one normal and one mutant X; and his sister (lane 4) and grandfather (lane 2) do not carry the mutation. Lane 1 shows undigested DNA as a control*

primer for each of the mutations tested for. Figure 6 shows the results of an ARMS assay developed by AstraZeneca Diagnostics to test for 20 different mutations in the cystic fibrosis gene. This multiple testing approach is especially important in a disease like cystic fibrosis, since

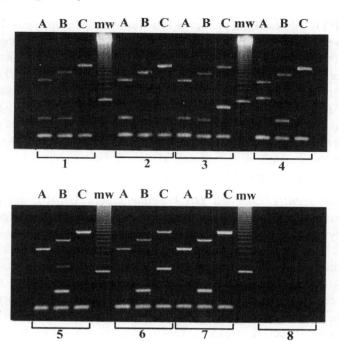

Figure 6 *Screen of cystic fibrosis mutations using the ELUCIGENE^{TM} CF20 kit from AstraZeneca Diagnostics. Each test comprises three multiplexes (A, B and C) and includes control primers as confirmation of amplification. Comparison of the diagnostic band position against an adjacent marker track (mw) allows identification of the mutation. The results shown are as follows: (1) delF508 het.; (2) delF508 homo.; (3) delF508/delI507; (4) G542X het.; (5) G551D het.; (6) 1078delT; (7) Normal DNA control; (8) Negative control*

more than 500 different mutations have been identified in patients with this condition.

2.3.4 Oligonucleotide Ligation. The oligonucelotide ligation technique is a recent development. The method is currently used in a kit developed and marketed by PE-Applied Biosystems, which is used for cystic fibrosis screening.[13] The kit screens for the presence of 31 mutations in the cystic fibrosis gene. The first step of the assay involves co-amplification of 15 exons of the cystic fibrosis gene using standard primers. The second stage of the assay involves hybridization and ligation of sequence-specific oligonucleotides/primers. Each mutation under test is probed by three oligonucleotides. One of the probes (in each set of three) hybridizes to the sequence common to the mutant and

---

[13] P. D. Grossman, W. Bloch, E. Brinson, C. C. Chang, F. A. Eggerding, S. Fung, D. A. Iovannisci, S. Woo, and E. S. Winn-Deen, *Nucleic Acids Res.*, 1994, **22**, 4527.

normal CFTR alleles. This primer is labelled with a fluorescent dye. One of the other primers in the trio is specific for the normal sequence the last is specific for the mutant sequence. These primers compete for binding to the amplicon. When the binding is perfect (without mismatches) ligation to the common probe will occur. The products of ligation are detected by electrophoresis on an automated DNA analyser. The PE-Applied Biosystems kit makes full use of the available fluorescent dyes and the specific primers have special size adjusting tags attached to permit separation of mutant and normal alleles during electrophoresis. An example electropherogram is shown in Figure 7.

2.3.5   Fluorescently Labelled DNA Sequencing.   Until quite recently DNA sequencing was rarely used for DNA diagnostics. The technique was mostly used in the research environment to identify disease-causing mutations. Once a disease causing mutation was identified in, say, a boy with Duchenne muscular dystrophy, a specific PCR-restriction digest assay was developed to screen other members of the family (Section 2.3.2). There have been many improvements to the dideoxy terminator sequencing technique and it is now possible to screen point mutations reliably using direct fluorescent DNA sequencing of PCR products. Furthermore, the method is used as the only detection technique in some circumstances. At the Guy's DNA laboratory DNA sequencing is used

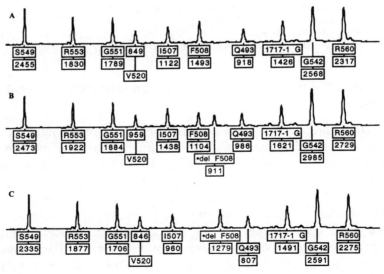

Figure 7   *Screen of cystic fibrosis mutation by the OLA. Each peak represents a signal for the labelled mutation (only a selection is shown). A different sized band indicates a given mutation is present: (A) Normal control; (B) delF508 het; (C) delF508 homo*

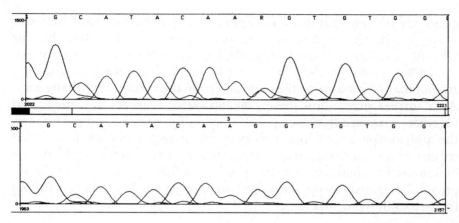

Figure 8 *Fluorescent DNA sequencing. Top panel shows sequence from an individual who is heterozygous for a G to A substitution (indicated by an R above G and A peaks superimposed). The lower panel is a normal control*

to screen for mutations in several genes including COL4A5, mutations in which are responsible for Alport syndrome. The quality of the sequence data and the standard of the analysis software permits the identification of heterozygote peaks whenever there is a mutation present (Figure 8).

3 INDIRECT DIGANOSIS WITH LINKED GENETIC MARKERS

In cases where the gene that caused the disease is known and the particular mutation identified, accurate diagnosis for other family members using one of the methods described above is straightforward. However, the mutation that causes the disease in a particular family is not always known. The fact that cystic fibrosis (CF) is genetically complex has already been mentioned. The disorder can be caused by any of at least 500 mutations. At present it is not feasible to test for all possible mutations in each family requiring genetic diagnosis. In practice an affected individual is tested for the more common mutations first. If these tests are negative but the diagnosis is confirmed by a number of clinical criteria then indirect methods can be employed to offer genetic testing to the family of an affected individual. Indirect testing is based on the use of normal variations of DNA sequence among individuals known as DNA polymorphisms. Polymorphic loci are used to 'mark' the normal and mutated copies of the gene in a family. The polymorphic markers used have to be located close to (tightly linked to) the gene so that the chance of the mutation and the marker being separated through a crossover (recombination) during sperm or egg production is low. For some diseases the markers used flank the gene being followed, but for CF, the markers used are intragenic, located in non-coding sections of the CFTR gene, *e.g.* IVS17bTA and IVS8CA,[14] therefore the risk of a recombination separating the mutation from the marker is virtually zero. As a further precaution these two intragenic markers are often used together. Standard fluorescent PCR methods are used to amplify across the polymorphic loci and products are detected on an automated sequencer as described before. In most cases it is possible to 'mark' the mutant and normal copies of the gene in a family in order to offer future prenatal diagnosis to the parents of an affected child and carrier testing to at-risk family members.

4 FUTURE PROSPECTS

There are always new technologies on the horizon that may one day be used in diagnostic laboratories. Some of these are competing methodologies and therefore may fall by the wayside; others may yet be pushed aside by exciting new breakthroughs in robotics and automated analysers. However, there are several procedures that are already being investigated in research laboratories and may soon find their way into the diagnostic tool cupboard: the first of these is denaturing high performance liquid chromatography (DHPLC) which can be used to detect unknown mutations in a given PCR product and has so far proven more sensitive than gel based systems, and, importantly, is largely automated;[15] the 5′ nuclease assay was first described some years ago, but has more recently found favour in its fluorescent version known as the TaqMan[TM] assay.[16] This technique is used for detecting known mutations in large numbers of samples and again its forte is in its automation. Although this is a proven technique, the main obstacle to its more widespread use is the high cost per sample. Lastly, screening for hundreds or thousands of mutations simultaneously is the ultimate goal of the diagnostic laboratory particularly for multigenic diseases and even this ideal may be addressed by advances in microchip arrays.[17] Essentially, it is now possible to synthesize thousands of oligonucleotides on a chip and use this chip for hybridization. Special scanning equipment can

[14] M. Morral, and X. Estivill, *Genomics*, 1992, **13**, 1362.
[15] M. C. O'Donovan, P. J. Oefner, S. C. Roberts, J. Austin, B. Hoogendoorn, C. Guy, G. Speight, M. Upadhyaya, S. S. Sommer, and P. McGuffin, *Genomics*, 1998, **52**, 44.
[16] J. K. Livak, S. J. A.Flood, J. Marmaro, W. Giusti, and K. Deetz, *PCR Methods Applic.*, 1995, **4**, 357.
[17] F. Ginot, *Hum. Mutat.*, 1997, **10**, 1.

determine from the pattern of signals which alleles a particular DNA has. Although still in its infancy, this technology could eventually replace many others and become the routine way to analyse DNA for any variations.

CHAPTER 11

DNA in Forensic Science

PAUL DEBENHAM AND PETER D. MARTIN

1 INTRODUCTION

Forensic scientists have never rested on the current state of forensic technology. There is a continuous evolution of both the scientific methods employed and in the levels of identification achieved. For many years the investigation of biological samples derived from scenes of crime were centred on the analysis of blood groups (red cell surface antigens and polymorphic enzymes). Depending on the nature of the sample a number of blood groups could be analysed and the results combined to give an overall result for comparative purposes. Although the analysis of blood groups has served the criminal justice system well for many years there were limitations to the use of these markers. Some of the groups were tissue specific and, in general, a considerable amount of blood or other body fluid stain was required to give a worthwhile result. An additional problem was the instability of many of the isoenzymes which meant that the stain had to be reasonably fresh.

It was always the ambition of forensic scientists to have systems which would approach individuality without suffering the difficulties encountered with the blood groups. DNA was the obvious choice because the same result would be obtained from any body fluid or tissue as long as cellular material was present. While it was appreciated that the whole genome in theory defines the individual, it was neither a practical nor possible option to determine the total DNA sequence of an individual. Therefore, it was necessary to have a methodology which would analyse small sections of the DNA in which there was sufficient variability to theoretically distinguish between individuals in the relevant population. Any such system had to be sufficiently robust so that results could be obtained from small amounts of starting material which might be old and subjected to adverse environmental conditions. Ideally the analysis time should be measured in hours rather than weeks in order that innocent suspects could be eliminated and released quickly so that the scientific work does not become the limiting factor in the criminal enquiry. The results should also be in a form which allows for the formation of computer searchable databases.

Surprisingly, given that our individuality is due to the expression of thousands of coding gene sequences, the methodology presently used around the world utilises DNA of the non-coding regions of the genome.

In the short time that forensic DNA analysis has been available the process has undergone a number of radical developments culminating in the present technology which allows for results to be rapidly and accurately obtained from very small amounts of cellular material. A national DNA database is now a reality. As well as methods for obtaining results from the analysis of genomic DNA, techniques for the examination of mitochondrial DNA have proved extremely valuable in a number of high profile investigations. There has also been considerable development in the analysis of markers from the Y chromosomes.

When DNA profiling was first used in crime cases there were aggressive court room criticisms of the laboratory methods. Some of these challenges were justified, whilst many just reflected the skills of the defence barristers manipulating a complex science before a lay jury. As a consequence of this intense scrutiny the science of DNA Profiling has evolved to a point where disputes with respect to methodology have largely disappeared and now the challenges tend to be directed to the significance of the results and the manner of presentation of the evidence.

2   MLP AND SLP TECHNOLOGY

DNA analysis came to the fore in 1985 when Professor Alec Jeffreys discovered that tandemly repeated sequences of nucleotides found in 'minisatellites' within the non-coding regions of the DNA could be exploited to determine individuality.[1]

Ninety per cent of human DNA is non-coding and the minisatellites which contain the repeating units are scattered throughout these areas. Within any particular minisatellite there is a repeating sequence of nucleotides. The sequence that is repeated is relatively invariant within the minisatellite and consists of approximately 7–30 nucleotides. For example the MS1 minisatellite has the sequence (GTGGAC/TAGG) (GTGGAC/TAGG) (GTGGAC/TAGG), *etc.* Thus it is not the

---

[1] A. J. Jeffreys, *Biochem. Soc. Trans.*, 1987, **15**, 309.

sequence *per se* that differs between individuals, it is in fact the number of times that the unit is repeated which varies between individuals. This variation in repeats is astonishing with individuals being found with about 100 repeats in a minisatellite, whilst others may have thousands and others with any number in between. This hypervariability is termed the variable number of tandem repeat (VNTR) polymorphism (Figure 1). These VNTRs can be assessed and provide the basis for individualization.

DNA can be readily isolated from a sample and when sufficiently pure can be subjected to digestion by restriction endonuclease enzymes which reduce the macromolecules of human DNA in a highly reproducible manner to many small fragments suitable for laboratory analysis. These can be size-fractionated by electrophoresis to produce a 'ladder' of fragments, some of which will contain minisatellites. The actual size of the DNA fragments containing minisatellites will be largely determined by the VNTRs contained. The larger fragments will have the slower electrophoretic mobilities and therefore the minisatellites with the greatest number of repeat units will have the slowest migration

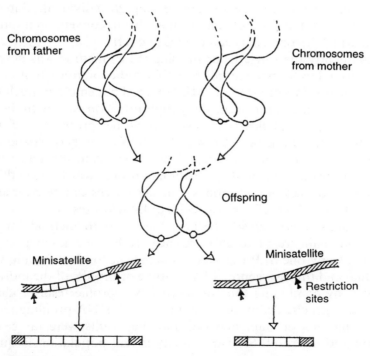

Figure 1   *Inheritance of VNTRs. The offspring has inherited one minisatellite which contains a VNTR of eight repeat units and another which has four repeat units*

speeds. The minisatellites can be detected by hybridization with a labelled probe (a piece of DNA with a sequence of nucleotides complementary to that of the repeat unit). The probe binds to the minisatellite DNA and exposes an X-ray film placed adjacent to the DNA causing a black band to develop where the minisatellite sequence is present. The physical position of the band(s) directly relates to the VNTRs for the individual and thus can be codified and compared between samples and individuals.

The original methodology developed by Alec Jeffreys involved a probe with a generic sequence complementary to a large number of minisatellites (*i.e.* Multi-Locus Probe known as a MLP) in human DNA. This MLP method of analysis therefore generated a result that contained numerous bands akin to a barcode and became known as a DNA fingerprint. While DNA fingerprint analysis provided the breakthrough in DNA individualization it did suffer some limitations for use in crime investigation. The method required a relatively large amount of cellular material, the bands produced could not be assigned to any particular chromosome and the results could not be used to prepare computer databases. Furthermore, whilst it was argued that the DNA fingerprint was a unique pattern of bands representing an individual, akin to the classic fingerprint, it was not readily amenable to conventional statistical analysis which hampered its acceptance in court.

Consequently the single locus probing (SLP) method was developed and was found to be much more amenable to the forensic situation.[2] The SLP system is very similar to the MLP method but employs much higher stringency conditions for the probing (hybridization) step with the result that only one minisatellite (locus) is detected per probe. Therefore just two bands (alleles) will be visualized at each locus (only one band will be seen if the person is homozygous). The alleles from other loci can be determined by the sequential application of probes which detect different repeating sequences of nucleotides. The allele sizes can be estimated by reference to a 'ladder' of control alleles of known sizes.

SLP analysis very quickly became the standard method for DNA analysis in mainstream forensic science. It became known as DNA profiling as any one SLP test was not sufficient to identify uniquely an individual and thus gave more of an outline or profile, of the individual. However, by combining four or more DNA profile tests produced a near-unique genetic definition of an individual. DNA profiling required smaller amounts of stain material and the results were far easier to compare and interpret. Statistical analysis and population frequencies

[2] Z. Wong, V. Wilson, I. Patel, S. Povey and A. J. Jeffreys, *Ann. Hum. Genet.*, 1987, **51**, 269.

were also easier to develop and defend. Importantly the results could be expressed in terms of band sizes for a given DNA locus and thus it was possible to produce computer searchable databases.

2.1 MLP/SLP Methods

2.1.1 Extraction and purification of the DNA. The majority of samples received in a forensic science laboratory are small and often old and it is therefore important that the maximum yield of DNA is recovered from the extraction. As the samples will be from different cellular material the extraction procedures will vary in detail but the overall method is essentially the same (Figure 2).

Initially the cells are usually lysed using a detergent like sodium dodecyl sulfate and general protein degradation is effected by a non-

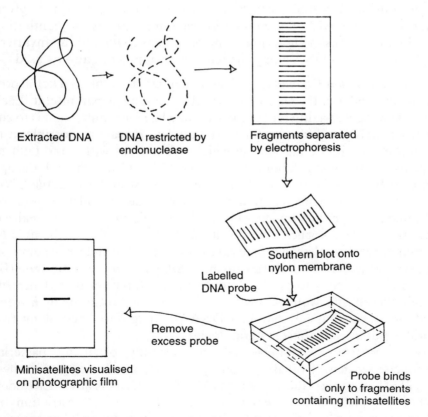

Extracted DNA   DNA restricted by endonuclease   Fragments separated by electrophoresis

Southern blot onto nylon membrane

Labelled DNA probe

Remove excess probe

Probe binds only to fragments containing minisatellites

Minisatellites visualised on photographic film

Figure 2 *SLP method. Diagrammatic representation of the steps involved with the production of single locus profiles*

specific proteinase. The resulting protein and cellular debris is partitioned into an organic phase such as phenol/chloroform whereby, as long as the pH is maintained at 7.8, the DNA will partition into the aqueous phase away from the debris. The DNA can then be recovered from the aqueous phase by precipitation with ethanol.

2.1.2 Quantitation. DNA absorbs strongly in the UV region with a maximum at the 260 nm wavelength (see Chapter 2). Thus historically the solution of extracted DNA could be quantitated by UV spectrophotometery. Given that both RNA and proteins also absorb in this region an assessment of the ratio of absorbance at 260 and 280 nm gave both an indication of DNA concentration and purity. This method was superseded by the use of DNA-intercalating dyes such as Hoechst 33258 or PicoGreen which have a high specificity for binding to DNA and which have a distinctive fluorescent emission spectra when bound to DNA. These latter methods have the advantage of being able to work with quite small samples which was of considerable importance given that one did not wish to use up a forensic sample purely in quantitating it. More laborious methods involving probing the DNA for an invarient section of human DNA alongside reference samples can also be used.

2.1.3 Restriction Endonuclease Digestion of DNA. This as a key aspect of the method and the choice of restriction enzyme is paramount. Each enzyme will cut at a different sequence and while the requirement is to cut in the flanking regions as near to the tandem repeats as possible, it is important to choose an enzyme which will not cut within the tandem repeats. In the main, laboratories in the UK and continental Europe selected *Hin*f1 as the enzyme of choice because it would restrict the DNA to produce fragments containing tandem repeats which would be identified by a predetermined set of probe. In this context it was realized that the forensic science laboratories in Europe would not be able to exchange results unless they all used the same enzyme and some common probes. However, laboratories in the USA chose an enzyme (*Hae*III) which cut the DNA to produce a different set of repeat units. Unfortunately this enzyme would cut within the minisatellites chosen by Europe and *vice versa* thus international DNA profiling developed along two parallel but incompatible methodologies.

The restriction endonucleases are naturally occurring bacterial products which will break the phosphodiester bonds at specific sites, *e.g. Hin*f1 recognizes the sequence GANTC (where N is any nucleotide). When complete, it is necessary to test for successful restriction by comparing samples before and after the operation by an assessment of DNA fragment size in a gel electrophoresis test.

2.1.4 Electrophoretic Separation.   In general, for MLP analysis 2–5 μg of DNA are required, whereas for SLP analysis 200 ng to 1 μg of DNA are used for the electrophoretic separation which is carried out at low voltage. The conditions for electrophoresis are arranged such that a spread of fragments ranging from approximately 1 kb to 20 kb can be detected. However, with such a range of fragment sizes resolution suffers particularly in the high molecular range. Control 'ladders' of known fragment sizes (prepared from a λ phage digest) are placed at intervals across the plate to aid in the size assessment of the sample alleles. Following electrophoresis the plate is treated with alkali to provide single stranded DNA. The fragments are then blotted onto a nylon membrane and fixed with short-wave UV light.

2.1.5 Hybridization.   At this point the MLP and SLP methods differ. Lower stringency conditions of temperature and ionic strength are used for the MLP technique to allow the probe to bind to sequences which are similar but not necessarily identical. For the SLP method high stringency is necessary to ensure that the probe will only bind to sequences of exact complementarity such that only one site on each of the homologous chromosomes is detected. In the original method the probes were ^{32}P labelled and detection was achieved through autoradiography of an X-ray film placed over the nylon membrane. The length of time taken to produce a worthwhile result depends to some extent on the amount of DNA which was loaded onto the plate. In a situation where only a small amount of DNA was available the time taken to produce a result on an X-ray film could approach 2 weeks. In a later development a chemiluminescence method was employed in which an alkaline phosphatase was bound to the probe.[3] When hybridization is complete the nylon is sprayed with a solution containing a phosphated luminescent substrate. In the areas where the phosphate is removed by the alkaline phosphatase an unstable intermediate is produced which emits light. Signal amplification occurs *via* fluorescein containing micelles and an image is produced on a photographic film. This latter method has two distinct advantages. Results are produced very quickly (typically 1–2 days) and the need for radioactivity is obviated.

In the SLP method once this process is complete and recorded, the probe can be removed with alkaline detergent and another probe, which will detect another locus, can be applied. This process can be continued with sequential probing until a sufficiently discriminating result is produced.

---

[3] A. F.Giles, K. J. Booth, J. R. Parker, A. J. Garman, D. T. Carrick, M. Akhaven and A. P. Schaap, *Adv. Forensic Haemogenet.*, 1990, 3, 40.

2.2   Analysis of Results

The analysis of MLP results is relatively straightforward as long as the samples to be compared are run simultaneously on the same plate. The tracks which contain the samples for comparison can be assessed visually to determine whether there is exact agreement between the two band patterns. The run conditions used will dictate how many bands are produced with sufficient separation for visual comparison. The DNA fingerprints to be compared needed to be positioned closely to allow for the pattern matching to be achieved by eye. It was generally considered appropriate to make the assessment of the bands greater than about 2 kb in size as below this point the bands became too numerous to be distinguishable.

In SLP analysis the first assessment is made visually (Figure 3), and it is usually obvious if there is no match. When the operator considers that there is a match confirmation is achieved by the use of automated scanning equipment. In this the size of the fragments (alleles) are estimated by reference to the control ladders which contain fragments of known molecular weight. By assaying known samples many times and calculating the variance within the results it is possible to obtain a window within which matches can be determined. The more probes that are used the greater the confidence with which a match can be made.

3   PCR TECHNOLOGY

With the development of the polymerase chain reaction (PCR) technique,[4] forensic science went through as big an upheaval as the discovery of DNA fingerprints had caused. While the SLP method had provided a system which gave extremely valuable information to the criminal justice system, the process time was unacceptably long and often there was not sufficient material to obtain a result. The PCR amplification procedures have provided systems which produce results in a very short timescale from minute amounts of starting material (see Chapter 2).

The PCR technology has become standard practice in so many branches of molecular biology and is, in essence, a method for amplifying short lengths of DNA using oligonucleotide primers. These primers are short lengths of nucleotides which will bind uniquely to the flanking region of the target sequences and initiate the amplification. It is there-

[4] K. B. Mullis, *Scientific American*, 1990, **262**, 36.

MS1   MS43A

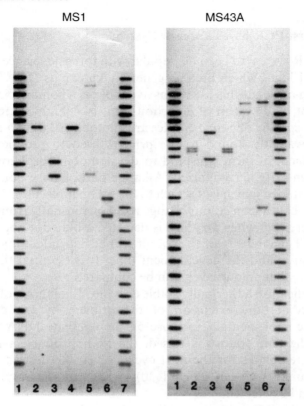

Figure 3   *Single locus probing in crime casework. The samples have been sequentially probed with MS1 (left) and MS43A (right). Tracks 1 and 7 are control allelic ladders. Track 2 is the crime scene sample and track 3 is the profile obtained from the victim. Tracks 4, 5 and 6 are three different suspects. Suspect 1 (track 4) matches the crime scene sample using both probes*

fore necessary to know the sequences within the flanking regions in order to prepare the primers. The amplification reaction requires a thermo-stable DNA polymerase. The primers bind to each of the homologous DNA strands acting like margins between which the DNA polymerase copies the target DNA. In one cycle of the process the DNA polymerase starts at the primers and copies the two strands of the target sequence. The DNA is then quickly heated to a near boiling temperature to separate the copied DNA from the template strands, and cooled again to allow fresh primers to anneal to the two copies of DNA now present. By sequential cycling of temperatures the polymerase can create millions of copies from the template DNA and the products can be analysed in a semi-automated system.

3.1 The First PCR-based Forensic System

The first PCR system to become popular with forensic science was that of HLA DQA1 locus which became generally known as DQα.[5] A method for detecting alleles at this locus was available as a commercial kit which used the 'dot blot' system of detection. This is a sandwich technique in which allele-specific oligonucleotides are immobilized onto a nylon strip. Biotin is covalently bound to the primers used to produce the PCR products which, in turn, will bind to the immobilized oligonucleotides if complementarity is recognized. Alkaline phosphatase bound strepta-vidin is then added which will attach to the biotin molecules. The alkaline phosphatase catalyses a colour change reaction normally from colourless to blue. Therefore, when an allele is detected a blue dot appears on the nylon strip. The intensity of the colour will be dependent upon the level of amplification and the development time. It is essential that controls are present so that colour levels can be compared.

For amplification a thermally stable enzyme, *Taq* DNA polymerase, is used to mediate the extension of the primers by the addition of nucleotides complementary to those on the template DNA. Following extension, the temperature is raised, the PCR products separate and become the templates for the next cycle. When the desired number of cycles (normally between 25 and 30) is reached the resulting allelic products can be detected.

Although the DQα system is still widely employed in forensic science because of its ease of use, it is a relatively uninformative locus with few alleles. However, in the absence of other systems this provided a very good starting point for forensic science laboratories to become acquainted with PCR technology. Other loci rapidly became available, many of which identified sequences which were often too large for reproducible amplification, but some survived which are still commonly used. The systems, however, which have become the technology of choice are those of the short tandem repeats (STR).

4 SHORT TANDEM REPEATS

STRs are a simple cousin of the VNTR in which the repeating sequence is very small, just two to six nucleotides long. By analogy to the mini-satellites the DNA regions containing STRs loci have been termed microsatellites. STRs occur extensively throughout the human genome

[5] C. T. Comey, B. Budowle, D. E. Adams, A. L. Baumstark, J. A. Lindsaey and L. Presley, *J. Forens. Sci.*, 1993, **38**, 239.

and are present in both coding and non-coding regions. The DNA region analysed by the PCR method must include the flanking sequence around the microsatellite for the primers to bind and typically involves DNA fragments with a total microsatellite length varying between 100 and 400 bases. STRs are not as naturally hypervariable in their number of repeats as that found in SLP minisatellites. In fact many of the commonly studied microsatellites only have about 10–20 alleles and consequently are not sufficient in their own right to provide unique identification. It is therefore necessary to examine a larger combination of loci to achieve a high level of discrimination. There are, however, so many different loci available that the choice will depend on the level of discrimination that is desired and which loci provide the basis for commonality between laboratories.

Like SLPs, STRs have a known chromosome location. This means that a package of STRs which are located on different chromosomes can be used. This allows much easier calculation of population frequencies because independence is assured.

Along with the introduction of STRs into casework came the concept of multiplexing in which the alleles from two or more loci are amplified simultaneously in the same analysis reaction tube (Figure 4). It is now commonplace to use a multiplex of seven or more loci and there are commercial kits which contain 13 STRs. When designing a multiplex it is important to ensure that there is no primer–primer interaction and that all alleles will amplify to approximately the same amount. This is not straightforward and it is often necessary to re-engineer the various primers so that their annealing temperatures (determined by their sequence) are comparable.

4.1 Method

The method for the analysis of STRs is essentially the same as that for the detection of DQα genotypes. The differences are associated with the separation and identification of the STR alleles. While there are many ways of determining the size of the PCR products, *e.g.* silver staining, ethidium bromide staining, radiolabelling, the method which has found the greatest popularity is that of fluorescent dye tagging with laser detection during electrophoresis on a DNA sequencer.[6]

4.1.1 Extraction of DNA. DNA can be extracted and denatured in one step using a chelex resin. For most body fluid stains it is advisable to

---

[6] C. J. Fregeau and R. M. Fourney, *Biotechniques*, 1993, **15**, 100.

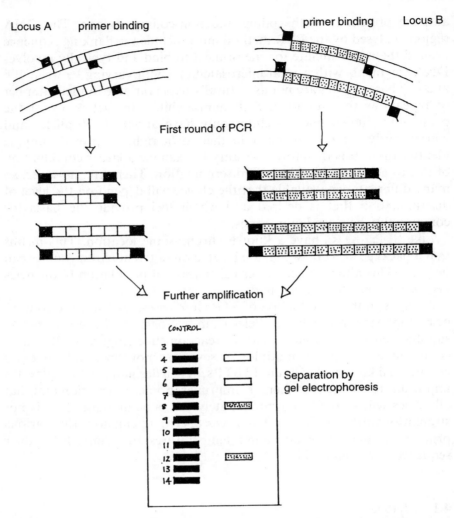

Figure 4 *Multiplexing of STR loci. PCR of locus A and locus B take place in the same tube which contains both primer sets. The control ladder consists of all known alleles from both loci*

perform a wash step before extraction especially if the stain has been subjected to unfavourable conditions. Extraction takes place by boiling in a 5% suspension of the chelex resin followed by centrifugation. The supernatant provides the DNA extract and an aliquot can be used directly for PCR. The combination of the alkalinity of the chelex and the high temperature is responsible for lysis of the cellular material and the denaturation of the DNA. However, as the analysis only requires short lengths of DNA, in general this method produces DNA of sufficient quality for amplification purposes.

4.1.2 Quantitation of DNA. In order to obtain successful results from the extracted DNA it is important to amplify an optimum amount. Too little starting material can result in the production of partial results and too much can give rise to non-specific amplified products. The method must be able to determine nanogram concentrations of human DNA in the presence of contaminants such as bacterial DNA. The slot-blot method is commonly used in which primate specific DNA is determined. This is a sandwich method in which the sample extract is immobilized onto a nylon strip. A 40 bp biotinylated probe, specific for primate sequences, is added. A streptavidin–peroxidase conjugate can then be used to enzymically produce a colour reaction using hydrogen peroxidase as the substrate.

The density of the colour reaction can be compared with a set of prepared controls to make an estimate of the DNA concentration in the extract. The PicoGreen method of quantitation indicated in Section 2.1.2 is also sufficiently sensitive to quantitate DNA concentrates found in the small samples used for STR analysis.

4.1.3 Amplification of DNA. A thermal cycler is employed to facilitate rapid and accurate changes in temperature. Initially the sample is denatured at high temperature (95°C for 60 s) to provide the single strands which become the template material. The temperature is then lowered (~ 54°C for 60 s) to allow the primers, one for each DNA strand, to anneal to the flanking regions of the target sequences. The reaction is then adjusted to the optimal temperature for the *Taq* DNA polymerase which catalyses the extension of the primers is 72°C. This is then maintained for 60 s.

The number of temperature cycles will depend on the loci being amplified and is usually adjusted by experiment. Likewise, the temperature for annealing will be influenced by the primer design and whether singleplex or multiplex analyses are being performed. When analysis of the multiplex STRs is to be performed using a fluorescent detection system such as the ABI 377, the primer sequences need to be tagged with a distinct dye to distinguish one loci's alleles from another. A combination of non-overlapping DNA size ranges and fluorescent labels allows the STR multiplex to provide interpretable products.

4.1.4 Separation of Products. The most common method for separating the PCR products is electrophoresis using automated fluorescent detection in an ABI 373 or 377 gene sequencer. The gene sequencer

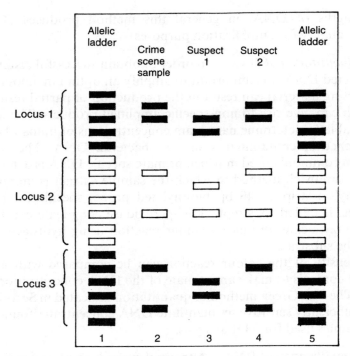

Figure 5   *Allele size measurement from three different loci. Each allele is measured against the control alleles in tracks 1 and 5. As primers for locus 1 and locus 3 have the same dye tag it is important to ensure that there are no overlapping alleles. Sample 2 shows matching alleles with the crime scene sample which is hetero- zygous in all three loci. Suspect 1 is eliminated*

technology has a laser that scans the PCR product DNA fragments as they are size-separated by electrophoresis. With the use of reference markers run at the same time and DNA molecular weight standards which are actually added into each sample for analysis (distinguished by a specific colour dye), it is possible to achieve extremely accurate measurement of the DNA fragment obtained (Figure 5). Accuracy is normally down to a fraction of a single base pair and this is essential, as it has been found that some variants of STRs can differ from the expected repeat lengths by just a base pair. Such variation is of course crucial for identification purposes.

5   DATABASES

For a number of years it has been the aim of forensic scientists to have a database which incorporates the various countries within Europe. The introduction of the STR technology has been instrumental in progressing this initiative. Organisations have been formed to produce a set of recommendations based around the use of STR loci. There are a number of criteria to be considered among which are the issues of the various legal systems and the standards which are necessary to ensure that accurate results are exchanged. The extent to which individual DNA profiles can be collected and stored on computer searchable databases is dependent upon local legislation and will decide the effectiveness of any pan-European database which might be developed.

Although databases have been formed from SLP results it was not until comprehensive legislation was enacted in the UK in 1995 that the first real nationwide DNA database was formed. The use of a multiplex of STRs to form the individual profiles was the basis for the formation of this index. When deciding on which component STRs to use, a number of criteria have to be considered:

(i) number of alleles at each locus.
(ii) ability to multiplex successfully (no preferential amplification).
(iii) minimal artefact production (stuttering and false peaks).
(iv) each locus should be on a different chromosome.
(v) population frequency data should be available.
(vi) results should be easy to interpret.
(vii) reagents must be readily available.

A recent study of European laboratories indicated that there are more than 50 STR loci within regular use and this does not include Y chromosome loci.

Initially, in the UK, the database was designed around a multiplex of four loci: THO1, VWA, FES, F13A. This, however, proved to be a false start and in the developments which followed, six loci were identified which fitted the criteria for forming a successful multiplex: THO1, VWA, FGA, D8S1179, D18S51, D21S11. In 1999 these tests were extended to include 10 loci.

Since the adoption of this multiplex for the UK National DNA Database it has been used routinely in casework and has become known as the Second Generation Multiplex (SGM).[7,8] It has also proved popular with many laboratories within Europe. In an Interpol initiative to register convicted sex offenders the following were identified as the core loci which should be used by all participating laboratories: THO1, VWA, FGA, D21S11. These four loci are commonly used in casework and inter-European laboratory exercises have shown these four loci to be robust

[7] R. Sparkes, C. Kimpton, S. Watson, N. Oldroyd, T. Clayton, L. Barnett, J. Arnold, C. Thompson, R. Hale, J. Chapman, A. Urquhart and P. Gill, *Int. J. Legal Med.*, 1996, **109**, 186.
[8] R. Sparkes, C. Kimpton, S. Gilbard, P. Carne, J. Andersen, N. Oldroyd, D. Thomas A. Urquhart, and P. Gill, *Int. J. Legal Med.*, 1996, **109**, 195.

and efficient in multiplexing. They have become known as the European core loci. In 1999 this was extended to include three more loci.

A more recent development (1998) has seen the introduction an STR based index in North America. Considerable research has culminated in the identification of thirteen loci to comprise the multiplex which will be used: THO1, VWA, FGA, TPOX, CSF1PO, D3S1358, D5S818, D7S820, D8S1179, D13S317, D16S539, D18S51, D21S11. With the interest shown by forensic science laboratories and the commercial potential, kits are now being marketed which offer a choice of loci both for the North American system and the European laboratories. One such kit contains loci which include the SGM and some loci identified in the North American multiplex: THO1, VWA, FGA, D21S11; D8S1179; D18S51; D2S1338, D6S477, D16S539, D19S433. A point has now been reached where there is sufficient overlap between the European laboratories for the effective exchange of information. This contrasts with the situation which existed with SLP profiling in which different enzymes and different loci were analysed and precluded the comparison of results.

6   INTERPRETATION OF THE RESULTS

As with all forensic identification evidence it is necessary to give some information as to the significance of a match. With SLP technology this proved to be the most contentious aspect of the use of DNA profiling.

Because of the broad range of band sizes (approximately 2 kb to 20 kb) and slight changes in electrophoretic mobility of different samples a tolerance 'window' had to be determined, based on experimentation, in order for a match to be declared. Once a match had been decided a numerical value for the total band pattern was required.

The frequency of occurrence of the various alleles was calculated with reference to population frequency databases. In order to multiply the calculated frequencies of each of the alleles, to obtain an overall estimate of the rarity of the profile, it was necessary to show independence within and between chromosomes. These aspects of the technology were aggressively challenged in court, especially in the USA. This culminated in the US National Academy of Sciences sponsoring a study to recommend the best way of dealing with the data for the courts.[9] In the UK it was generally agreed that when there was doubt regarding population substructure the calculation should be modified to give a more conservative value.[10]

[9] National Research Council USA, National Academy Press, Washington DC., 1994.
[10] A. R. Nicholls and D. J. Balding, *Heredity,* 1991, **66**, 297.

In the past it had been normal practice to give straightforward probability evidence when matching profiles were obtained. Modern statistical preferences, however, were for the use of a likelihood ratio, based on a Beyesian statistical approach,[11] in which the posterior odds (chance of guilt) were calculated from the prior odds in which all pieces of evidence were assessed. The DNA result would form one part of the prior odds. However, in a recent appeal court judgement it was considered that forensic scientists should not calculate such an assessment of guilt, but rather they should stick to identification statistics given the observation of specific genetic types.

With the introduction of the STR technology the calculations became more straightforward and the basis of their determination less contentious. The resolution of the system was sufficient to distinguish differences of one base pair and therefore it was much easier to give an accurate estimate of the STR allele by reference to a control set. In addition, STR loci on different chromosomes could be chosen to ensure genetic independence.

Present forensic practice is to present the DNA evidence given the two opposing hypotheses that the DNA found at the scene of the crime came from the defendant *versus* a totally unrelated man of the same ethnic origin. Even though more population studies have now been performed it is still common practice to modify the value of the profile if there is any doubt about population stratification.

7 MITOCHONDRIAL DNA

Mitochondrial DNA (mtDNA) exists as a single circular chromosome and can be used for certain aspects of forensic science. There is considerable sequence variation between individuals with the hypervariable regions, HV1 and HV2 in the D loop, being the areas commonly examined within forensic science (Figure 6).[12]

In general terms PCR methodology is used in which each hypervariable region has two primer sets, the products of which overlap. Sequencing of the PCR products indicates the points at which differences occur. The process, especially the interpretation of the results, requires some considerable skill and experience and the methodology is generally very labour intensive. The system is not as useful as other methods for

[11] D. J. Balding and P. Donnelly, *Crim. Law Rev.*, 1994, **October**, 711.
[12] S. Anderson, A. T. Bankier, B. G. Barrell, M. H. L. deBrujin, A. R. Coulson, J. Drouin, I. C. Eperon, D. P. Nierlich, B. A. Roe, F. Sanger, P. H. Schreier, A. J. H. Smith, R. Staden and I. G. Young, *Nature*, 1981, **290**, 457.

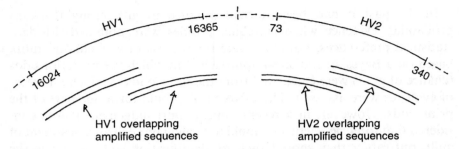

Figure 6   *Diagram of mtDNA showing the variable regions (HV1 and HV2) of the D loop. When amplified the PCR products are sequenced. The overlap allows for confirmation of the procedure*

determining differences between individuals but because there is such a high copy number in each cell there is a much greater chance of detecting mtDNA from old and degraded material and particularly hairs. Hair shafts, for example, have very little or no genomic DNA and, in these circumstances, the forensic scientist can only use mtDNA analysis,[13] A simplified mtDNA 'profiling' method has been developed which assesses polymorphisms at ten key hypervariable sites in the D loop.[14]

Inheritance of mtDNA is *via* the maternal line and this is both a strength and a weakness. For mainstream forensic science it precludes the separation of siblings but in other investigations, where comparison with relatives are required, this can be an advantage. When trying to identify the badly burnt bodies from the Waco siege in Texas, forensic scientists were able to use mtDNA sequences and compare the results with living relatives. The Armed Forces in the USA have also found this method extremely useful for identifying, for example, the remains of soldiers killed in battle, particularly the Vietnam war.

Unfortunately, because some mtDNA hypervariable sites are highly mutable, the system is found to contain some heteroplasmy, where slightly different sequences are found within and between samples from tissues obtained from the same person. Reported differences in mtDNA analysis of hairs taken from one individual indicate that caution must prevail when either eliminating or including an individual in a crime investigation.

---

[13] M. R. Wilson, D. Polanskey, J. Butler, J. A. DiZinno, J. Replogle and B. Budowle, *Biotechniques*, 1995, **18**, 662.
[14] G. Tully, K. M. Sullivan, P. Nixon, R. E. Stones and P. Gill, *Genomics*, 1996, **34**, 107.

8 Y CHROMOSOME ANALYSIS

There are also many STRs in the non-coding regions of the Y chromosome. In contrast to the mtDNA markers the Y-linked STRs follow a male inheritance. Furthermore, as there is minimal recombination at meiosis these genetic markers are inherited in a haplotypic fashion which, barring mutations, will be identical for successive generations of males. Although a considerable number of Y-linked STRs have been described,[15] most are not very polymorphic and therefore not very useful for identifying individuals in forensic science. The most informative of these markers, so far described, is DYS385 which is unusual in that it is derived from two tandemly repeated regions of the chromosome. Following amplification, two products are identified with variable number repeat units and these are interpreted as a haplotype rather than as single alleles.

These Y-linked STRs have already found a place in forensic examinations[16] especially in sexual assault cases. An example is a rape case in which a suspect was eliminated, using a multiplex of six non-sex-linked STR loci, by a mismatch of two alleles. Analysis using Y-linked STRs, however, showed an exact match indicating that there was a possibility that the true perpetrator could be a brother of the suspect. In the event a brother matched in all STR loci used. In a mixture of body fluids, semen is often a minor component and, with the examination of autosomal STRs, it is not unusual to experience difficulty in uniquely identifying those alleles originating from the rapist. This situation does not occur with Y-linked STRs.

9 THE FUTURE

9.1 Capillary Electrophoresis

In the main, forensic science laboratories are now working towards the compilation of databases which hold the profiles of criminals and results obtained from the analysis of crime scene samples. It has also been agreed that the systems used to compile these databases should be common to all laboratories to allow for exchange of information. At the present time laboratories are holding large numbers of sample results

---

[15] M. Kayser, A. Caglia, D. Corach, N. Fretwell, C. Gehrig, G. Graziosi, F. Heidorn, S. Herrmann, B. Herzog, M. Hidding, K. Honda, M. Jobling, M. Krawczak, K. Leim, S. Meuser, E. Meyer, W. Oesterreich, A. Pandya, W. Parson, G. Penacino, A. Perez-Lezaun, A. Piccinini, M. Prinz, C. Schmitt, P. M. Schneider, R. Szibor, J. Teifel-Greding, G. Weichold, P. de Knijff, and L. Roewer, *Int. J. Legal Med.*, 1997, **110**, 125.

[16] M. A. Joblin, A. Pandya, and C. Tyler-Smith, *Int. J. Legal Med.*, 1997, **110**, 118.

together with population data based on a defined set of STRs. In order to get this far it has been necessary to invest considerable resources and, consequently, any progress which is made in DNA profiling is only ideal if it can utilize the results which are currently held. For this reason capillary electrophoresis appears to provide a means of performing the same analyses in a more efficient way.[17]

Capillary electrophoresis offers greater efficiency due to fully automated sample loading procedures and faster analysis time. This is a high resolution technique incorporating silica-fused capillaries containing a semi-liquid non-sieving polymer which allows for high voltage separations. Although the separated DNA fragments could be detected by UV absorption, laser induced fluorescence is far more sensitive and therefore the use of dye-tagged primers is also employed in this methodology. In a short time there have been many improvements which include better separation polymers and automated matrix exchange between each run to avoid the chance of contamination. Interlaboratory exercises have shown that results obtained from capillary electrophoresis and conventional slab-gel methods produce identical results.

Multi-capillary equipment is now available and if this proves to be reliable for the accurate fragment sizing of STR multiplexes it could well replace the slab-gel systems in current use.

9.2   DNA Chip Technology

The rapid rise in the demand for systems which will use genetic information in clinical diagnostics has been responsible for the commercial interest in producing a technology which is sensitive, fast and relatively inexpensive. Although there are a number of variations, chip technology, in general, involves the immobilization of very small amounts of target DNA on a solid support and the use of an automated reader (see Chapter 2). Thousands of microdots of different genetic sequences can be robotically deposited and fixed onto a 2 cm square of solid support (glass, plastic, *etc.*). For diagnostic purposes large numbers of permutations of sequence variations for a particular clinical trait can be used for matching against samples from individuals. Primer dye tagging of the patients sample allows for automated optical reading of hybrids which are formed with the target sequences. Theoretically this technology is extremely sensitive and highly specific with a fast analysis time. As the whole procedure can be automated it is also claimed to be very cost effective.

[17] Y. Wang, J. M. A. Wallin, J. Ju, G. F. Sensabaugh, and R. A. Mathies, *Electrophoresis*, 17, 1485.

A variation on this theme involves the use of matrix assisted laser desorption/ionization time-of-flight mass spectrometry (MALDI-TOF-MS) for the measurement of the deposited DNA sequences. The DNA sample is combined in an organic matrix prior to deposition and subjected to laser irradiation for analysis. The DNA/matrix is volatilized and ionized by the laser pulse and the gas phase ions are analysed by the measurement of the time-of-flight in the mass spectrometer.

Although it is claimed that microsatellites can be analysed using these methods the major thrust of chip technology is aimed at the detection of single nucleotide deletions or substitutions and this is a considerable departure from the systems used at present.

Single nucleotide polymorphisms (SNPs) present particular problems to forensic science not least of which is the fact that the results are not compatible with current STR profiles which constitute national databases. Furthermore, as the necessity in forensic science is individualisation, many hundreds of these SNPs may need to be analysed from each sample because each, in its own right, is likely to be only dimorphic. Intriguingly, the statistical chance that a random man might have the same SNP profile as that from the scene of crime material may actually be quite high because many dimorphic loci will be commonly found in the population.

It has always been an agreed principle that only non-coding regions of the DNA would be investigated for forensic casework because of the ethical consideration associated with inadvertent knowledge of a person's predisposition to a particular disease. At the moment SNP databases are being generated for clinical diagnostic use and there might be the temptation for forensic scientists to use these existing databases. This has the potential for creating a challenging ethical dilemma in the future. There is an inevitability in the way that progress is leading towards a fully automated laboratory and it can be envisaged that a completely robotic process from reception through extraction, quantitation, amplification and analysis to interpretation is only a short time away. Further still, point of testing, where the laboratory process is miniaturised into a mobile system, is an obvious sequel for the future. The challenge for such systems will be to separate PCR product-rich environments from the sample input area. Any deviation from the established principles could result in serious challenge to the creditability of the results. The criminal justice system is going to have a difficult task in coming to terms with the emerging technologies and it is important that evidential safeguards are maintained.

CHAPTER 12

Vaccination and Gene Manipulation

MICHAEL MACKETT

1 INFECTIOUS DISEASE – THE SCALE OF THE PROBLEM

Vaccination against viral and bacterial diseases has been one of the success stories of human and veterinary medicine. Probably the most outstanding example of the effectiveness of vaccination is the eradication of smallpox. In 1967 between 10 and 15 million cases of smallpox occurred annually in some 33 countries. By 1977 the last naturally occurring case was reported in Somalia. Polio too has been controlled in developed countries, for example, the number of cases in the USA was reduced from over 40 000 per year in the early 1950s before a vaccine was available, to only a handful of cases in the 1980s. In 1998 polio was declared eradicated from the Americas. Eradication is not an option for some pathogenic agents such as *Cornybacterium diphtheriae* and *Clostridium tetanii* which are commensal agents widely distributed in nature and, unlike smallpox, which only had a human reservoir for re-infection, are unlikely to be eliminated from the environment. Consequently vaccination against these bacteria will be required for the foreseeable future. The dire effect of ceasing vaccination can be seen in the Newly Independent States of the former Soviet Union where between 1991 and 1996 some 52 000 cases of diphtheria were reported resulting in 1700 deaths. Despite these tremendous achievements infectious disease is still a major global issue. Table 1, compiled from the 1999 WHO World Health Report[1] (see http://www.who.int/dsa/cat98/world8.htm), shows the enormous scale of the problem, estimates for 1998 suggest that over 12.5 million deaths each year are attributable to an infectious disease.

The International Community has responded vigorously to the chal-

[1] The World Health Report, 1999, 'Making a difference', WHO, Geneva, ISBN 92 4 156194 7.

Table 1 *Illness and mortality burden for selected infectious diseases*

Condition	DALYs[a]	Deaths ($\times 1000$)
Respiratory disease[b]	85 085	3507
HIV/AIDS	70 930	2285
Diarrhoea[c]	73 100	2219
Tuberculosis	28 189	1498
Malaria	39 267	1110
Measles	30 255	888
Tetanus	12 979	410
Pertussis	13 226	346
Syphilis	4967	159
Tropical diseases[d]	10 984	106
Hepatitis B	1700	92

[a] Disability adjusted life years.
[b] This includes influenza viruses and a number of other viruses as well as a bacterial pathogens.
[c] This includes rotaviruses and a host of bacterial and parasitic pathogens.
[d] This includes trypanosomiasis, leishmaniasis, shistosomiasis, onchoceriasis and filariasis.

lenge and the World Health Organization (WHO) expanded programme of immunization, now called the Children's Vaccine Initiative (CVI), has immunized the majority of the world's children against six major infectious diseases. Vaccines included in the WHO children's vaccine initiative are: diphtheria, pertussis, tetanus, poliomyelitis, measles, tuberculosis, Japanese encephalitis in parts of the Far East, and yellow fever in parts of Africa. However, if the CVI were completely successful less than a quarter of the diseases in Table 1 would have been dealt with. Thus there is compelling need for new vaccines to viral, bacterial and parasitic pathogens.

Even in developed countries where many of the diseases listed in Table 1 are not a major problem or have been controlled by vaccination, new vaccines are being introduced regularly. For example, from 1993 all children under 6 months old in the UK receive vaccinations against *Haemophilus influenzae B* to prevent small numbers of cases of meningitis associated with infection of this bacterium. Multivalent capsular polysaccharide vaccines to prevent pneumonia due to pneomococcus infection are currently being introduced for at risk groups and it is likely that in the near future vaccination against meningococcus will become more widespread because of recent outbreaks of meningitis in schools and colleges in the UK.

Not only are new vaccines required but more effective and safer vaccines than those currently used are also still needed. For example, a more effective vaccine against cholera is desirable. The current vaccine is at most effective in only 50% of vaccinees and the duration of immunity is relatively short. Similarly inactivated virus vaccines for influenza confer protection in only 30% of vaccinees although about 70% have reduced disease when exposed to influenza. Reversion to virulence by the live, attenuated poliovirus vaccine is responsible for poliomyelitis at a frequency of 1 to 2 cases per million vaccinations.

2 CURRENT VACCINATION STRATEGIES

Table 2 summarizes the standard vaccines available in the UK and illustrates the widespread use of both live, attenuated, killed and subunit vaccines.[2] These are examples from human infectious disease, however there are equally successful vaccines from the point of view of veterinary medicine. Leptospirosis, rinderpest and foot and mouth in cattle, clostridial diseases in sheep, cattle and pigs, diarrhoea caused by enterotoxic *Escherichia coli* in piglets, and Newcastle disease and Marek's disease in poultry are all examples of diseases in animals for which highly effective vaccines are available.

2.1 Inactivated Vaccines

Inactivated vaccines are made from virulent pathogens by destroying their infectivity usually with β-propiolactone or formalin to ensure the retention of full immunogenicity. Vaccines prepared in this way are relatively safe and stimulate circulating antibodies against the pathogens surface proteins, thereby conferring resistance to disease. Two or three vaccinations are usually required to give strong protection and booster doses are often required a number of years later to top up flagging immunity. Subunit vaccines can be seen as a subcategory of inactivated vaccines because similar considerations apply to subunits and whole organisms. Doses, routes, duration of immunity and efficacy of these vaccines are all very comparable. In this case a part of the pathogen, such as a surface protein, is used to elicit antibodies that will neutralize the pathogenic agent. The widespread use of hepatitis B virus surface antigen purified from the blood of carriers[3] or more recently from recombinant yeast[4] shows that this can be a very effective way to immunize. Hepatitis

---

[2] Compiled from 'Immunisation against Infectious Disease', HMSO, London, 1996, ISBN 0-11-321815-X.
[3] 'Hepatitis B Vaccine', INSERM Symposium 18, ed. P. Maupas and P. Geusry, Elsevier, Amsterdam, 1981.
[4] P. Valenzuela, A. Medina, W. J. Rutter, G. Ammerer, and B. D. Hall, *Nature (London)*, 1982, **298**, 347.

Table 2   *Vaccines intended for use in the UK.*
(Compiled from 'Immunisation against Infectious Disease', HMSO,
London, 1996 ISBN 0-11-321815-X)

Vaccination practice in the UK			
Vaccine	*Type*	*Age/reason given*	*Route*
Anthrax	Inactivated whole bacterium	High risk	Intramuscular
BCG (tuberculosis)	Attenuated *M. bovis*	10–14 years	ID
Diphtheria Pertussis Tetanus	Toxoid	2, 3 and 4 months, diphtheria and pertussis before school entry, prior to school leaving	Intramuscular or deep subcutaneous
Haemophilus influenzae B	Capsular polysaccharide conjugated to protein (diphtheria, tetanus, *etc.*)	2, 3 and 4 months (with diphtheria, pertussis and tetanus)	Intramuscular or deep subcutaneous
Hepatitis A	Inactivated whole virus	Travel/high risk	Intramuscular
Hepatitis B	Recombinant HBsAg	Increased risk, lifestyle, occupation, *etc.*, three doses	Intramuscular
Influenzae	Trivalent, inactivated whole virus or purified HA/N	Increased risk	Intramuscular or deep subcutaneous
Japanese encephalitis	Inactivated whole infected cell	Travellers to South East Asia/Far East. Three dose	Subcutaneous
Measles/mumps/ rubella (MMR)	Live attenuated	12–15 months, pre-school	Intramuscular or deep subcutaneous
Meningococcal	Capsular polysaccharide (A + C)	High risk/travel	Intramuscular or deep subcutaneous
Pneumococcal	Capsular polysaccharide (23/84 capsular serotypes)	High risk	Intramuscular or subcutaneous
Poliomyelitis (OPV/ IPV)	OPV	2, 3 and 4 months, before school entry, prior to school leaving	Oral
Rabies	Inactivated whole virus	High risk/post-exposure. Three to five doses	Intramuscular or subcutaneous
Smallpox and vaccinia	Live attenuated	Laboratory workers?	Scarification
Tick borne encephalitis	Inactivated whole virus	High risk/travel. Two doses	Intramuscular
Typhoid	(1) Monovalent whole cell (2) Vi capsular polysaccharide (3) Ty21A	Travel/laboratory workers	(1) Intramuscular (2) Intramuscular deep subcutaneous (3) 3 × oral
Yellow fever (17D)	Live attenuated	Travel	Deep subcutaneous

B virus surface antigen, the product of a single gene, assembles into a highly antigenic 22 nm particle which if used in three 40 μg doses at 0, 15 and 6 months gives virtually complete protection against infection with hepatitis B virus. Another example that can be included in the subunit vaccine class is the use of bacterial toxoids. Many bacteria produce toxins which play an important role in the development of the disease caused by a particular organism. Thus vaccines against some agents, for example tetanus and diphtheria, consist of the toxin inactivated with formaldehyde conjugated to an adjuvant. Immunization protects from disease by stimulating antitoxin antibody which neutralizes the effects of the toxin. A further type of vaccine included in the subunit category is the capsular polysaccharide vaccines, for example, those against *Haemophilus influenzae* and meningococcal meningitis. In this case an extract of the polysaccharide outer capsule of the bacterium is used as a vaccine and is sometimes conjugated to protein to improve immunogenicity. Antibody persists for several years and is able to protect against the bacterium.

2.2 Live Attenuated Vaccines

Table 2 also shows that over half of the standard vaccines used in the UK are live attenuated derivatives pathogenic organisms. In effect live vaccines mimic natural infection, yet produce subclinical symptoms often eliciting long lasting immunity. Most of today's attenuated vaccine strains have been derived by a tortuous, rather empirical, route involving passage in culture until they were found to have lost the virulence of the parental pathogen. This virulence is tested in animal model systems before being tested in human volunteers. For example, the vaccine used to immunize against tuberculosis was derived after 13 years' passage in bile-containing medium by Calmette and Guerin (hence the name BCG – bacille Calmette–Guerin).

An example of the fact that many approaches to a vaccine can be taken is that of the current typhoid vaccine. In the UK three different vaccines have been licensed. One is a killed whole cell vaccine, a second is based on a capsular polysaccharide extract of typhoid, and the third is a live attenuated strain of *Salmonella typhi* (Ty21a).

2.3 The Relative Merits of Live *versus* Killed Vaccines

There has been much debate over the last 45 years as to the relative merits of live and killed vaccines, often generating more heat than light! The evidence is that either route will give adequate vaccines that can be

Table 3   *Relative merits of live versus killed vaccines*

		Live	Killed/subunit
Production	Purification[a]	Relatively simple	More complex
	Cost	Low[b]	Higher
Administration	Route	Natural or injection	Injection
	Dose	Low, often single	High, multiple
	Adjuvant	None	Required[c]
	Heat lability	Yes	No
	Need for refrigeration[d]	Yes	Yes
Efficacy	Antibody response	IgG; IgA	IgG
	Duration of immunity	Many Years	Often Less
	Cell mediated response	Good	Poor
Safety	Interference	Occasional OPV only[e]	No
	Reversion to virulence	Rarely[f]	No
	Side effects	Low Level[g]	

[a] Increasing safety standards mean that for new vaccines some of the older methodologies would not be acceptable. [b] The price for new vaccines will approach that of killed subunit vaccines as safety standards are increased. [c] Very few adjuvants for human are acceptable. [d] The need for refrigeration increases the costs significantly. [e] Especially in the developing countries. [f] At very low levels, 1–2 cases per 10^6 vaccinations. [g] This varies from occasional mild symptoms with rubella and measles vaccines to poliomyelitis.

used to protect against disease under the appropriate conditions. Table 3 shows some of the major points of debate which are discussed more fully in reference 5. Many factors including cost, safety, number of immunizations, ease of access to vaccines, politics and social acceptance will determine whether there is a high uptake of a particular vaccine and whether it is ultimately successful in eradicating the target disease. Even if a perfectly viable, relatively safe vaccine is available uptake may be limited. For example, it has been estimated that vaccination against measles up to 1993 by WHO CVI prevented over 60 million cases and 1.37 million deaths. Despite these efforts there are still many millions of cases of measles annually resulting in nearly 1 million deaths, mostly in children.

The WHO acknowledges these shortcomings and at one congress[6] adopted the goals of increasing immunization coverage, improving surveillance, developing laboratory services and improving vaccine quality. Other goals included improved training, promoting social

[5] C. Mimms and D. O. White, 'Viral Pathogenesis and Immunology', Blackwell Scientific Publications, Oxford, 1984.
[6] 42nd World Health Assembly, Expanded Programme on Immunisation, WHO, 42, 32, WHO, Geneva, 1989.

mobilization and developing rehabilitation services, as well as improving research and development.

These goals serve to illustrate the importance of factors other than the efficacy of the vaccine itself in disease prevention. Health education, improved living conditions, nutrition and sanitation conditions all impact on the effectiveness of a vaccine. Indeed in many cases vaccines would not be needed if these factors could be addressed.

The single most important issue in developed countries is the safety of a vaccine, a single death in a million vaccinations for a new vaccine would be unacceptable (except possibly if it were an effective AIDS vaccine). While this is obviously important in a third world country other issues such as cost and how to deliver the vaccine are of paramount importance.

3 THE ROLE OF GENETIC ENGINEERING IN VACCINE IDENTIFICATION, ANALYSIS AND PRODUCTION

3.1 Identification and Cloning of Antigens with Vaccine Potential

Many pathogens are virtually impossible to culture outside their natural host and this makes it unlikely that conventional approaches to vaccination would be successful. For example, hepatitis B virus (HBV), the agent of human syphilis (*Treponema pallidum*) and the bacterium that causes leprosy (*Mycobacterium leprae*) have never been grown *in vitro* although they can be propagated in animal models. Consequently it is not possible to generate live attenuated or inactivated vaccines by culturing the agents. Recombinant DNA technology allows the transfer of genetic information from these fastidious organisms to more amenable hosts such as *E. coli*, yeast, or mammalian cells. Not all protective antigens are as simple to identify, clone and express as the surface antigen gene of hepatitis. The entire sequence of the HBV genome became available[7] and was found to be less than 10 kb in size. Consequently it was relatively simple to establish which open reading frame to express. It has been known for many years that irradiated malarial sporozoites can protect against malaria.[8] As the sporozoite stage in the life-cycle of the malarial parasite can only be grown in small quantities it was left to recombinant DNA technology to identify, clone and express components of the sporozoite that might be of use in vaccine production. The genome of the malarial parasite is many thousands of times larger than the genome

[7] Y. Ono, H. Onda, R. Sasada, K. Igarashi, Y. Sugino, and Nishioka, *Nucleic Acids Res.*, 1983, **11**, 1747.
[8] V. Nussenzweig and R. S. Nussenzweig, *Cell*, 1985, **42**, 401.

of HBV and therefore provides a different scale of problem. Not only was there little sequence data available but there was also no idea of which gene products may be protective. See Section 3.1.3 for detail of the initial cloning of the malarial sporozoite surface antigen. The starting point of any recombinant DNA work is to generate a library of DNA in *E. coli* which is representative of the organism under study. Once having a cDNA bank or a genomic library there are a number of basic ways of identifying and isolating a gene of interest. With advances in sequencing technology and the routine nature of the generation of cDNA and genomic libraries sequence information derived from whole genomes or individual chromosomes is becoming increasingly important.

3.1.1 DNA/Oligonucleotide Hybridization.   If there is some pre-existing knowledge of the nucleic acid sequence, or where purified mRNA is available, it is possible to detect recombinant clones by hybridization of ^{32}P labelled DNA or RNA to bacterial colonies or bacteriophage plaques. Often a protein has been purified and some amino acid sequence is available which allows a corresponding nucleic acid sequence to be synthesized. Due to the degeneracy of the genetic code a complex mixture of oligonucleotides is required to ensure that all possible sequences are represented. Labelling this mixture of oligonucleotides yields a probe that can be used to screen a cDNA (or possibly genomic) library that might be expected to contain the gene of interest.

3.1.2 Hybrid Selection and Cell-free Translation.   A second approach is to use hybrid selection of mRNA coupled with cell-free translation. DNA clones from a library, either individually or in pools of clones can be immobilized by binding to a solid support and mRNA hybridized to them. Only the mRNA that corresponds to the clones will bind and this can then be eluted and translated to protein in a cell-free system. The protein can then be immunoprecipitated with antisera to the gene product of interest or assayed for activity. An example that encompasses both this approach and the sequence route is in the development of a vaccine for Epstein–Barr virus (EBV). It had been known since 1980[9] that antibody to the major membrane antigen of the virus (gp350/220) would neutralize the virus. Around 1983 a fragment of the virus genome was cloned and sequenced; using computer predictions the gp340/220 gene was identified.[10] The experimental evidence that confirmed this prediction was published in 1985[11] and came from experimental work

[9] J. R. North, A. J. Morgan, and M. A. Epstein, *Int. J. Cancer*, 1980, **26**, 231.
[10] M. Biggin, P. J. Farrell, and B. G. Barrell, *EMBO J.*, 1984, **3**, 1083.
[11] M. Hummel, D. Thorley-Lawson, and E. Keiff, *J. Virol.*, 1984, **49**, 413.

that managed to hybridize select EBV mRNA using genomic DNA clones. This was followed by cell-free translation of the eluted mRNA and immunoprecipitation of gp340/220 with a high titre antibody. The DNA clone that hybridized with the gp340/220 mRNA was then mapped to an open reading frame which agreed with the one predicted by computer analysis. The hybrid selection approach is rather labour intensive and has for the most part been superseded by one of the forms of expression cloning.

3.1.3 Expression Cloning. This approach is invaluable when the only means of identification is an antisera against the protein or pathogen of interest.

Probably the most laborious form of this approach is its use in conjunction with a biological assay. cDNA libraries are cloned into a plasmid that will allow expression in eukaryotic cells, *e.g.* SV40 or EBV vectors. Clones or pools of clones are then transferred to appropriate cell types, *e.g.* COS cells for SV40 vectors and cell extracts or cell supernatant is assayed for biological activity. If a pool of clones gives the biological activity then the individual clones can be re-assayed and the desired cDNA clone identified. This methodology, although tedious, has allowed many of the interleukin genes to be cloned probably because the assays for these proteins are very sensitive.

Other gene products or vaccine antigens may require an enrichment step. For example, many genes expressed on the cell surface, *i.e.* receptors, adhesion molecules, *etc.*, have been cloned by 'panning' techniques where the cells expressing the gene of interest are selected out either with antibody or by interaction with other cells. cDNA libraries are constructed in *E. coli* and the library is transferred to eukaryotic cells. Those cells expressing the gene of interest are enriched for and the library transferred back to *E. coli*. This can be done for several rounds of expression and eventually individual clones conferring the selected phenotype will be isolated.

The most extensively used form of expression cloning involves the use of plasmid or bacteriophage vectors in *E. coli* and identification of DNA clones using antisera to the protein of interest. Here a vector such as the bacteriophage λgt11 is set up so that when cDNA fragments are cloned into sites adjacent to the β-galactosidase gene bacteria will express a β-galactosidase fusion protein containing epitopes present in the cDNA. Recombinant bacteriophage are detected with antisera.[12] The cDNA

[12] C. K. Stover, V. F. de la Cruz, T. R. Fuerst, J. E. Burlien, L. A. Benson, L. T. Bennet, G. P. Bansal, J. F. Young, M. H. Lee, G. F. Hatfull, S. B. Snapper, R. G. Barletta, W. R. Jacobs, and B. R. Bloom, *Nature (London)*, 1991, **351**, 456.

insert is then sequenced and the whole gene can then be isolated in a more traditional way. The antisera used can be a monoclonal antibody, polyclonal monospecific antisera or even polyclonal antisera with many antibody specificities present. A variation in this method allowed the initial cloning of the malarial sporozoite surface antigen.[13] Malarial sporozoite stage cDNAs were introduced into the ampicilin resistance gene of the plasmid pBR322. Low levels of expression of the sporozoite surface antigen were detected by solid phase radioimmunoassay using a monoclonal antibody specific for the protein. In this way a cDNA clone coding for the antigen was isolated and subsequently sequenced. This information was then used to design peptide vaccines based on a repeat found in the antigen. Several human clinical trials been carried using this information. The λgt11 system is a more utilized and sophisticated system than that described above and has been used to isolate various antigens from stages in the life-cycle of the malarial parasite using human immune sera.

3.1.4   Genomic Sequencing. With the improvement in sequencing technologies, particularly as a result of automated fluorescent dye technology, there is an explosion of DNA sequence information available from a wide number of pathogens. Table 4 illustrates what is fast becoming the tip of an iceberg. The institute of genome research (TIGR) database is available (http://www.tigr.org/tdb/mdb/mdb.html) and lists 23 complete sequences of genomes or chromosomes, with a further 88 underway. Many of these sequences are from pathogens for which there is no vaccine available and many others for which more effective vaccines would be desirable. Although an entire field of functional genomics has developed around the use of sequence data for drug discovery, much less thought has been given to their use in the production of vaccines. At present there is no algorithm that can be used to identify the targets of protective antibody or T-cell responses from genomic sequences, although some useful approximations have been used. For antibody responses products predicted to reside on the cell surface or be secreted are presumed to be accessible to antibody and therefore potential targets. For T-cell responses (see Section 3.2.2) the subcellular location is less important, the crucial factor being a processed antigens ability to interact with HLA binding epitopes in the sequence. Hoffman and colleagues[14] have suggested that one route to identifying important antigens for B-cell responses is to clone an open reading frame (ORF)

---

[13] J. Ellis, L. S. Ozaki, R. W. Gawdz, A. H. Cochrane, V. Nussenzweig, R. S. Nussenzweig, and G. N. Godson, *Nature (London)*, 1983, **302**, 536.
[14] S. L. Hoffman, W. O. Rogers, D. J. Carucci, and J. C. Ventner, *Nature Medicine*, 1998, **4**, 1351.

Table 4 *Completed genomes and chromosomes of selected pathogens*

Genome	Associated disease/comment	Size (Mb)	Reference
Haemophilus influenzae Rd	Pnuemonia	1.83	a
Helicobacter pylori strain 26695	Stomach ulcers/cancer	1.66	b
Helicobacter pylori strain J99		1.64	c
Borrelia burgdorferi	Lyme disease	1.44	d
Mycobacterium tuberculosis	Tuberculosis	4.40	e
Saccharomyces cerevisiae	Highly attenuated yeast	13	f
Treponema palladium	Syphilis	1.14	g
Chlamidia trachomatis	Trachoma, STD	1.05	h
Chlamidia pneumoniae	Pneumonia	1.23	i
Plasmodium falciparum (chromosome 2)	Malaria	1.00	j
Leishmania major (chromosome 1)	Leishmaniasis	0.27	k

[a] Fleischmann *et al.*, 1995, *Science*, **269**, 496.
[b] Tomb *et al.*, 1997, *Nature (London)*, **388**, 539.
[c] Alm *et al.*, 1999, *Nature (London)*, **397**, 176.
[d] Fraser *et al.*, 1997, *Nature (London)*, **390**, 580.
[e] Cole *et al.*, 1998, *Nature (London)*, **393**, 537.
[f] Goffeau *et al.*, 1997, *Nature (London)*, **387**, (suppl.) 5.
[g] Fraser *et al.*, 1998, *Science*, **281**, 375.
[h] Stephens *et al.*, 1998, *Science*, **282**, 754.
[i] Kalman *et al.*, 1999, *Nature Genetics*, **21**, 385.
[j] Gardner *et al.*, 1998, *Science*, **282**, 1126.
[k] Myler *et al.*, 1999, *Proc. Natl. Acad. Sci. USA*, **96**, 2902.

into a plasmid and immunize mice with the construct (see Section 5.1). Sera from the immunized mice can then be tested for their ability to react with or neutralize the pathogen of interest. For T-cell responses, HLA binding epitopes can be predicted with some level of success. These epitopes may then be validated as targets of CTL by analysing naturally infected individuals for reactivity to the predicted epitope. A number of peptide epitopes might then be joined together rather like a 'string of beads' and used to generate a vaccine.

While the fruits of this approach are a long way off it is clear that it will become increasingly important for complex pathogens such as parasites and is likely to remain so for some time to come.

3.2 Analysis of Vaccine Antigens

3.2.1 B-cell Epitopes. The structural analysis of a potential vaccine antigen can yield valuable information in the development of a vaccine, for example, a knowledge of the location of epitopes against which neutralizing antibodies can be raised will allow the suitability of peptide vaccines to be investigated (see Section 5.2). Epitopes are usually referred to as continuous or discontinuous. Continuous epitopes are peptides that are recognized in their random coiled form so that antisera to the epitope will react with the whole molecule from which the sequence is derived. Discontinuous epitopes are made up of molecules that are brought together due to the secondary structure of a protein or arise from constraints imposed by the structure of a particular infectious agent. Some neutralizing epitopes are continuous while others are discontinuous.

An example demonstrating the variety of techniques that can be used to analyse vaccine antigens is the VP1 protein of foot and mouth disease virus (FMDV). As long ago as 1973 it was shown that of the four virus structural proteins, only VP1 could induce FMDV neutralizing antibodies.[15] It was not until 1982 that immunogenic epitopes were identified on VPI and this was achieved by the classical method of cyanogen bromide (CNBr) cleavage.[16] CNBr cleaves at methionines and fragments the VP1 protein. Comparison of CNBr cleaved isolated VPl and enzymatically digested virus particles, suggesting that the immunogenic epitopes were contained within amino acid sequences 146–152 and 200–213.

A second, more indirect approach was based on sequence analysis of four isolates of FMDV. It was argued that the most variable regions of VP1 would be those subject to the most immunological pressure for mutation and would thus be the sites of greatest antigenicity. The sequence showed 80% conservation and three highly variable regions, amino acids 42–61, 138–160 and 193–204. As antigenic sites are likely to be on the surface of the virus and therefore hydrophilic in nature, it was assumed (and later shown) that the 42–61 hydrophobic sequence was not antigenic.[17] Subsequently the Pepscan method was used to identify immunogenic sites on VP1. This method is a 'sledgehammer' approach and involves synthesis of overlapping hexapeptides corresponding to the amino acid sequences 1–6, 2–7, 3–8, 4–9, *etc.* These peptides are synthesized on polyethylene rods in a microtitre plate format. Each pin is reacted with neutralizing antisera. Those peptides to which antibody attached are detected by a secondary antibody conjugated to an indicator enzyme. Such an analysis of the protein identified the l46–152

[15] J. Laprte, J. Grosclaude, J. Wantyghem, S. Bernard, and P. Rouze, *C. R. Acad. Sci., Ser. D*, 1973, **276**, 399.
[16] K. Strohmaier, R. Franke, and K. H. Adam, *J. Gen. Virol.*, 1982, **59**, 295.
[17] J. L. Bittle, R. A. Houghten, H. Alexander, T. M. Shinnick, J. G. Sutcliffe, R. A. Lerner, D. J. Rowlands, and F. Brown, *Nature (London)*, 1982, **298**, 30.

region of the molecule as antigen. Several groups went on to show that the 200–213 region of the molecule was also antigenic and, if coupled to the 141–160 peptide gives an improved antibody response compared to either of the two individual peptides.[18]

Although unsuccessful with this particular protein, a further approach is the use of bacteria to express fragments of the protein of interest. Naturally occurring antibody and monoclonal antibodies can then be assessed for reactivity with the bacterially produced proteins. VP1 was found to be extremely difficult to produce in bacterial expression systems, but other virus and bacterial antigens have been analysed in this way.

3.2.2 T-cell Epitopes. T lymphocytes recognize foreign antigens as processed peptides in association with the extracellular portion of the MHC molecule. Helper (CD4 +) T-cells recognize antigen in conjunction with MHC class II whereas cytotoxic (CD8 +) T-cells (CTLs) recognize antigen in association with MHC class I molecules. The genetic polymorphism of Class I and II MHC molecules determines the specificity and affinity of peptide binding in T-cell recognition. This means that individuals with different HLA haplotypes will vary in the specificity of the cell-mediated immune they make to the same antigen. It is therefore clearly important, particularly if protection against disease is mediated by T-cell responses, that the peptides recognized by T-cells in a potential vaccine antigen are identified.

Vaccinia recombinants (see Section 4.2.1) have been particularly useful in identifying targets of CTLs.[19] Processing of a foreign antigens expressed in recombinant vaccinia infected cells appears to take place authentically. CTL generated *in vitro* (or occasionally directly from patients) can be examined by using a panel of vaccinia recombinants to create targets for the CTL. Lysis of a specific recombinant infected cell reveals the target of the cytotoxic activity. Even cells infected with a recombinant expressing a 15 amino acid peptide from the nucleoprotein of influenza A could be lysed by CTL specific for this peptide.[20] Thus in theory it is possible to map the specificity of CTL clones by using a panel of recombinants expressing overlapping peptides to create targets. In practice, however, it is more common to use synthetic peptides in conjunction with computer prediction. Other assays used for detection and prediction of CTL epitopes involve binding of peptides to isolated HLA molecules[21] or stabilizing HLA expression in cell lines such as T2.[22]

[18] H. M. Geysen, R. H. Meleon, and Barteling, *Proc. Natl. Acad. Sci. USA*, 1984, **81**, 3998.
[19] J. R. Bennink and J. W. Yewdell, *Curr. Top. Microbiol. Immunol.*, 1990, **163**, 153.
[20] K. Gould, J. Cossins, J. Bastin, G. G. Brownlee, and A. J. Townsend, *J. Exp. Med.*, 1989, **170**, 1051.
[21] G. S. Ogg and A. J. McMichael, *Immunol. Lett.*, 1999, **66**, 77.
[22] Y. Deng, J. W. Yewdell, L. C. Eisenlohr, and J. R. Bennink, *J. Immunol.*, 1997, **158**, 1507.

Table 5 *Characteristics of major protein expression systems used to express vaccine antigens*

Characteristics[a]	Expression systems				
	Bacterial	Yeast	Mammalian	Baculovirus (insect)	Transgenic animals
Proteolytic cleavage	Possibly	Possibly	Yes	Yes	Yes
Glycosylation	No	Yes[b]	Yes	Yes[c]	Yes
Secretion	Some antigens	Yes	Yes	Yes	Yes
Folding	Possibly	Possibly	Yes	Yes	Yes
Assembly of proteins	Yes	Yes	Yes	Yes	Yes
Phosphorylation	No	Yes	Yes	Yes	Yes
Myristylation	No[d]	Yes	Yes	Yes	Yes
% yield (dry weight)	1–5	1	< 1	30	> 10

[a] Other post-translational modifications could have been included in this table, *e.g.* palmitoylation, acylation, 5′ methionine removal and signal cleavage.
[b] *S. cereviseae* hyperglycosylates some antigens.
[c] Insect cell glycosylation differs from mammalian cell glycosylation. Complex sugars are added only inefficiently to the core glycosylated protein.
[d] When the yeast enyme *N*-myristyl transferase is expressed in *E. coli* it will myristylate foreign genes.

Considering the diversity of the MHC class I and class II molecules and the fact that each haplotype can recognize different peptides, it is surprising to find that algorithms have been developed which have some predictive value.[23] Faced with a molecule where nothing is known about potential CTL epitopes the most straightforward approach for analysis would be to use an algorithm for prediction of CTL epitopes, synthesize the predicted peptides and use these to stimulate CTL *in vitro* or create targets for CTL.

3.3   Generation of Subunit Vaccines

3.3.1   Expression of Potential Vaccine Antigens.   Other chapters in this book refer to cloning and expression of genes in bacteria, yeast and mammalian cell culture. Any of these technologies can be applied to vaccine production (Table 5). In general, however, mammalian cell culture is likely to be the method of choice for vaccines against pathogens that replicate in eukaryotic cells. *E. coli* are unable to carry out some post-translational modifications of some vaccine candidates. For example, bacterial systems cannot add carbohydrate which is important in the antigenicity and structure of many protective antigens from viruses.

[23] J. R. Schafer, B. M. Jesdale, J. A. George, N. M. Kouttab, and A. S. De Groot, *Vaccine*, 1998, **16**, 1880.

One of the best examples of the power of recombinant DNA technology is seen in the development of the current hepatitis B virus (HBV) subunit vaccine.[3,4] About 300 million people worldwide have been infected with the virus. Infection in adults leads to a short acute phase associated with viral replication. In about 10% of cases patients develop a carrier state and are at a high risk from liver cirrhosis and hepatocellular carcinoma. Infection in children leads to much higher carrier frequencies. It has been estimated that 800 000 deaths per year are due to HBV infection and its sequelae. In order to eradicate the virus large quantities of an efficient and affordable vaccine are required, however, traditional approaches were not tenable because, as mentioned previously, the virus cannot be grown *in vitro*. The first vaccine for HBV was licensed in 1984 and consisted of HBV surface antigen protein, which self-assembles into 22 nm particles, purified from the blood of HBV carriers. With the advent of HIV, manufacturers turned to recombinant DNA technology as an alternative source of HBV surface antigen to avoid using blood products. Table 6 shows some of the systems used to express the HBsAg gene along with the date they were reported and with a number of comments.

Interestingly, fairly poor yields of surface antigen in bacteria and a complete failure of the protein to fold properly and assemble into 22 nm particles meant that the most appropriate system for expression proved to be in the yeast *Saccharomyces cereviseae*.[4] Not only were yields of protein reasonable but 22 nm particles were also formed. This, together with experience of scale-up from fermentation technology, made the system particularly attractive. Highly purified preparations of yeast-derived HBsAg particles were shown to be innocuous and have a high protective efficacy in humans and a licence for general use of the vaccine was granted in the USA in 1986. However, this vaccine is not totally effective as about 1% of vaccinees do not respond. In the more developed countries this has meant that individuals who receive the vaccine need to be monitored to ensure the vaccine has 'taken'. This increases the cost of vaccination and as a result vaccine companies have investigated alternative vaccines. It was thought that incorporation of the whole Pre S region of the HbsAg gene or inclusion of an epitope from tetanus toxoid might improve take rates. It is not clear yet whether this will be successful but efforts to improve yeilds and take rates along with increasing levels of hepatitis B infection in North America will ensure continued activity in this area.

Following the success of the hepatitis vaccine and because of the availability of established technology to provide large amounts of antigen the subunit approach is one of the most favoured methods for

Table 6 *Expression of the hepatitis B virus surface antigen in a variety of systems*

Gene/peptide	Vector/host	Year reported/Comment
HBs-β-galactosidase	Plasmid/*E. coli*	1980
HBs-β-lactamase	Plasmid/*E. coli*	1979
HBs	Plasmid/*E. coli*	1983
HBs peptide in *salmonella fimbrae*	Plasmid/*S. typhimurium* attenuated strains	1989; Intended as a prototype for oral immunization against HBV
HBs	Plasmid/*S. cerevisiae* (yeast)	1982; Licensed in 1986 as RECOMBIVAX HB and ENERGEX B
HBs	Chromosomal insertion/*P. pastoris* (yeast) and *H. polymorpha* (yeast)	1987; Methylotropic yeast with five-fold increase over the *S. cerevisiae* system
HBs-HSV gD fusion	Plasmid/*S. cerevisiae* (yeast)	1985
HBs	Adenovirus/mammalian cells	1985
HBs -Tetanus toxoid epitope	Adenovirus/mammalian cells	1999; Tetanus toxoid epitope may improve take rates
HBs	Vaccinia virus	1983
HBs	Varicella Zoster virus	1992
HBs	Alexander cell line (derived from a liver cancer patient)	Investigated as a potential source of HBs but low levels and safety concerns over the use of a tumour cell line limited use
HBs	Baculovirus/insect cells	1987
HBs	SV40/Cos cells	1984
HBs	Bovine papillomavirus (BPV)/ NIH3T3 or C127 cells	1983; Clinical trial of C127/BPV derived material (1990) comparable to other sources of HBs
HBs	Vero (monkey kidney) cells	1984; Cell line licensed for some vaccine preparation
HBs + preS1 region	Chinese hamster ovary cells	1986; Licensed as GENEHEVAC B

the development of new vaccines. Had glycosylation been important in the immunogenicity of HbsAg then mammalian cell culture may have been more appropriate. Indeed one of the licensed hepatitis B vaccines is produced in mammalian cells.

4 IMPROVEMENT AND GENERATION OF NEW LIVE ATTENUATED VACCINES

4.1 Improving Current Live Attenuated Vaccines

Molecular biological techniques allow the analysis of virulence and antigenicity at the molecular level. This enables a more rational approach to be adopted in the generation of attenuated organisms eventually enabling engineering of live vaccines with desired properties. DNA containing viruses and other microorganisms can be engineered directly where the DNA is infectious, *e.g.* adenoviruses and herpes simplex virus, and indirectly in other cases where plasmids are used to transfer information by recombination into the genome. RNA viruses are somewhat more problematical and, although there has been some success with poliovirus (see Section 4.2.4) and influenza virus, the compactness of the virus genome and packaging constraints of virus particles makes it unlikely that vaccines to many RNA viruses will be achieved by engineered attenuation. It should be noted that as the incidence of immune suppression due to HIV infection increases, the number of vaccine-associated complications from live vaccines is also likely to increase. Vaccine manufacturers are by nature very cautious, particularly when litigation from vaccine associated complications is possible. It is therefore likely that they will only favour the live approach under well defined circumstances, particularly if viable subunit or peptide vaccines are available.

4.1.1 New vaccines for Pseudorabies Virus.[24] The first live viral vaccine on the open market produced by recombinant DNA technology was licensed in the USA in January 1986, for use in pigs to combat pseudorabies virus (PrV), a member of the herpesvirus family. The disease is a serious difficulty for swine production because of reproductive problems, death of piglets and increased secondary respiratory disease. Vaccination is used to reduce economic losses and aid in preventing reactivation and shedding of the virus. The recombinant vaccine was constructed by introducing a plasmid with a 148 bp deletion in the PrV thymidine kinase (TK) gene into cells infected with PrV. At a low level homologous recombination takes place between PrV genomic DNA at the TK locus and the deleted *tk* gene carried by the plasmid. The recombinant virus is TK negative and can be distinguished from parental virus by its ability to grow unimpaired in the presence of 5-bromodeoxyuridine (BdUR). The *tk* gene lesion reduces the ability of the virus to

[24] S. Kit, *Vaccine*, 1990, **8**, 420.

establish neuronal replication leading to a latent infection. Since the initial system was worked out other gene-deleted pseudorabies vaccines have been licensed and used in both the USA and Europe. Use of these live attenuated vaccines can lead to problems in diagnosis of natural infections. Is any particular PrV outbreak in pigs due to wild-type virus or reversion of the vaccine? To answer this question antibody based assays have been developed that are specific for the wild-type protein deleted from the vaccine strain. Genetic deletion of genes for specific glycoproteins has also enabled the development of diagnostic kits to distinguish an antibody response to the vaccine from the response to concomitant infection with field strains. A further sophistication has also been incorporated in some vaccines where added non-natural DNA sequences have been used to trace and distinguish vaccine virus strains. Such markers in any recombinant vaccine would be useful for clear identification of the vaccines should they be suspected of circulating in the environment or reverting to virulence. Recombination of live vaccines with field viruses could also be followed if modified viruses were genetically marked in this way.

4.1.2   Improving Attenuation in Vibrio cholerae.   As mentioned previously the currently available cholera vaccine is fairly ineffective. A great deal of time and effect has been expended in analysing the molecular basis for the pathogenicity of *V. cholerae* and in engineering specific deletions for testing as potential vaccines. Cholera is a severe diarrhoeal disease of humans caused by infection of the small intestine by virulent strains of *V. cholerae*. Although these virulent strains of *V. cholerae* may possess a number of virulence determinants the clinical manifestations of cholera are primarily due to the action of the holotoxin on the intestinal epithelia. Cholera toxin is comprised of two distinct subunits, termed A and B, which are encoded by genes located on the bacterial chromosomes. The B subunit specifically binds to the GM 1 ganglioside located on the surface of eukaryotic cells, whereas the A subunit possesses ADP-ribosylation activity resulting in the ribosylation of the Gs protein of the adenylate cyclase system in host cells. Virulent strains of cholera are non-invasive, and therefore stimulation of humoral immunity by the currently available inactivated Vibrio strains is likely to provide only limited protection against the disease. Because the B subunit of the holotoxin is immunogenic, it has been investigated as a vaccine candidate to stimulate mucosal immunity at the site of infection. The production of specific immunoglobulins at the site of infection may lead to neutralisation of holotoxin activity by inhibiting adsorption to target cells.

Recombinant plasmids have been described in which the ctxB determinant is placed under the control of the strong bacterial tac promoter resulting in over expression of the B subunit by strains possessing this plasmid.[25] The introduction of these plasmids into nontoxigenic strains of cholera results in transformants analogous to a current B subunit/whole cell vaccine that has undergone field trails in Bangladesh.[26]

The use of recombinant DNA technology has made it possible to develop live attenuated strains of cholera that can be used as oral vaccines.[27,28] This was achieved by an *in vivo* marker exchange procedure which involved the use of a plasmid in which the toxin A and B subunits were replaced by a mercury resistance gene. This gene and DNA flanking the toxin genes were mobilized into cholera on a plasmid belonging to the P incompatibility group. Homologous recombination occurred with the net result of transferring the mercury resistance gene to the genomic site and eliminating the toxin genes. The recombination was detected by introducing a second IncP plasmid containing sulfur resistance and selecting for both mercury and sulfur resistance. The requirement for sulfur resistance would maintain the second plasmid while eliminating the now incompatible mercury resistance plasmid. If recombination had taken place then a dual resistant Vibrio would have the toxin genes deleted.

The advantages of using attenuated strains of *V. cholerae* as vaccine material relies upon the observation that nontoxigenic strains of *V. cholerae*, when ingested by volunteers, produce a mild diarrhoea.[29] These results indicate that the extreme purging observed in severe cases of cholera is due to the action of the cholera toxin but that other virulence factors produced by the bacteria contribute towards infection. Such a hypothesis is strengthened by the fact that attenuated strains of *V. cholerae* that also possess no *ctx* genes can be used to stimulate a significant degree of protection in volunteers challenged with virulent strains. Virulence determinants other than the holotoxin that have been implicated in mediating infection by *V. cholerae* include colonization antigens and secondary toxins. Therefore, the strategy for the development of a safe efficient cholera vaccine is currently focused upon the use

[25] J. Sanchez and J. Holmgren, *Proc. Natl. Acad. Sci. USA*, 1989, **86**, 481.
[26] J. Clemens, D. A. Sack, J. R. Harris, J. Chakrabarti, M. R., Kahn, B. F. Stanton, B. A. Kay, M. U. Kahn, M. D. Yunus, W. Atkinson, A.-M. Svennerholm, and J. Holmgren, *Lancet*, 1986, **ii**, 124.
[27] J. B. Kaper, H. Lockman, M. M. Baldini, and M. M. Levine, *Nature (London)*, 1984, **308**, 655.
[28] M. M. Levine, J. B. Kaper, D. Herrington, J. Ketley, G. Losonsky, C. O. Tacket, B. Tall, and S. Cryz, *Lancet*, 1988, **ii**, 476.
[29] M. M. Levine, J. B. Kaper, D. Herrington, G. Losonsky, J. G. Morris, M. L. Clements, R. E. Black, B. Tall, and R. Hall, *Infect. Immunol.*, 1988, **56**, 161.

of live, attenuated strains of Vibrio that possess a functional *ctxB* gene but lack the *ctxA* determinant.

4.1.3 Improving Stabilily—Poliovirus. The possibility of producing engineered polioviruses is dependant on the observation that infectious viruses can be rescued from cDNA clones when the appropriate sequence is put under the control of a strong eukaryotic promoter and transfected into susceptible cells.[30] Thus by standard molecular technology it is possible to introduce defined mutations or alterations into the cDNA and rescue a recombinant virus. This technology, coupled with the knowledge of the crystal structure of the virus[31] and a knowledge of the major antigenic epitopes of the virus, will allow more stable polioviruses to be generated and opens the door on generating polioviruses expressing foreign gene epitopes (see Section 4.2.4). The three Sabin attenuated strains of poliovirus have been used successfully for many years to protect against paralytic polio. The strains are very stable but at low frequency it has been shown that the type 2 and 3 strains can revert to virulence and cause vaccine associated paralytic polio. Sabin type[1] however, is much more stable. Two possible approaches to the generation of more stable viruses have been investigated. The first is to replace the type 1 major virus neutralizing epitopes with type 3 epitopes; this would hopefully give as stable a virus as Sabin type 1 and have some of the antigenic characteristics of type 3.[32] A second approach was based on the observation that virulence was in part associated with the 5′ noncoding region of the virus RNA. The 5′ non-coding region of the type 1 strain was used in place of the 5′ non-coding region of the type 3 strain. Here the chimaeric virus RNA codes for a completely type 3 virus particle but hopefully would have a more stable phenotype.[33] These theoretical improvements may not be taken up by vaccine manufacturers for a variety of reasons, not least of which is being able to show conclusively that a complication rate in humans is reduced from 1 in 10^6 vaccinations to an even smaller number.

4.2 Recombinant Live Vectors

This strategy has been pioneered by vaccinia virus recombinants and uses currently available live attenuated vaccines as hosts for foreign

[30] B. L. Semler, A. J. Dorner, and E. Wimmer, *Nucleic Acids Res.*, 1984, **12**, 5123.
[31] J. M. Hogle, M. Chow, and D. J. Filman, *Science*, 1985, **229**, 1358.
[32] K. L. Burke, G. Dunn, M. Ferguson, P. D. Minor, and J. W. Almond, *Nature (London)*, 1988, **332**, 81.
[33] G. Stanway, P. J. Hughes, G. D. Westrop, D. M. A. Evans, G. Dunn, P. D. Minor, G. C. Schild, and J. W. Almond, *J. Virol.*, 1985, **57**, 1187.

genes. With careful choice of the protective antigen gene the immune response to the carrier vaccine and to the foreign gene product can be sufficient to protect against the original target of the host vaccine vector and the pathogen the foreign gene is derived from.

4.2.1 Vaccinia Virus Recombinants. Vaccinia virus has been used for over 150 years as a live attenuated vaccine for the control of smallpox. The cheapness and simplicity of the vaccine to manufacture and administer without refrigeration, its stability with a single inoculation, and stimulation of both cell-mediated and antibody responses are all advantages traditionally associated with vaccinia. These advantages should also be enjoyed by vaccinia recombinants that express foreign genes. Over 1000 different vaccinia recombinants[34] expressing genes from viral, bacterial and parasitic pathogens have been described. Many of them have been shown to protect in animal model systems against challenge with the appropriate pathogen. A vaccinia recombinant expressing the HIV1 envelope glycoprotein gp160 has been tested in humans and shown to induce immunological responses to HIV gp160.[35] However complications associated with vaccination and increasing numbers of individuals with immunodeficiencies (a contraindication for vaccination with vaccinia) may limit the usefulness of recombinants for human vaccination. Plans for use of two vaccinia-based vaccines in animals are, however, well advanced. One vaccine will protect cattle against rinderpest,[36] the other protects wildlife (and hence indirectly the human population) from rabies infection.[37] In fact a recombinant expressing the rabies virus glycoprotein has now been used extensively. It has been shown to induce neutralizing antibody and cytotoxic T-cells in vaccinated animals. More impressively it protects foxes, fox cubs, skunks and racoons against challenge with wild-type rabies virus, even when the vaccine is presented as baited food. Field trials in Belgium, dropping baited food from the air over large areas, have demonstrated that wildlife can be protected from rabies virus and that it appears to be safe with very little spread of the recombinant virus in the environment. Probably the most extensive use of the recombinant is shown in Table 7, where 11 million doses of baited food have been used in Texas. Other states and countries are also using this vaccinia recombinant extensively.

Poxviruses have been found in many species of animal and they often have a limited host range. For instance, fowlpoxvirus will only replicate

[34] B. Moss, *Proc. Natl. Acad. Sci. USA*, 1996, **93**, 11341.
[35] E. L. Cooney, A. C. Collier, P. D. Greenberg, R. W. Coombs, J. Zarling, D. E. Arditti, M. C. Hoffman, S.-L. Hu, and L. Corey, *Lancet*, 1991, **337**, 567.
[36] T. Yilma, *Devel. Biol. Stand.*, 1995, **84**, 201.
[37] P. P. Pastoret, B. Brochier, *Vaccine*, 1999, **17**, 1750.

Table 7   *Texas Department of Health Oral Rabies Vaccine Program (ORVP): Programme Summary 1995– 1999. See http://www.tdh.state.tx.us/zoonosis/orvp/*

Year	Bates dropped	Area covered (square miles)
February 1995	850 000	South Texas: 15 000
January 1996	1 300 000	South Texas: 21 000
	1 200 000	West Texas: 20 000
January 1997	1 500 000	South Texas: 23 000
	1 100 000	West Texas: 18 000
January 1998	1 500 000	South Texas: 24 000
	1 100 000	West Texas: 17 000
January 1999	1 000 000	South Texas: 15 000
	1 700 000	West Texas: 18 100
Totals	11 250 000	(171 100)

in avian species. Thus it is well suited for a vector for expression of vaccine antigens to immunize poultry against pathogens such as Newcastle disease virus. Indeed recombinants expressing genes from Newcastle disease virus will protect poultry against the disease[38] (see also Section 4.2.5).

4.2.2   Recombinant BCG Vaccines.   BCG is an avirulent bovine tubercle bacillus that is the most widely used vaccine in the world. Since 1948 over 5 billion vaccinations have been carried out. It is possible to introduce foreign DNA into BCG[12,39] to express antigens from other organisms, for example, the envelope glycoprotein of human immunodeficiency virus (HIV).[40] Recombinant BCG has a number of distinct advantages over other approaches for multivalent vaccines primarily due to experience gained with the parent BCG vaccine. Advantages include the fact that BCG and oral polio vaccine are the only two vaccines WHO recommend to be given at birth, the younger the age at which vaccination can begin the better the chances of success in vaccination programmes. A single immunization with BCG gives long-lasting cell-mediated immunity to tuberculosis, it can be given repeatedly, is very safe with less than one complication per million vaccinations, and is a highly potent adjuvant in its own right. Although phage and plasmid vectors have been used with some success a significant amount of development is still required both to achieve higher levels of expression and to allow the system to be more readily manipulated.

[38] J. Taylor, C. Edbauer, A. Rey-Senelonge, J.-F. Boquet, E. Norton, S. Goebel, P. Desmettre, and E. Paoletti, *J. Virol.*, 1990, **64**, 1441.
[39] W. R. Jacobs Jr, M. Tuckman, and B. R. Bloom, *Nature (London)*, 1987, **327**, 532.
[40] A. Aldovini and R. A. Young, *Nature (London)*, 1991, **351**, 497.

4.2.3 Attenuated Salmonella Strains as Live Bacterial Vaccines. It is possible to introduce totally defined mutations or deletions in a variety of bacterial strains in order to attenuate them. These rationally designed attenuated vaccines can also be used as carriers for antigens cloned from other pathogenic organisms. Attenuated salmonella strains seem good candidates for this approach because they can be used as oral vaccines to stimulate secretory and cellular immune responses in the host. For example, the gene for heat labile B subunit of enterotoxic *E. coli* was introduced into the attenuated AroA strain of salmonella. This recombinant salmonella was able to induce IgG and IgA antibodies to the enterotoxic B subunit (as well as salmonella) in vaccinated animals.[41] A further modification of this strategy is to incorporate peptides into the flagellin gene of salmonella.[42] A potential HBV vaccine was constructed by incorporating synthetic oligonucleotides coding for sequences from the HBV surface antigen and from the pre-S2 antigen into the flagellin gene followed by introducing the hybrid gene into a flagellin negative salmonella strain.[43] The recombinant salmonella expressed the hybrid flagellin gene and when used to vaccinate mice, guinea pigs or rabbits induced antibodies that reacted with native HBV surface antigen. In addition isolated T-cells from immunized mice, proliferated in response to the hepatitis peptide contained in the flagellin gene showing that T-cell mediated immune responses can also be generated by recombinant salmonella. The antibody responses were greater in mice immunized by intramuscular inoculation rather than those vaccinated by the oral route. However, continued efforts to improve oral immunization are important, not only because of the ease of administration (syringes and needles are not required) and the reduced costs but also because mucosal immune responses to antigens after oral vaccination may offer more protection against pathogens that have their initial replicative cycle on similar mucosal surfaces.

4.2.4 Poliovirus Chimaeras. The live attenuated poliovirus type 1 Sabin strain has proved to be a very safe and effective vaccine stimulating good secretory and circulating antibody responses. A knowledge of the crystal structure of the virus together with the ability to generate virus from cDNA molecules has allowed antigenic domains from other pathogens to be incorporated precisely into the virus particle at the most antigenic sites. For example, the DNA coding for the major

---

[41] G. Dougan and J. Tite, *Semin. Virol.*, 1990, **1**, 29.
[42] S. M. C. Newton, C. O. Jacob, and B. A. D. Stocker, *Science*, 1989, **244**, 70.
[43] Y. J. Wu, S. Newton, A. Judd, B. Stocker, and S. W. Robinson, *Proc. Natl. Acad. Sci. USA*, 1989, **86**, 4726.

antigenic site of the Sabin type 1 strain was replaced by a peptide sequence from HIV1. Antisera to the peptide recognized the recombinant poliovirus particle and it was found that in immunization studies the recombinant virus could induce broadly neutralizing anti-HIV antibodies.[44]

These new poliovirus chimaeras suffer some of the same limitations as peptide vaccines but do not require potent adjuvants to work.

4.2.5   Cross-species Vaccination, 'Live-dead' Vaccines.   Debate over the safety of vaccinia virus has led to the suggestion that poxviruses of other species might be used to immunize humans. Canarypoxvirus replicates in cells of avian origin but is blocked in its ability to replicate in human cells. The idea is to express a vaccine antigen in a recombinant canarypoxvirus and use this as an immunogen. As it cannot replicate in the vaccinee there is no danger of virus spread, however, the virus does enter cells and produce the antigen. This 'live-dead' vaccine may prove to have the advantages of live vaccines with authentic antigenic presentation without the possible complications associated with live viruses. Indeed a canarypoxvirus recombinant expressing the rabies virus glycoprotein induces rabies virus neutralizing antibody in vaccinated animals and has been tested in phase I clinical trials.[45]

Modified vaccinia ankara (MVA) is an attenuated vaccinia virus that has been used to immunize against smallpox in about 200 000 people, including some immunosuppressed individuals, without adverse effects.[46] As the virus will not replicate in human cells it can be considered in the 'live-dead' category. The virus can, however, be propagated in rodent cell lines allowing ease of use and safe handling.[47] Recently it has been used in a variety of settings as a vector for foreign genes and it has proved particularly successful in a number of vaccination related settings.[48] High levels of cytotoxic T-cells can be detected after immunization (up to 4% of total peripheral blood T-cells).[49] In a protocol involving boosting after DNA vaccination an MVA recombinant protected against malaria challenge in mice where all other proto-

[44] D. J. Evans, J. McKeating, J. M. Meredith, K. L. Burke, K. Katrak, A. John, M. Ferguson, P. D. Minor, R. A. Weiss, and J. W. Almond, *Nature (London)*, 1989, **339**, 385.
[45] M. Cadoz, A. Strady, B. Meignier, J. Taylor, J. Tartaglia, E. Paoletti, and S. Plotkin, *Lancet*, 1992, **339**, 1429.
[46] A. Mayr and K. Danna, *Devel. Biol. Stand.*, 1978, **41**, 225.
[47] M. W. Carroll and B. Moss, *Virology*, 1997, **238**, 198.
[48] A. P. Durbin, L. S. Wyatt, J. Siew, B. Moss, and B. R. Murphy, *Vaccine*, 1998, **16**, 1324.
[49] A. Seth, I. Ourmanov, M. J. Kuroda, J. E. Schmitz, M. W. Carroll, L. S. Wyatt, B. Moss, M. A. Forman, V. M. Hirsch, and N. L. Letvin, *Proc. Natl. Acad. Sci. USA*, 1998, **95**, 10112.

cols had failed.[50] Recombinants have been used as vaccines in tumour models too and can even break tolerance to self-antigens.[51]

4.2.6 Other Virus Vectors. A number of other virus vectors which are appropriate for vaccination have been described, some mostly in the context of gene therapy. These include both replication deficient and replication competent recombinant adenoviruses. Adenovirus types 4 and 7 have been used to immunize military personnel for some time and as it is now possible to engineer adenoviruses to express vaccine antigens they have some potential as live recombinant vaccines.[52] For example, adenoviruses expressing HBsAg engineered as a fusion gene with a tetanus toxoid epitope may be effective in immunizing vaccinees who fail to respond to conventional sub-unit HbsAg.[53] At present there are reservations about these viruses from several standpoints: large amounts of virus are needed to achieve any effective vaccination. For replication competent viruses there are concerns over the Ela and Elb gene products which have oncogenic potential and for the replication defective variants there is concern over the cell-lines that are required for growth of the recombinants. The Oka strain of *Varicella zoster* virus (chicken poxvirus) is highly attenuated and can be engineered to express foreign vaccine antigens, for example, the hepatitis B virus surface antigen has been expressed and immunization with the recombinant was shown to induce antibody to hepatitis B surface antigen.[54]

4.2.7 Recombinant E. coli *Strains.* Enterotoxigenic *E. coli* strains (ETEC) cause diarrhoeal diseases in young pigs and under some circumstances in man (*E. coli* serotype O 157). These bacteria adhere to the intestine of the host *via* surface-associated fimbrae and secrete toxins which can be classified into heat stable (ST-toxin) and heat labile (LT-toxin). The fimbrae are highly antigenic and the first vaccines against ETEC consisted of whole cells or acellular extracts enriched for fimbrae. Vaccines prepared from ETEC strains gave significant levels of adverse reactions due to high levels of lipopolysaccharide and capsular antigens on the surface of the wild-type ETEC strains. *E. coli* K12 gave far fewer adverse reactions and was used as a vector for plasmid constructs that

[50] J. Schneider, S. C. Gilbert, T. J. Blanchard, T. Hanke, K. J. Robson, C. M. Hannan, M. Becker, R. Sinden, G. L. Smith, and A. Hill, *Nature Medicine*, 1998, **4**, 397.

[51] W. W. Overwijk, D. S. Lee, D. R. Surman, K. R. Irvine, C. E. Touloukian, C.-C. Chan, M. W. Carroll, B. Moss, S. Rosenberg, and N. P. Restifo, *Proc. Natl. Acad. Sci. USA*, 1999, **96**, 2982.

[52] A. Ballay, M. Levrero, M.-A. Buendia, P. Tiollais, and M. Perricaudet, *EMBO J.*, 1984, **4**, 3861.

[53] M. V. Chengalvala, R. A. Bhat, B. M. Bhat, S. K. Vernon, and M. D. Lubeck, *Vaccine*, 1999, **7**, 1035.

[54] R. S. Lowe, P. M. Keller, B. J. Keech, A. J. Davidson, Y. Whang, A. J. Morgan, *et al.*, *Proc. Natl. Acad. Sci. USA*, 1987, **84**, 3896.

expressed one or more different antigenic types of fimbrae.[55] However, to produce a vaccine with a wider spectrum of protection an anti-toxin component was introduced. Plasmid vectors were constructed using a strong prokaryotic promoter that expressed the LT toxin B subunit at high levels.[56] Cetus corporation now market a pig vaccine which consists of an *E. coli* Kl2 strain that expresses high levels of the LT-B subunit and contains fimbrae from ETEC strains. This engineered *E. coli* was the first licensed vaccine produced by recombinant DNA technology to be used in the USA.

5   OTHER APPROACHES TO VACCINES

5.1 DNA Vaccines (Genetic Immunization)

Rather surprisingly, plasmids carry genes specifying one or more antigenic proteins can be used for immunization. The plasmids delivered by injection, often into muscle, put genes directly into some cells and leads to uptake by cells in the vicinity of the inserted needle. Alternatively the plasmids can be delivered by a gene gun which propels plasmids into cells near the surface of the body – typically those of the skin or mucous membranes. Once inside cells, some of the recombinant plasmids make their way to the nucleus and because the gene is under the control of a strong eukaryotic promoter, the cells synthesize the plasmid-encoded antigens.

Thus, for example, intramuscular injection of plasmid DNA encoding influenza A virus genes generates both humoral and cell-mediated immunity in mice and ferrets which can protect against a challenge with influenza virus.[57] Many studies have described the use of this technology to induce antibody and cell mediated immunity to a variety of pathogens. An indication of the interest in this area is shown by the fact that a web site dedicated to DNA immunization has compiled a list of over 900 journal references on the subject, all published since the original observation in 1993 (http://www.genweb.com.Dnavax/dnavax.html).

Part of the surprise that genetic immunization worked was the 'feeling' that it was not possible for the antigens to be presented along with the appropriate co-stimulatory molecules. It now appears that some of the plasmid DNA does make its way to professional antigen-presenting cells, thereby inducing a response.[58]

[55] M. Kehoe, M. D. Winther, P. Morrisey, G. Dowd, and G. Dougan, *FEMS Microbiol. Lett.*, 1982, **14**, 129.
[56] S. Attridge, J. Hackett, R. Marona, and P. Whyte, *Vaccine*, 1988, **6**, 387.
[57] J. J. Donnelly, A. Friedman, J. B. Ulmer and M. A. Liu, *Vaccine*, 1997, **15**, 865.
[58] A. J. Ramsay, I. A. Ramshaw, and G. L. Ada, *Immunol.–Cell–Biol*, 1997, **75**, 360.

5.1.1 Optimising Responses. Responses to DNA vaccines often require significant amounts of plasmid DNA ($> 50\ \mu$g per injection) and are not as efficient at generating immunity as natural infection with the pathogen or that stimulated by live attenuated vaccines. However, responses can be improved by a variety of methods including the use of the gene gun where DNA is absorbed onto gold particles and propelled into cells.

The immune response to an antigen is influenced by the plasmid DNA surrounding the gene. This effect appears to stem from the fact that plasmid DNA, derived from bacteria, has a greater frequency of CG sequences than does the DNA in vertebrates. Moreover, the CG units in bacterial plasmids tend to have no methyl group attached, whereas those in vertebrates generally are methylated. It has been proposed that the vertebrate body interprets a high frequency of unmethylated CG pairs as a danger signal. In response, a relatively primitive part of the immune system (one not dependent on antigen recognition) attempts to destroy or wall off the foreign intruder. This phenomenon is most potent when the CG in plasmid DNA is flanked by two purines to their 'C' side and two pyrimidines to their 'G' side. In mice, plasmids containing such immunostimulatory sequences induced more vigorous antibody and cytotoxic T-cell activity than did an otherwise identical vaccine.[59] Hence, increasing the number of immunostimulatory sequences in plasmids might well amplify the immunogenicity of the antigens coded in a DNA vaccine.

Interestingly incorporation of IL-12 DNA along with the antigen appears to bias the response in favour of a Th1 cell-mediated response while fusion with other cytokines such as GM-CSF appears to enhance the Th2 humoral response. This is a distinct advantage, as the ability to generate an appropriate response may deal with a pathogen more effectively then a conventional vaccine could. Thus all that is required for the hepatitis B vaccine is high levels of antibody while the elimination of intracellular organisms, *e.g.* Leishmania, will require strong cell-mediated immunity. As well as these cytokines, genes for chemokines might be incorporated into constructs. Chemokines are small molecules that attract both antigen-presenting cells and T-cells to damaged or infected tissues. Like cytokines, these substances differ in the mix of cells on which they act and in the precise effects they exert. As their individual actions are better understood, carefully combining specific chemokine genes with selected cytokine genes could better customize both the type and the extent of immune responses elicited.

[59] M. Roman, E. Martin-Orozco, J. S. Goodman, M. D. Nguyen, Y. Sato, A. Ronaghy, R. S. Kornbluth, D. D. Richman, D. A. Carson, and E. Raz, *Nature Medicine*, 1997, 3, 849.

DNA vaccines could even side-step the need for classical antigen-presenting cells to prime T-cell responses. If a gene for an antigen were co-expressed with a co-stimulatory molecule, then inoculated skin, muscle or other cells could themselves display both the antigen and the crucial 'second signal', thereby facilitating both the priming and the activation of cytotoxic T-cells, thus obviating the need for a professional antigen-presenting cell.

Another alternative to improve responses is to target a gene product to the proteosome by fusion of the gene with the ubiquitin gene.[60] Unconjugated peptide epitopes delivered by DNA vaccination failed to stimulate an immune response while the same epitopes conjugated to ubiquitin-induced cytotoxic T-cells.

There is also considerable interest in combining DNA vaccination with other protocols. In some instances priming with DNA vaccination followed by boosting with recombinant virus has achieved better results than either DNA or recombinant virus on its own (see Section 4.2.5).

5.1.2 RNA immunization. Vaccines composed of RNA are also being tested. Once inside the cell, RNA is readily translated resulting in the synthesis of any encoded vaccine candidate. The technology being developed to introduce DNA into cells could be used for RNA as well. However, RNA is less stable than DNA, presenting a problem for vaccine manufacture and distribution. This difficulty is probably surmountable and current approaches to gene therapy with RNA may provide an appropriate solution.

DNA (and possibly RNA derived) vaccines have the potential to preserve the positive aspects of existing vaccines while avoiding some of their risks. In addition to activating both arms of the immune system they can be biased to stimulate an appropriate type of immunity. As a bonus, they are easy to design and to generate in large quantities and they are as stable as other vaccines when stored. They should therefore be relatively inexpensive to manufacture and to distribute widely. Furthermore, because they can be engineered to carry genes from different strains of a pathogen, they can potentially provide immunity against several strains at once, something that should be very helpful when the microorganism is highly variable, as in the case of HIV and hepatitis C. DNA vaccination has an inherent attractiveness because of the flexibility of the approach and this will ensure it is an active area of research for many years to come.

---

[60] F. Rodriguez, J. Zhang, and J. L. Whitton, *J. Virol.*, 1997, **71**, 8497.

5.2 Peptide Vaccines

This approach is the ultimate in the reductionist approach to vaccines. As mentioned in Sections 3.2.1 and 3.2.2 it is possible to identify the epitopes within a protein that can induce neutralizing antibodies or epitopes that are important in T-cell responses to vaccines. Chemical synthesis of these epitopes is relatively straightforward and with an appropriate adjuvant or conjugation to carrier proteins, they can induce antibody or T-cell mediated responses to the synthesized epitope. These immune responses are in some cases sufficient to give protection against the organism the protein epitope was derived from. For example, 100 μg of a 20 amino acid synthetic peptide to amino acids 141–160 of VP1, the major coat protein of foot and mouth disease virus (FMDV) will protect guinea pigs against a severe challenge with FMDV.[61] The cloned sporozoite surface antigen of the malarial parasite (see Section 3.1.3) was shown to contain a repeat sequence within the molecule. This repeat seemed an ideal target for the development of an anti-malarial peptide vaccine. Indeed several human clinical trials of vaccines based on this sequence have been carried out with some success.[62] Modifications to the basic technique of coupling a peptide to a protein and using it as a vaccine include: (i) Incorporation of a helper T-cell epitope within the peptide and use of the peptide on its own. (ii) Cyclization of the peptide to improve antigenicity. (iii) Incorporation of the peptide into antigenic regions of other proteins such as HBV surface (HBsAg) or core (HBcAg) antigens. The advantage of incorporation into HBsAg or HBcAg rather than other proteins is that they both form particles which by their very nature are highly immunogenic. (iv) Construction of higher order structures with multiple copies of the peptide. This has been achieved in a number of ways including polymerization and branching using a lysine core. A good example of the polymerization approach is attempts to make an anti-malarial vaccine by linking peptide sequences from three different asexual stages of the parasite with NANP a repeat sequence present in the major protein on the surface of the sporozoite. The polymer known as Spf66 has shown some efficacy in phase III clinical trials.

The advantages of this approach include the fact that the product is stable and chemically defined without the presence of an infectious agent and can be designed to stimulate specific T- and B-cell responses. Many other approaches require large-scale production plants and complex

[61] F. Brown, *Semin. Virol.*, 1990, **1**, 67.
[62] F. Nosten, C. Luxemburger, D. E. Kyle, W. R. Ballou, J. Wittes, E. Wah, T. Chongsuphajaisiddhi, D. M. Gordon, N. J. White, J. C. Sadoff and D. G. Heppner, *Lancet*, **348**, 701.

downstream processing peptide, vaccines are relatively simple and require limited work-up and purification. The major disadvantage often cited for peptide vaccines is based on the fact that often only one peptide is used. Many pathogens are characterized by the fact that there is extreme variation in the antigenic proteins of the agent. Could a single epitope or even multiple epitopes be found that protect in all cases in the face of the extreme variation of an agent such as HIV or hepatitis C virus? There is also the possibility that neutralization escape mutants can arise, particularly for RNA viruses, and hence allow growth of the pathogen because the monospecific sera, generated by a single peptide, does not recognize the mutant. In general a peptide from a pathogen will only be recognized by a specific HLA haplotype. Consequently a further problem arises from the need to provide sufficient diversity of recognition sites to immunize in an outbred population. Despite these reservations the ease of production, stability and safety of peptide vaccines make them an approach that will receive much attention in the future.

In recent years it has become possible to isolate dendritic cells from bone marrow or peripheral blood and culture them *in vitro*.[63] These cells can then be primed with peptide (or other source of antigen) and given back to the individual the cells were originally derived from. These professional antigen presenting cells then efficiently generate an immune response that might not otherwise have been generated. This immunotherapeutic approach is under intensive investigation for cancer treatment and might be relevant for some types of infectious agent. The main drawback is that this high-tech approach is patient specific and therefore very costly.

5.3 Anti-idiotypes

It has been suggested that anti-idiotype (anti-Id) antibodies may make effective vaccines. This is based on the finding that antibodies themselves can act as immunogens. An immune response raised against the unique antigen combining site of an antibody is termed an anti-idiotypic response and may bear a structural resemblance to the original antigen. When this occurs the anti-idiotypic antibody (monoclonal or polyclonal) may be able to induce an antibody response that recognizes the original antigen and hence act as a vaccine. Anti-idiotypes have been shown to give protection in a variety of animal model systems; probably the best demonstration of the potential of this approach was the protection of chimpanzees from HBV-associated disease by previous immunization

---

[63] J. Banchereau and R. M. Steinman, *Nature (London)*, 1998, **392**, 245.

with anti-Id.[64] There are several advantages of anti-Ids over traditional approaches, most of which are true of other subunit/peptide vaccine approaches. These advantages are apparent: (i) Where antigen is difficult to obtain, *i.e.* when the infectious agent is hazardous or cannot be grown *in vitro*. (ii) Where attenuated vaccines have high reversion frequencies or possess genes that may be involved in oncogenesis. Any problems vaccinating immunocompromised individuals with live vaccines would also be avoided. (iii) Where a single epitope can confer protection but other epitopes of the whole molecule might induce autoimmunity. (iv) When organisms display wide genetic diversity but a single cellular receptor. An anti-Id response could theoretically induce a serological response that mimics the receptor and binds the infectious agent at its receptor binding site. (v) Probably the most important advantage is the ability of anti-Id to mimic non-proteinaceous epitopes such as carbohydrate, lipid or glycolipid, all of which cannot yet be produced easily to act as subunit vaccines.

There are a number of disadvantages to this approach, in particular the restriction of the vaccine to a single epitope or a few epitopes administered together which may not be enough to protect against some organisms and the limitations associated with multiple use of anti-Id preparations. Over time it is assumed that with repeated anti-Id immunization, antibodies to constant region immunoglobin determinants will arise and this might then prejudice subsequent immunizations with anti-Id. Despite these limitations and although this approach offers little advantage over other recombinant sources of antigen other than for non-proteinaceous epitopes it may prove to be an important adjunct to other strategies for immunization.

5.4 Enhancing Immunogenicity and Modifying Immune Responses

5.4.1 Adjuvants, Carriers and Vehicles. Many purified proteins are poorly immunogenic and require measures to enhance their antigenicity. The most simple way of achieving this is by addition to the protein of an adjuvant. Adjuvants work by a variety of mechanisms, including enhancing antigen deposition and persistence of antigen, recruiting inflammatory cells and altering the balance between T-cell subsets. The only artificial adjuvants widely used in man are aluminium salts, often referred to as 'alum' and although they are very safe, they have a number of limitations, the most important being that, even at best, their immunostimulating properties are limited particularly with respect to the

[64] R. C. Kennedy, J. W. Eichberg, R. E. Lanford, and G. R. Dreesman, *Science*, 1986, **232**, 220.

production of cell-mediated immunity. Consequently the design and testing of systems to improve immune responses is an area of intense activity.

Edelman and Tacket[65] have approached the question of immuno-stimulation in a useful way by classifying immunostimulatory molecules into three categories: adjuvants, carriers for antigens and vehicles. Adjuvants include the aluminium salts previously mentioned; *B. pertussis*, as well as some of its active molecules such as muramyl di- and tri-peptides; monophosphoryl lipid A; saponin; and various cytokines such as IL-12 and GM-CSF. Carriers include the bacterial toxoids of tetanus, cholera or diphtheria as well as live vectors such as poliovirus chimaeras and hybrid proteins that form particulate structures such as yeast TY retrotransposons or HBV core antigen fusions. The current conjugate vaccines for *Haemophilus influenzae*, where capsular polysaccharide is conjugated to tetanus or diphtheria toxoids or group B meningococcal outer membrane protein, are examples of the use of carriers to improve immunity.

Part of most modern synthetic vaccines is the adjuvant formulation which comprises of a vehicle and an adjuvant. The vehicle often has independent immunostimulatory properties and can consist of any of the following, mineral oil emulsions (Freund's), vegetable oil emulsions, non-ionic copolymer surfactants, squalene or squalane, liposomes or biodegradable polymer microspheres. An adjuvant is not licensed 'on its own' but as part of an adjuvant formulation. However the widespread use of alum has placed it in a special category from a regulatory standpoint, all other adjuvants being considered experimental.

5.4.2 Carriers. Carriers are a diverse group of agents ranging from the licensed toxoid conjugates with bacterial capsular polysaccharide to higher order structures such as liposomes. Immune responses to virus particles tend to be vigorous compared with an equivalent amount of isolated antigen. A deliberate design feature of many of the carriers is to present numerous copies of an antigen or epitope in a particulate structure. Antigens have incorporated into various types of liposomes have been tested over the past 25 years with promising results. Other higher order structures that have shown promise include ISCOMs (immune stimulating complexes) which consist of a 35 nm cage-like structure formed between Quil A, added lipids and the antigen. High levels of antibody and CTL in response to a variety of antigens have been demonstrated.

---

[65] R. Edelman and C. O. Tacket, *Int. Rev. Immunol.*, 1990, **7**, 51.

Another rather interesting concept in delivery of antigen is the use of biodegradable polymers such as poly(D,L-lactide-co-glycolide) used in soluble stitching. They are known for their safe use in humans and can be used to encapsulate antigen or even DNA in microparticles. Different formulations of polymer can be used to allow variation in the time taken to deliver the antigen or even change the delivery profile to give a pulse of antigen followed by a trickle. Microparticles between 5 and 10 μm, containing toxoids administered orally are taken up in Peyer's patches in the gastrointestinal tract and prime anti-toxoid IgG and IgA. The deposition and slow release of antigen acts in a similar manner to that of classical adjuvants. Further clinical studies will be required to establish the efficacy of this approach and determine if low doses will generate unwanted side effects such as tolerance or hypersensitivity.

5.4.3 Mucosal Immunity. Most vaccines are delivered by injection which requires trained personnel to give repeated immunizations. Oral vaccination, if developed, could be administered without the cost of needles, syringes and training of personnel and might well decrease the difficulty in the logistics of delivering a course of vaccination in remote areas. As a consequence a large amount of effort is being made to develop oral vaccination. Many infections are acquired at the mucosal surfaces of the respiratory, genitourinary or gastrointestinal tracts. It has been argued that protection against pathogens that infect *via* these sites is best achieved by stimulating IgA at a mucosal surface. Interestingly it appears that stimulation of immunity at one mucosal surface can lead to IgA being present at distal sites *via* trafficking immune cells. This means that stimulation in the gastrointestional tract can lead to local immunity in the lungs and genitourinary tract. This *common mucosal immune system* increases the attractiveness of oral vaccination. Oral immunization has been attempted with live vectors such as *Salmonella* (see Section X) and live attenuated cholera strains, the latter by linking protein to the cholera toxin B subunit (or *E. coli* LTB), either as a gene fusion or by conjugation.

5.4.4 Modulation of Cytokine Profile. Most adjuvants act by eliciting an appropriate mixture of cytokines needed to generate an immune response. Cytokines can have a profound effect on the balance between cellular and humoral response as well as the isotype of immunoglobin produced. As our understanding of the action of cytokines increases we may be able to modulate the type of immunity produced in favour of the mechanism of immunity required to protect against a particular pathogen. As mentioned in Section 5.1.1, IL-12 can bias immune responses to a Th1 phenotype while GMCSF appears to bias towards Th2 antibody type responses. These cytokines might be added to a vaccine either as protein or as gene fusions with the antigen of interest. Incorporation of chemokines in a similar way may have profound effects by attracting, for instance, dendritic cells to the site of immunization thereby improving presentation of antigen.

Live vectors can also be engineered to express cytokines or chemokines thus modulating the immune response in favour of the desired type of immunity.

5.4.5   Modulation by Antigen Targeting.   There are many cell surface molecules which might be used to target antigen to specific cell populations, for example, MAd CAM recognizes the receptor $\alpha_{IEL}\beta7$ located on cells residing in the mucosal epithelium. Thus fusion of antigens to MAdCAM may direct their delivery to mucosal epithelium.

Another approach is to incorporate signalling sequences within the antigen that target it to the appropriate compartment of antigen presenting cells. For example, a polyepitopic synthetic protein presented as a DNA vaccine failed to produce CTL responses. However, when it was tagged with ubiquitin, which presumably targeted the protein into the proteosome, CTL responses could be detected.

5.4.6   Modulation of Signalling.   Live attenuated infectious vectors might also be used to modulate signals received by important immune effector cells. For instance, co-stimulatory molecules such as B7.1 have been incorporated into vaccinia recombinants and shown to improve responses. Incorporation of other molecules involved in cell–cell contact, *e.g.* ICAM I, LFA3, *etc.*, may also enhance immune responses.

6   SUMMARY AND CONCLUSIONS

The increasing knowledge of basic immunological processes and the contribution of different types of immune responses in the prevention and control of infectious diseases has clarified the immunological requirements for a vaccine to give long-lasting immunity. The ideal vaccine should, immunolgically speaking, generate large numbers of memory T and B lymphocytes, be processed to peptides which associate with MHC and induce T-cell responses to sufficient numbers of T-cell epitopes to overcome genetic variability between hosts. It should also result in the persistence of antigen so that memory B-cells are continually recruited to produce circulating antibody.

Not only do these immunological criteria need to be taken into account but also a series of other basic criteria need to be considered when designing new vaccines. For example, target epitopes must be clearly identified and characterized if possible and these target epitopes must remain conserved across any variant population and expressed to sufficient levels to allow immune mechanisms to function. Immune effector mechanisms should be identified and the direct involvement of the target epitope confirmed. New immunostimulating (adjuvants) agents need to be identified and approved for use. The safety of the immunogen must be established, *i.e.* minimize side effects such as immunosuppression, autoimmunity, or excessive inflammation.

In practice this is a difficult set of criteria to meet for many organisms. For example, the identification of a protective antigen can be difficult. It is not enough to assume that a major surface protein will be a good vaccine antigen because of antigenic variability of the molecule concerned. It is also very clear that even if a good vaccine antigen can be identified new adjuvants will need to be used. At present aluminium hydroxide gel (alhydrogel) is the only widely licensed adjuvant for human use and it is not effective enough for some antigens. For example, although there are a number of difficulties associated with the use of HIV envelope protein gp160 as a vaccine for HIV (antigenic variation being the most serious problem) it has been used for human trials. These trials have indicated that alhydrogel is not a very effective adjuvant and consequently large trials have been set up to assess a series of experimental adjuvants in conjunction with HIV gp160 produced by recombinant DNA technology in mammalian cells.

Once a candidate antigen has been identified it is pertinent to ask which of these multitude of approaches should be taken? There is no simple answer; the strategy used will vary from case to case. A simple guide for the developed countries is that subunit vaccines produced by eukaryotic cells will probably receive the widest acceptance because of their perceived safety. Peptide vaccines, if shown to be effective, will also be valuable. In the third world many more secondary requirements come into play such as those listed below.

(i) Cost. The six childhood vaccines supplied to the WHO global programme are supplied through UNICEF for a cost of about 20 pence per schedule of administration. Any new vaccines must cost a similar amount for them to be afforded on a global scale. There is a difference between what developed countries feel is appropriate and what third world countries can afford.

(ii) Many vaccines are administered parenterally which adds to the cost of administration requiring skilled personnel, needles and syringes. Delivery *via* mucosal surfaces, preferably orally, offers many advantages.

(iii) Thermal stability, particularly for those vaccines used in tropical countries, is vital unless there are monies available for setting up a cold chain.

(iv) A single immunization, if effective, would also reduce administration costs.

(v) Long-lived immunity. Live vaccines tend to be more effective at stimulating long-lasting immunity than non-infectious vaccines without adjuvant.

For the successful application of this new technology and in particular for the choice of which route to use another important consideration is the portal of entry of the pathogen. It may be that the most effective immunity to a particular agent will be generated by immunity at the site of entry of that pathogen. The disease would then dictate to some degree the most effective means of vaccination. It is probable that the immune response required for protection against a blood borne pathogen such as a virus or a parasite would be quite different from a virus or bacterium that infects at a mucosal surface. Part of the overall assessment requires an answer to the question of whether protection is best achieved by circulating antibody or secretory antibody or a vigorous T-cell response. It should also be borne in mind that neither immunization nor recovery from natural infection always protects a person against infection with the same organism. This principle holds true for diseases that have been successfully dealt with by immunization, for example, polio, measles and rubella. Control is achieved not by inducing sterile immunity, *i.e.* no infection, but rather by limiting replication and spread of the organism; *e.g.* poliovirus is prevented from reaching nervous tissue but a limited replication does occur. Immunity to respiratory infections is often poor following natural infection and it is probably unrealistic to expect any vaccine to produce sterile immunity. However, it may be possible to protect against serious pulmonary disease yet not prevent rhinitis thus limiting serious disease but not completely eliminating the pathogen may be a valuable approach to decreasing the burden of infectious disease.

Ultimately medical, political, economic and social considerations will determine the use of any vaccine. It is salutary to realize that the only vaccine generally licensed for use in humans produced by molecular technology is the hepatitis B surface antigen produced in yeast. The question remains how long will it take to have, for example, a vaccine for HIV. In 1983, soon after HIV was discovered, an official from the American National Institutes of Health said we will have a vaccine within 5 years. In 1999 there are only two Phase III clinical trials of an HIV vaccine planned or taking place anywhere in the world. It is generally believed that these will not show efficacy. It is clearly too soon to say if any one avenue of vaccine research will provide a global answer but with the plethora of new approaches to vaccines it is to be hoped that before too long many of the problems associated with infectious disease will be tackled.

7 FURTHER READING AND SOURCES OF INFORMATION

M. Mackett and J. D. Williamson, 'Human Vaccines and Vaccination', 1995, BIOS Scientific Publishers.

Science series on vaccines: 1994, Vol. 285, issue 5177. Includes: Vaccines for varicella-zoster virus and cytomegalovirus: recent progress. S. A. Plotkin, *Science*, **265**, 1383–5. Pneumococcal disease: prospects for a new generation of vaccines. G. R. Siber, *Science*, **265**, 1385–7. Measles vaccine: do we need new vaccines or new programs? S. L. Katz and B. G. Gellin, *Science*, **265**, 1391–2.

Lancet series on vaccination: 1997, Vol. 350, various issues from 9086, *e.g.* Host immunobiology and vaccine development, G. Nossal, *Lancet*, **350**, 1316–19. Rotavirus vaccines against diarrhoeal disease, T. Vesikari, *Lancet*, **350**, 1538–41. Influence of disease burden, public perception, and other factors on new vaccine development, implementation and continued use, M. M. Levine and O. S. Levine, *Lancet*, **350**, 1386–92. What is required of an HIV vaccine? C. R. Bangham and R. E. Phillips, *Lancet*, **350**, 1617–21.

Clinical trial of virus recombinant (1996) expressing 7 foreign proteins: NYVAC-Pf7: a poxvirus-vectored, multiantigen, multistage vaccine candidate for *Plasmodium falciparum* malaria, 1996, J. A. Tine *et al.*, *Infection Immunity*, **64**, 3833–44.

A peptide vaccine Phase III trial (1996): Randomised double-blind placebo-controlled trial of SPf66 malaria vaccine in children in north-western Thailand. Shoklo SPf66 Malaria Vaccine Trial Group, 1996, F. Nosten *et al.*, *Lancet*, **348**(9029), 701–7. [Comment in: *Lancet* 1996, **348**(9029), 695.]

Immunisation Infectious Diseases (1996): Department of Health, Welsh Office, Scottish Office Home and Health *against* Department, DHSS (Northern Ireland), HMSO, London. ISBN 0-11-321815-X

Issues around a vaccine for HIV: A Vaccine for HIV: The antibody perspective, 1997, D. R. Burton, *PNAS USA* **94**, 10018–23 and references therein. Escape of human immunodeficiency virus from immune control, 1997, A. J. McMichael. *Ann. Rev. Imunol.*, **15**, 271–96. Cyto-toxic T-cells – protection from disease progression – protection from infection, 1996, Gotch *et al.*, *Imunol. Let.*, **51**, 125–8. An HIV vaccine: How long must we wait? *Lancet*, **352**, 1323. Is an HIV vaccine possible? A. J. McMichael and T. Hanke, *Nature Medicine*, **5**, 612–4.

Hepatitis B virus, the vaccine, and the control of primary cancer of the liver: 1997, B. S. Blumberg, *PNAS USA*, **94**, 7121–5 and references therein.

WEB sites for General Virology and Vaccines:

'All the Virology on the WWW' is a large site with many useful topics and links to other sites: http://www.tulane.edu/~dmsander/garryfavweb.html. It has a UK mirror site which might provide faster access at certain times of the day. See vaccines page on http://www-micro.msb.le.ac.uk/garryfavweb/garryfavweb.html

Site designed for schools with some useful basics: http://www.path.ox.ac.uk/dg/

Virus databases on line: http://www.res.bbsrc.ac.uk//mirror/auz/welcome.html. UK mirror of Australian site.

A good collection of virus resources is maintained at The University of Wisconsin: http://www.bocklabs.wisc.edu/Welcome.html. Among these are many useful pictures of virion structures: http://www.bocklabs.wisc.edu/virusviz.html

The University of Cape Town Medical Microbiology Department has a good website of electron micrograph images: http://www.uct.ac.za/depts/mmi/stannard/emimages.html

DNA vaccines website: http://www.genweb.com/Dnavax/dnavax.html

Clinician's Handbook of Preventive Services – including a great deal of information on vaccines and immunization: http://indy.radiology.uiowa.edu/Providers/ClinGuide/PreventionPractice/TableOfContents.html

The Centers for Disease Control in the US monitors infectious diseases in the US and has weekly updates on the incidence of viral diseases: http://www.cdc.gov/ncidod/ncid.htm

The World Health Organization in Geneva runs virus vaccination and health education programs all across the world and their website provides access to many online factsheets, reports and press releases.

http://www.who.int/ WHO programmes were recently reorganized. The old Global Programme for Vaccines and Immunization (GVI) and The Expanded Program for Immunization which has been so successful are not spcifically identified. For information on a variety of infectious diseases see http://www.who.int /inf-dg/home/combatting_ill_health/ communicable_diseases/index.html

Web sites for specific vaccines and diseases:

Measles, *etc*.: http://www.who.int/gpv-dvacc/research/virus1.htm
Infuenza A and B: http://www.cdc.gov/ncidod/diseases/flu/fluvirus.htm and pages inside or http://www.who.int/emc/diseases/flu/index.html. Also see FluNet a WHO sponsored influenza outbreak monitoring site http://oms.b3e.jussieu.fr/flunet/. Hepatitis B: http://www.who.int/gpv-dvacc/diseases/hepatitis_b.htm, http://www.cdc.gov/ncidod/diseases/hepatitis/hepatitis.htm
Polio: http://whqsabin.who.int:8002/ The beginning of the end – see also pages inside.

CHAPTER 13

Transgenesis

LINDA J. MULLINS AND JOHN J. MULLINS

1 INTRODUCTION

Transgenesis may be defined as the introduction of exogenous DNA into the genome, such that it is stably maintained in a heritable manner. Over the last 15 years, the introduction of transgenes into the mammalian genome has become a routine experimental tool and is gaining increasing importance in the biotechnology industry. Traditionally, DNA (the transgene) is introduced into the one-cell embryo by micro-injection and surviving embryos are subsequently reimplanted into a pseudopregnant female and allowed to develop to term. In a proportion of the embryos, provided that the DNA integrated into the genome prior to the first cell division, the transgene will be passed on to subsequent generations through the germline.

Transgenic techniques have far-reaching research applications. At the molecular level they allow the identification of *cis*-acting DNA sequences important in directing developmental and/or tissue-specific gene expression, and the specific manipulation of gene expression *in vivo*. One can equally ask broad developmental questions concerning cell or organ function with a degree of finesse not possible with other techniques. With the advent of embryo stem (ES) cell technology, and the development of strategies for achieving homologous recombination, the researcher now has the ability to question the function of specific genes and to ascertain the *in vivo* effects of precise alterations to gene function. These innovations have important implications for many areas of biomedical research including the design of disease models, the use of mammals as bioreactors for the production of human therapeutic proteins and, ultimately, the correction of inborn errors of metabolism by gene targeting.

357

2 THE PRODUCTION OF TRANSGENIC ANIMALS BY MICROINJECTION

The first transgenic mice were produced almost 20 years ago, and for both scientific and practical reasons, the mouse is still the animal of choice in the majority of transgenic experiments. During recent years, however, transgenic techniques have been extended to other species including the rabbit, the rat, and also a range of commercially important animals, notably the cow, the pig, the goat and the sheep.

2.1 Transgenic Mice

The technique by which transgenic mice are produced is schematically outlined in Figure 1. The first stage in development of transgenic animals is the isolation of sufficiently large numbers of fertilised eggs for micro-injection. This is achieved by the superovulation of young virgin females (approximately 4–5 weeks of age), which are injected with a source of follicle-stimulating hormone (pregnant mare's serum gonadotrophin). Forty-eight hours later, the females are given an artificial surge of leutinizing hormone by administration of human chorionic gonadotrophin and are paired with proven stud males. The following day, females which have mated (as identified by the presence of a vaginal plug) are sacrificed and fertilized eggs are removed from the swollen ampullae of the Fallopian tubes by dissection. Using such a protocol, up to 30 zygotes can be isolated per female, depending on the strain used. The zygotes are freed from attached cumulus cells by brief incubation in the presence of hyaluronidase, transferred to appropriate medium, and are stored in a CO_2 incubator at 37°C prior to micro-injection.

Typically 20–25 fertilized eggs are processed at a time. One by one, they are picked up by gentle suction onto a holding pipette and injected with the micro-injection needle. The movement of both holding and micro-injection pipettes is controlled by micromanipulators (Figure 2) which are either mechanically or pneumatically controlled. Suction to the holding pipette is applied through oil-filled tubing *via* a micrometer-controlled syringe.

The micro-injection needle, which has an internal tip diameter of approximately 1 μm and contains DNA at a typical concentration of 1–2 μg ml^{-1}, is manipulated gently but firmly until both the zona pellucida, and the nuclear membrane of one of the pronuclei have been pierced. Care must be taken to ensure that the highly elastic nuclear membrane is punctured, and that the needle does not touch the nucleoli which causes

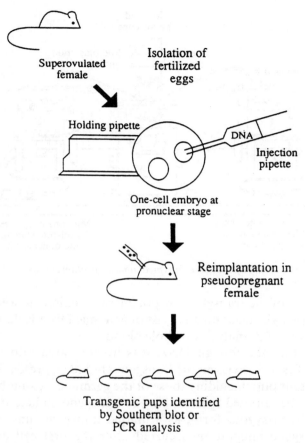

Figure 1 *Generation of transgenic animals by micro-injection of the one-cell embryo*

blockage of the needle and damage to the egg. DNA is injected into the pronucleus using a pneumatic pump or a hand-operated syringe/micrometer and successful injection is indicated by swelling of the pronucleus prior to removal of the pipette tip.

Eggs that have been successfully injected are returned to the incubator and may be left to develop to the two-cell stage overnight. This allows one to check that the eggs are still viable, following injection. (A proportion of the injected eggs do not survive the ordeal.) Whether at the one- or two-cell stage, embryos are reimplanted into the oviduct of anaesthetized pseudopregnant females (experienced mothers which have been mated the previous night with vasectomized or genetically infertile males). The females are allowed to recover from the anaesthetic and the pregnancy is continued to term. At weaning, progeny are tested for

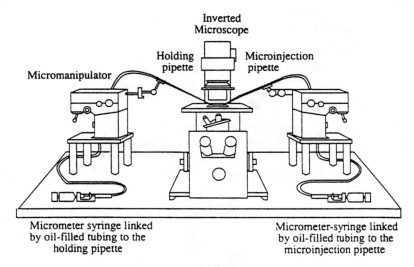

Figure 2 *Schematic representation of a typical micro-injection set-up*

incorporation of the transgene, by polymerase chain reaction (PCR)[1] or Southern blot hybridization analysis[2] of genomic DNA isolated from tail biopsies or by PCR analysis of whole blood.[3]

Provided that the foreign DNA was incorporated into the genome prior to the first cell division, the transgene should be present in every cell of the resultant pup, including those of the germline. A suitable breeding strategy can be initiated to maintain the transgene, in heterozygous and eventually homozygous form, as a unique transgenic line. Sometimes, where transgene integration occurred after the first cell division, the founder is chimaeric and, depending upon the representative proportion of cells populating the germline, a fraction of its progeny may carry the transgene, allowing it to be rescued. If the transgene is not represented in cells of the germline, then it is not possible to generate a transgenic line from the founder. For a more technically detailed description of transgenesis see Hogan *et al.*[4]

2.2 Transgenic Rats

Transgenic rats are generated using the same strategy as outlined for transgenic mice, but with some important modifications. Firstly, virgin

[1] K. B. Mullis and F. Faloona, *Methods Enzymol.*, 1987, **155**, 335.
[2] E. M. Southern, *J. Mol. Biol.*, 1975, **98**, 503.
[3] A. J. Ivinson and G. R. Taylor, in 'PCR – A Practical Approach', ed. M. J. McPherson, P. Quirke, and G. R. Taylor, Oxford University Press, New York, 1991, Chapter 2, p. 15.
[4] B. Hogen, R. Beddington, F. Costantini, and E. Lacy, 'Manipulating the Mouse Embryo – A Laboratory Manual', 2nd Edn, Cold Spring Harbor Laboratory Press, 1986.

females (approximately 30 days of age) have been found to be most responsive to superovulation. On that day, a highly purified source of follicle-stimulating hormone ('Foltropin', Vetrepharm, Canada; or 'Ovagen', Synergy Prod. Ltd, Melksham, Wiltshire, UK) is administered using a subcutaneously implanted osmotic minipump.[5] Following induction with human chorionic gonadotrophin 48 h. later, and mating, between 50 and 100 fertilized eggs can be isolated from a single female (depending on the strain used). The pronucleii take several hours longer to develop in the rat embryo than those of the mouse, and it has been found that the zona pellucida and pronuclear membranes are much more elastic, making them slightly more difficult to micro-inject.

2.3 Choice of Animal

As previously stated, the mouse is traditionally the animal of choice for transgenic research. A transgenic animal programme requires significant animal breeding facilities, first to ensure the regular production of large numbers of eggs for micro-injection and pseudopregnant females to receive the injected eggs, and, secondly, to maintain breeding colonies for the various transgenic lines generated from each micro-injection series. For some physiological studies such as cardiovascular research, neurobiology and pharmacology, however, rat transgenics may be preferable because of size constraints in the mouse[6] or the historical use of this species within a particular discipline.

From mouse studies, it is apparent that F1 cross-bred animals are superior to inbred strains because they produce higher numbers of eggs on superovulation, have larger litter sizes, and are generally better mothers. Factors such as characteristics of endogenous gene *versus* the transgene, however, may play a part in the decision to use inbred rather than F1 hybrids.

If larger animals are to be used, it is practical to carry out initial transgenic studies with a given construct in rodents prior to costly trials in large domestic animals. It must be noted, however, that the response of one species to a gene construct may vary from that of another.[6,7]

2.4 Application of Micro-injection Techniques to Other Animals

Transgenic technology has been extended to include a number of agriculturally important animals. Details regarding superovulation vary

[5] D. T. Armstrong and M. A. Opavsky, *Biol. Reprod.*, 1988, **39**, 511.
[6] J. J. Mullins, J. Peters, and D. Ganten, *Nature*, 1990, **344**, 541.
[7] R. E. Hammer, S. D. Maika, J. A. Richardson, J.-P. Tang, and J. D. Taurog, *Cell*, 1990, **63**, 1099.

from species to species and the reader is referred to other publications for specific information (*e.g.* production of transgenic sheep,[8,9] goats,[10] and pigs[11]). There are significant limitations regarding the application of pronuclear injection to non-murine animals, not least being the time and cost due to longer gestation and generation times, reduced litter sizes and higher maintenance costs. Additionally, large numbers of fertilized eggs (and donor animals) are required for micro-injection, and the cost of carrying non-transgenic offspring to term is much more significant. These considerations resulted in significant departures from the standard procedures for the development of bovine transgenics,[12] with use of *in vitro* embryo production in combination with gene transfer technology. Micro-injected embryos can be developed through to the morula/ blastocyst stage in recipient rabbits or sheep, allowing time for sexing, and transgene screening to take place before reintroduction into the natural host, providing that the screening methods are reliable.[13] Such techniques are also now used in the goat.[14]

2.5 Animal Cloning

The much publicized animal cloning is a means of increasing the numbers of scientifically or commercially important animals. Initially, full development of the mouse was only obtained when a pronucleus or nucleus (karyoplast) from early two-cell stage mouse embryos were transferred to enucleated oocytes.[15] Nuclei from later stages were considered incapable of reprogramming since they failed to support embryogenesis. It has since been shown that artificial activation (by electrical stimulation) allows nuclei from eight-cell mouse embryos to be transferred to MII-staged cytoplasts,[16] and that serial transfer (growing reconstituted eggs in medium containing cytochalasin B, to produce two 'pronuclei-like nuclei', and subsequent transfer of these to previously enucleated fertilized one-cell embryos) improves the pro-

[8] J. P. Simons, I. Wilmut, A. J. Clark, A. L. Archibald, J. O. Bishop, and R. Lathe, *Bio/Technology*, 1988, **6**, 179.
[9] G. Wright, A. Carver, D. Cottom, D. Reeves, A. Scott, P. Simons, I. Wilmut, I. Garner, and A. Colman, *Bio/Technology*, 1991, **9**, 830.
[10] K. M. Ebert, J. P. Selgrath, P. DiTullio, J. Denman, T. E. Smith, M. A. Memon, J. E. Schindler, G. M. Monastersky, J. A. Vitale, and K. Gordon, *Bio/Technology*, 1991, **9**, 835.
[11] R. E. Hammer, V. G. Pursel, C. E. Rexroad, R. J. Wall, D. J. Bolt, K. M. Ebert, R. D. Palmiter, and R. L. Brinster, *Nature*, 1985, **315**, 680.
[12] P. Krimpenfort, A. Rademakers, W. Eyestone, A. van der Schans, S. van den Broek, P. Kooiman, E. Kootwijk, G. Platenburg, R. Strijker, and H. de Boer, *Bio/Technology*, 1991, **9**, 844.
[13] J. J. Cooeau, Y. Heyman, and J. P. Renard, *Prod. Animales*, 1998, **11**, 41.
[14] E. A. Amoah and S. Gelaye, *J. Animal Sci.*, 1997, **75**, 578.
[15] J. McGrath and D. Salter, *Science*, 1983, **220**, 1300.
[16] H. T. Cheong, Y. Takahashi, and H. Kanagawa, *Biol. Reprod.*, 1993, **48**, 958.

duction of mouse clones.[17] Using nuclear transfer, sheep have been cloned from embryo-derived cell lines,[18] and also, significantly from fetal and mammary gland-derived *somatic* cell-lines.[19] More recently, cloning of bovine embryos by multiple nuclear transfer has also been reported.[20]

3 EMBRYO STEM CELL TECHNOLOGY, HOMOLOGOUS RECOMBINATION AND TRANSGENESIS

An alternative and powerful strategy for transgenesis involves the introduction of foreign DNA into embryonic stem (ES) cells (Figure 3a). To establish an ES cell-line, cells are removed from the inner cell mass of the developing blastocyst and are passaged either on feeder layers, or in the presence of differentiation-inhibiting activity (DIA),[21] to maintain their undifferentiated state. Foreign DNA can be introduced into the ES cells by a number of means: electroporation, transfection or micro-injection. Selected cells are then reintroduced into a blastocyst and are reimplanted into a pseudopregnant female and allowed to develop to term. An important distinction between pups obtained in this way, and those resulting from micro-injection of the one-cell embryo, is that they will, by definition, be chimaeras, since cells harbouring the transgene only constitute a proportion of the inner cell mass of the blastocyst. However, providing transgene-containing cells have contributed to the germline, then a suitable breeding strategy will allow the establishment of a transgenic line.[22]

The introduction of foreign DNA into the ES cell has significant advantages over micro-injection of the one-cell embryo by virtue of the fact that the foreign DNA can be designed to homologously recombine with its endogenous counterpart (Figure 3b).[23] Following selection, correctly targeted clones can be identified, either by Southern blot[2] or PCR[1] analysis, and reintroduced into host blastocysts to generate the desired transgenic chimaeras. A number of strategies have been devised for the positive selection of homologous recombination events combined with negative selection against random integration. Positive selection can be achieved by interrupting homologous sequences in the targeting

[17] O. Y. Kwon and T. Kono, *Proc. Natl. Acad. Sci.*, 1996, **93**, 13010.
[18] K. H. S. Campbell, J. McWhir, W. A. Ritchie, and I. Wilmut, *Nature*, 1996, **380**, 64.
[19] I. Wilmut, A. E. Schnieke, J. McWhir, A. J. Kind, and K. H. S. Campbell, *Nature*, 1997, **385**, 810.
[20] H. Takano, C. Kozai, S. Shimizu, Y. Kato, and Y. Tsunoda, *Theriogenology*, 1997, **47**, 1365.
[21] A. G. Smith, J. Nichols, M. Robertson, and P. D. Rathjen, *Dev. Biol.*, 1992, **151**, 339.
[22] J. Nichols, E. P. Evans, and A. G. Smith, *Development*, 1990, **110**, 1341.
[23] M. J. Evans, *Mol. Biol. Med.*, 1989, **6**, 557.

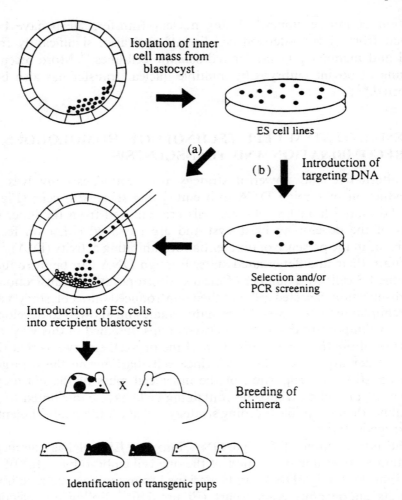

Isolation of inner
cell mass from
blastocyst

ES cell lines

(a)

(b) Introduction of
targeting DNA

Selection and/or
PCR screening

Introduction of ES cells
into recipient blastocyst

X   Breeding of
chimera

Identification of transgenic pups

Figure 3 *(a) Generation of transgenic animals by means of ES cell manipulation.
(b) Following introduction of DNA, ES cells carrying the homologous recombination event can be selected prior to introduction into the blastocyst*

vector with a selectable marker, such as the bacterial neomycin-resistance
gene *neo*^r (using G418 selection). If the vector integrates into the genome
in a homologous fashion, then the *neo*^r gene will be incorporated into the
germline. Any additional vector sequences without the region of homology will be lost. By placing the HSV-1-*tk* gene in the non-homologous
region of the targeting vector, any cells retaining this gene through
random integration events will be killed in the presence of appropriate
synthetic nucleosides. Such an approach was used to inactivate the *Wnt*-1

proto-oncogene[24,25] and the retinoblastoma gene,[26-28] however toxigenic genes can equally be used as a negative selection.[29]

Homologous integration can be indirectly selected if expression of the selectable marker is dependent on correct gene targeting. Here, the marker, lacking its translational start, is fused, in frame, to coding sequences of the targeted gene. It is essential that the target gene be either active or inducible in the ES cell for this approach to work. The strategy was used successfully to sequentially inactivate both allelles of the *pim*-1 proto-oncogene.[30] In some cases, generation of chimaeras using homozygous ES cells lacking gene functions essential to early development may be informative where breeding to homozygosity fails because of early lethality. To investigate expression patterns of a target gene, one can replace it with a reporter gene such as the *Escherichia coli lacZ* gene.[31] Developmental expression can then be followed by staining for enzymatic activity of the gene product, β-galactosidase, using X-gal which is potentially more sensitive than *in situ* hybridization.

All the above techniques generate null mutations through insertional inactivation of the target gene. By using the so-called 'hit and run' or double replacement strategies, more subtle gene alterations can be introduced into the target gene. The 'hit and run' procedure introduces a site-specific mutation into a non selectable gene by a two-step recombination event.[32] In the first step, the vector containing the desired mutation within sequences homologous to the target gene, together with selectable markers for monitoring the integration and reversion events, integrates into the target gene by single reciprocal recombination. The resultant duplication is resolved, by single intrachromasomal recombination to yield clones either restored to wild-type or carrying the desired mutation. The double-replacement strategy requires two targeting constructs. The first introduces a functional hypoxanthine phosphoribosyl transferase or *hprt* minigene into the target gene,[33] whilst the second

[24] K. R. Thomas and M. R. Capecchi, *Nature*, 1990, **346**, 847.

[25] A. P. McMahon and A. Bradley, *Cell*, 1990, **62**, 1073.

[26] E. Y.-H. P. Lee, C.-Y. Chang, N. Hu, Y.-C. J. Wang, C.-C. Lai, K. Herrup, W.-H. Lee, and A. Bradley, *Nature*, 1992, **359**, 288.

[27] T. Jacks, A. Fazeli, E. M. Schmitt, R. T. Bronson, M. A. Goodell, and R. A. Weinberg, *Nature*, 1992, **359**, 295.

[28] A. R. Clarke, E. R. Maandag, M. van Roon, N. M. T. van der Lugt, M. van der Valk, M. L. Hooper, A. Berns, and H. te Riele, *Nature*, 1992, **359**, 328.

[29] T. Yagi, Y. Ikawa, K. Yoshida, Y. Shigetani, N. Takeda, I. Mabuchi, T. Yamamoto, and S. Aizawa, *Proc. Natl. Acad. Sci. USA*, 1990, **87**, 9918.

[30] H. Riele, E. R. Maandag, A. Clarke, M. Hooper, and A. Berns, *Nature*, 1990, **348**, 649.

[31] H. L. Mouellic, Y. Lallemand, and P. Brulet, *Proc. Natl. Acad. Sci. USA*, 1990, **87**, 4712.

[32] P. Hasty, R. Ramirez-Solis, R. Krumlauf, and A. Bradley, *Nature*, 1991, **350**, 243.

[33] R. Ratcliff, M. J. Evans, J. Doran, B. J. Wainwright, R. Williamson, and W. H. Colledge, *Transgen. Res.*, 1992, **1**, 177.

removes the *hprt* gene and replaces it with a subtly altered target gene. Selection both for and against the incorporation of *hprt* can be achieved using appropriate media.

Although gene targeting strategies are becoming increasingly elegant, the design of replacement vectors is somewhat empirical. Progress towards defining the extent of homology required between targeting vector and target locus, for high-fidelity recombination to occur, is now being made.[34,35] For optimal targeting efficiency the transgene should be isogenic, *i.e.* isolated from the same strain as that from which ES cells were derived.[35,36] It appears that sequence differences in non-isogenic DNA significantly reduce recombination efficiency.

One major limitation in the widespread application of homologous recombination is that ES cells have, as yet, only been isolated from the mouse. Initially, ES cells were only isolated from the 'permissive' 129 strain, however, more recently apparently pure ES cell cultures have been isolated from primary explants of 'non permissive' embryos by selective ablation of differentiated cells.[37] Additionally, ES cell-lines have been isolated from the early epiblast by culture on STO feeder cells.[38] When primary embryonic fibroblasts were used as feeders in the presence of leukemia inhibitory factor (LIF), ES cell-lines were readily obtained from refractory strains such as CBA/Ca.[38] Putative ES cells in rat[39] and other species have not so far proved to go germline, but the development of these alternative strategies for ES cell derivation may hold promise for the future.

4   GENERAL CONSIDERATIONS

4.1   The Construct

Correct tissue-specific or developmental expression of a micro-injected transgene requires the incorporation of all necessary flanking control elements within the construct, but this is not always sufficient to ensure correct expression. The major problem regarding pronuclear micro-injection is the random integration of the transgene into the chromosome. Position effect variegation[40] results from influences of flanking

[34] K. R. Thomas, C. Deng, and M. A. Capecchi, *Mol. Cell Biol.*, 1992, **12**, 2919.
[35] C. Deng and M. A. Capecchi, *Mol. Cell Biol.*, 1992, **12**, 3365.
[36] H. te Riele, E. Robanus Maandag, and A. Berns, *Proc. Natl. Acad. Sci. USA*, 1992, **89**, 5128.
[37] J. McWhir, A. E. Schnieke, R. Ansell, H. Wallace, A. Coleman, A. R. Scott, and A. J. Kind, *Nature Genet.*, 1996, **14**, 223.
[38] F. A. Brook and R. L. Gardner, *Proc. Natl. Acad. Sci. USA*, 1997, **94**, 5709.
[39] P. M. Iannaccone, G. U. Taborn, R. L. Garton, M. D. Caplice, and D. R. Brenin, *Dev. Biol.*, 1994, **163**, 288.
[40] G. H. Karpen, *Curr. Opin. Genet. Devel.*, 1994, **4**, 281.

elements at the site of integration, and can have consequences ranging from complete silencing of the transgene to inappropriately high, copy number-independent expression or loss of developmental or cell specificity. If feasible, constructs should be transfected into appropriate cell-lines prior to introduction into the germline, to determine whether the construct can be expressed *in vitro*. This can often give valuable information about the extent of promoter sequences necessary for expression, prior to *in vivo* studies. One could consider the relative benefits of linking promoter sequences to a reporter gene such as the *lacZ* gene,[31] or green fluorescent protein (GFP) gene[41] if expression might be masked by the endogenous copy of the gene.

Position-independent, copy number-related expression may be achieved if the transgene includes locus control regions such as those identified at the human globin gene cluster,[42] the chicken β-globin gene,[43] adjacent to the human red/green visual pigment genes[44] and flanking the chicken lysozyme gene,[45] or matrix attachment regions. Such elements have been shown to function across species barriers, and appear to maintain an open chromatin structure for heterologous gene expression, independent of integration site. Identification of other similar regulatory elements and transcription factor binding sites,[46,47] may give the researcher a greater degree of control over construct design, limiting expression of the transgene to specific cells and yet ensuring good expression levels.

Since the inclusion of extensive flanking sequences may be sufficient to overcome position effects and should greatly improve the chances of incorporating important regulatory elements, researchers are turning to P1,[48] bacterial artificial chromosome[49] and yeast artificial chromosome[50] libraries as a source of transgenes. Because of the large size of such clones, modification (*e.g.* incorporation of reporter sequences or deletion/mutation of control elements) by standard restriction enzyme 'cut and ligate' strategies is not readily achieved. Techniques for homologous

[41] A. K. Hadjantonakis, M. Gertsenstein, M. Ikawa, M. Okabe, and A. Nagy, *Mechan. Devel.*, 1998, **76**, 79.
[42] P. Collis, M. Antoniou, and F. Grosveld, *EMBO J.*, 1990, **9**, 233.
[43] M. Reitman, E. Lee, H. Westphal, and G. Felsenfeld, *Nature*, 1990, **348**, 749.
[44] Y. Wang, J. P. Macke, S. L. Merbs, D. J. Zack, B. Klaunberg, J. Bennett, J. Gearhart, and J. Nathans, *Neuron*, 1992, **9**, 429.
[45] C. Bonifer, M. Vidal, F. Grosveld, and A. E. Sippel, *EMBO J.*, 1990, **9**, 2843.
[46] S. Philipsen, D. Talbot, P. Fraser, and F. Grosveld, *EMBO J.*, 1990, **9**, 2159.
[47] D. Talbot, S. Philipsen, P. Fraser, and F. Grosveld, *EMBO J.*, 1990, **9**, 2169.
[48] J. C. Pierce, B. Sauer, and N. Sternberg, *Proc. Natl. Acad. Sci. USA*, 1992, **89**, 2056.
[49] H. Shizuya, B. Birren, U.-J. Kim, V. Mancino, T. Slepak, Y. Tachiiri, and M. Simon, *Proc. Natl. Acad. Sci. USA*, 1992, **89**, 8794.
[50] Z. Larin, A. P. Monaco, and H. Lehrach, *Proc. Natl. Acad. Sci. USA*, 1991, **88**, 4123.

recombination in yeast,[51] and bacteria,[52-55] mean that modified transgenes upwards of 100 kb can now routinely be used for micro-injection.
Homologous recombination has been demonstrated in murine zygotes,
where a large functional gene (33 kb) was reconstructed from micro-
injected DNA fragments.[56] A significant proportion of resulting transgenic mice contained the correctly reconstituted human serum albumin
gene. With the advent of BAC and P1 technology, this method has
largely been superseded with few recent applications in the literature.[57]

4.2 Aberrant Expression

As already stated, a micro-injected transgene integrates into the genome
in an entirely random fashion. If the transgene integrates within an
endogenous gene, it may alter or insertionally inactivate that gene
function and result in a totally unexpected phenotype.[58-60] Ethical
concerns regarding the generation of transgenic animals, which have
been engineered specifically for pharmaceutical, medical or nutritional
reasons, lie outside the scope of this review. However, careful consideration should be given to the likely consequences of expressing a transgene
in vivo. Despite all these cautions, transgenesis is a very powerful tool
with many applications, some of which are outlined below (Section 5).

5 DESIGN OF THE TRANSGENIC EXPERIMENT

5.1 Investigating Gene Expression

Increasing the expression of a gene is often informative in evaluating the
role that the gene product plays in normal development or physiology. It
is possible to determine the effect of overexpression of a gene by placing
it under the control of a strong, heterologous promoter.[61] Alternatively,

[51] C. Huxley, *Methods Companion Methods Enzymol.*, 1998, **14**, 199.
[52] X. W. Yang, P. Model, and N. Heintz, *Nature Biotech.*, 1997, **15**, 859.
[53] J. Boren, I. Lee, M. J. Callow, E. M. Rubin, and T. L. Innerarity, *Genome Res.*, 1996, **6**, 1123.
[54] J. R. Jessen, A. Meng, R. J. McFarlane, B. H. Paw, L. I. Zon, G. R. Smith, and S. Lin, *Proc. Natl. Acad. Sci. USA*, 1998, **95**, 5121.
[55] Y.Zhang, F. Buchholz, J. P. P. Muyrers, and F. Stewart, *Nature Genetics*, 1998, **20**, 123.
[56] F. R. Pieper, I. C. M. de Wit, A. C. J. Pronk, P. M. Kooiman, R. Strijker, P. J. A. Krimpenfort, J. H. Nuyens, and H. A. de Boer, *Nucleic Acids Res.*, 1992, **20**, 1259.
[57] S. Ali, C. M. G. A. Fontes, G. P. Hazelwood, B. H. Hirst, A. J. Clark, H. J. Gilbert, and J. Hall, *Gene*, 1997, **202**, 203.
[58] R. P. Woychik, T. A. Stewart, L. G. Davis, P. D'Eustachio, and P. Leder, *Nature*, 1985, **318**, 36.
[59] A. K. Ratty, L. W. Fitzgerald, M. Titeler, S. D. Glick, J. Mullins, and K. W. Gross, *Mol. Brain Res.*, 1990, **8**, 355.
[60] U. Karls, U. Muller, D. J. Gilbert, N. G. Copeland, N. A. Jenkins, and K. Harbers, *Mol. Cell Biol.*, 1992, **12**, 3644.
[61] M. E. Steinhelper, K. L. Cochrane, and L. J. Field, *Hypertension*, 1990, **16**, 301.

expression can be rendered constitutive by linking the transgene to a housekeeping gene promoter, such as that of the phosphoglycerate kinase gene (PGK), or inducible by using, for example, the metallothionine promoter[62] or Cyp1a1.[63-65]

In order to identify and define the control elements in and around a gene which effect its tissue-specific and developmental pattern of expression, one can design a series of constructs with nested deletions around the promoter region, the 3' end of the gene, and if necessary, within introns. By analysing expression patterns of the transgenes in the resultant transgenic lines, with ontogeny studies and tissue surveys, one can build up a map of the sequences which are absolutely required for correct gene expression.

If expression of the transgene is likely to be masked or affected by expression of its endogenous counterpart then it is possible to link the promoter sequences to a reporter gene, such as the simian virus 40 large tumour antigen (SV40 TAg),[66] bacterial chloramphenicol acetyltransferase (CAT),[67] *lacZ*[31] or green fluorescent protein.[41] One can then ascertain whether or not the promoter directs expression to the appropriate cell types. Using the SV40 TAg reporter, tumours often develop in the target cells and these have proved to be invaluable in generating new cell lines which retain highly differentiated phenotypes.[68,69] β-Galactosidase staining of the embryos/tissue slices readily identifies promoter-directed sites of expression *in situ*. However, because GFP is a 'living stain', it provides the potential for real time imaging of gene expression and intracellular protein transport.[70,71]

5.2   Reduction of Gene Function

Abolishing or reducing the function of a gene can be equally informative regarding its role *in vivo*, though it should be noted that, due to the

[62] R. R. Behringer, R. L. Cate, G. J. Froelick, R. D. Palmiter, and R. L. Brinster, *Nature*, 1990, **345**, 167.
[63] S. J. Campbell, F. Carlotti, P. A. Hall, A. J. Clark, and C. R. Wolf, *J. Cell Sci.*, 1996, **109**, 2619.
[64] S. N. Jones, P. G. Jones, H. Ibarguen, C. T. Caskey, and W. J. Craigen, *Nucl. Acids Res.*, 1991, **19**, 6457.
[65] J. D. Smith, E. Wong, and M. Ginsberg, *Proc. Natl. Acad. Sci. USA*, 1995, **92**, 11926.
[66] C. D. Sigmund, K. Okuyama, J. Ingelfinger, C. A. Jones, J. J. Mullins, C. Kane, U. Kim, C. Wu, L. Kenny, Y. Rustum, V. J. Dzau, and K. W. Gross, *J. Biol. Chem.*, 1990, **265**, 19916.
[67] M. L. Lui, A. L. Olson, W. S. Moye-Rowley, J. B. Buse, G. I. Bell, and J. E. Pessin, *J. Biol. Chem.*, 1992, **267**, 11673.
[68] J. J. Windle, R. I. Weiner, and P. L. Mellon, *Mol. Endocrinol.*, 1990, **4**, 597.
[69] P. L. Mellon, J. J. Windle, P. C. Goldsmith, C. A. Padula, J. L. Roberts, and R. I. Weiner, *Neuron*, 1990, **5**, 1.
[70] X. G. Zhu, J. A. Hanover, G. L. Hager, and S. Y. Cheng, *J. Biol. Chem.*, 1998, **273**, 27052.
[71] J. Ellenberg, J. Lippincott Schwartz, and J. F. Presley, *Biotechniques*, 1998, **25**, 838.

extreme complexity of mammals, genes of the same family or pathway may be functionally redundant and that a given gene often has multiple roles. There are a number of strategies by which gene expression can be reduced. The introduction of a dominant negative mutation, by gene targeting, can reduce or inhibit the function of the wild-type endogenous protein in a manner equivalent to the gene knock-out procedure.[72,73] Alternatively, gene expression can be downregulated by the production of antisense transcripts. Inverting a DNA sequence, relative to its promoter, produces transcripts which are complimentary to the target gene transcript. Since the transgene is driven by the promoter sequences of its endogenous counterpart, expression should be limited to those sites where it can reduce the amount of targeted gene product. The mechanism by which antisense inhibition occurs remains obscure, but it is probably brought about by RNA–DNA interactions interfering with transcription, general interference with nuclear RNA processing and transport, inhibition of translation, and/or rapid degradation of sense–antisense RNA hybrids. There are numerous applications in the literature.[74-76]

A related strategy involves the use of a ribozyme, an RNA molecule with enzymic activity which is capable of cleaving specific target RNA molecules. If such a sequence is placed within an appropriate antisense sequence, one would predict a much more efficient inhibition of targeted gene expression, though few studies have rigorously tested this prediction (for review see Sokol & Murray[77]). Sokol *et al.*[78] using an array of antisense transgenes, with and without ribozyme motifs, found no evidence that ribozymes enhance the effectiveness of antisense mRNA directed against CAT gene transcripts. Despite this, there are many examples of ribozyme applications in the literature.[79-81]

[72] T. Bowman, H. Symonds, L. Gu, C. Yin, M. Oren, and T. van Dyke, *Genes Devel.*, 1996, **10**, 826.
[73] R. M. Friedlander, V. Gagliardini, H. Hara, K. B. Fink, W. W. Li, G. MacDonald, M. C. Fishman, A. H. Greenberg, M. A. Moskowitz, and J. Y. Yuan, *J. Exp. Med.*, 1997, **185**, 933.
[74] M.-C. Pepin, F. Pothier, and N. Barden, *Nature*, 1992, **355**, 725.
[75] R. P. Erickson, L. W. Lai, and J. Grimes, *Devel. Genet.*, 1993, **14**, 274.
[76] C. M. Moxham, Y. Hod, and C. C. Malbon, *Science*, 1993, **260**, 991.
[77] D. L. Sokol and J. D. Murray, *Transgen. Res.*, 1996, **5**, 363.
[78] D. L. Sokol, R. J. Passey, A. G. Mackinlay, and J. D. Murray, *Transgen. Res.*, 1998, **7**, 41.
[79] P. J. Huillier, S. Soulier, M. G. Stinnakre, L. Lepourry, S. R. Davis, J. C. Mercier, and J. L. Vilotte, *Proc. Natl. Acad. Sci. USA*, 1996, **93**, 6698.
[80] A. S. Lewin, K. A. Dresner, W. W. Hauswirth, S. Nishikawa, D. Yasamura, J. G. Flannery, and M. M. LaVail, *Nature Med.*, 1998, **4**, 967.
[81] R. Morishita, S. Yamada, K. Yamamoto, N. Tomita, I. Kida, I. Sakurabayashi, A. Kikuchi, Y. Kaneda, R. Lawn, J. Higaki, and T. Ogihara, *Circulation*, 1998, **98**, 1898.

5.3 Cell Ablation

To answer questions about the lineage, fate or function of a cell, it can be informative to observe the effects of removing that cell.[82] By introducing genes encoding cytotoxins, such as the catalytic subunits of Diphtheria toxin (DT-A)[83,84] or ricin (RT-A),[85] under appropriate cell-specific promoters, one can selectively ablate cell types which might be difficult or impossible to remove by physical means. Toxigenic ablation is potentially very powerful, since the toxic gene products may act at very low concentrations. However, it is important that the toxin should be confined to the cells where it is expressed, or damage to neighbouring cells may occur. To overcome this problem, toxigenes often lack the signal sequences of their native counterparts. Obviously, some promoter–toxigene constructs are likely to be lethal to the developing embryo, if the ablated cells are essential for viability. To give the researcher more control over the degree of cell ablation or timing of the event during development, a number of strategies have been devised. The first is the use of an attenuated DT-A gene.[86] Although the degree of penetrance of such a gene may be variable, it should prove more versatile in achieving cell ablation with a broad range of cell- and tissue-specific promoter sequences.

An alternative strategy makes ablation dependent on the administration of drugs. This was first achieved by introducing the herpes simplex virus-1 thymidine kinase (HSV-1-*tk*) transgene under the control of appropriate promoter elements. Cells expressing the gene are rendered susceptible to drugs such as gancyclovir.[87] The power of this strategy is that the timing and degree of cell ablation is controlled by the investigator. Evidence suggests that both actively dividing and non-dividing cells may be susceptible to drug-induced ablation indicating a wide application for this strategy.

Recently, transgenic mice specifically expressing the *E. coli* nitroreductase (NTR) gene in the luminal epithelial cells of the mammary gland have been reported.[88,89] Administration of the anti-tumour pro-drug

[82] C. J. O'Kane and K. G. Moffat, *Curr. Op. Genet. Devel.*, 1992, **2**, 602.
[83] R. D. Palmiter, R. R. Behringer, C. J. Quaife, F. Maxwell, I. H. Maxwell, and R. L. Brinster, *Cell*, 1987, **50**, 435.
[84] M. L. Breitman, S. Clapoff, J. Rossant, L.-C. Tsui, L. M. Glode, I. H. Maxwell, and A. Bernstein, *Science*, 1987, **238**, 1563.
[85] C. P. Landel, J. Zhao, D. Bok, and G. A. Evans, *Genes Dev.*, 1988, **2**, 1168.
[86] M. L. Breitman, H. Rombola, I. H. Maxwell, G. K. Klintworth, and A. Bernstein, *Mol. Cell Biol.*, 1990, **10**, 474.
[87] E. Borrelli, R. A. Heyman, C. Arias, P. E. Sawchenko, and R. M. Evans, *Nature*, 1989, **339**, 538.
[88] A. J. Clark, M. Iwobi, W. Cui, M. Crompton, G. Harold, S. Hobbs, T. Kamalati, R. Knox, C. Neil, F. Yull, and B. Gusterson, *Gene Therapy*, 1997, **4**, 101.
[89] B. Gusterson, W. Cui, M. Iwobi, M. R. Crompton, G. Harold, S. Hobbs, T. Kamalati, R. Knox, C. Neil, F. Yull, B. Howard, and A. J. Clark, *Endocrine-Related Cancer*, 1997, **4**, 67.

CB1954 (5-aziridin-1-yl-2,4-dinitrobenzamide), which requires enzymatic activation by NTR, rapidly and selectively killed the cells. This model should allow the examination of mammary carcinogenesis and its modulation through manipulation of target cell population size. The strategy is potentially applicable to the investigation of complex cellular interactions during development and in the adult organism, and because of the ease of control, it is predicted that this system will supersede HSV-1-*tk* ablation.

5.4   Conditional Gene Alteration

To circumvent problems related to complete inactivation of a gene early in development masking its full spectrum of functions, many novel strategies have been developed which place temporal and spatial restrictions on the gene alteration. (For detailed reviews, see Sauer,[90] Lobe and Nagy[91] and Porter[92]).

5.4.1   Inducible Gene Targeting Using the Cre-lox System.   By exploiting the site-specific DNA recombination systems of bacteria and yeast, one can, through gene targeting or transgenesis, effect *in vivo* manipulation of DNA in ES cells or living animals.

Cre recombinase (Cre) from the temperate phage P1, recognizes a 34-bp site called *loxP*, and efficiently catalyses reciprocal conservative DNA recombination between pairs of *loxP* sites. The *loxP* site consists of two 13-bp inverted repeats flanking an 8-bp non-palindromic core region, which gives the site directionality. Cre-mediated recombination between two directly repeated *loxP* sites leads to excision of the intervening sequences. If the *loxP* sites are inverted, then recombination will result in inversion of the intervening sequences.[93] A similar system involving FLP recombinase and the FRT recognition site exists in yeast.[94] The applications of such switch mechanisms are almost unlimited and numerous examples are given below.

If introduction of a transgene into the mouse results in lethality or reduced viability, it is advantageous to maintain the transgene in a dormant state until the transgenic line has become established. This can be achieved by placing a lox^2STOP cassette between the potentially toxic transgene and its promoter (Figure 4a). By intercrossing the mouse to a second (well characterized) line which expresses Cre, the transgene can

[90] B. Sauer, *Methods Companion Methods Enzymol.*, 1998, **14**, 381.
[91] C. G. Lobe and A. Nagy, *BioEssays*, 1998, **20**, 200.
[92] A. Porter, *Trends Genet.*, 1998, **14**, 73.
[93] B. Sauer, *Molec. Cell Biol.*, 1987, **7**, 2087.
[94] N. J. Kilby, M. R. Snaith, and J. A. Murray, *Trends Genet.*, 1993, **9**, 413.

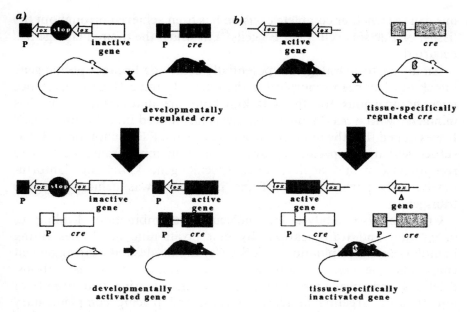

Figure 4 *Applications of the* cre-*lox technology. (a) The target gene is inactivated because of the insertion of a loxP-flanked stop cassette between the gene and its promoter. On crossing to a second transgenic mouse which carries a developmentally regulated cre recombinase, developmental activation of the target gene is achieved. (b) The target gene is flanked by loxP sites through homologous recombination in ES cells, and the resultant mouse (following blastocyst injection, etc.) is bred to homozygosity. On crossing to a second transgenic mouse which carries a tissue-specifically regulated* cre *(in addition to null alleles for the target gene) tissue-specific ablation of the target gene is achieved*

be activated as desired.[95] (It should be noted that, following Cre excision of the inhibitory cassette, a 34-bp *loxP* site remains, and it is essential that this does not interfere with transgene expression.) If the transgene would potentially be expressed universally, then conditional expression of Cre, using a developmental- or tissue-specific promoter, will limit transgene expression. One can envisage using a whole panel of *Cre* transgenics to analyse specific aspects of transgene function. In a similar way, if one modifies the target gene by homologous recombination in ES cells, so that it is flanked by *loxP* sites, and then crosses the resultant line to a second transgenic line conditionally expressing Cre, one will achieve tissue-specific gene ablation (Figure 4b).[96] If transgene excision is designed to simultaneously activate a reporter gene such as GFP, then

[95] M. Lakso, B. Sauer, J. B. Mosinger, E. J. Lee, R. W. Manning, S.-H. Yu, K. L. Mulder, and H. Westphal, *Proc. Natl. Acad. Sci. USA*, 1992, **89**, 6232.
[96] H. Gu, J. D. Marth, P. C. Orban, H. Mossmann, and K. Rajewsky, *Science*, 1994, **265**, 103.

one can see whether complete excision has been achieved throughout the tissue, or Fac sort cells (*e.g.* ES cells[97]) in which the desired change has occurred.

An alternative strategy for potentially achieving lineage-specific gene knock-out has been reported for the retinoblastoma (RB) locus. Since mice homozygous for the RB knockout die mid-gestation, an RB minigene, with a *loxP*-flanked exon was introduced into the embryos.[98] It was hoped that the minigene would rescue the RB phenotype and that subsequent tissue-specific removal of the functional transgene by Cre recombinase would achieve tissue-specific gene knockout. Unfortunately, only partial rescue of the phenotype was achieved by the transgene.

Cre can also catalyse intermolecular recombination, leading to integration and translocation. By first positioning *loxP* sites (using homologous recombination in ES cells) at the desired rearrangement endpoints one can orchestrate precise chromosome translocations, deletions and inversions. This can be achieved using two rounds of homologous targeting in *Hprt* ES cells, and placing complementary halves of a *loxP–Hprt* fusion at each target.[99] Successes to date suggest that the engineering of tissue-specific chromosomal translocations[99,100] and deletions[101] will provide mouse models for human diseases caused by specific monosomies.[102]

5.4.2 Tetracycline/Tamoxifen. An attractive alternative to the use of native promoters is to place the transgene under the control of a synthetic inducer as exemplified by the tetracycline-regulated transcription system. The transgene is placed downstream of a transcriptionally silent minimal promoter carrying a repeated *tet* operator sequence. In the presence of tetracycline (Tc) the Tc-controlled transactivator is unable to bind to the operator, and the expression unit is turned off.[103] Conversely, construction of an altered transactivator, which only binds to the *tet* operator in the presence of tetracycline (reverse Tc-controlled transactivator or rtTa) puts gene *activation* in the hands of the

[97] S. Gagneten, Y. Le, J. Miller, and B. Sauer, *Nucleic Acids Res.*, 1997, **25**, 3326.
[98] E. Zackenhaus, Z. Jiang, D. Chung, J. Marth, R. Phillips, and B. Gallie, *Genes Devel.*, 1996, **10**, 3051.
[99] A. J. H. Smith, M. A. de Sousa, B. Kwabi-Addo, A. Heppell-Parton, H. Imprey, and P. Rabbitts, *Nature Genetics*, 1995, **9**, 376.
[100] R. Ramirez-Solis, P. Lui, and A. Bradley, *Nature*, 1995, **378**, 720.
[101] Z.-W. Li, G. Stark, J. Gotz, T. Rulicke, U. Muller, and C. Weissmann, *Proc. Natl. Acad. Sci. USA*, 1996, **93**, 6158.
[102] M. Lewandoski and G. Martin, *Nature Genetics*, 1997, **17**, 223.
[103] M. Gossen, S. Freundlieb, G. Bender, G. Muller, W. Hillen, and H. Bujard, *Science*, 1995, **268**, 1766.

researcher.[104] Although initial attempts at tetracycline control of Cre-mediated recombination in transgenic animals were only partially successful due to mosaicism of transgene expression,[105] it is becoming widely used. Both systems act as true genetic switches, and potential interactions with cellular transcription factors and potential epitopes which might have elicited a cellular immune response should be eliminated by recent refinements.[106]

An alternative strategy for conditionally regulating somatic mutagenesis was accomplished by tissue-specifically expressing Cre as a fusion protein with a mutant oestrogen receptor ligand binding domain which is insensitive to the endogenous hormone β-oestradiol, but responsive to the synthetic oestrogen antagonist 4-OH-tamoxifen. The required Cre modification was thus limited to a particular tissue, and was only effected on administration of 4-OH-tamoxifen to the transgenic animals.[107]

6 COMMERCIAL APPLICATIONS

6.1 Biopharmaceuticals in Transgenic Animals

Transgenic livestock hold the promise of being able to produce large quantities of important therapeutic proteins. The important distinction between 'pharming' and protein production by large-scale mammalian or microbial cell culture is that proteins produced *in vivo* can be post-translationally modified in a manner identical to that of the native product.

Early attempts to generate transgenic farm animals were hampered by low frequencies of transgene integration, low numbers of animals expressing the recombinant protein, and reproductive and physiological problems. Some of these problems can be bypassed by directing expression of the transgene to the mammary gland, an exocrine organ in which the expressed transgene remains separate from the animal's blood circulation. Typically, the transgene is fused to the regulatory sequences of a milk protein.[10,108,109] The yield of heterologous protein

[104] A. Kistner, M. Gossen, F. Zimmermann, J. Jerecic, C. Ullmer, H. Lubbert, and H. Bujard, *Proc. Natl. Acad. Sci. USA*, 1996, **93**, 10933.
[105] L. St-Onge, P. A. Furth, and P. Gruss, *Nucleic Acids Res.*, 1996, **24**, 3875.
[106] U. Baron, M. Gossen, and H. Bujard, *Nucleic Acids Res.*, 1997, **25**, 2723.
[107] F. Schwenk, R. Kuhn, P.-O. Angrand, K. Rajewsky, and A. J. Stewart, *Nucleic Acids Res.*, 1998, **2**, 1427.
[108] A. L. Archibald, M. McClenaghan, V. Hornsey, J. P. Simons, and A. J. Clark, *Proc. Natl. Acad. Sci. USA*, 1990, **87**, 5178.
[109] A. J. Clark, H. Bessos, J. O. Bishop, P. Brown, S. Harris, R. Lathe, M. McClenaghan, C. Prowse, J. P. Simons, C. B. A. Whitelaw, and I. Wilmut, *Bio/Technology*, 1989, **7**, 487.

has been found to be extremely variable,[110] but can account for a significant proportion of the total milk proteins. Poor expression of the ovine promoter in the mouse may reflect species differences in recognizing heterologous *versus* homologous promoters, and the predictive value of mouse models must be cautioned. The best example, to date, is the transgenic sheep which expresses 35 g l^{-1} (50% of the total milk protein) of glycosylated human alpha-1-antitrypsin under the direction of the ovine β-lactoglobulin promoter.[9] Taking into account the milk yield of goats, sheep and cows, such levels of transgene expression would represent a substantial yield per year compared to that from mammalian or microbial cell culture. The development of suitable purification methods and the use of transgenically produced proteins in clinical trials are well advanced. The range of proteins which can be produced in milk may not be limitless, however, since high expression of certain proteins may adversely affect the physiology of the mammary gland.[111] However, one can anticipate, for example, the future development of human milk substitutes.[112]

6.2 Xenografts

The shortage of human organs for transplantation has prompted developments in the use of animal organs. The major barrier to successful xenogeneic organ transplantation is the phenomenon of complement-mediated hyperacute rejection (HAR), which involves the classical complement cascade. The strategy used to address HAR in porcine-to-primate xenotransplantation has been to produce transgenic pigs expressing human complement inhibitors such as decay accelerating factor (DAF)[113] and hCD59.[114] After transplantation, pig hearts survived in recipient baboons for prolonged periods without rejection.[114] If ES cells can be isolated from the pig, it may be possible to knockout the antigenic determinants to which antispecies antibodies bind, as a further strategy for eliminating HAR. Having overcome the problems associated with organ rejection, two further hurdles have

[110] A. S. Carver, M. A. Dalrymple, G. Wright, G. S. Cottom, D. B. Reeves, Y. H. Gibson, J. L. Keenan, J. D. Barrass, A. R. Scott, A. Colman, and I. Garner, *Bio/Technology*, 1993, **11**, 1263.

[111] R. J. Wall, V. G. Pursel, A. Shamay, R. A. McKnight, C. W. Pittius, and L. Henninghausen, *Proc. Natl. Acad. Sci. USA*, 1991, **88**, 1696.

[112] S. J. Kim, D. Y. Yu, Y. M. Han, C. S. Lee, and K. K. Lee, *Adv. Exp. Med. Biol.*, 1998, **443**, 79.

[113] A. M. Rosengard, N. R. B. Cary, G. A. Langford, A. W. Tucker, J. Wallwork, and D. J. G. White, *Transplantation*, 1995, **59**, 1325.

[114] K. R. McCurry, D. L.Kooyman, C. G. Alvarado, A. H. Cotterell, M. J. Martin, J. S. Logan, and J. L. Platt, *Nature Medicine*, 1995, **1**, 423.

come to light, namely acute vascular rejection and cell-mediated rejection (for a review see Platt[115]).

6.3 Toxicological Applications

Transgenic animals have been designed which allow the researcher to screen for genotoxic or carcinogenic compounds. There are now commercially available transgenic mice, including Mutamouse and Big Blue,[116] which contain the *E. coli lacZ* and *lacI* genes, respectively, cloned into bacteriophage lambda vectors that are integrated into the genome. Following treatment of the mice with a test chemical, the integrated bacteriophage vectors are rescued from genomic DNA, and mutant phage are recognized by their ability to grow on susceptible *E. coli* host strains and by resulting plaque colour. Genotoxic agents which cause large deletions can be detected by transgenic mice carrying a plasmid-based *lacZ* system.[117]

Classically, carcinogenicity testing has depended on the chronic rodent bioassay which uses a large number of animals (400 to 500 per compound), and since the development of a malignant clone requires several genetic changes in the affected cells, animals must be exposed to the test compound for considerable lengths of time. Transgenic animals have now been generated with a predisposition to carcinogen-induced tumorigenesis, by introducing an activated oncogene,[118] an inactivated tumour suppresser (p53) gene,[119,120] or an inactivated DNA repair (XPA) gene[121] into the genome. It is essential that the transgene does not cause an increase in spontaneous tumours, but simply allows shorter periods of exposure to test compounds and a considerable reduction in the numbers of animals required. An important consideration when using these transgenic lines is that the effects of mutations in certain genes may be subject to species variation.

6.4 Immortomouse

The study of cell-lines has greatly improved our understanding of many important biological systems. However, the limited number of available

[115] J. L. Platt. *Nature*, 1998, **392** (suppl), 11.
[116] R. Forster in 'Environmental Mutagenesis', ed. D. H. Phillips and S. Venitt, Bios Scientific Publishers, Oxford, UK, 1995, p. 291.
[117] J. A. Gossen, H.-J. Martus, J. Y. Wei, and J. Vijg, *Mutation Res.*, 1995, **331**, 89.
[118] E. D. Kroese, H. van Steeg, C. T. van Oostrom, P. M. Dortant, P. Wester, H. J. van Kranen, A. de Vries, and C. F. van Kreijl, *Carcinogenesis*, 1997, **19**, 975.
[119] R. W. Tennant, J. E. French, and J. W. Spalding, *Environ. Health Perspect.*, 1995, **103**, 942.
[120] S. Yamamoto, *Carcinogenesis*, 1996, **17**, 2455.
[121] A. de Vries, C. T. M. van Oostrom, P. M. Dortant, R. B. van Beems, C. F. van Kreijl, P. J. A. Capel, and H. van Steeg, *Mol. Carcinogen.*, 1997, **19**, 46.

cell-lines and, more pertinently, the difficulty in obtaining new ones, have impeded many areas of study. Traditionally, cell-lines were obtained only as tumour cells or spontaneously immortalized variants of cells in culture. Transfection and retroviral-mediated gene insertion of immortalizing genes have facilitated the production of cell-lines from various tissues, but such genes can alter the normal physiology of the cell. This problem has been overcome by the introduction of a conditional immortalizing gene, a temperature sensitive variant of the simian virus 40 (SV40) large tumour antigen (TAg), into transgenic mice under the mouse major histocompatibility complex H-$2K^b$ promoter.[122] The promoter is widely active and is inducible by γ-interferon, meaning that cell-lines can potentially be derived from a wide number of tissues grown at the permissive temperature (33°C),[123] but the thermolabile TAg has only minimal activity *in vivo* (39°C), so that tumours should not develop in the immortomouse.

7 FUTURE PROSPECTS

Transgenesis is having and will continue to have far-reaching affects on the fields of animal model production, gene therapy strategies, new therapeutic drug treatments, and the commercial production of biologically important molecules. Improvements in the efficiency and success of nuclear transfer as a means of increasing numbers of scientifically or commercially important animals will have many important applications. Additionally, with refinements in gene targeting, and the extension of ES cell technology to include commercially important animals, the potential for subtle gene alteration promises to yield exciting future developments.

[122] P. S. Jat, M. D. Noble, P. Ataliotis, Y. Tanaka, N. Yannoutsos, L. Larsen, and D. Kioussis, *Proc. Natl. Acad. Sci. USA*, 1991, **88**, 5096.
[123] R. H. Whitehead, P. E. VanEeden, M. D. Noble, P. Ataliotis, and P. S. Jat, *Proc. Natl. Acad. Sci. USA*, 1993, **90**, 587.

CHAPTER 14

Protein Engineering

JOHN R. ADAIR

1 INTRODUCTION

Protein engineering is the process of constructing novel protein molecules, by design from first principles or by altering an existing structure. There are two main reasons for wishing to do this. First, there is the desire to understand for its own sake how proteins are assembled and what elements of the primary sequence contribute to folding, stability and function. These features can be probed by altering one or more specific amino acids in a directed manner within a protein and observing the outcome after production of the altered version. Often related proteins with similar but not identical sequences exist in nature that have slightly different properties and these differing sequences can be used as guides for the alterations.

A second reason for wishing to change a protein is that the protein may be suitable, in principle, for a particular technology purpose but the version found in nature does not have the optimal properties required for the task. For example, an enzyme may be considered as part of an industrial process but a feature of the protein, such as the temperature stability or pH optimum for the catalytic activity, or the need for a co-factor may not be compatible with the process. Amino acid changes can be made that can tailor the enzyme so that it functions better in the new environment. There are many other examples of how proteins can be altered to make them better suited to commercial and technological activities and some of these are noted in Section 3 below.

To engineer a protein implies an understanding of protein structural principles, knowledge of proteins as a material, and an appreciation of the limits of the material so that rational design, or alteration, of the properties can be achieved. In addition the engineer should have to hand the tools to produce and analyse the desired protein. The tools and the underlying principles have been developing in a parallel but intertwined manner over the past two decades.

This chapter will provide a background to protein engineering and summarize some recent developments.

1.1   Protein Structures

Our current knowledge of how proteins look, how they behave and the principles that determine how a primary structure folds derives from two sources. First, studies on the structures of proteins using physical techniques along with biochemical studies on the properties and physical interactions of the amino acids within proteins and with their environment have shown that the amino acids of proteins adopt particular secondary structures such as the α-helix or β-strand, and these secondary structures in turn are folded into tertiary structure motifs. Small proteins may comprise only one such motif, however larger proteins are often comprised of a number of motifs, or domains, that are themselves folded into a particular arrangement with specific interactions between the domains. In some cases separate proteins are organized into larger complexes by inter-domain interactions to form quaternary structures. Within the tertiary structures particular amino acid groupings combine to provide the function of the protein. For example, certain amino acid side chains spread throughout the primary amino acid sequence may be brought into close association by the tertiary structure to generate a catalytic function or a specific ligand-binding surface.

Second, and drawing on the structural studies, theorizing and calculation have provided insights into the ways proteins may assemble and function. For example, early observations on the likelihood of particular amino acids to participate in different secondary structure motifs provided the initial impetus for protein structure prediction algorithms (*e.g.* reference 1).

Many basic questions about the properties of proteins remain unresolved, for example the mechanics of protein folding is still a topic of some debate (*e.g.* reference 2). This may not be unexpected, given that the number of sequences that have currently been obtained, and structures that have been determined, are a very small fraction of the available repertoire occurring in nature. Fortunately, *via* selection,

[1] P. Y. Chou and G. D. Fasman, *Biochemistry*, 1974, **13**, 222.
[2] C. M. Dobson and M. Karplus, *Curr. Opin. Struct. Biol.*, 1999, **9**, 92.

nature repeats successful structural motifs and this redundancy provides an insight into the way that different amino acid sequences can adopt similar structures. As more new sequences become available and structures determined, and as methods become more efficient at placing new sequences within existing sequence and structure families, then the 'who, what and when' of protein structures may become more accessible. The 'why' and the 'how' still require a lot more effort. Protein engineering seeks to accelerate our understanding of the 'why' and the 'how'.

In the next section the basic tools that are required for protein engineering are described, followed by a series of examples of successful protein engineering

2 TOOLS

2.1 Sequence Identification

Sequence identification, by protein or gene sequencing is now a relatively straightforward process. Large databases now exist containing many thousands of sequences.[3] In addition genome sequencing projects are providing new sequences and open reading frames at an ever increasing rate. The functions of many of these sequences are known, from biochemical or genetic data, or by sequence homology to other known sequences. This gives the protein engineer an approach to rational design.

2.2 Structure Determination and Modelling

High-resolution structural information, determined by X-ray or electron crystallography[4-6] or NMR techniques,[7] is at the core of understanding of protein biochemistry. The number of proteins that have high-resolution structures is increasing rapidly, but remains well below the number of newly identified sequences. However, proteins with little primary sequence homology have been observed to adopt similar folds and can be grouped into protein superfamilies. These groupings may reflect ancient evolutionary relationships (but care needs to be taken in inferring functional activities to regions that may be conserved for purely physical reasons). It has been estimated that there exist in nature

[3] A. Bairoch and R. Apweiler, *Nucleic Acids Res.*, 1997, **25**, 31.
[4] J. R. Helliwell, *Meth. Enzymol.*, 1997, **276**, 203.
[5] T. Walz and N. Grigorieff, *J. Struct. Biol.*, 1998, **121**, 142.
[6] J. P. Wery and R. W. Schevitz, *Curr. Opin. Chem. Biol.*, 1997, **1**, 365.
[7] K. Wuthrich, *Nat. Struct. Biol.*, 1998, **NMR Supplement**, 492.

a relatively small number (1000–3000) of unique fold topologies,[8–13] although there is some comment about how to distinguish between closely related types (*e.g.* reference 14). Given the rate at which new folds appear in new structure determinations it has been estimated that only(!) in the order of 10 000–15 000 additional structures may be required to be determined to cover 'structure space'.[15,16] Deciding how to, and who will, fill the gaps in the structure database will presumably require some international co-ordination,[16,17] but it is estimated that this will be done in the next decade. In the meantime, while this comprehensive structural database is being assembled, attempts are being made to predict protein structures by a variety of means.

Prediction of protein structures from primary sequence has a long history (*e.g.* reference 1). Jones,[18] Sanchez and Šali,[19] and Westhead and Thornton[20] have reviewed recent developments. Significant progress has been made in predicting structures by modelling new sequences onto the known structures of homologous sequences or sub-sequences, or, where there is little homology to known sequences, by using 'threading' or fold-recognition techniques, in which a new sequence is compared directly to known structural types.[21,22] *Ab initio* calculation is still problematic. A question that recurs is how close does a model approach reality and what confidence can be placed in the detail, before protein engineering experiments can begin. As will be seen below, some of the manipulation methods seek to circumvent this problem.

2.3 Sequence Modification

Modification of an existing protein by alteration of the gene sequence, from whole domains down to single amino acids, is now a routine process and this area is now the least difficult part of the engineering

[8] S. E. Brenner, C. Chothia, and T. J. Hubbard, *Curr. Opin. Struct. Biol.*, 1997, **7**, 369.
[9] A. V. Efimov, *Structure*, 1994, **2**, 999.
[10] J.-F. Gibrat, T. Madej, and S. H. Bryant, *Curr. Opin. Struct. Biol.*, 1996, **6**, 377.
[11] A. G. Murzin, S. E. Brenner, T. Hubbard, and C. Chothia, *J. Mol. Biol.*, 1995, **247**, 536.
[12] M. B. Swindells, C. A. Orengo, D. T. Jones, E. G. Hutchinson, and J. M. Thornton, *BioEssays*, 1998, **20**, 884.
[13] C. A. Orenga, D. T. Jones, and J. M. Thronton, *Nature*, 1994, **372**, 631.
[14] L. Holm and C. Sander, *Structure*, 1997, **5**, 165.
[15] T. Gaasterland, *Nat. Biotechnol.*, 1998, **16**, 625.
[16] B. Rost, *Structure*, 1998, **6**, 259.
[17] T. C. Terwilliger, G. Waldo, T. S. Peat, J. M. Newman, K. Chu, and J. Berendzen, *Protein Sci.*, 1998, **7**, 1851.
[18] D. T. Jones, *Curr. Opin. Struct. Biol.*, 1997, **7**, 377.
[19] R. Sanchez and A. Šali, *Curr. Opin. Struct. Biol.*, 1997, **7**, 206.
[20] D. R. Westhead and J. M. Thornton, *Curr. Opin. Biotechnol.*, 1998, **9**, 383.
[21] D. T. Jones, W. R. Taylor, and J. M. Thornton, *Nature*, 1992, **358**, 86.
[22] M. J. Sippl and H. Flöckner, *Structure*, 1996, **4**, 15.

process. Since the early 1980s methods for directed protein modification by oligonucleotide-directed site specific mutagenesis (see below), by whole gene synthesis from oligonucleotides (*e.g.* reference 23), and by the use of the polymerase chain reaction (PCR[24]) have become routine (see references 25–27 for reviews).

These methods are restricted to gene-coded amino acids. Sequences may also be generated by chemical protein synthesis *in vitro* (*e.g.* Holford and Muir, 1998[28]; Muir, 1995[29]) which also allows the incorporation of non-coded amino acids into the protein sequence.[30]

2.3.1 Site-directed Mutagenesis Methods. There are two basic methods of alteration of the existing coding sequence of a protein at the DNA level. Both involve annealing of one or more oligonucleotides to a region of (at least temporarily) single stranded DNA (ssDNA) followed by *in vitro* DNA polymerase directed extension of the oligonucleotide(s).

2.3.1.1 Non-PCR Methods. In the older non-PCR procedures, the DNA sequence to be modified is linked to a replication origin (*e.g.* in a plasmid, bacteriophage or 'phagemid), allowing *in vivo* amplification of the modified and parental genotypes. A synthetic oligonucleotide harbouring the required mutation is annealed either to circular ssDNA template, *e.g.* the genome of a ssDNA 'phage such as M13 or ΦX174 or the ssDNA form of a phagemid (Figure 1a) or to a partially single-stranded dsDNA template (Figure 1b). Partial ssDNA templates can be generated by a variety of enzymatic methods.

The annealed oligonucleotide acts as a primer for *in vitro* DNA synthesis using DNA polymerase, usually T4 or T7 DNA polymerase, and in the presence of DNA ligase, closed circular double-strand DNA (dsDNA) molecules are generated (Figure 1c). The dsDNA is introduced into a suitable host cell where replication and hence segregation of parental and mutant daughter strand occurs. The desired modification is then identified by one of a number of screening or selection procedures (summarized in reference 25).

[23] M. D. Edge, A. R. Green, G. R. Heathcliffe, P. A. Meacock, W. Schuch, D. B. Scanlon, T. C. Atkinson, C. R. Newton, and A. F. Markham, *Nature*, 1981, **292**, 756.
[24] R. K. Saiki, S. Scharf, F. Faloona, K. B. Mullis, G. T. Horn, H. A. Erlich, and N. Arnheim, *Science*, 1985, **230**, 1350.
[25] J. Adair and T. P. Wallace in 'Molecular Biomethods Handbook', ed. R. Rapley and J. Walker, Humana Press Inc, Totowa, NJ, USA, 1998, 347.
[26] C. R. Newton and A. Graham, 'PCR', Bios Scientific Publishers, Oxford, UK, 1994.
[27] M. Smith, 'Synthetic DNA and Biology' ©The Nobel Foundation, 1993.
[28] M. Holford and T. W. Muir, *Structure*, 1998, **6**, 951.
[29] T. W. Muir, *Structure*, 1995, **3**, 649.
[30] J. Wilken and S. B. H. Kent, *Curr. Opin. Biotechnol.*, 1998, **9**, 412.

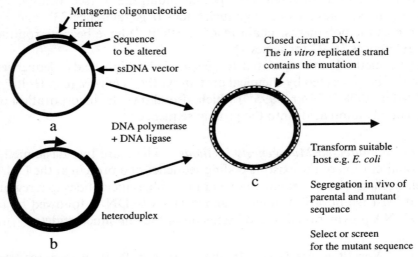

Figure 1 *Strategy for non-PCR mutagenesis of ssDNA. A single-strand DNA (ssDNA) template containing the sequence to be mutated is obtained. This is usually achieved by cloning the required sequence into a phage or phagemid vector and generating the ssDNA form. For a single-point mutation an oligonucleotide of between 15–20 nucleotides with the proposed mutant sequence located centrally in the sequence is synthesized and is mixed in molar excess with, and annealed to, the ssDNA template. For more complex mutations it is generally useful to have 15–18 nucleotides at either side of the mutation that will form a perfect match with the template sequence. A DNA polymerase is added along with DNA ligase and the remaining ssDNA regions are converted to dsDNA and closed by the ligase (Figure 1c). T7 DNA polymerase is often used which has efficient 3′–5′ proofreading activity and good processivity but which lacks the detrimental 5′–3′ exonuclease activity. (b) As an alternative to the procedure in (a) the vector containing the parental is rendered partially single stranded across the region to be mutated. A very simple way to achieve this is by cloning the required sequence into a phage or phagemid vector and generating the ssDNA form. Then the dsDNA vector without the insert is annealed to the ssDNA form generating a heteroduplex in which the cloned sequence is exposed as ssDNA. The mutation is then generated using a synthetic oligonucleotide, DNA polymerase and DNA ligase as in (a). (c) The closed circular DNA that is generated has the mutation coded in the in vitro generated strand and the parental sequence in the vector strand. At the site of the mutation there is a DNA mismatch. The mutant and parental sequences are separated by transformation into a suitable host and allowing replication to occur. The mutant sequence can then be identified by one of a number of screening or selection procedures (reviewed in reference 25). In either procedure multiple mutations can be incorporated within one oligonucleotide or by annealing several oligonucleotides to the ssDNA region. (Figure adapted from reference 25 with permission of the publisher.)*

2.3.1.2 PCR-based Methods. The non-PCR methods are well known and have been optimized over a period of many years. However, the more recent PCR-based methods have become more widespread, particularly for rapid rearrangements of protein domains where convenient common restriction sites are often not available.

In the PCR-based approaches the desired modification is generated and amplified *in vitro* by annealing to denatured target DNA a synthetic oligonucleotide harbouring the required mutation along with an oligonucleotide that can act as a primer for replication of the complementary DNA strand (Figure 2).

There are a large variety of variants to the generalized method outlined in Figure 2, some of which involve including restriction sites into the mutagenic olignucleotides. This can reduce the number of reactions involved. (For example, if in Figure 2 primers A and D have useful restriction sites in the sequences and in D the site is 3′ to the mutagenic sequence, the first reaction products can then be directly cloned.) The amplified DNA fragment is then linked to a bacterial replication origin in a plasmid or 'phage and cloned by introduction into bacteria.

A variant of the procedure allows the linking together of separate sequences which may code for domains from different proteins or may allow the reorganization of domains within a protein. In this procedure, outlined in Figure 3, the primers C and D are hybrids containing sequences that can anneal to both of the domains in question.

The first reactions then generate dsDNA fragments that now overlap in sequence and the desired product can then be generated in a third reaction using the primers A and B.

In all of these cases the design of the oligonucleotide sequences and the reaction conditions need to be considered carefully, along with the correct choice of thermostable DNA polymerase (summarized in reference 25). However these procedures are very rapid and highly efficient.

2.4 Molecular Evolution

The difficulty with protein engineering has, and will continue to be, knowing what to alter. In many circumstances directed modification of a sequence is not a suitable procedure for obtaining a desired outcome, because it is often not clear where the true target amino acid(s) lie(s) and what to alter it/them to. A number of strategies have been developed to produce and test large libraries or repertoires of variants of a particular sequence.

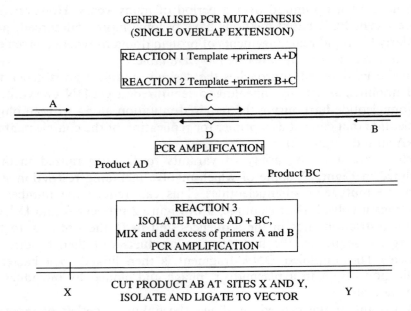

GENERALISED PCR MUTAGENESIS
(SINGLE OVERLAP EXTENSION)

REACTION 1 Template +primers A+D

REACTION 2 Template +primers B+C

A   C

D

B

PCR AMPLIFICATION

Product AD   Product BC

REACTION 3
ISOLATE Products AD + BC,
MIX and add excess of primers A and B
PCR AMPLIFICATION

X   CUT PRODUCT AB AT SITES X AND Y,   Y
ISOLATE AND LIGATE TO VECTOR

Figure 2   *PCR mutagenesis by single overlap extension. In this method the mutation to be introduced is coded into two oligonucleotides shown as C and D. These are designed to be complementary to each of the DNA strands at the site to be mutated and are capable of stably annealing. For a single-point mutation the oligonucleotides are likely to be around 15–20 nucleotides. This usually allows the selection of an incubation temperature for the PCR that maintains the initial stability of the annealed oligonucleotide. Oligonucleotides A and B are often complementary to sequences within the vector and can be used for DNA sequence confirmation of the mutation when it is made. In a first round of parallel reactions oligonucleotides A and D are mixed and annealed in molar excess over the dsDNA template and similarly for B and C. PCR is performed and new dsDNA fragments AD and BC are produced that each have the mutation. After purification of the fragments an aliquot of each is mixed with molar excess of oligonucleotides A and B and PCR again performed. The result is the fragment AB. This fragment can then be cloned into a suitable vector using sites within the sequence, or quite frequently, sites introduced to the sequence in the oligonucleotides A and B. This method has been termed single-overlap extension (SOE) PCR. (S. N. Ho, H. D. Hunt, R. M. Horton, J. K. Pullen, and L. R. Pease, Gene, 1989, 77, 51). (Figure adapted from reference 25 with permission of the publisher.)*

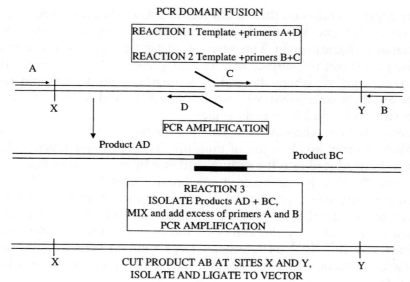

REACTION 3
ISOLATE Products AD + BC,
MIX and add excess of primers A and B
PCR AMPLIFICATION

X   CUT PRODUCT AB AT SITES X AND Y,   Y
ISOLATE AND LIGATE TO VECTOR

Figure 3   *PCR domain fusion. The domain fusion PCR is similar to the operation described in Figure 2. In this case the design of the central primers is more complex. Oligonucleotide A has the sequence of the sense strand at the N-terminal end of the first domain and so can anneal to the antisense strand of the coding sequence of the first domain. Oligonucleotide B has the sequence of anti-sense strand at the C-terminal end of the second domain. Oligonucleotides A and B are usually 18– 20 nucleotides in length and these lengths can be adjusted to ensure that the oligonucleotides remain annealed at the temperature of the PCR extension reaction. Oligonucleotide C is a hybrid of two sequences. At the 5' end the first 20 (approximately) nucleotides are the same as the coding sequence at the C-terminus of the first domain. This part of the oligonucleotide sequence ends at the point where the fusion will occur. The second portion of the oligonucleotide sequence (also often around 20 nucleotides) has the sequence of the sense strand at the N-terminal end of the second domain. Therefore oligonucleotide C can anneal to the antisense strand of the coding sequence of the second domain, leaving a non-annealed 'tail'. Oligonucleotide D is a similar hybrid sequence but can anneal to the sense strand of the first domain, also leaving a non-annealed 'tail'. Two PCR reactions are done. In the first oligonucleotides A and D are in molar excess over the coding sequence for the first domain. In the second reaction oligonucleotides B and C are in molar excess over the coding sequence for the second domain. At the end of the reactions products AD and BC are formed. Product AD codes for the first domain and also codes for a number of amino acids from the N-terminal region of the second domain, with the point of fusion exactly as dictated by the sequence in oligonucleotide D. In product BC, the DNA sequence codes for the C-terminal amino acids of the first domain and the amino acids of the second domain. Importantly the products AD and BC now have a significant nucleotide sequence homology. Products AD and BC are purified and an aliquot of each is mixed with molar excess of oligonucleotides A and B in a third PCR reaction. The resultant product AB now has the coding sequence of both domains in-frame and joined exactly at the desired point. This fragment can then be cloned into a suitable vector using sites near the ends of the sequence, or quite frequently, using sites introduced to the sequence in the oligonucleotides A and B. (Figure adapted from reference 25 with permission of the publisher.)*

The approach relies on three main features: first, that the nucleic acid that codes for the protein sequence of interest remains physically associated with the protein. This association can be achieved by presenting the protein on the surface of a bacteriophage,[31,32] or a bacteria or eukaryotic cell[33–36] where the coding sequence is resident within the 'phage or cell, or on a polysome[37] where the mRNA and newly translated protein are still linked by the ribosome.

Second, that a method exists to generate the large number of variants; this method often involves use of multiply degenerate oligonucleotides which are introduced into the coding sequence by cassette insertion or a PCR procedure, or by an *in vitro* mutagenesis method. Even with such procedures only small segments of a protein can be altered at one time because of the number of possible combinations which are involved. Libraries are limited by the ability of the host to take up individual members, and 10^{12}–10^{14} are considered large numbers here, however the possible combinations of all residues in even small protein domains far exceeds this.

Third, these methods require a screening or selection strategy that will enrich from among the library those novel protein sequences that have the phenotype of interest. Screens which involve binding to a ligand can easily be devised.[38] Screens which involve catalysis and which result in the presented protein being captured on a solid support have been described.[39] Selection systems often involve complementation of an essential function within the host organism,[40] but more novel *in vivo* selection systems have also been proposed[41] where selection takes place within a whole organism.

The key to the process is that once useful proteins have been isolated they can be amplified by propagation of the 'phage or cell containing the gene sequence, or by direct amplification of the gene sequences themselves. Then, multiple rounds of screen/selection and amplification can be done to enrich the desired sequence. Methods of this kind are rapidly becoming the workhorse for the protein engineer and are termed directed

[31] K. T. O'Neil and R. H. Hoess, *Curr. Opin. Struct. Biol.*, 1995, **5**, 443.
[32] C. Rader and C. F. Barbas III, *Curr. Opin. Biotechnol.*, 1997, **8**, 503.
[33] E. T. Boder and K. D. Wittrup, *Nat. Biotechnol.*, 1997, **15**, 553.
[34] Y. Boublik, P. DiBonito, and I. M. Jones, *Bio/Technology*, 1995, **13**, 1079.
[35] G. Georgiou, C. Stathopoulos, P. S. Daugherty, A. R. Nayak, B. L. Iverson, and R. Curtiss III, *Nat. Biotechnol.*, 1997, **15**, 29.
[36] S. Ståhl and M. Uhlén, *TIBTECH*, 1997, **15**, 185.
[37] L. Jermutus, L. A. Ryabova, and A. Pluckthun, *Curr. Opin. Biotechnol.*, 1998, **9**, 534.
[38] H. Zhao and F. H. Arnold, *Curr. Opin. Struct. Biol.*, 1997, **7**, 480.
[39] K. D. Janda, L.-C. Lo, C.-H. Lo, M.-M. Sim, R. Wang, C.-H. Wong, and R. A. Lerner, *Science*, 1997, **275**, 945.
[40] P. Kast and D. Hilvert, *Curr. Opin. Struct. Biol.*, 1997, **7**, 470.
[41] R. Pasqualini and E. Ruoslahti, *Nature*, 1996, **380**, 364.

evolution.[42-46] Tawfik and Griffiths[47] have recently demonstrated a completely *in vitro* approach to directed evolution.

2.5 *de novo* Sequence Design

There are different challenges for those interested in *de novo* protein design. In principle, for any protein of n residues there are 2×10^n different possible sequences. The structures and sequences databases demonstrate that there are a range of sequences that can be accommodated into similar protein folds, and of sequences which can fulfil similar functions. Therefore the reverse approach of deciding what fold is appropriate and then identifying what sequence would be needed to generate the desired fold and function may be more relevant.[48,49] Dahiyat and Mayo[50] and Harbury *et al.*[51] have recently described computational procedures for design of peptide domain sequences. Regan[52] has summarized the recent developments in this area. Once the peptide sequence has been devised the protein can be built either by peptide synthesis, if the sequence is of a reasonable size (see earlier Section 2.3) or by gene synthesis. Currently PCR amplification methods for gene synthesis are the most often used, to fill in and amplify partially overlapping oligonucleotides (Figure 4, summarized in reference 25).

2.6 Expression

Once a novel sequence has been identified and made it must be expressed to determine that the function is as required. Expression of recombinant proteins is a vast enterprise and host expression systems range from bacteria (*Escherichia coli* being the most obvious), to yeasts (*e.g. Saccharomyces cerevisiae*, and *Pichia pastoris*) to insect cells and mammalian cell cultures to transgenic animals and plants. The scale of production also varies. In the early stages of development of a novel protein, for example, where a protein has been identified among a number from a display library, only small quantities (micrograms) of

[42] F. H. Arnold, *Nat. Biotechnol.*, 1998, **16**, 617.
[43] L. P. Encell and L. A. Loeb, *Nat. Biotechnol.*, 1998, **16**, 234.
[44] P. A. Patten, R. J. Howard, and W. P. C. Stemmer, *Curr. Opin. Biotechnol.*, 1997, **8**, 724.
[45] W. P. C. Stemmer, *Nature*, 1994, **370**, 389.
[46] W. P. C. Stemmer, *Bio/Technology*, 1995, **13**, 549.
[47] D. S. Tawfik and A. D. Griffiths, *Nat. Biotechnol.*, 1998, **16**, 652.
[48] K. E. Drexler, *Proc. Natl. Acad. Sci. USA*, 1981, **78**, 5275.
[49] C. Pabo, *Nature*, 1983, **301**, 200.
[50] B. I. Dahiyat and S. L. Mayo, *Science*, 1997, **278**, 82.
[51] P. B. Harbury, J. J. Plecs, B. Tidor, T. Alber, and P. S. Kim, *Science*, 1998, **282**, 1462.
[52] L. Regan, *Structure*, 1998, **6**, 1.

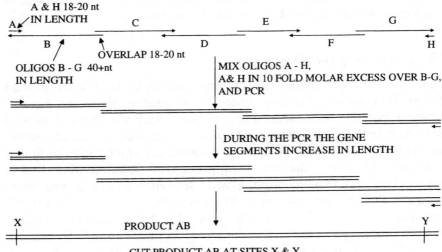

CUT PRODUCT AB AT SITES X & Y,
ISOLATE SEQUENCE AND LIGATE TO VECTOR

Figure 4 *Gene assembly PCR. The required amino acid sequence is determined and a suitable nucleotide coding sequence is generated. The sequence may incorporate codon bias to assist with expression in a host cell and will usually also take advantage of the redundancy of the genetic code to include several useful restriction sites for later manipulation of the sequence. The sequence is then marked off into segments as shown in the figure (A–H), in which the sequences used alternate from the top (usually the sense strand) and bottom (antisense) strand. Oligonucleotide A has the sequence of the 5' end of the top strand of the planned nucleotide sequence and will anneal to the 3' region of oligonucleotide B. Oligonucleotide H has the sequence of the 5' end of the bottom strand planned nucleotide sequence and will anneal to the 3' region of oligonucleotide G. Oligonucleotides A and H are usually short oligonucleotides and will be present in the PCR in molar excess to drive the amplification of the full-length sequence once this has assembled in the early stages of the PCR. Oligonucleotides B to G in the figure cover the whole sequence to be assembled and can be of any length desired; the shorter the length the more that are required. With ever increasing fidelity of long oligonucleotide synthesis, lengths in excess of 100 nucleotides can be used. The 5' region of oligonucleotide B can anneal to the 5' region of oligonucleotide C and the 3' region of oligonucleotide C can anneal to the 3' region of oligonucleotide D, and so on, until the terminal oligonucleotide is reached. Oligonucleotides B to G are mixed with molar excess of A and H and PCR is performed. In the early rounds of the reaction individual pairs of sequence anneal, e.g. C to D and are extended. Similarly A anneals to B and converts it to a dsDNA sequence. In the next and subsequent rounds the CD product can now anneal to the AB product to give product covering the AD region. The CD product can also anneal to the EF product, and so on. Similar annealing and extension events occur during subsequent rounds until a full-length sequence is assembled. The presence of excess A and H then ensures that this product is rapidly amplified. In the nucleotide sequence design the terminal oligonucleotides (B and G) in the figure will encode restriction sites (X and Y in the figure) suitable for insertion into a cloning vector. (Figure adapted from reference 25 with permission of the publisher.)*

the protein may be needed to confirm a biological property. In the second stage larger quantities (10s to 100s of milligrams) of purer material are required, typically to obtain structural information and to perform more demanding *in vitro* and *in vivo* bioassays. In some situations larger quantities (grams to multi-kilograms) of material purified by rigorous (and usually costly) procedures may be required if the protein is required for commercial purposes (for example for industrial enzymes).

Often a number of approaches may have to be tried to find a suitable expression host, as it cannot always be assumed that an altered protein will express in the same manner as the parental sequence (sometimes it is better, often it is not), and in the case of novel sequences, no precedent exists. These factors are becoming of increasing importance as more strategies rely on library display systems, where members of a library may be lost or underrepresented because of poor expression or folding properties. Much effort has gone into a better understanding of the expression and protein folding environment of the host systems, particularly of *E. coli*.[53,54] To overcome some of these problems *in vitro* transcription–translation systems have also been used and are being optimized for efficient small-scale production of novel proteins[37,55,56] in amounts suitable for analysis.

2.7 Analysis

Methods must exist for the analysis of the properties of the modified proteins. Where a function is being altered (introduced, modified or removed) a biological assay can usually be devised to measure the ability of the newly produced protein. The issue is often of ensuring that the assay can function with very small amounts of materials, such as that obtained from library amplification procedures. Methods that include physical adsorption or binding may be better suited to this than those that rely on a detection of a catalytic product.

In many cases acquiring and demonstrating the new function may be all that is required. However, in addition to functional assays, interpretation of the results often requires a structural understanding of the alteration, particularly to gain further insights into protein folding. Assuming a reasonable quantity of the novel protein can be prepared,

[53] H. Bothmann and A. Pluckthun, *Nat. Biotechnol.*, 1998, **16**, 376.
[54] J.G. Wall and A. Pluckthun, *Curr. Opin. Biotechnol.*, 1995, **6**, 507.
[55] E. A. Burks, G. Chen, G. Georgiou, and B. L. Iverson, *Proc. Natl. Acad. Sci. USA*, 1997, **94**, 412.
[56] L. A. Ryabova, D. Desplancq, A. S. Spirin, and A. Pluckthun, *Nat. Biotechnol.*, 1997, **15**, 79.

gross properties can readily be detected by spectroscopic or other large-scale measurements (*e.g.* circular dichroism, turbidity, enthalpic, sedimentation or chromatographic properties). However, detailed structural information about the end product is still the rate-limiting step in rational protein engineering and design. Recently, Casimiro *et al.*[57] have shown that PCR-based gene synthesis, expression in *E. coli* of milligram quantities of isotopically labelled protein and NMR spectroscopy can be achieved in a reasonably short space of time, with first NMR spectra available for a novel sequence as little as two months after the initiation of the gene synthesis steps.

3 APPLICATIONS

Applications of protein engineering are everywhere in biology, as a means of producing novel molecules and as a means of understanding the basic properties of proteins. Therefore to attempt a comprehensive survey is fruitless. However, it is useful to note some specific examples of the different types of molecules that can be achieved, from the simplest point mutations and domain rearrangements to more demanding multi-site alterations and *de novo* designs. The examples given in the next sections reflect a personal bias, but hopefully will prove of general interest.

3.1 Point Mutations

Individual point mutations in proteins can be readily achieved once the gene sequence is available, using the techniques noted earlier. Numerous examples of amino acid sequence variants are known. A few examples are given below.

3.1.1 Betaseron/Betaferon (Interferon β-1b). One of the earliest examples of pharmaceutical protein engineering was the production of interferon β-1b. This novel protein was generated by the substitution of a cysteine (Cys) for serine (Ser) at residue 17 of the 154 amino acid interferon β.[58] This substitution reduces the possibility of incorrect disulphide bridge formation during synthesis in *E. coli*, and also removes a possible site for post-translational oxidation. The resultant protein is expressed in *E. coli* at a specific activity approaching that of native fibroblast-derived interferon β. The molecule has been licensed since

[57] D. R. Casimiro, P. E. Wright, and H. J. Dyson, *Structure*, 1997, **5**, 1407.
[58] D. F. Mark, S. D. Lu, A. A. Creasey, R. Yamamoto, and L. S. Lin, *Proc. Natl. Acad. Sci. USA*, 1984, **81**, 5662.

1993 for use for the reduction of frequency and degree of severity of relapses in ambulatory patients with relapsing remitting multiple sclerosis.

3.1.2 Humalog (Lispro Insulin). Humalog is an engineered form of human insulin in which two residues at the *C*-terminus of the B chain, proline (Pro) and lysine (Lys) at residues 28 and 29, respectively, have been reversed in their order. Humalog is a fast acting analogue of insulin, designed to mimic the body's natural rate of insulin response to food. The *C*-terminal alterations were designed based on structural and sequence homology to insulin-like growth factor 1 (IGF-1) and the sequence reversal reduces the dimerization of the B subunit. This reduced self-association means that the monomer is more readily available to function after food uptake and therefore can be administered very shortly before meals. Humalog has been licensed for use since 1996.[59]

3.1.3 Novel Vaccine Adjuvants. Vaccines have until recently been one of the more successful but underrated pharmaceutical classes.[60,61] However, recently, novel approaches to vaccine design have brought this area of medicine to the forefront. Vaccines are now the major class of biotechnology products in development.[62] Protein engineering is being used to construct novel protein molecules that provide immunological help (adjuvants) to stimulate an immune response to a co-administered antigen. These novel protein adjuvants are of particular interest for stimulating mucosal immune responses after oral immunization so as to avoid the use of injections for vaccinations. The most well characterized of these are based on *V. cholerae* toxin (CT) and the heat labile toxin of *E. coli* (LT). The crystal structures of LT and CT have been solved.[63–65] Site-directed mutagenesis experiments based on the LT structure have led to an understanding of the relationship of the single enzymatic A subunit to the five B subunits of these heterohexameric toxins, the location of the enzymatic and cofactor binding sites in the A subunit, and of the activating cleavage points in the A subunit. These experiments have led to the construction of mutant toxins that are still able to assemble into the holotoxin, but that have minimal toxicity *in vivo* while

[59] M. R. Burge, A. G. Rassam, and D. S. Schade, *Trends Endocrinol. Metabol.*, 1998, **9**, 337.
[60] B. R. Bloom and R. Widdus, *Nat. Med. (Vaccine Suppl.)*, 1998, **4**, 480.
[61] R. J. Saldarini, *Nat. Med. (Vaccine Suppl.)*, 1998, **4**, 485.
[62] A. Persidis, *Nat. Biotechnol.*, 1998, **16**, 1378.
[63] T. K. Sixma, S. E. Pronk, K. H. Kalk, E. S. Wartna, B. A. M. van Zanten, B. Witholt, and W. G. J. Hol, *Nature*, 1991, **351**, 371.
[64] T. K. Sixma, K. H. Kalk, B. A. M. van Zanten, Z. Dauter, J. Kingma, B. Witholt, and W. G. J. Hol, *J. Mol. Biol.*, 1993, **230**, 890.
[65] R.-G. Zhang, D. L. Scott, M. L. Westbrook, S. Nance, B. D. Spangler, G. G. Shipley, and E. M. Westbrook, *J. Mol. Biol.*, 1995, **251**, 563.

retaining their immunological and adjuvant properties. For example, a mutation of Ser to Lys at residue 63 in the A subunit of LT inhibits the binding to the toxin of the NAD cofactor for the ADP ribosyltransferase activity. This S63K mutant exhibits a six orders of magnitude reduction in toxicity whilst retaining immunological and adjuvant properties.[66,67] Alternatively, a mutation at residue 192 (arginine (Arg) to glycine (Gly)) in the A subunit of LT has been made. This mutation is at the point where the A1 subdomain is proteolytically cleaved from the A2 sub-domain and which is the first step in activation of the enzymatic function of the A subunit. The mutation leads also to an enzymatically inactive toxin that retains adjuvant properties.[67,68] These mutant toxins are currently being tested in preclinical and clinical studies to assess their utility.

3.2 Domain Shuffling (Linking, Swapping and Deleting)

Most large proteins are composed of smaller independently folding domains, generally corresponding to the protein folds mentioned earlier (reviewed by Heringa and Taylor[69]). These domains are often linked together by short peptide sequences and in many, but not all, cases the domains are identifiable as separate exons in the gene sequence. Using standard molecular biology techniques it is possible to add, remove or swap domains from one protein to another to rebuild proteins.

3.2.1 Linking Domains.
3.2.1.1 Domain Fusions for Cell Targeting. Amongst the very earliest examples of protein engineering the binding site region of an antibody was genetically linked to an enzyme. The basic antibody unit is a Y or T shaped structure comprising two identical 50 kDa 'heavy' chains and two identical 25 kDa 'light' chains. The antigen-binding region, known as the Fab region, is comprised of the *N*-terminal half of one heavy chain and one light chain. The *C*-terminal halves of the two heavy chains associate together to form the Fc region. This latter region is involved with interactions with various cells of the immune system and with complement, and is also important in determining the serum half-life of antibodies, at least of the IgG type.[70]

[66] M. Pizza, M. R. Fontana, M. M. Giuliani, M. Domenighini, C. Magagnoli, V. Giannelli, D. Nucci, W. Hol, R. Manetti, and R. Rappuoli, *J. Exp. Med.*, 1994, **180**, 2147.

[67] V. Giannelli, M. R. Fontana, M. M. Giuliani, D. Guangcai, R. Rappuoli, and M. Pizza, *Infect. Immun.*, 1997, **65**, 331.

[68] B. L. Dickinson and J. D. Clements, *Infect. Immun.*, 1995, **63**, 1617.

[69] J. Heringa and W. R. Taylor, *Curr. Opin. Struct. Biol.*, 1997, **7**, 416.

[70] J. V. Ravetch, *Curr. Opin. Immunol.*, 1997, **9**, 121.

The antibody–enzyme gene fusions were done by taking the DNA sequence that codes for the heavy chain component of Fab of the antibody and linking this either to staphylococcal nuclease[71] or to *E. coli* DNA polymerase coding sequences.[72] Introduction of these fusion genes in a cell that produces the antibody light chain, and expression of the fusion gene, reconstituted both antigen binding and enzymic activity. These early examples provided the foundation for many other examples where a binding function (antibody, cytokine, growth factor or extra-cellular ligand binding domain (ECD) of a receptor) is linked to an effector function (toxin, enzyme, cytokine). A non-antibody binding domain can be attached to the Fc region of an antibody to take advantage of the long serum half-life of antibodies to improve the pharmacokinetic properties of a designed molecule. Examples of these domain fusions have been reviewed.[73–77]

One such example, Enbrel™ (etanercept), has recently been licensed for human use. Enbrel™ consists of the extracellular domains of the receptor for the cytokine tumour necrosis factor α (TNFα) genetically fused to the Fc regions of IgG and is used to block the activity of the TNFα.[78]

Enbrel™ is currently licensed for the reduction in signs and symptoms of moderately to severely active rheumatoid arthritis in patients who have an inadequate response to one or more disease-modifying anti-rheumatic drugs.

3.2.1.2   Fused Cytokines.   Different cytokines often have overlapping functions, or can act on the same cell synergistically to cause a physiological change to the cell. Linking cytokines by fusing the genes in-frame tethers the two functional groups together and can in principle lead to the desired biological effect at lower doses than if administered separately. Fusions of this type were described for the interferons more than a decade ago.[79] More recently PIXY321, a fusion of granulocyte

[71] M. S. Neuberger, G. T. Williams, E. B. Mitchell, S. S. Jouhal, J. G. Flanagan, and T. H. Rabbitts, *Nature*, 1985, **314**, 268.
[72] G. T. Williams and M. S. Neuberger, *Gene*, 1986, **43**, 319.
[73] I. Benhar and I. H. Pastan in 'Antibody Therapeutics', ed. W. J. Harris and J. R. Adair, CRC Press Inc, Boca Raton, FL, USA, 1997, p. 73.
[74] S. M. Chamow and A. Ashkenazi, *TIBTECH*, 1996, **14**, 52.
[75] R. J. Kreitman and I. Pastan in 'Antibody Therapeutics', ed. W. J. Harris and J. R. Adair, CRC Press Inc, Boca Raton, FL, USA, 1997, p. 34.
[76] J. R. Murphy, *Curr. Opin. Struct. Biol.*, 1996, **6**, 541.
[77] G. A. Pietersz and I. F. C. McKenzie, in 'Monoclonal Antibodies: The Second Generation', ed. H. Zola, BIOS Scientific Publishers Ltd., Oxford, UK, 1995, p. 93
[78] K. M. Mohler, D. S. Torrance, C. A. Smith, R. G Goodwin, K. E. Stremler, V. P. Fung, H. Madani, and M. B. Widmer, *J. Immunol.*, 1993, **151**, 1548.
[79] L. D. Bell, K. G. McCullagh, and A. G. Porter, 'Covalently Linked Polypeptide Cell Modulators', US Patent No. 4935233.

macrophage-colony stimulating factor (GM-CSF) and interleukin 3 (IL-3), has been described.[80,81] In this case the GM-CSF and IL-3 genes were modified to remove mammalian *N*-glycosylation sites and were then linked by addition of a 15 amino acid flexible linker between the *C*-terminus of GM-CSF and the *N*-terminus of IL-3. Yeast expressed PIXY321 exhibited enhanced receptor affinity, proliferative activity and colony stimulating activity compared with the either of the starting monomeric proteins.

3.2.1.3 Fusions to Stabilize Dimeric Proteins. In other situations it is useful to genetically fuse together domains which are normally associated by non-covalent interactions or disulfide bridges. This allows more convenient production and ensures that the domains remain in proximity even at the low concentrations that would be observed during *in vivo* dosing. The binding domains of antibodies have been joined together with a linker peptide to form single chain Fvs (scFvs). The sequence and length of this peptide can be varied (reviewed by Huston *et al.*[82]). Similarly, cytokine subunits can also be linked together. The heterodimeric IL-12 has been linked together and shown to retain function.[83]

3.2.2 Swapping Protein Domains. Another simple procedure used in protein engineering is whole domain swapping, in which analogous domains from different sources are swapped into a multi-domain protein to provide a novel functionality.

3.2.2.1 Chimaeric Mouse-Human Antibodies. Here the antigen binding domains from a mouse monoclonal antibody are linked to the constant regions (which provide immune effector functions and dictate biological half-life) from a human antibody.[84] This switching can markedly reduce unwanted immunogenicity compared to the original mouse antibody.[85] To date four chimaeric antibody products have been

[80] B. M. Curtis, D. E. Williams, H. E. Broxmeyer, J. Dunn, T. Farrah, E. Jeffrey, W. Clevenger, P. deRoos, U. Martin, D. Friend, V. Craig, R. Gayle, V. Price, D. Cosman, C. J. March, and L. S. Park, *Proc. Natl. Acad. Sci. USA*, 1991, **88**, 5809.
[81] S. Vadhan-Raj, H. E. Broxmeyer, M. Andreeff, J. C. Bandres, E. S. Buescher, R. S. Benjamin, N. E. Papadopoulos, A. Burgess, S. Patel, C. Plager, W. N. Hittelman, I. McAlister, L. Garrison, and D. E. Williams, *Blood*, 1995, **86**, 2098.
[82] J. S. Huston, J. McCartney, M.-S. Tai, C. Mottola-Hartshorn, D. Jin, F. Warren, P. Keck, and H. Oppermann, *Intern. Rev. Immunol.*, 1993, **10**, 195.
[83] G. J. Lieschke, P. K. Rao, M. K. Gately, and R. C. Mulligan, *Nat. Biotechnol.*, 1997, **15**, 35.
[84] S. L. Morrison, M. J. Johnson, L. A. Herzenberg, and V. T. Oi, *Proc. Natl. Acad. Sci. USA*, 1984, **81**, 6851.
[85] A. F. LoBuglio, R. H. Wheeler, J. Trang, A. Haynes, K. Rogers, E. B. Harvey, L. Sun, J. Ghrayeb, and M. B. Khazaeli, *Proc. Natl. Acad. Sci. USA*, 1989, **86**, 4220.

licensed as pharmaceuticals, including the anti-platelet compound Reo-Pro^{TM}, the lead drug in its therapeutic class (Table 1).

3.2.2.2 Polyketide Synthases (PKSs). Polyketides are a class of chemical compounds which include many pharmaceutical compounds, including antibiotics, anti-fungals and immunosuppressants, and which together account for billions of pounds sterling of sales per annum.[86,87] These chemicals are produced in various soil micro-organisms and fungi. Their synthesis involves the regulated action of a sequence of enzymes, the PKSs. The PKSs occur either as small proteins that have a small number of distinct non-repeating catalytic domains (iterative PKSs), or as large multi-domain polypeptides. In the latter, while each domain has a specific enzyme function, the same function may be present a number of times (modular PKSs). In both cases the product of one catalytic domain forms the substrate for the reaction of a neighbouring domain, the modular polyketides producing the more complex molecules. By selectively adding, deleting and rearranging the order of the catalytic domains to generate novel PKSs, a novel order of catalytic reactions is established and hence new products can be generated (*e.g.* reference 88, and references therein). The opportunity for combinatorial biosynthesis, perhaps in a cell-free system, is being examined.

3.2.3 Deleting Domains. Tissue plasminogen activator (tPA) is a serine protease secreted by endothelial cells. Following binding to fibrin tPA activates plasminogen to plasmin which then initiates local thrombolysis. Reteplase is a variant form in which three of the five domains of tPA have been deleted. One of the domains that confers fibrin selectivity, and the catalytic domain are retained.[89] Reteplase is licensed as Retavase for the treatment of acute myocardial infarction to improve blood flow in the heart.

3.3 Whole Protein Shuffling

Many proteins exist as multi-member families in which the homologues display slightly different biological activities determined by the sequence variation. In the early days of protein engineering hybrid genes were generated by swapping sequence stretches using convenient common

[86] C. W. Carreras and D. V. Santi, *Curr. Opin. Biotechnol.*, 1998, **9**, 403.
[87] C. Khosla and R. J. X. Zawada, *TIBTECH*, 1996, **14**, 335.
[88] M. A. Alvarez, H. Fu, C. Khosal, D. A. Hopwood, and J. E. Bailey, *Nat. Biotechnol.*, 1996, **14**, 335.
[89] S. Noble and D. McTavish, *Drugs*, 1996, **52**, 589.

Table 1 *Examples of licensed chimaeric antibodies*

Name (international non-proprietary name)	Format	First licensed	Antigen	Indication
ReoPro® (abciximab)	Mouse–human chimaeric antibody Fab fragment	1994	Recognizes an epitope on the gpIIb/IIIa antigen on the surface of platelets and inhibits blood clotting by inhibiting the aggregation of platelets.[a]	ReoPro® is used as adjunctive therapy to prevent cardiac ischaemic complications in a broad range of patients undergoing percutaneous coronary intervention (PCI), as well as in unstable angina patients not responding to conventional medical therapy when PCI is planned within 24 h. PCI includes balloon angioplasty, atherectomy and stent placement.
Remicade™ (infliximab)	Mouse–human chimaeric IgG	1998	It is believed that Remicade™ reduces intestinal inflammation in patients with Crohn's disease by binding to and neutralizing tumour necrosis factor α (TNF-α) on the cell membrane and in the blood and by destroying TNF-α producing cells.[b]	Remicade™ is approved for treatment of moderately to severely active Crohn's disease for the reduction of the signs and symptoms, in patients who have an inadequate response to conventional therapy. It is also indicated as a treatment for patients with fistulizing Crohn's disease for reduction in the number of draining enterocutaneous fistula(s).

Rituxan™ (also marketed as MabThera® (Rituximab)	Mouse–human chimaeric IgG	1998	CD20 antigen found on the surface of normal and malignant B lymphocytes.[c] The Fab domain of Rituxan™ binds to the CD20 antigen on B lymphocytes and the Fc domain recruits immune effector functions to mediate B-cell lysis *in vitro*. Possible mechanisms of cell lysis include complement-dependent cytotoxicity (CDC) and antibody-dependent cell-mediated cytotoxicity (ADCC).	Rituxan™ is approved for the treatment of patients with relapsed or refractory low-grade or follicular, CD20 positive, B-cell non-Hodgkin's lymphoma.
Simulect® (basiliximab)	Mouse–human chimeric IgG	1998	Recognizes an epitope on CD25 and functions by blocking the binding of IL-2 to CD25.[d]	Simulect® is approved for prophylaxis of acute rejection during the first several weeks following renal transplantation. Because it is selective for T lymphocytes activated by IL-2, Simulect® has a mode of action that complements the immunosuppressive cyclosporin A, and is expected to increase its efficacy without sacrificing specificity of action.

[a] H. F. Weisman, T. F. Schaible, R. E. Jordan, C. F. Cabot, and K. M. Anderson, *Biochem. Soc. Trans.*, 1995, **23**, 1051.

[b] H. M. van Dullemen, S. J. van Deventer, D. W. Hommes, H. A. Bijl, J. Jansen, G. N. Tytgat, and J. Woody, *Gastroenterology*, 1995, **109**, 129.

[c] D. G. Maloney, T. M. Liles, D. K. Czerwinski, C. Waldichuk, J. Rosenberg, A. Grillo-Lopez, and R. Levy, *Blood*, 1994, **84**, 2457.

[d] J. Kovarik, P. Wolf, J. M. Cisterne, G. Mourad, Y. Lebranchu, P. Lang, B. Bourbigot, D. Cantarovich, C. Gerbeau, A. G. Schmidt, and J. P. Soulillou, *Transplantation*, 1997, **64**, 1701.

restriction sites (*e.g.* reference 90), or by *in vivo* recombination between homologous genes in a more random manner.[91] These methods produced small numbers of novel genes that could be examined individually. More recently protein homologues have been used to generate very large libraries of novel variants by sequence shuffling. In this procedure the genes for two or more members of the family are fragmented randomly by DNase I. The fragments are then used as PCR primers to generate new gene sequences formed by random hybridization of the fragments. New variants can then be identified by selection or screening procedures as noted earlier. Patten *et al.*[44] have reviewed examples generated by this procedure, which include enzymes, cytokines and antibody binding sites.

3.4 Protein–Ligand Interactions

3.4.1 Enzyme Modifications. Many examples of protein engineering involve changes to enzymes to examine and modify enzyme–substrate interactions. These experiments can be traced back to the earliest attempts at protein engineering, using tyrosyl-tRNA synthetase[92,93] (reviewed in reference 94).

The types of changes include: enhancing catalytic activity; modification of substrate specificity, including the *de novo* generation of novel catalytic functions; alterations to pH profiles so that an enzyme can function in non-physiological conditions; improving oxidation resistance by replacing oxidation sensitive amino acids such as Cys, tryptophan (Trp) or methionine (Met) by sterically similar non-oxidizable amino acids such as Ser, phenylalanine (Phe) or glutamate (Glu), respectively; improving stability to heavy metals by replacing Cys and Met residues and surface carboxyl groups; removing protease cleavage motifs; removing sites at which catalytic product might otherwise bind to induce allosteric feedback inhibition.

Numerous examples are now available, perhaps the most obvious being in the modification of the range and tolerances of enzymes used in industrial processes. Improvements to a number of commercially useful enzymes, including proteases, amylases, lipases, cellulases and xylanases

[90] A. G. Porter, L. D. Bell, J. Adair, G. H. Catlin, J. Clarke, J. A. Davies, K. Dawson, R. Derbyshire, S. M. Doel, L. Dunthorne, M. Finlay, J. Hall, M. Houghton, C. Hynes, I. Lindley, M. Nugent, G. J. O'Neil, J. C. Smith, A. Stewart, W. Tacon, J. Viney, N. Warburton, P. G., Boseley, and K. G. McCullagh, *DNA*, 1986, **5**, 137.
[91] H. Weber and C. Weissmann, *Nucleic Acids Res.*, 1983, **11**, 5661.
[92] G. Winter, A. R. Fersht, A. J. Wilkinson, M. Zoller, and M. Smith, *Nature*, 1982, **299**, 756.
[93] A. J. Wilkinson, A. R. Fersht, D. M. Blow, P. Carter, and G. Winter, *Nature*, 1984, **307**, 187.
[94] A. Fersht and G. Winter, *Trends Biochem. Sci.*, 1992, **17**, 292.

have been recently reviewed.[95-97] As one example, the subtilisin BPN' from *Bacillus amyloliquefaciens* has been modified to increase its calcium-independent stability by 1000-fold. This improves its ability to function in a chelating environment.[98]

3.4.2 Hormone Agonists. Increasing the affinity of binding of hormones to their receptors can lead to the development of super-agonists with markedly increase bioactivity. Recently Grossmann *et al.*[99] have described the rational design of variants of human thyroid-stimulating hormone (TSH) with up to 1300-fold increase in activity. The variants were designed based on homology to human chorionic gonadotrophin (HCG), another glycoprotein hormone that shares a common α subunit. Substituting positively charged groups into one loop region of TSH led to synergistic increases in activity.

3.4.3 Substitution of Binding Specificities. A number of examples are available where the specificity for a ligand has been transferred from one protein background to another. Some of the examples require small changes, some require large-scale transfer of many residues. For example, the specificity of the hormone prolactin has been modified by substitution of eight amino acids at the receptor binding surface so that the modified hormone now binds to the receptor for growth hormone.[100]

Protein domains often have loop regions that span between other secondary structure motifs (α-helix, β-strand). In some cases the functionality of the protein resides in these loops and they can be the target for relatively simple substitution experiments. For example, the specificity of basic fibroblast growth factor was altered to that of acidic fibroblast growth factor by the substitution of one particular loop region.[101]

On a more adventurous scale antibody humanization involves transfer of up to six loops that form the antigen binding surface, from the β-sheet frameworks of the antigen binding domains of a non-human antibody to that of a human antibody. This feat has taken some years to become routine, since the first description in 1986.[102] However it has now

[95] W. Aehle and O. Misset, in 'Biotechnology. A Multi-volume Comprehensive Treatise, Volume 5a: Recombinant Proteins, Monoclonal Antibodies and Gene Therapy', ed. A. Mountain, U. Ney, and D Schomberg, WILEY-VCH, Weinheim, FRG, 1999, p. 189.

[96] D. N. Rubingh, *Curr. Opin. Biotechnol.*, 1997, **8**, 417.

[97] C. Vita, *Curr. Opin. Biotechnol.*, 1997, **8**, 429.

[98] S. L. Strausberg, P. A. Alexander, D. T. Gallagher, G. L. Gilliland, B. L. Barnett, and P. N. Bryan, *Bio/Technology*, 1995, **13**, 669.

[99] M. Grossmann, H. Leitolf, B. D. Weintraub, and M. W. Szkudlinski, *Nat. Biotechnol.*, 1998, **16**, 871.

[100] B. C. Cunningham, D. J. Henner, and J. A. Wells, *Science*, 1990, **247**, 1461.

[101] A. P. Seddon, D. Aviezer, L.-Y. Li, P. Bohlen, and A. Yayon, *Biochemistry*, 1995, **34**, 731.

[102] P. T. Jones, P. H. Dear, J. Foote, M. S. Neuberger, and G. Winter, *Nature*, 1986, **321**, 522.

progressed to the point where a number of humanized antibodies are marketed drugs (Table 2).

More recently more ambitious projects have been described that involve the transfer of functional loops between different structural backgrounds (reviewed by Vita[97]). This leads on to the beginnings of *de novo* design in which a required function is grafted into a particular structural scaffold.

3.5   Towards *de novo* Design

An intermediate step between experiments of the kind described above and 'true' *de novo* design is to take a particular protein scaffold and modify it so as to introduce new functions. This is often done by taking a small compact domain of known structure and using phage display procedures to modify part of the surface of the protein and to screen or select for the introduction of a new function. For example, Martin *et al.*[103] have described how a 'minibody' scaffold derived from an immunoglobulin binding domain and retaining only two loops and six β-strands has been used to obtain a novel antagonist for IL-6. Nygren and Uhlen[104] have reviewed a number of other similar examples.

In an alternative approach the structural plasticity of a protein fold has been examined by redesigning the *N*- and *C*-termini within the sequence of the IL-4 cytokine, circularly permuting the sequence without affecting the overall fold.[105]

3.5.1 de novo Design.   Attempts to design protein domains from first principles have been underway for some time (*e.g.* reference 106). Substantial progress has been made in constructing small proteins which fold in the planned fashion (reviewed in references 107–109). However, for some of these designed proteins side-chain interactions in the domain core led to non-native 'molten globule' interiors. Much work has been done on examining the packing of protein interiors by mutagenesis to determine the destabilizing effects of side chain modifica-

[103] F. Martin, C. Toniatti, A. L. Salvati, G. Ciliberto, R. Cortese, and M. Sollazzo, *J. Mol. Biol.*, 1996, **255**, 86.
[104] P.-A. Nygren and M. Uhlen, *Curr. Opin. Struct. Biol.*, 1997, **7**, 463.
[105] R. J. Kreitman, R. K. Puri, and I. Pastan, *Proc. Natl. Acad. Sci. USA*, 1994, **91**, 6889.
[106] L. Regan and W. F. deGrado, *Science*, 1988, **241**, 976.
[107] C. Sander, M. Scharf, and R. Schneider, in 'Protein Engineering: A Practical Approach', ed. A. R. Rees, M. J. E. Sternberg, and R. Wetzel, IRL Press, Oxford, UK, 1992, p. 88.
[108] J. W. Bryson, S. F. Betz, H. S. Lu, D. J. Suich, H. X. Zhou, K. T. O'Neil, and W. F. DeGrado, *Science*, 1995, **270**, 935.
[109] C. E. Schafmeister and R. M. Stroud, *Curr. Opin. Biotechnol.*, 1998, **9**, 350.

Table 2 *Examples of licensed humanized antibodies*

Name (international non-proprietary name)	Format	First licenced	Antigen	Indication
Zenapax® (daclizumab)	Humanized IgG[a,b]	1997	CD25, a subunit of the high affinity interleukin-2 (IL-2) receptor, expressed on the surface of activated lymphocytes.	Zenapax® is approved for prophylaxis of acute organ rejection in patients receiving renal transplants, to be used as part of an immunosuppressive regimen that includes cyclosporine and corticosteroids.
Synagis™ (palivizumab)	Humanized IgG[c]	1998	Respiratory syncytial virus (RSV) F protein.	Synagis™ is approved for the prevention of serious lower respiratory tract disease caused by RSV in paediatric patients at high risk of RSV disease.
Herceptin® (trastuzumab)	Humanized IgG[d]	1998	HER2 (human epidermal growth factor receptor 2) protein.	Herceptin® is approved for use in patients with metastatic breast cancer who have tumours that overexpress HER2. It is indicated for treatment of patients both as first line therapy in combination with paclitaxel and as a single agent in second- and third-line therapy.

[a] C. Queen, W. P. Schneider, H. E. Selick, P. W. Payne, N. F. Landolfi, J. F. Duncan, N. M. Avdalovic, M. Levitt, R. P. Junghans, and T. A. Waldmann, *Proc. Natl. Acad. Sci. USA*, 1989, **86**, 10029.
[b] T. A. Waldmann and J. O'Shea, *Curr. Opin. Immunol.*, 1998, **10**, 507.
[c] S. Johnson, C. Oliver, G. A. Prince, V. G. Hemming, D. S. Pfarr, S. C. Wang, M. Dormitzer, J. O'Grady, S. Koenig, J. K. Tamura, R. Woods, G. Bansal, D. Couchenour, E. Tsao, W. C. Hall, and J. F. Young, *J. Infect. Dis*, 1997, **176**, 1215.
[d] P. Carter, L. Presta, C. M. Gorman, J. B. B. Ridgway, D. Henner, W. L. T. Wong, A. M. Rowland, C. Kotts, M. E. Carver, and H. M. Shepard, *Proc. Natl. Acad. Sci. USA*, 1992, **89**, 4285.

tions.[110–112] More recently computational approaches have been used to design interiors of proteins.[113,114] These have led to computational methods for design of protein sequences compatible with a fixed[50] or flexible[51] protein backbone. Malakauskas and Mayo[115] recently described the use of such computer algorithms for the design of a thermostable variant of a domain from a streptococcal Ig binding protein. Computations identified a number of locations for point mutagenesis. The resultant protein has a melting temperature in excess of 100°C.

4 CONCLUSIONS AND FUTURE DIRECTIONS

Protein engineering is now a mature technology. Through the pragmatic approach allowed by molecular evolution methods novel proteins are being produced for research and commercial applications at an ever-increasing rate. In the near future there will be an increasing effort to exploit the technology on areas that have been less tractable to date, for example, in understanding the structure and function of membrane proteins (*e.g.* references 116 and 117), that are the focus for many pharmaceutical strategies and which may play a part in future biosensor, drug delivery and computing devices.

Protein design is still at an early stage and it appears still to be the domain of experts. However, much the same could be said for protein engineering even 10 years ago. It remains to be seen what the next decade will produce.

[110] W. S. Sandberg and T. C. Terwilliger, *TIBTECH.*, 1991, **9**, 59.
[111] E. P. Baldwin and B. W. Matthews, *Curr. Opin. Biotechnol.*, 1994, **5**, 396.
[112] B. Lee and G. Vasmatzis, *Curr. Opin. Biotechnol.*, 1997, **8**, 423.
[113] J. R. Desjarlais and N. D. Clarke, *Curr. Opin. Struct. Biol.*, 1998, **8**, 471.
[114] H. W. Hellinga, *Nat. Struct. Biol.*, 1998, **5**, 525.
[115] S. M. Malakauskas and S. L. Mayo, *Nat. Struct. Biol.*, 1998, **5**, 470.
[116] H. Bayley, *Curr. Opin. Biotechnol.*, 1999, **10**, 94.
[117] I. Mingarro, G. von Heijne, and P. Whitley, *TIBTECH*, 1997, **15**, 432.

CHAPTER 15

Bioinformatics

PETER M. WOOLLARD

1 INTRODUCTION

Bioinformatics can be defined as the application of information technology to biology. It aids in the chromosomal localization of genes, rapid searching to find sequences similar to ones of interest, gene identification and searching of scientific literature.[1] In recent years there has been a huge growth in the amount of biological data being generated. In just a few years we will have the entire 3 000 000 000 base pairs of the human genome sequence. The achievement of all the sequencing is the easy part; the difficult task will be trying to interpret this vast array of sequence data and determine its functions. This should ultimately help us to understand how the human and other organisms work, and provide us with insights into disease mechanisms. An important function of bioinformatics is to store biological data in databases and provide computing tools to access and analyse this data.

Various model organisms (mouse, *Saccharomyces cerevisiae* (yeast), *Caenorhabditis elegans* (nematode worm), *Arabidopsis thaliana* (plant), rice, *Escherichia coli*) are also being completely sequenced. These model organisms have been experimentally well characterized. The knowledge from all this work will aid the interpretation of the human genome as well as being of much interest in their own right. There are many major mapping and sequencing projects for animals, plants and microorganisms of agricultural and medical interest, either completed or under way. These make extensive use of the bioinformatics data and techniques developed from the human and model organism

---

[1] 'DNA and Protein Sequence Analysis – A Practical Approach', ed. M. J. Bishop and C. J. Rawlings, IRL Press, OUP, 1997. (General reading with many useful chapters.)

sequencing programmes, especially sequencing technology, sequence assembly software and homologous sequences (in assembly and annotation).

Comparing homologous genes within or between organisms (comparative genomics), particularly if they are distantly related by evolution, will help sequence analysis for intron/exon boundaries, functionally important parts of a translated protein, and regions of expression control (*e.g.* promoter sites). Preliminary analysis of genomic regions in different organisms has shown conservation of homologous genes in the same relative areas (synteny), although the gene order may vary.[2] If a gene of interest has been found and chromosomally located in one organism, synteny could point to a chromosomal location for investigation in another organism. It should be noted that it is often easier to find a gene in a compact genome, *e.g.* the Fugu fish,[3] rather than in a human, mouse or Zebra fish. This is because the Fugu fish has essentially the same number of genes as in the human genome, but generally has far less 'junk' DNA between genes and shorter introns. Biochemical pathways vary between species and computer programs are used to attempt to deduce pathways present. Bioinformatics plays a key role in all of this.

The work of developmental biologists studying model organisms, sequence data, protein structure and gene expression data will greatly help our understanding of how organisms work at the molecular level. Biotechnology and pharmaceutical companies have major projects studying expression analysis using microarray technologies. Hundreds or even thousands of probes (*e.g.* cDNA) can be attached to tiny chips. Various experiments are performed using these microarrays to determine which genes are expressed, when and at what level, *e.g.* in Arabidopsis.[4] All this information needs to be stored in computer databases and researchers need simple ways of querying the data to answer questions or stimulate new ones. See the sequence analysis section below for a discussion of some of these methods.

Bioinformatics is a young, exciting and rapidly expanding field. Many computer programs have been written by or for biologists, particularly during the last 10 years, and these have a produced a comprehensive, but *ad hoc* toolkit. You will find bioinformatics throughout biological or medical research (Figure 1). *In silico,* is a term sometimes used when experiments are done on a computer using bioinformatics. Many of the

[2] The 1st International Workshop on Comparative Genome Organisation. Comparative Genome Organisation of Vertebrates, *Mamm. Genome*, 1996, **7**, 717.
[3] S. Brenner, *et al.*, *Nature*, 1993, **366**, 265.
[4] T. Desprez, J. Amselem, M. Caboche, and H. Höfte, Differential gene expression in Arabidopsis monitored using cDNA arrays. *Plant J.*, 1998, **14**, 643.

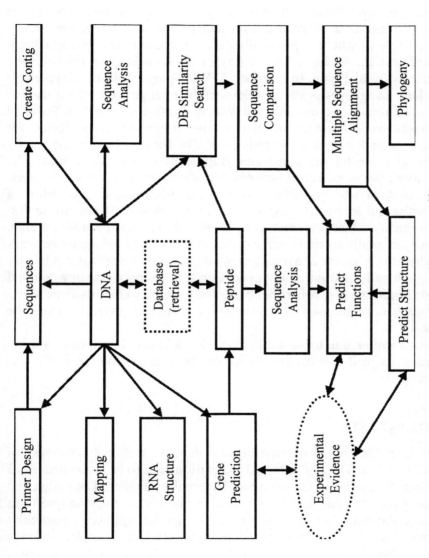

Figure 1 *Provides an overview of some of the uses bioinformatics in molecular biology*[35]

early programs were difficult to use, having their own particular data formats and tied to particular computer systems. Fortunately the quality and ease of use of bioinformatics programs though is constantly improving.

The World Wide Web (WWW) has been a real boon to biologists wanting to access bioinformatics resources. One of the biggest benefits is that vast amounts of related data can be connected together almost irrespective of where it physically exists. Hypertext is a piece of high-lighted text containing a link to further data; clicking on this with the mouse will take you to another WWW page. HTML (HyperText Markup Language) is the simple computer language used to display text, pictures and hyperlinks on WWW browsers. The WWW-based browsers have allowed intuitive mouse driven interface forms for programs and database accessing. WWW forms often transparently wrap up powerful programs and graphically present the results. The increasing use of powerful network orientated programming languages, such as Java and possibly ActiveX, are promising an explosion of powerful and interactive applications on the WWW. There are limita-tions such as being restricted by a WWW form/application design, security firewalls around your site and, in particular inadequate network bandwidth. What the above means in practice is that you may not be able to significantly customize an application that nearly suits your needs; if local computing controllers are excessively security anxious some appli-cations cannot be used, and interactive graphical programs could be slow.

This chapter has to be fairly brief; for a fuller description you are directed to read a dedicated molecular biology computing book, *e.g.* reference 1.

2 DATABASES

Storage of the results of experimental data is an important aspect of bioinformatics. The raw data and annotation need to be integrated into a database in a reliable, consistent and accessible manner. The design and maintenance of databases is of critical importance for asking questions of the data; databases need to be easy to update and access particular information of a database entry (see Figure 2). For example, you may wish to find all the complete, eukaryotic serine proteinase genomic sequences deposited in the last three months. Databases are increasingly

[5] Overview of Bioinformatics chart, by Y. Karavidopoulou, http://www.hgmp.mrc.ac.uk/MANUAL/faq/chart.gif

```
ID   THRR_HUMAN      STANDARD;       PRT;    425 AA.
AC   P25116;
DT   01-MAY-1992 (REL. 22, CREATED)
DT   01-MAY-1992 (REL. 22, LAST SEQUENCE UPDATE)
DT   01-NOV-1997 (REL. 35, LAST ANNOTATION UPDATE)
DE   THROMBIN RECEPTOR PRECURSOR.
GN   F2R OR PAR1 OR TR.
OS   HOMO SAPIENS (HUMAN).
OC   EUKARYOTA; METAZOA; CHORDATA; VERTEBRATA; TETRAPODA; MAMMALIA;
OC   EUTHERIA; PRIMATES.
RN   [1]
RP   SEQUENCE FROM N.A.
RX   MEDLINE; 91168254.
RA   VU T.-K.H., HUNG D.T., WHEATON V.I., COUGHLIN S.R.;
RL   CELL 64:1057-1068(1991).
CC   -!- FUNCTION: RECEPTOR FOR ACTIVATED THROMBIN.
CC   -!- SUBCELLULAR LOCATION: INTEGRAL MEMBRANE PROTEIN.
CC   -!- TISSUE SPECIFICITY: PLATELETS AND VASCULAR ENDOTHELIAL CELLS.
CC   -!- PTM: IT IS THOUGHT THAT CLEAVAGE AFTER AA 41 BY THROMBIN LEADS TO
CC       ACTIVATION OF THE RECEPTOR. THE NEW AMINO TERMINUS FUNCTIONS AS
CC       A TETHERED LIGAND AND ACTIVATES THE RECEPTOR.
CC   -!- SIMILARITY: BELONGS TO FAMILY 1 OF G-PROTEIN COUPLED RECEPTORS.
DR   EMBL; M62424; G339677; -.
DR   PIR; A37912; A37912.
DR   HSSP; P00734; 1NRN.
DR   GCRDB; GCR_0088; -.
DR   MIM; 187930; -.
DR   PROSITE; PS00237; G_PROTEIN_RECEPTOR; 1.
KW   G-PROTEIN COUPLED RECEPTOR; TRANSMEMBRANE; GLYCOPROTEIN; SIGNAL;
KW   BLOOD COAGULATION.
FT   SIGNAL        1     26       POTENTIAL.
FT   PROPEP       27     41       REMOVED FOR RECEPTOR ACTIVATION.
FT   CHAIN        42    425       THROMBIN RECEPTOR.
FT   DOMAIN       42    102       EXTRACELLULAR (POTENTIAL).
FT   TRANSMEM    103    128       1 (POTENTIAL).
FT   DOMAIN      129    137       CYTOPLASMIC (POTENTIAL).
FT   TRANSMEM    138    157       2 (POTENTIAL).
FT   DOMAIN      158    176       EXTRACELLULAR (POTENTIAL).
FT   TRANSMEM    177    198       3 (POTENTIAL).
FT   DOMAIN      199    218       CYTOPLASMIC (POTENTIAL).
FT   TRANSMEM    219    239       4 (POTENTIAL).
FT   DOMAIN      240    268       EXTRACELLULAR (POTENTIAL).
FT   TRANSMEM    269    288       5 (POTENTIAL).
FT   DOMAIN      289    311       CYTOPLASMIC (POTENTIAL).
FT   TRANSMEM    312    334       6 (POTENTIAL).
FT   DOMAIN      335    350       EXTRACELLULAR (POTENTIAL).
FT   TRANSMEM    351    374       7 (POTENTIAL).
FT   DOMAIN      375    425       CYTOPLASMIC (POTENTIAL).
FT   CARBOHYD     35     35       POTENTIAL.
FT   CARBOHYD     62     62       POTENTIAL.
FT   CARBOHYD     75     75       POTENTIAL.
FT   CARBOHYD    250    250       POTENTIAL.
FT   CARBOHYD    259    259       POTENTIAL.
FT   SITE         41     42       CLEAVAGE (BY THROMBIN).
FT   DOMAIN       57     60       ASP/GLU-RICH (ACIDIC).
FT   DISULFID    175    254       BY SIMILARITY.
SQ   SEQUENCE   425 AA;  47410 MW;  E9A485AE CRC32;
     MGPRRLLVA ACFSLCGPLL SARTRARRPE SKATNATLDP RSFLLRNPND KYEPFWEDEE
     KNESGLTEYR LVSINKSSPL QKQLPAFISE DASGYLTSSW LTLFVPSVYT GVFVVSLPLN
     IMAIVVFILK MKVKKPAVVY MLHLATADVL FVSVLPFKIS YYFSGSDWQF GSELCRFVTA
     AFYCNMYASI LLMTVISIDR FLAVVYPMQS LSWRTLGRAS FTCLAIWALA IAGVVPLVLK
     EQTIQVPGLN ITTCHDVLNE TLLEGYYAYY FSAFSAVFFF VPLIISTVCY VSIIRCLSSS
     AVANRSKKSR ALFLSAAVFC IFIICFGPTN VLLIAHYSFL SHTSTTEAAY FAYLLCVCVS
     SISSCIDPLI YYYASSECQR YVYSILCCKE SSDPSSYNSS GQLMASKMDT CSSNLNNSIY
     KKLLT
//
```

Figure 2 *An example of a SwissProt entry accessed using SRS.[8] The leftmost two letters, code for the field name, e.g. FT – feature table. Note the highlighted clickable hyperlinks, which take us to pertinent WWW pages for more information* being linked together and this creates a very powerful and flexible resource. Sequence annotation is text describing features of the sequence, such as the exon positions and the gene function. In the WWW applications of Entrez[6] and SRS[7] for example, when looking at a genomic sequence and its associated annotation there are clickable hyperlinks to the literature abstracts, relevant information in genome database, mutation databases, protein products, *etc.*

2.1 Sequence Databases

There will soon be raw sequence data of 3 000 000 000 bases for humans alone. Sequence annotation is derived from experimental evidence, either directly from molecular biology or by comparison with homologous sequences. Substantial efforts are made to provide useful annotation about the sequence and references to literature, by the sequence submitter and the database annotators;[8] an example sequence with annotation can be seen in Figure 2. Ultimately we need to rely on primary experimental data for annotation, rather than that derived from predicted homologous sequences. This is an increasing problem, particularly as increasing numbers of sequences are added to the databases. When one also considers the continual need to update annotation, just the technical database side of this presents interesting computing challenges.

The WWW has greatly simplified the accessing of sequence and annotation information. Sequence databases are generally searched in two ways, by keywords in the annotation and by database scans looking for similar sequences. These are discussed under sequence analysis (Section 3).

2.1.1 Nucleic Acid Sequence Databases. The results of the sequencing efforts of individual researchers and sequencing factories are compiled and collated into sequence databases. Most of the sequence databases have full releases (every 2–3 months) and then updates (daily or weekly) of new or amended entries. These are automatically picked up by bioinformatics centres and indexed for all the required programs. Research scientists should thus always have access to newly released sequences. To obtain a sequence that is expected to be released soon and/

---

[6] NCBI (Information about BLAST, databases, Entrez , SNPs) http://www.ncbi.nlm.nih.gov/
[7] SRS; http://srs.ebi.ac.uk/ Thure Etzold, Anatoly Ulyanov, and Patrick Argos, SRS: Information Retrieval System for Molecular Biology Data Banks, *Meth. Enzymol.*, 1996, **266**, 114.
[8] The EMBL Nucleotide Sequence Database; http://www.ebi.ac.uk/ebi_docs/embl_db/ebi/database home.html

or the latest known homologues, frequent databases scans are required; if this is for a particular genome or chromosomal region, then the laboratory which is sequencing this is a wise place to check regularly. The NCBI[6] maintains a list of who is sequencing what. Major sequencing centres like the Sanger Centre allow WWW searching of their latest sequences.

Entries are submitted to one of the four major sequence database centres and then annotated entries are rapidly distributed to the other centres (Table 1). The sequence databases are split into many divisions of taxonomy (*e.g.* bacteria and plants) and sequence type (*e.g.* EST, STS and GSS). It is sometimes useful to search only particular database sections.

ESTs (expressed sequence tags) are cDNAs derived from mRNA sequences. They are comparatively cheap to produce and provide a useful resource, even if only containing part of the coding section of expressed genes. Most of the ESTs are from human and mouse, but there is increasing creation of ESTs from other organisms. STSs (sequence tagged sites) are short stretches of genomic sequences of much use in mapping. Their physical position on chromosomes is known, so it is a great benefit if your sequence contains or is linked to an STS. GSSs (genomic survey sequences) are usually short stretches of sequences distributed over a genome and are a useful resource for gene hunting, *e.g.* as has been demonstrated with the Fugu (Puffer fish) landmark

Table 1 *Shows the major nucleic acid sequence databases.*
Databases 1–4 act as joint repositories for all nucleic acid sequence data, going to their WWW sites will show the database release notes and allow searching of these databases. The following (databases 5–7) are sections from these databases, but with extra genome mapping annotation. The remainder (databases 8–9) sets of EST clusters

Database	Location	Description
1 GenBank	http://www.ncbi.nlm.nih.gov/	These are the four primary repositories
2 EMBL	http://www.ebi.ac.uk/	of all nucleic acid sequence data, they
3 DDBJ	http://www.ddbj.nig.ac.jp/	regularly exchange all sequences.
4 GSDB	http://www.ncgr.org/gsdb/	(Genome Sequence DataBase)
5 dbEST	http://www.ncbi.nlm.nih.gov/	Expressed sequence tags
6 dbGSS	http://www.ncbi.nlm.nih.gov/	Genome survey sequences
7 dbSTS	http://www.ncbi.nlm.nih.gov/	Sequence tagged sites
8 Stack	http://www.sanbi.ac.za/stack/	Sequence Tag alignment and consensus knowledgebase
9 UniGene	http://www.ncbi.nlm.nih.gov/	Unique gene sequence collection for human and mouse

mapping project.[9] As it is expensive to completely sequence organisms, GSSs help to locate a gene/region of interest, so that only this region can be sequenced. The EST, STS and GSS sections also exist as separate databases, with different annotation more relevant to mapping.

The EST section consists of a redundant collection of ESTs of varying sequence quality. Many of the ESTs are relatively short fragments, but there are longer ones. Valiant attempts have been made to cluster ESTs (Stack and UniGene (Table 1)) and the results are longer, better quality non-redundant consensus regions, often with links to genetic and physical maps. Hence, if database searches are carried out against these consensus sequences there will be fewer, but probably more informative results.

2.1.2 Protein Sequence Databases. Most genes code for polypeptides rather than rRNA or tRNA, *etc.* To best understand the function of a gene, analysis is often most useful at the protein level. Protein sequence database entries are usually better annotated than the nucleic ones; some even have invaluable experimental information. The emphasis in protein sequence databases (Table 2) is on gene families and protein function.

Protein sequences are being generated *via* the genome projects faster than they can be well characterized and annotated. SwissProt is the single best quality protein sequence database, with excellent annotation. Databases like SPTREMBL make available the protein sequence whilst their annotation is being brought up to the SwissProt standard. There are also useful rival protein database organisations such as PIR and PRF (Protein Resource Foundation). Most databases have some redundancy within them and, although there is a significant overlap between them (*e.g.* PIR will contain most of the protein sequences found in SwissProt), some will have collected unique sequences.

OWL was an early attempt at creating a non-redundant protein sequence database collated from different protein databases, but SPTR and NCBIs nr (non-redundant) National Centre for Biotechnology Information are now the most useful ones encountered.

2.1.3 Protein Family and Motif Databases. Information from the alignment of protein sequences, as well as from structural biology and biochemical studies, reveals areas of protein with distinct functions (domains). There are often distinctive sequence patterns (motifs or signatures) that are structurally or functionally important. These motifs are useful in attempting to determine the function of a protein. Table 3

---

[9] HGMP; UK MRC Human Genome Mapping Project Resource Centre, http://www.hgmp.mrc.a-c.uk/

Table 2 *Major protein sequence databases. SwissProt and PIR are the two major protein sequence databases. TREMBL and GenPept are translations of the polypeptide, encoded by nucleic acid sequence entries. The non-redundant protein sequence databases almost allow searching of only one protein database*

Database	Location	Description
SwissProt	http://www.ebi.ac.uk/	A quality protein sequence database
PIR	http://nbrfa.Georgetown.Edu/pir/	The largest annotated protein sequence database
TREMBL	http://www.ebi.ac.uk/	Translation of EMBL coding sequences (CDS); SPTREMBL will go into SwissProt, REMTREMBL is the remainder.
GenPept	http://www.ncbi.nih.gov/	Translation of GenBank coding sequences (CDS)
NRL_3D	http://nbrfa.Georgetown.Edu/pir/	The actual protein sequences from 3-D protein structure (PDB) entries
PRF	http://www.prf.or.jp/	The Protein Resource Foundation database. Includes many unique entries from the literature
SPTR	http://www.ebi.ac.uk/	SwissProt + SPTREMBL + TREMBLNEW
OWL	http://www.uk.embnet.org/	SwissProt + PIR + GenPept + NRL_3D
NCBI nr	http://ncbi.nlm.nih.gov/	SwissProt + PIR + GenPept + PDB + PRF

Table 3 *Protein family and motif databases which can be very useful in identifying the function of specific regions of a protein*

Database	Location	Description
Pfam	http://www.sanger.ac.uk/Software/Pfam/	Multiple alignments of protein domains or conserved protein regions
BLOCKS	http://www.blocks.fhcrc.org/	Based around automatic ungapped alignments
PRINTS	http://www.biochem.ucl.ac.uk/bsm/dbbrowser/PRINTS/PRINTS.html	Fingerprints of a series of small conserved motifs making up a domain
PROSITE	http://expasy.hcuge.ch/sprot/prosite.html	Regular expression patterns, but now also includes profiles

lists some of the most useful resources. The prediction of patterns/ domains will result in some false hits, so care has to be taken. Pfam is useful for finding domains and for determining protein families. The consensus approach of using various programs and databases means that known protein domain signatures are less likely to be missed and there is more predictive evidence for those that are found.

2.2 Genome Databases

Mapping, sequence, mutation, expression and functional assignments data is maintained in an array of databases (Table 4). Databases such as GeneCards are showing the way forward with the integration of useful data from many resources and then appropriate links to specific databases for more information. Comparative database resources will become increasingly important and there are already human/mouse comparative databases at the NCBI and MGI (Mouse Genome Informatics).

A running list of the amazing number of genome sequencing projects can be found at Terri Gaasterland's http://www.mcs.anl.gov/home/ gaasterl/genomes.html page. This also provides links to all the publicly available microorganism genome databases.

Applications like ENSEMBL,[10] GeneQuiz[10a] and MAGPIE (Automated Genome Project Investigation Environment)[11] attempt to automatically analyse whole genomes using bioinformatics. The results of this analysis can be searched on the WWW. Automated analysis is a very difficult task and improvements are continually being made as our biological knowledge increases, other genomes are sequenced and analysed, and computational power increases. It is necessary to be wary of the results of automatic analysis and evaluate where the experimental evidence for functional assignment has been 'inherited' from. At the very least, such evidence can provide insights into genes and help scientists decide upon wet laboratory experiments.

2.3 Enzyme Databases

The enzyme classification database is a repository for each characterized enzyme and can be found at Expasy.[12] It is very useful for finding out about enzymatic activities for particular proteins, diseases associated

[10] ENSEMBL – a baseline annotation of the human genome; http://www.ensembl.org/

[10a] GeneQuiz; http://www.sander.ebi.ac.uk/genequiz/genequiz.html

[11] MAGPIE; T. Gaasterland and C. W. Sensen, http://www-fp.mcs.anl.gov/~gaasterland/magpie. html

[12] EXPASY (excellent site for protein and enzymes) http://www.expasy.ch/

Table 4   *Selected genome databases*

Database	Location	Description
GeneCards	http://bioinformatics.weizmann.ac.il/cards/	An excellent database of human genes, their products and their involvement in diseases
GDB	http://www.gdb.org/ http://www.hgmp.mrc.ac.uk/gdb/gdbtop.html	The Genome Database holds data on human gene loci, polymorphisms, mutations, probes, genetic maps, GenBank, citations and contacts
GeneMap	http://www.ncbi.nlm.nih.gov/genemap/	Radiation hybrid map of the human genome
HGMD	http://www.uwcm.ac.uk/uwcm/mg/hgmd0.html	The Human Gene Mutation Database collate the majority of known (published) gene lesions responsible for human inherited disease
OMIM	http://www3.ncbi.nlm.nih.gov/Omim/searchomim.htmlhttp://www.hgmp.mrc.ac.uk/omim/	On-line Mendelian Inheritance in Man – catalogue of human genes and genetic disorders.
MGI	http://www.informatics.jax.org/ http://mgd.hgmp.mrc.ac.uk/	Mouse Genome Informatics – provides integrated access to various sources for information on the genetics and biology of the laboratory mouse
UK CropNet	http://synteny.nott.ac.uk/	The UK Crop Plant Bioinformatics Network – access to WWW plant genome databases (especially *Arabidopsis thaliana*)
AGIS	http://probe.nalusda.gov:8300/plant/index.html	The Agricultural Genome Information Server – access to plant genome databases (especially rice)
FlyBase	http://flybase.bio.indiana.edu/ http://www.ebi.ac.uk/flybase/	A database of the Drosophila genome
ACeDB	http://www.sanger.ac.uk/Projects/C_elegans/	Access to *C. elegans* genomic sequence data with maps, genetics and bibliographic information
Yeast	http://speedy.mips.biochem.mpg.de/mips/yeast/	Genome of *Saccharomyces cerevisiae*
EcoCyc	http://ecocyc.PangeaSystems.com/ecocyc/	Encyclopaedia of *E. coli* genes and metabolism

with particular enzymes, *etc*. At the same site, searches can be made for enzymes by keyword and by viewing of the appropriate section(s) of Boehringer Mannheim's Biochemical Pathways.

2.4 Literature Databases

There is a huge wealth of experimental and other information in journals; computing resources have made this far easier to access, particularly for abstracts. Two of the most accessed literature resources in UK academia are Medline and the Elsevier databases (at BIDS[13]), which together provide a wide coverage. Many journals allow access to entire articles electronically over the WWW, such as *Nature* and *New Scientist*, although subscriptions are often required.

2.4.1 Medline. Medline covers about 3400 journals, published in 70 countries, and chapters and articles from selected monographs. It contains over eight million records, with over 300 000 added annually. Medline can be accessed from University libraries. Entrez/Pubmed are available on the WWW from the NCBI.[7] This allows keyword searches and there are copious links to protein and nucleic acid sequences.

2.4.2 BIDS Embase. The Excerpta Medica database contains major pharmacological and biomedical literature. It covers approximately 3500 journals from 110 countries, plus some book reviews and conference proceedings. It has strong coverage of European journals.

2.4.3 BIDS ISI Citation Indexes and Index to Scientific and Technical Proceedings (ISTP). SciSearch is the science Citation Index with additional journal coverage from the Current Contents series of publications. The Index to Scientific and Technical Proceedings contains details of papers published at over 4000 conferences per year.

3 SEQUENCE ANALYSIS

Evolution has kindly provided us with a variety of sequence information to help us predict gene structure and function. Once proteins or nucleic acid regions have been sequenced extensive computational analysis is possible. Genes from sequencing projects are characterized and the coding regions translated into peptide sequences. This is a rapidly expanding area of bioinformatics.

---

[13] BIDS bibliographic services for the UK higher education community: http,//www.bids.ac.uk/

3.1 Sequence Database Searching

Once some sequence data, either protein or nucleic acid is available, database scans can be carried out to identify similar sequences. Scan results may contain matches to sequences which are related by evolution or occur by chance.

Two or more sequences are said to be homologous if they share a common evolutionary ancestor. Orthologues are homologous sequences with the same function in different organisms. Paralogues are produced when a gene is duplicated and subsequent mutations produce gene products with different functions. Being aware of the above is important when trying to assign a function to a gene. A homologous sequence may have lots of experimental data and may even already have had its structure solved. Pairwise or multiple sequence alignments of homologous, particularly using divergent sequences will give insights into the protein structure and hopefully function of the protein sequence under investigation.

3.1.1 Keyword Searching. WWW applications such as the Sequence Retrieval System (SRS)[7] and Entrez[6] provide an excellent means of searching annotation in sequence and many other molecular biology databases. There is a wealth of information contained in these databases and hyperlinks to many other databases; this is illustrated in Figures 2 and 3.

3.1.2 Database Scanning. The usual aim of database scanning is to find homologues. There is trade off between speed and sensitivity in the program used to search a sequence database; some of the protein sequence searching programs are shown in Table 5. The speed itself will vary depending upon the computer hardware, the database being scanned and the length of target sequence. BLAST could take between a few seconds and an hour, but usually takes just a couple of minutes.

In a typical database scan, the sequence under investigation is aligned against each database entry. The algorithm (method) used depends upon the program and the alignment score, which will be program dependent. A statistical analysis is essential to understand what this score means and for comparing results between programs. The Expect (E) score is the predicted number of alignment scores at least this good occurring by chance in a database of this size. The P value is the probability of scores at least this good occurring by chance. At low values the E score and P values converge. E values of less that 1×10^{-3} are statistically signifi-

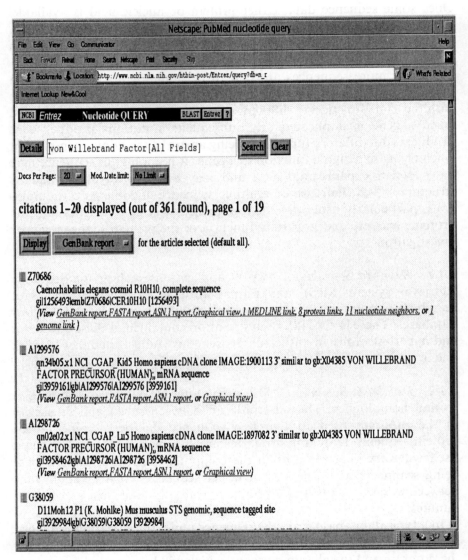

Figure 3   *Illustration of a keyword search for 'von Willebrand Factor' in the GenBank nucleic acid sequence database, using Entrez on the WWW*

Table 5   *Comparison of speed and sensitivity of some protein sequence searching programmes*

Package	Speed	Sensitivity
BLAST[6]	Very fast	Fairly
FASTA[13]	Reasonably	Fairly good
ssearch	Slow	Good
BLITZ	Very fast (parallel)	Good

cant, particularly for E values greater than 1×10^{-3} biological knowledge should be used.

For protein comparisons such as in database scanning, it should be noted that physiochemically similar amino acid residues are generally conserved; this needs to be considered during the alignment. Programs allow the selection of the appropriate substitution matrix.[1] A substitution matrix is basically a look-up table used by the alignment program when comparing two amino acid residues in order to measure how biologically similar they are; see the Section 3.2.1 for more detail on pairwise comparison.

Dayhoff and colleagues studied closely related proteins and observed residue replacements; they constructed the PAM (point accepted mutation) model of molecular evolution.[1,14] A 'PAM' corresponds to an average change in 1% of all amino acid residue positions. A series of PAM matrices have been calculated using this model for different evolutionary distances. Other studies have been done since, using the vast amounts of protein family sequence data we now possess, but these newer PAM matrices are not greatly different from the originals. Henikoff and Henikoff[1] studied multiple alignments of distantly related protein regions; using blocks of conserved residues they created a series of BLOSUM matrices. These BLOSUM matrices have been shown to be generally superior to PAM matrices for detecting biological relationships.

In the future, better matrices will be derived from protein tertiary structure information as more structures are known. It can be important to use a matrix that will best characterize the similarity between two sequences and thus most likely to be a 'correct' alignment. The evolutionary relationship is not generally known in advance so a small range of matrices should be used, *e.g.* several of the BLOSUM series.

[14] M. J. E. Sternberg, 'Protein Structure Prediction – a Practical Approach', ed. by M. J. E. Sternberg, IRL Press at Oxford University Press, Oxford, 1996, ISBN 0 19 963496 3 (Includes useful information about BLAST, FASTA, matrices and database scanning in general, see the above reference http://barton.ebi.ac.uk/barton/papers/rev93_1/rev93_1.html).

The choice of parameters is important, especially gap creation and extension penalties for the alignments; these are discussed in Section 3.2.1. Filters should be used to mask out areas of sequences containing repeats or low complexity regions. *i.e.* replace amino acid residues by Xs or nucleic acid bases by Ns. With filtering sequence searches hits are more likely to be biological rather than simply statistically significant, although information is lost.

The BLAST[6] and FASTA[15] packages contain programs for searching DNA against DNA, protein against protein, and DNA against protein. For sensitive DNA against DNA searches, all forward and reverse reading frames can be automatically translated.

There are other biologically more sensitive methods which try to find distant members of the same gene family. PSI BLAST (Position-Specific Iterated BLAST) begins with a normal database scan, the significant alignments are compiled into a position-specific score matrix and this is then scanned against the database. These steps are repeated until no new entries are found. PHI BLAST (Pattern Hit Initiated BLAST) is similar, but here the input is the protein sequence and specified pattern. This is integrated with PSI BLAST and the search results in sequences similar to the protein sequence and contains the pattern. There are now also programs like HMMer[16] which make use of Hidden Markov Models (HMMs) (a branch of statistics much used in the study of speech patterns and now applied with great effect to many areas of bioinformatics).[16]

Increasingly, graphical result viewers are available and these also provide links to the sequence annotation. A good example is that found when doing BLAST searches at the NCBI.

In conclusion, database scanning is most efficient if one starts with an initial scan using a simple BLAST search, and then experiments with parameters and matrices accordingly. Remember that usually evolutionary relationships are being analysed and that searches at the protein level are the most sensitive.

3.2 Pairwise and Multiple Sequence Comparisons and Alignments

The aligning of two sequences is termed a pairwise alignment and can be done by all database scanning programs but, for computational speed reasons, these are usually not as rigorous as they could be. Alignments with different parameters and matrices can be used to explore the possible relationships between two sequences. Even mathematically

[15] FASTA and align, ftp://ftp.virginia.edu/pub/fasta/
[16] R. Durbin, S. Eddy, A. Krogh, and G. Mitchison, 'Biological Sequence Analysis: Probabilistic Models of Proteins and Nucleic Acids', Cambridge University Press, Cambridge, 1998.

rigorous alignments between two sequences are practical, but this is not computationally feasible for aligning more that a small number of sequences, hence heuristic approaches have been used. The techniques described below are particularly useful in comparative genomics.

3.2.1 Pairwise Comparisons. Graphical displays such as dotplots,[1] are excellent for obtaining an overview of how two or more sequences compare; regions of similarity appear as diagonal runs of dots (Figure 4). Comparing a sequence against itself with dotplots will reveal simple, tandem and inverted repeats. Sequence alignment (Figure 5) shows the comparison at the detailed sequence level. The best single regions of alignment are most clearly identified using local alignments programs *e.g.* lalign[18] and GCG's[17] bestfit. Global alignments should be used to align entire sequences, *e.g.* align[19] and GCG's gap. The alignment programs work by aligning two sequences to obtain the highest score: positive scores for identical (or physiochemically similar) matches, and negative ones for introducing or extending gaps. Alignments do vary depending upon the parameters chosen. Programs have different scoring mechanisms, so it is useful to quote percentage identity/similarity along with the alignment length. Some programs also allow alignments of protein against DNA and this is useful such as to show exon positions.

3.2.2 Multiple Sequence Alignments. These highlight conserved and diverged regions between sequences and notably reveal functionally important residues. Residue conservation in proteins helps provide structural insights. It is difficult to automate biologically reliable alignment because it is more complex than simply comparing strings of letters. A variety of approximate methods for finding 'good' multiple alignments exist; clustering and the Hidden Markov Model (HMM)[16] technique are used the most.

The initial step in clustering is to calculate all pairwise comparison scores. The algorithm now creates a clustering tree which is used to guide the alignment:

- Pair off the two 'closest' sequences (> score) and create a cluster.
- Repeat for the next two sequences or clusters.

[17] Wisconsin Package Version 9.1, Genetics Computer Group (GCG), Madison, Wisconsin, http://www.gcg.com/.
[18] lalign: X. Huang and W. Miller, *Adv. Appl. Math.*, 1991 **12** 337.
[19] E. L. L. Sonnhammer and R. Durbin, *Gene*, 1995, **167**, GC1–10, http://www.sanger.ac.uk/Software/Dotter

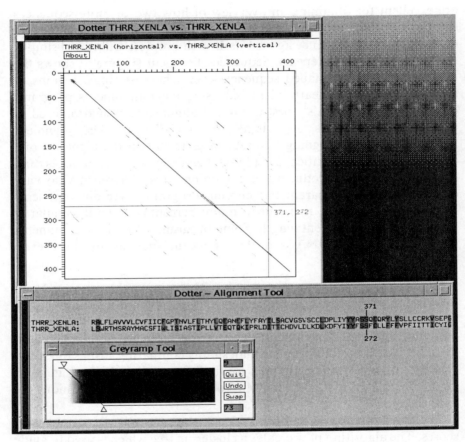

Figure 4 *Pairwise dotplot comparisons of two protein sequences using the dotter program*[19]

GAP of: THRR_HUMAN and THRR_XENLA

Symbol comparison table: /packages/gcg9/gcgcore/data/rundata/blosum62.cmp
CompCheck: 6430

```
        Gap Weight:      12      Average Match:   2.912
     Length Weight:       4      Average Mismatch: -2.003

           Quality:    1112            Length:    428
             Ratio:   2.648              Gaps:      2
Percent Similarity: 61.391    Percent Identity:  52.278
```

```
        Match display thresholds for the alignment(s):
                       | = IDENTITY
                       : =      2
                       . =      1
```

THRR_HUMAN x THRR_XENLA   December 16, 1998 14:42 ..

```
                   .         .         .         .         .
   1 ...MGPRRLLLVAACFSLCGPLLSARTRARRPESKATNATLDPRSFLLRN  47
      :     |||.       | | |       | | | |:
   1 MMELRVLLLLLLLLTLLGAMGSLCLANSDTQAKGAHSNNMTIKTFRIFDDS  50

               .         .         .         .         .
  48 PNDKYEPFWEDEEKNESGLTEYRLVSINKSSPLQKQLPAFISEDASGYLT  97
      .:   |    |::  :..    |    :    ||   .   |:.:     |..:|  ||.
  51 ESEFEEIPWDELDESGEGSGDQAPVSRSARKPIRRN....ITKEAEQYLS  96

             .         .         .         .         .
  98 SSWLTLFVPSVYTGVFVVSLPLNIMAIVVFILKMKVKKPAVVYMLHLATA 147
      |  |||  ||||.|| ||:| ||||::||::|: ||||:||||||||.||  |
  97 SQWLTKFVPSLYTVVFIVGLPLNLLAIIIFLFKMKVRKPAVVYMLNLAIA 146

             .         .         .         .         .
 148 DVLFVSVLPFKISYYFSGSDWQFGSELCRFVTAAFYCNMYASILLMTVIS 197
      ||  ||||||||||.|:   ||.|| ||   ||| |||||| |:||.  ||
 147 DVFFVSVLPFKIAYHLSGNDWLFGPGMCRIVTAIFYCNMYCSVLLIASIS 196

             .         .         .         .         .
 198 IDRFLAVVYPMQSLSWRTLGRASFTCLAIWALAIAGVVPLVLKEQTIQVP 247
      :|||||||||  |||||:|  ||   : || :-||   :||..  |||  .:|
 197 VDRFLAVVYPMHSLSWRTMSRAYMACSFIWLISIASTIPLLVTEQTQKIP 246

             .         .         .         .         .
 248 GLNITTCHDVLNETLLEGYYAYYFSAFSAVFFFVPLIISTVCYVSIIRCL 297
      |.|||||||.      |. :| ||||.|  .||||| ||.|:||: |||  |
 247 RLDITTCHDVLDLKDLKDFYIYYFSSFCLLFFFVPFIITTICYIGIIRSL 296

             .         .         .         .         .
 298 SSSAVANRSKKSRALFLSAAVFCIFIICFGPTNVLLIAHYSFLSHTSTTE 347
      |||.:|    ||.||||||.  | |:||||||||||| : ||         |
 297 SSSSIENSCKKTRALFLAVVVLCVFIICFGPTNVLFLTHYL....QEANE 342

             .         .         .         .         .
 348 AAYFAYLLCVCVSSISSCIDPLIYYYASSECQRYVYSILCCKESSDPSSY 397
      ||||:|   || |:|  |:||||||||||||:||||.||:|||:.  |:|  |
 343 FLYFAYILSACVGSVSCCLDPLIYYYASSQCQRYLYSLLCCRKVSEPGSS 392

             .         .         .
 398 NSSGQLMASKMDTCSSNLNNSIYKKLLT 425
      |  |  | ||.|  .||||||||
 393 TGQLMSTAMKNDNCSTNAKSSIYKKLLA 420
```

BESTFIT of: THRR_HUMAN and THRR_XENLA

Symbol comparison table: /packages/gcg9/gcgcore/data/rundata/blosum62.cmp
CompCheck: 6430

```
        Gap Weight:      12    ˙ Average Match:   2.912
     Length Weight:       4      Average Mismatch: -2.003
```

Figure 5 *Shows the alignment of two protein sequences using local (bestfit) and global*
(gap) alignment programs. Identities are shown between the two positions by a
'|' symbol, similarities by ':' or '.' and gaps are represented by a '.' at the sequence
level (continued overleaf)

```
          Quality:    1118              Length:    419
            Ratio:    2.720               Gaps:      2
Percent Similarity: 62.044    Percent Identity: 53.041

     Match display thresholds for the alignment(s):
                    | = IDENTITY
                    : =    2
                    . =    1

THRR_HUMAN x THRR_XENLA   December 16, 1998 14:42  ..

                    .                   .                   .
    6 LLLVAACFSLCGPLLSARTRARRPESKATNATLDPRSFLLRNPNDKYEPF  55
      |||.        | |   | .   .  . | |:      . .:  |
    9 LLLLLTLLGAMGSLCLANSDTQAKGAHSNNMTIKTFRIFDDSESEFEEIP  58

                    .                   .                   .
   56 WEDEEKNESGLTEYRLVSINKSSPLQKQLPAFISEDASGYLTSSWLTLFV 105
      |::  :..    |   :     ||  .    |:.:      |..:| ||.| ||| ||
   59 WDELDESGEGSGDQAPVSRSARKPIRRN....ITKEAEQYLSSQWLTKFV 104

                    .                   .                   .
  106 PSVYTGVFVVSLPLNIMAIVVFILKMKVKKPAVVYMLHLATADVLFVSVL 155
      ||.||  ||:| ||||::||::|: ||||:|||||||||.|| ||| |||||
  105 PSLYTVVFIVGLPLNLLAIIIFLFKMKVRKPAVVYMLNLAIADVFFVSVL 154

                    .                   .                   .
  156 PFKISYYFSGSDWQFGSELCRFVTAAFYCNMYASILLMTVISIDRFLAVV 205
      ||||.|: ||.|| ||  :|| ||| ||||||| :||.  |:|||||||
  155 PFKIAYHLSGNDWLFGPGMCRIVTAIFYCNMYCSVLLIASISVDRFLAVV 204

                    .                   .                   .
  206 YPMQSLSWRTLGRASFTCLAIWALAIAGVVPLVLKEQTIQVPGLNITTCH 255
      ||| |||||| ||   |  || :.||  :||.. ||| .:|  |.|||:|
  205 YPMHSLSWRTMSRAYMACSFIWLISIASTIPLLVTEQTQKIPRLDITTCH 254

                    .                   .                   .
  256 DVLNETLLEGYYAYYFSAFSAVFFFVPLIISTVCYVSIIRCLSSSAVANR 305
      |||.     |. :| |||||.| .|||| ||:|||:|: |||.| |
  255 DVLDLKDLKDFYIYYFSSFCLLFFFVPFIITTICYIGIIRSLSSSIENS 304

                    .                   .                   .
  306 SKKSRALFLSAAVFCIFIICFGPTNVLLIAHYSFLSHTSTTEAAYFAYLL 355
      ||.|||||.  | |:|||||||||| : ||      |  ||||:|
  305 CKKTRALFLAVVVLCVFIICFGPTNVLFLTHYL....QEANEFLYFAYIL 350

                    .                   .                   .
  356 CVCVSSISSCIDPLIYYYASSECQRYVYSILCCKESSDPSSYNSSGQLMA 405
      || |:| |:|||||||||||:|||| .||:||:. :| |         |
  351 SACVGSVSCCLDPLIYYYASSQCQRYLYSLLCCRKVSEPGSSTGQLMSTA 400

  406 SKMDTCSSNLNNSIYKKLL 424
      | | ||.|  .|||||||
  401 MKNDNCSTNAKSSIYKKLL 419
```

Figure 5 *Continued*

Two commonly used clustering programs are Clustal[20] and pileup[15]; Clustalx[20] is a graphical front end to Clustal and can be seen in Figure 6.

A more recent method has been to use HMM techniques. Here one provides a seed alignment of half a dozen sequences which has been manually edited and checked, and the program is then allowed to automatically align further sequences to this seed alignment. The major advantages of this method are that it allows the use of biological

[20] Clustal, J. D. Thompson, D. G. Higgins, and T. J. Gibson, *Nucl. Acids Res.*, 1994, **22** 4673, http://www-igbmc.u-strasbg.fr/BioInfo/ClustalW/

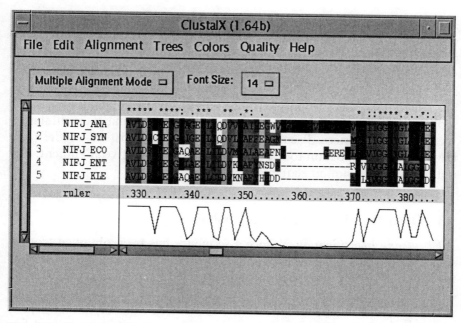

Figure 6 *Shows multiple sequence alignment of five flavodoxin sequences using Clustal*[20]

knowledge to get the seed alignment right and align far more sequences than is practical with most other techniques.

3.2.3 Improving the Alignment. Multiple sequence alignments can be improved greatly by applying both biological knowledge and specific information that investigators may have about the sequences. Computer programs largely treat sequences as groups of letters, although there are algorithmic improvements. It is important to make use of any available structural information; proteins may have a known structure or at least use a secondary structure prediction program.[21,22] Consider, for example, whether gaps are sensible, *e.g.* in the middle of a likely alpha helix, rather than in a loop region. It is wise to concentrate on aligning domains, not loops, as the former are better conserved. If there are biologically active residues, then it is likely that these too are conserved. Reflect upon which sequences are being used: full sequences will be better than fragments, and check that the sequences really are homologous. If both nucleic acid and peptide sets of sequences are available, aligning both would provide insights. It is also important to try different programs and parameters.

[21] PHD; B. Rost, C. Sander, and R. Schneider, *J. Mol. Biol.*, 1994, **235**, 13, http://www.embl-heidelberg.de/predictprotein/
[22] Jpred, http://circinus.ebi.ac.uk:8081/

3.2.4   Profile Searching.   A sensitive method for finding distant family members is to use the information in a multiple sequence alignment. A consensus sequence consists of the residues shared by the majority of sequences in an alignment. A database scan with just the consensus sequence would probably miss many homologous sequences. A profile is a mathematical description of a multiple sequence alignment, either as a score matrix (*e.g.* GCG's profile programs, where the matrix is similar to the Dayhoff matrices discussed earlier) or as a statistical model (*e.g.* HMM's). It is important to use a balanced, representative alignment, *e.g.* not to have very similar sequences.

3.3   Other Nucleic Acid Sequence Analysis

3.3.1   Gene Identification.   Once a sequence has been obtained from a sequencer, ideally the gene products, control of expression and ultimately the functions of the gene need to be determined. Careful interpretation of database scanning results often greatly aids gene identification. Gene structure, such as promoter sites, exons, *etc.*, can be usefully predicted.[23,24] Many of these programs provide parameters optimized for different species. tRNA genes can be very accurately predicted[25] because they share sequence (structural) motifs. Consensus approaches which run many types of analysis and present the results graphically together, greatly aid the interpretation, *e.g.* using the NIX interface.[9]

3.3.2   Restriction Mapping.   A restriction site is the sequence pattern recognized and cut by a restriction enzyme. Given a nucleotide sequence, all the cutting sites of the desired restriction enzymes and the fragment sizes can be easily found; useful programs are tacg[26] and map.[17] Restriction maps are useful in physical and genetic mapping and there are also programs for interpreting the results of restriction enzyme digests.[17]

3.3.3   Single Nucleotide Polymorphismis (SNPs).   Single nucleotide polymorphisms (SNPs) have sparked considerable interest particularly in medical and pharmaceutical research.[7] SNPs are single base pair differences between two sequence regions. A key aspect of research in

[23] GenScan, C. Burge, 'Identification of Genes in Human Genomic DNA', PhD thesis, Stanford University, Stanford, California, 1997.
[24] fgenes: V. V. Solovyev, and A. A. Salamov, The Gene-Finder computer tools for analysis of human and model organisms genome sequences. In 'Proceedings of the Fifth International Conference on Intelligent Systems for Molecular Biology', ed C. Rawling, D. Clark, R. Altman, L. Hunter, T. Lengauer, and S. Wodak.), AAAI Press, Halkidiki, Greece, 294–302.
[25] T. M. Lowe, and S.R. Eddy, *Nucleic Acids Res.*, 1997, **25**, 955.
[26] tacg, by H. Mangalam, http://hornet.bio.uci.edu/%7Ehjm/projects/tacg/tacg.main.html

genetics is associating sequence variations with heritable phenotypes. SNPs are the most common and occur approximately once every 500 bases. They can be especially useful when occurring in promoters or exons, *i.e.* affecting gene expression. SNPs are expected to facilitate large-scale disease-gene association studies.

4 PROTEIN STRUCTURE

Knowing at least some information about a protein's structure is very useful in determining possible functions. Homology is often better conserved in structure than at the sequence level. Protein secondary and tertiary structure prediction is often difficult to do correctly and indeed in reality an active protein may have a range of conformations. Books such as *Protein Structure Prediction – A Practical Approach*[14] and protein prediction courses are much recommended for those wanting to know more about this interesting and challenging area.

Secondary structure prediction rates from a single sequence are of the order of 70–75% accurate, this can be improved greatly when using multiple sequences, *e.g.* PHD[21] and a consensus approach (jpred).[22] The programs usually correctly predict the structure of regions of a structural type, but have difficulties with the boundaries.

Reliable tertiary structure prediction is difficult, but there have been advances. Ideally one or more homologous proteins of known structure should be found (*e.g.* database scanning) and then used as a template. Such sequences can be found by performing database scans of the sequences from PDB.[27] Another approach is to thread the protein sequence onto known protein folds, although incorrect protein conformations can still get high scores. Once a basic model has been achieved, it can be analysed and obvious structural or biological problems corrected, *e.g.* torsion angles and large hydrophobic residues exposed in water on the outside of the molecule.

In nature, proteins with different evolutionary ancestors may undergo convergent evolution to possess similar functions. It can be very useful to the understanding of function to compare two such structures.

There are many commercial and academic software packages and programs for protein structure prediction. Our prediction ability will improve as more protein structures are experimentally determined and our knowledge of how proteins fold improves.

---

[27] Protein DataBank (PDB), http://www.pdb.bnl.gov/

5 MAPPING

5.1 Introduction

Genetic and physical mapping has undergone an explosive growth over the past 10 years. The traditional approach to analysing eukaryotic genomes has been to create a map of markers on the chromosomes of an organism. Bioinformatic techniques are used to assemble such maps. Scientists who wish to investigate an unknown region of DNA (usually a gene) try to predict the location relative to markers on this map. Markers are any identifiable position on the DNA (*e.g.* microsatellites, RFLPs (restriction fragment length polymorphisms)) and the type(s) of marker used is dependent upon the organism, resources available, *etc*. The scientist can then attempt to sequence this region of DNA, assemble the resultant sequences into a contig (consensus) and analyse the gene using sequence analysis techniques. These maps are also used for the large-scale sequencing projects.

5.2 Linkage Analysis

Genetic mapping relies on whether or not a target gene locus is co-segregated at meiosis with the genetic markers. There is a resolution of about 1 Mb. In investigations of genetic diseases, family tree (pedigree) information can be invaluable. Each member of the pedigree is medically assessed and diagnosed with respect to the disease in question and their genome analysed to see which markers they have inherited. The statistical analysis of predicting the likely locations with respect to the markers is achieved using specialized programs, *e.g.* Fastlink[28] and Vitesse.[29] There is now a WWW interface to the linkage programs: GLUE (Genetic Linkage User Environment).[9]

5.3 Physical Mapping

This requires the detection of overlaps between cloned DNA fragments. The resolution can be far better than genetic mapping, *i.e.* the genetic (sequence) problem is pinpointed. Complete sequencing is the ultimate physical map.

---

[28] fastlink, R. W. Cottingham Jr., R. M. Idury, and A. A. Schaffer, *Am. J. Hum. Genetics*, 1993, **53**, 252.
[29] J. R. O'Connell, and D. E. Weeks, *Nature Genetics*, 1995, **11**, 402.

5.4 Radiation Hybrids

Radiation hybrid mapping combines aspects of the genetical and physical approaches. A target organism's DNA is randomly fragmented by radiation and fused with a donor organism's cells. Most target DNA is ejected, but some cells retain fragments of DNA. The aim is to have the genome of the target organism distributed over a panel of 80–100 cell-lines. STS (sequence tagged site) markers are typed against this panel, for their presence or absence. Map building can then be achieved with the help of programs like RHMAPPER.[30] The results of typing a target marker against this panel can be easily and reliably analysed by RH servers on the WWW.[30] The analysis uses statistics to predict the likely position of the target marker with respect to the map framework STS markers.

5.5 Primer Design

Designing primers for PCR is an easy task for computer programs. The program is supplied with a sequence region to design primers within; the program will try a range of primer pairs to see if they meet all the criteria. Important parameters include primer and PCR product sizes, whether the primers will anneal to themselves or each other, and melting temperature of the primers to the target region. Primer[31] and prime[17] are well used programs.

If a multiple sequence alignment (either protein or nucleic acid) has been created, programs can be used to select primers for conserved or divergent regions.[32] This can be useful for identifying gene families within the same or different organisms.

6 BIOINFORMATICS SITES AND CENTRES

Bioinformatics programs and expertise are available from many universities, specialized government and academic bioinformatics centres. In addition there are numerous other useful resources on the WWW. Introductory and specialized bioinformatics courses are available, particularly from EMBnet nodes.[33] Increasing use is now made of distance learning and such material is available.

[30] Human Radiation Hybrid Mapping http://www.sanger.ac.uk/HGP/Rhmap/
[31] S. Primer and H. J. S. Rozen (1996, 1997), Primer3 Code available at, http://www-genome.wi.mit.edu/genome_software/other/primer3.html
[32] PRIMEGEN is available from Scientific Computing Services (7554 Jones Ave. N.W., Seattle, WA 98117).
[33] Embnet, http://www.uk.embnet.org/brochure/

Many sites offer complementary resources. If time is limited, it is recommended to make much use of local bioinformatics expertise, get an account on an EMBnet node (details are on the WWW[33]) and become familiar with resources at several of the major specialized sites.

6.1 Local Bioinformatics Services

The local bioinformatics supplied at a University or other institution can vary from an excellent and broad-ranging service to just one or two staff with bioinformatics expertise. Having local bioinformatics help is invaluable and should be investigated.

6.2 National EMBnet Nodes

EMBnet[33] is a science-based group of bioinformatics nodes throughout Europe and an increasing number of associated ones overseas (in Australia, China, *etc.*). The EMBnet national nodes are molecular biology computing centres which have been appointed by their governmental authorities to provide databases and software tools for molecular biology and biotechnology to their scientific community. National nodes offer software and on-line services covering a diverse range of research fields. The information they can provide includes sequence analysis, protein modelling, genetic mapping, phylogenetic analysis, user support and training in their local language. The EMBnet nodes benefit from external, local and collaborative research and development in these and other areas.

6.3 Specialised Sites

Numerous sites on the WWW offer access to specialized areas of bioinformatics, but there is insufficient space to list them here. Suitable specialized sites are soon found by following the links from any major bioinformatics site, such as EMBnet nodes or the centres listed below; alternatively a net search can be done.

The National Center for Biotechnology Information (NCBI)[6] in the USA offers access to Medline, sequence database access and searching, and much more. The European Bioinformatics Institute (EBI)[34] in the UK offers mirrors of many databases, a catalogue of bioinformatics programs, *etc.* The NCBI and EBI collaborate in compiling the major primary sequence databases. Major sequencing sites, such as the Sanger

[34] EBI, http://www.sanger.ac.uk/

Centre[35] and The Institute for Genome Research[36] allow access to their sequences, as well as to new bioinformatic tools they have developed.

7 CONCLUSION AND FUTURE PROSPECTS

Over the next decade, the genomes of humans, model organisms, disease microorganisms and many agricultural species will be mapped and sequenced. We are living in exciting times, particularly with the human genome destined to be fully sequenced within a few years. Proteomics, the study of proteins coded by the genome, will become increasingly important and bioinformatics will have a large part to play in protein structure and function determination. There will be huge amounts of gene expression and biochemical data. Bioinformatics is an essential component in the generation, storage and, more importantly, for the interpretation of the data. This is creating a healthy demand for bioinformaticians and improvements to bioinformatics software. WWW browser applications are becoming better designed and more useful. A good understanding of bioinformatics will greatly help research work in molecular biology and biotechnology.

[35] The Sanger Centre, http://www.sanger.ac.uk/
[36] TIGR, http://www.tigr.org/

CHAPTER 16

Immobilization of Biocatalysts

GORDON F. BICKERSTAFF

1 INTRODUCTION

A vital component in many areas of biotechnology is the utilization of a biocatalyst(s) to produce a required biotransformation(s). While many biocatalysts can be used as free enzymes or as whole cells, immobilization of biocatalysts provides additional features that can significantly improve the catalysed reaction. The technology required for immobilization of biological catalysts has expanded greatly in the past 30 years as the advantages of immobilization have been utilized in a wide range of analytical, biotransformation and medical applications.[1] A consequence of the explosion of this technology is that there is now a bewildering array of permutations for immobilization of biological material.[2]

2 BIOCATALYSTS

One reason for the great volume of immobilization methodology lies in the subject material to be immobilized. Biological catalysts have a high degree of individual variability, and while many immobilization techniques have wide applicability, it is impossible for one or even a few methods to cater for the great diversity of requirements inherent in biological material. This is especially so when the aim is to produce an optimum system in which an immobilized biocatalyst can function at high levels of efficiency, stability, economy, *etc.*

The range of biological catalysts available for industrial or medical use

[1] G. F. Bickerstaff, *'The Genetic Engineer and Biotechnologist'*, 1995, **15**, 13.
[2] G. F. Bickerstaff, Immobilization of Enzymes and Cells, Humana Press, New Jersey, 1997, pp. 365.

is increasing as advances in biological sciences reveal new levels of understanding of the component parts in living organisms. Enzymes/cells/organelles that were previously overlooked because of high cost/poor stability have benefited from advances in immobilization methodology and genetic technology. New catalyst types such as abzymes, ribozymes and multicatalytic complexes have added to the pool of biocatalysts, and stimulated new ideas on the future development of biocatalysts.

2.1   Enzymes

Enzymes are a specific group of proteins that are synthesized by living cells to function as catalysts for the many thousands of biochemical reactions that constitute the metabolism of a cell. More than 2500 different enzymes are known, and there are still many organisms that have not been screened for novel enzymes or metabolic pathways. It is likely that there are many enzymes awaiting discovery. Living cells require enzymes to drive metabolic pathways because at physiological temperature and pH, uncatalysed reactions would proceed at too slow a rate for the vital processes necessary to sustain life.[3]

In common with all catalysts, enzymes are subject to the normal laws concerning the catalysis of reactions. A catalyst cannot speed up a reaction that would not occur in its absence, because it is not thermodynamically possible. A catalyst is not consumed during the reaction, and so relatively few catalyst molecules are capable of catalysing a reaction many times. Lastly, a catalyst cannot alter the equilibrium position of a given reaction. The vast majority of reactions eventually proceed to a state of equilibrium, in which the rate of the forward reaction is equal to the rate of the reverse reaction. At equilibrium the substrate and product have specific equilibrium concentrations that are a special characteristic of the reaction. For example, the isomerization of glucose to produce fructose is catalysed by the enzyme glucose isomerase.

$$\text{Glucose} \xrightleftharpoons[\hphantom{glucose isomerase}]{\text{glucose isomerase}} \text{Fructose}$$

From a starting solution containing 100% glucose, the reaction proceeds to equilibrium, and for this reaction the relative proportions at equilibrium are; 45% fructose and 55% glucose.[3] The catalyst cannot change the equilibrium position of the reaction, but it can reduce the time that

[3] G. F. Bickerstaff, Enzymes in Industry and Medicine, Edward Arnold, London, 1987, pp. 93.

the reaction normally takes to reach equilibrium. In these respects enzymes are no different from other catalysts. However, enzymes do possess two special attributes that are not found to any great extent in non-biological catalysts, and these are specificity and high catalytic power.

2.1.1 Specificity. Perhaps the most distinctive feature of enzyme-based catalysis is specificity. Chemical catalysts display only limited selectivity, while enzymes have strong specificity for reactants (substrates), and also for the susceptible bond involved in the reaction. The degree of specificity can vary from absolute to fairly broad. For example, the blood enzyme urease is absolutely specific for its substrate urea (NH_2CONH_2), and structural analogues such as thiourea (NH_2SONH_2) are not hydrolysed. In contrast, the enzyme hexokinase is less specific, and shows group specificity for a small set of related sugar molecules.

$$\text{glucose} + \text{ATP} + \text{Mg}^{2+} \xrightarrow{\text{hexokinase}} \text{glucose-6-phosphate} + \text{ADP} + \text{Mg}^{2+}$$

Glucose is the principal substrate, but hexokinase will catalyse phosphorylation of several other sugars such as mannose and fructose, but not galactose, xylose, maltose or sucrose.

In addition to substrate specificity, enzymes display remarkable product specificity, which ensures that the final product is not contaminated with by-products. Thus, in the above phosphorylation of glucose, the product is exclusively glucose-6-phosphate, and no other phospho-glucose (*e.g.* glucose-1-phosphate, or glucose-2-phosphate) is produced during the reaction. The formation of by-products by side reactions is a significant problem associated with most of the less specific catalysts. In many enzymes, specificity also extends to selective discrimination between stereoisomers of a substrate molecule. This stereospecificity is shown by the enzyme D-amino acid oxidase, which is specific for D-amino acids only, and will not catalyse the oxidation of the L-amino acid stereoisomers.

Specificity is an inherent feature of enzymes because catalysis takes place in a particular region of the enzyme, which is designed to accommodate the substrates involved in the reaction (see Figure 1). This region is the active site, and it is normally a small pocket, cleft or crevice on the surface of the enzyme. It is designed to bring a few of the enzyme amino acid residues into contact with the substrate molecule. The site has strong affinity for the substrate because the site amino acid residues are primed for interaction with groups or regions on the

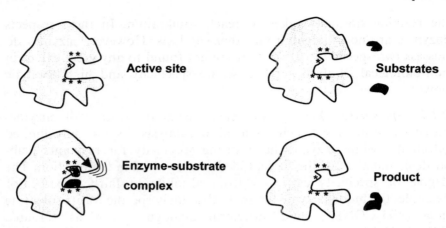

Figure 1 *Schematic representation of an enzyme active site, and its importance in determining enzyme specificity*

substrate molecule. Consequently the substrate molecule must have the correct shape and/or functional groups to fit into the active site, and participate in the interactions (see Figure 1). Enzymes with absolute specificity have very precise shape/interaction requirements, which are only found in a particular substrate molecule. Enzymes with broad specificity have more flexible active site requirements, and therefore accept a wider range of substrate molecules. The active site amino acid residues participate directly in the catalytic reaction, and are largely responsible for the high catalytic power associated with enzyme reactions.

2.1.2 Catalytic Power. During any reaction the reactants briefly enter a state in which the susceptible substrate bonds are not completely broken, and the new bonds in the product are not completely formed. This transient condition is called the transition state, and it is energy dependent because energy is needed to make and break chemical bonds (350 kJ mol^{-1} for each covalent bond). This represents an energy barrier to successful reaction, and is the reason why most reactions proceed extremely slowly in the absence of external help.

Substrates can be helped towards the transition state by heat energy, high pressure or extreme pH to weaken bonds or by the addition of catalysts. Enzyme catalysts are more effective than chemical catalysts at reducing the energy barrier to enable transition state formation and increased the rate of a reaction. The efficiency of enzyme catalysis varies, but most enzymes can enhance the rate of an uncatalysed reaction by a factor in the range of 10^5 to 10^{14}. One of the most efficient enzymes is the enzyme carbonic anhydrase, which catalyses the hydration of up to 600 000 molecules of CO_2 per second under optimal conditions.

$$CO_2 + H_2O \xrightarrow{\text{carbonic anhydase}} H_2CO_3$$
$$\text{carbon dioxide} \quad \text{water} \qquad\qquad\qquad \text{carbonic acid}$$

2.2 Ribozymes

Ribozymes are RNA molecules with catalytic activity, and are probably the evolutionary forerunners of enzymes. A variety of ribozymes have been discovered, and the hammerhead ribozyme, which is one of the smallest ribozymes, can catalyse the site-specific hydrolysis of a phosphodiester bond in RNA.[4] Production of these new biological catalysts, which can be tailored to suit a specific biotransformation, is a major development that will increase the usefulness of biological catalysts. Currently, the biocatalysts available are those existing in nature, and applications revolve around the existing set of available biocatalysts. New applications will be possible if ribozymes can be engineered to suit a designed application.

Ribozymes have attracted great attention because comparative studies on the properties of enzymes and ribozymes will provide new fundamental information on the chemical/physical principles of biological catalysis.[5] Also ribozymes recognize their target RNA (substrate) in a highly sequence-specific manner, and this degree of specificity should enable development of important therapeutic applications for both inherited and acquired diseases.[6] For example, ribozymes may be used to counteract harmful gene expression by attacking gene products (mRNA) produced by faulty, damaged or uncontrolled genes.[7] Such targets might be the mRNA produced by oncogenes, viral genomes and mRNA from human immunodeficiency virus-type 1 (HIV-1).

2.3 Abzymes

Antibodies are proteins and, like enzyme and receptor proteins, they have binding domains (regions) where they bind other molecules. Antibodies bind molecules called antigens, and can bind small molecules (haptens) or small parts (determinants) of large molecular weight

[4] G. J. Narlikar and D. Herschlag, *Ann. Rev. Biochem.*, 1997, **66**, 19.
[5] W. G. Scott and A. Klug, *Trends Biochem. Sci.*, 1996, **21**, 220.
[6] J. J. Rossi, *Biodrugs*, 1998, **9**, 1.
[7] H. A. James and I. Gibson, *Blood*, 1998, **91**, 371.

antigens, which may be attached to cells. Catalytic antibodies or abzymes are an additional class of biocatalyst because they have distinct catalytic properties.[8] They catalyse a range of reactions including hydrolysis, cyclization reactions, elimination reactions, synthetic reactions, *etc.*

Comparative studies between enzymes and abzymes indicate that enzymes have varying levels of specificity built into their structure, while antibodies have much greater recognition qualities. It seems likely that enzymes and antibodies are molecular cousins that evolved from a common ancestor with catalytic and recognition properties. Enzymes probably evolved to concentrate on catalysis, and their recognition property adapted to provide specificity. Antibodies evolved the recognition property to a very high degree, and retained useful elements of catalysis. Comparison of the properties of enzymes and abzymes will provide new understanding of the selectivity processes inherent in enzyme specificity.[9]

Abzymes could have exciting applications in many areas of biotechnology, medicine, analytical biochemistry and molecular biology. In particular, enzyme specificity is limited to that already provided by nature, and often the limitation restricts the range of possible catalysis. With protein engineering it will be possible to alter and improve the *specificity* of key enzymes, and improve the *catalytic power* of useful abzymes. Antibodies have a vast repertoire of highly specific binding functions, which can be used to significantly increase the total portfolio of biocatalysis.

Biosensors are an obvious example where abzymes could provide new improvement in the recognition of analyte (molecule to be measured), and thereby increase the range of analytes that can be detected and measured. Enzyme-based biosensors are limited in scope, and tend to discriminate between small molecular weight molecules. Abzyme-based biosensors, on the other hand, have considerable scope, and will be able to discriminate between toxins, viruses, microbial cells, native and foreign proteins, nucleic acids, human cells, cancer cells, *etc.* Evolution has shaped antibodies for a particular role in nature so it is not surprising that abzymes have less catalytic power than enzymes, and have a tendency not to release product quickly after catalysis. However, unlike enzymes, antibodies have similar gross protein structural features so an immobilization method may be readily applied to different abzymes.

[8] B. S. Green and D. S. Tawfik, *Trends Biotechnol.*, 1989, **7**, 304.
[9] S. J. Benkovic and A. Ballesteros, *Trends Biotechnol.*, 1997, **15**, 385.

2.4 Multienzyme Complex

Enzyme-based catalysis has mostly been concerned with single enzymes performing single step reactions and simple biotransformations. While there is still much work to be done with single enzyme biotransformations, enzyme technologists are keen to move onto the next phase of biocatalysis, *i.e.* sequential multiple-step biotransformations as performed in biochemical pathways by cells and organelles. Multienzyme complexes are ordered assemblies of functionally and structurally different enzymes and proteins that catalyse successive steps in a biochemical reaction pathway. Increased understanding of such complexes has provided opportunities to take enzyme-based catalysis beyond single-step reaction processes to more sophisticated reaction schemes. A number of multienzyme complexes have been evaluated, and the examples below represent the use of multienzyme complexes in handling highly reactive intermediates, rapid recycling and waste utilization.

2.4.1 PDC. The pyruvate dehydrogenase complex (PDC) catalyses the oxidative decarboxylation of pyruvate in the formation of acetyl CoA. The complex from *Escherichia coli* has a molecular mass of 5000 kDa, and consists of 60 polypeptide chains tightly organized in a polyhedral structure with a diameter of 30 nm. The reaction process generates and uses highly reactive intermediates, and the complex is structured so that the reactive intermediates can be passed directly from one enzyme to the next with the minimum of deviation to ensure that reaction intermediates do not become side-tracked or lost in fruitless side reactions.

2.4.2 Proteosome. This is a large 2000 kDa multienzyme complex consisting of more than 25 different proteins, which range in molecular weight from 22 to 110 kDa. It is found in both the cytoplasm and the nucleus of eukaryotic cells, and structurally it appears (side-on-view) as a cylindrical stack of four rings with six or seven proteins in each ring.[10] The function of the complex is rapid hydrolysis of intracellular protein to enable fast turnover of proteins. This function is vital and mutations introduced into yeast cells that knockout the proteosome are lethal to the cells.[11] Fast recycling of protein may be important for cell differentiation, cell-cycle control, response to environmental stress and removal of damaged/faulty protein.

Other more specialized functions (in mammalian cells) may include

[10] J. M. Peters, *Trends Biochem. Sci.*, 1994, **19**, 377.
[11] A. J. Rivett, *Biochem. J.*, 1993, **291**, 221.

contribution to cell-mediated cytotoxicity in interleukin-2 activated natural killer cells. The complex contains both endopeptidase (cut the inside of protein to produce large fragments) and exopeptidase (cut at the ends of polypeptide fragments to produce peptides and amino acids), and is therefore capable of complete hydrolysis of protein to amino acids.[12]

2.4.3 Cellulosome. Cellulose is the most abundant source of carbon and chemical energy on the planet. However, it is a very stable polymer of glucose, which is usually intermixed with other polysaccharides such as xylan and lignin to produce a complex structure that is resistant to simple mechanisms of chemical or biological degradation. Cellulolytic microorganisms have evolved a multifunctional multienzyme complex called a cellulosome, which has been designed to provide a systematic hydrolysis of cellulose to release glucose sugars for growth and energy supply.[13] The cellulosome is therefore a biocatalytic tool used by microorganisms to crack a difficult problem, which is the utilization of cellulose for energy and growth. Such is the complexity of cellulose that its hydrolysis requires a concerted co-operation between various micro-organisms to unpick the complex polysaccharides such as lignin and xylan, and remove the degradation products of that disassembly before the power of the cellulosome can be unleashed on cellulose.

Typically the cellulosome is composed of a central protein skeleton onto which are attached link proteins. Attached to the link proteins are various enzymes associated with cellulose hydrolysis, and also attached are binding proteins that attach to the cellulose substrate and physically locate the complex onto the cellulose. A raft of enzymes have been identified including endoglucanases, exoglucanases, β-glucosidase, xylanase, ligninase and pectinase, and the amount and pattern of enzymes can be varied by a given microorganism to suit the different forms of cellulose produced by a given plant. Flexibility and adaptability are key features that are highly desirable in biocatalysis, and it is likely there are some very clever principles that can be derived from detailed studies on the cellulosome.

Given the abundance of cellulose on the planet it is not surprising that several thousand cellulolytic strains have been found among micro-organisms. Much work remains to be done on the fine detail of the structure and function of the cellulosome, and any new versions that might be awaiting discovery in other cellulolytic microorganisms.

[12] W. Hilt and D. H. Wolf, *Trends Biochem. Sci.*, 1996, **21**, 96.
[13] E. A. Bayer, E. Morag, and R. Lamed, *Trends Biotechnol.*, 1994, **12**, 379.

2.4.4 Multienzyme Complexes and Immobilization Technology. A particularly exciting prospect is that studies on multienzyme complexes such as proteosomes and cellulosomes will undoubtedly shape future methods of biocatalyst immobilization technology. It will be possible to make immobilization a more active participant in biotransformation reactions by introducing specificity to support materials *via* immobilized specific binding domains that ensure accurate binding to particular substrate materials so increasing substrate specificity.

It will be possible to mimic the cellulosome and immobilize an array of specific enzymes and regulatory proteins together for sophisticated multi-step biotransformations. The future will provide one modular universal support material onto which can be attached any number of binding, regulatory and catalytic protein modules that are required for a given biotransformation (Figure 2). In the future, successful development of these systems will enable production of cell-free complex processes, and will allow more sophisticated opportunities for enzyme-based biocatalysis such as versatile fuel cells, recycling of complex waste materials, multianalyte analytical biosensors, and cell-free biosynthesis of biopolymers such as proteins and carbohydrates (see Figure 2). A new development that will support this approach to biocatalysis is nanobiotechnology, and the availability of highly defined nanoparticles (see Section 3.1.1).

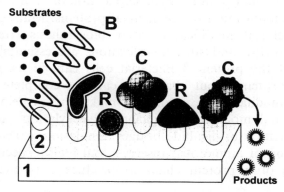

Figure 2   *Illustration of an immobilized modular multienzyme complex. The scheme shows a universal support material (1). Universal linker proteins (2) with flexible binding characteristics (similar to antibodies) bind and immobilize various catalytic (C) enzyme modules, regulatory (R) modules and substrate binding (B) modules to the support material, and create a programmed reaction pathway for multi-step conversion of substrates to products*

2.5   Cells

Cells represent a different class of biocatalyst in that they contain considerable arrays of molecular biocatalysts (such as enzymes), which they harness in sequential step reactions to catalyse extensive biotransformations (*e.g.* glucose $\rightarrow CO_2 + H_2O$). Therefore the range and scale of biocatalysis within a particular cell is vastly greater than that with particular enzymes, ribozymes, *etc.*[14] Cell biocatalysts have been obtained from animal, plant and microbial sources, and cell-based systems are particularly advantageous for multiple-step reactions, and reactions that involve energy transduction such as ATP hydrolysis.[15] At first sight it may seem that the need for enzymes is questionable. However, cells are living organisms, and their priority is life support, rather than completion of any external biotransformation. Consequently, biochemical resources and energy are used by cells for growth, and so they are not efficient biocatalysts for many simple biotransformations.[16] Cells are important biocatalysts for complex and multistep reactions, and the merits of various cells as biocatalysts are discussed briefly below.

2.5.1   Animal Cells.   Commercial applications for animal cells are increasing, and they are routinely used for expression of proteins from recombinant DNA. They are also used as hosts for attenuated strains of important viruses in the production of vaccines for foot-and-mouth disease, polio, rabies, measles and rubella, *etc.*[17]

A particularly important catalytic application of animal cells is post-translational modification of commercial proteins. Many key proteins receive biochemical modification after biosynthesis (translation of mRNA to protein), and such modifications are essential for proper functioning of the protein.[18] Bacterial cells can be used to express commercial proteins, but bacteria cannot complete the post-translational modifications needed to make the protein functional. Post-translational modifications include glycosylation (adding sugar residues), formation of disulfide bridges, amidation, carboxylation, or phosphorylation of amino acid residues, and highly specific protease-based cutting of a protein chain to produce a particular protein

[14] M. J. Núñez and J. M. Lema, *Enzyme Microb. Technol.*, 1987, **9**, 642.
[15] A. Groboillot, D. K. Boadi, D. Poncelet, and R. J. Neufeld, *Crit. Rev. Biotechnol.*, 1994, **14**, 75.
[16] S. Norton and J.-C. Vuillemard, *Crit. Rev. Biotechnol.*, 1994, **14**, 193.
[17] H. E. Swaisgood, in Enzymes and Immobilized Cells in Biotechnology, ed. A. I. Laskin, Benjamin/Cummings, London, 1985, pp. 1–24.
[18] S. Bidley, in Immobilized Cells and Enzymes: A Practical Approach, ed. J. Woodward, IRL Press, 1985, pp. 147–181.

shape.[19] However, animal cells are fragile, have special growth requirements, low product yields, and are susceptible to infection by bacteria and viruses. Animal cell culture is expensive and normally reserved for high-value medical products.

2.5.2 Plant Cells. In addition to crops, many plants produce compounds that have commercial value. Over three-quarters of the 32 000 + known natural products are derived from plants.[20] These include medicines and drugs such as atropine, morphine and digoxin; essential oils and fragrances such as menthol, strawberry, vanilla and camphor; pigments such as anthocyanin, betacyanin and saffron; and speciality products such as enzymes, fungicides, pesticides, peptides, vitamins and pigments.[21] Perhaps the most notorious plant products are the narcotics such as opium and morphine (which is an extract from opium).

However, development of plant biocatalysis has been slow due to the disadvantages associated with whole plant cultivation, such as weather requirements, low product yield, geographical complications, pesticide/herbicide requirements and expensive extraction processes. A particular problem relates to plant vacuoles that accumulate waste products. During cell disruption to release plant protein products, these waste materials are also released and contaminate the desired product causing disruption/inactivation of the plant protein product.[22]

Advances in plant genetic engineering have enabled a whole range of products, which were previously difficult to obtain from plants, to be produced in quantity from domesticated crop plants. Transgenic plants are beginning to compete with microbial cell systems for bulk production of biomolecules. Other improvements in plant cell tissue culture have enabled processes that provide a suitable alternative to whole plant cultivation for speciality low volume products.[23]

2.5.3 Microorganisms (Bacteria, Yeast and Filamentous Fungi). Microorganisms are without question the most versatile and adaptable forms of life, and this capability has enabled them to survive on this planet for over three billion years. A key feature in this remarkable success is the immense capacity that microorganisms have for biocatalysis. Man has made extensive use of the biocatalytic properties of

[19] K. Nilsson, *Trends Biotechnol.*, 1987, **5**, 73.
[20] P. Brodelius, in Enzymes and Immobilized Cells in Biotechnology, ed. A. I. Laskin, Benjamin/Cummings, London, 1985, pp. 109–148.
[21] A. C. Hulst and J. Tramper, *Enzyme Microb. Technol.*, 1989, **11**, 546.
[22] A. K. Panda, S. Mishra, V. S. Bisaria and S. S. Bhojwani, *Enzyme Microb. Technol.*, 1989, **11**, 386.
[23] O. J. M. Goddijn and J. Pen, *Trends Biotechnol.*, 1995, **13**, 379.

microorganisms in the production of beverages and foodstuff, and this is a common factor shared by all communities in the world.

Advances in microbiological sciences over the past 60 years have revealed the enormous biocatalytic potential inherent in microorganisms, and this has stimulated new contributions to medicine, agriculture, waste (water and hazardous) management, and animal feed. It is readily apparent that microorganisms have considerably more potential for biotransformation of a wide range of organic and biochemicals, and current opinion suggests that less than 2% of the total range of microorganisms on the planet have been thoroughly characterized in terms of biocatalytic potential. As biocatalysts, some microorganisms are producers of organic material (autotrophs), while some are consumers of organic material (heterotrophs), and although bacteria are prokaryotic (*i.e.* they lack nuclear membrane, mitochondria, endoplasmic reticulum, Golgi apparatus and lysosomes) they possess sophisticated and flexible catalytic systems for both biosynthesis and biodegradation. Although easier to cultivate than animal or plant cells, they are not without some drawbacks as biocatalysts. A critical concern is safety, and assurance that products are free from residual bacterial/fungal toxins/cells.

2.6 Biocatalyst Selection

The choice of catalyst for a particular biotransformation will be influenced by a number of factors, but three are fundamental. The first factor will centre on the availability of the biocatalyst, and this may range from pure enzyme, partially purified enzyme, crude enzyme preparation, cell/organelle extract, dead whole cells, to living whole cells, which may be growing or resting cultures. The next important factor is the bioreactor, and in particular the configuration and operating conditions. Many configurations are available, and these considerations will be influenced by biocatalyst characteristics, properties of the substrate and requirements for the finished product. The third key factor is the need or not of the additional features provided by immobilization, and this will depend on the choices made for the first two factors.

3 IMMOBILIZATION

In solution, biocatalysts behave as any other solute in that they are readily dispersed in the solution or solvent, and have complete freedom of movement in the solution (Figure 3). Immobilization may be considered as a procedure specifically designed to limit the freedom of

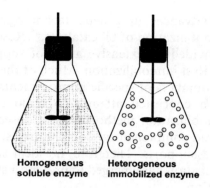

Homogeneous   **Heterogeneous**
soluble enzyme   **immobilized enzyme**

Figure 3 *Two-phase system generated by immobilization*

movement of a biocatalyst. Immobilization normally involves attachment of the biocatalyst to, or location within, an insoluble support material.[3] In effect the biocatalyst is separated from the bulk of the solution to create a heterogeneous two-phase system (Figure 3). Immobilization provides three basic advantages over soluble-based systems.

- Repeated use of biocatalyst in batch reactions.
- Simple separation of biocatalyst from product after reaction.
- Continuous use of biocatalyst in a continuous reactor system.

An immobilized biocatalyst is easily recovered after a batch reaction, and this facility allows repeated use of the biocatalyst in a fresh batch. This advantage can improve the commercial viability of a reaction process that was previously too expensive in terms of biocatalyst costs. Removal of the biocatalyst from the product solution after the reaction is complete means that the final product is not contaminated with biocatalyst, and extraction procedures are not required in the finishing steps. Immobilization of a biocatalyst will increase the options for bioreactor selection. The immobilized biocatalyst can be incorporated into cylindrical flow-through columns, or other devices to produce continuous operation bioreactors in which a continuous flow of substrate enters at one end of the bioreactor, and a continuous flow of product emerges from the other end.[24] Such systems can operate continuously for weeks or months.

Recent expansion of biotechnology, and expected developments that

[24] H. E. Swaisgood and F. M. L. Passos, in Immobilization of Enzymes and Cells, ed. G. F. Bickerstaff, Humana Press, New Jersey, 1997, pp. 237–242.

will accrue from advances in genetic technology, have stimulated enthusiasm for immobilization of biocatalysts.[25] Research and development work has provided an extensive array of support materials and methods for biocatalyst immobilization. Much of the expansion may be attributed to large numbers of specific improvements for given applications rather than a concerted attempt to build a set of industrial standards for immobilization methodology. Consequently there have been few detailed and comprehensive comparative studies on immobilization methods and supports. Therefore, no ideal support material or method of immobilization has emerged to provide a standard for a given type of immobilization.

Selection of support material and method of immobilization is made by weighing the various characteristics and required features of the biocatalyst application against the properties/limitations/characteristics of the combined immobilization/support. A number of practical aspects should be considered before embarking on experimental work to ensure that the final immobilized biocatalyst is fit for the planned purpose or application, and will operate with optimum effectiveness.[26]

3.1 Choice of Support Material

A decision on the type of support material for immobilization will require careful evaluation of the proposed use of the biocatalyst and characteristics of the intended application.[27] Virtually any inorganic, organic, or biological material can be used or modified for use as a support material for immobilization of biocatalysts (Table 1).[28] The choice of support will be influenced by some of the support properties listed below, and the evaluation process should help to inform which of the properties are important for the immobilized biocatalyst under consideration.

- Various physical properties of the support may be important such as strength, non-compression of particles, available surface area, shape/form (beads/sheets/fibres, *etc.*), non-porous, porous (degree of porosity), pore volume, permeability, density, space for increased biomass, flow rate and pressure drop.

[25] W. H. Scouten, J. H. T. Luong and R. S. Brown, *Trends Biotechnol.*, 1995, **13**, 178.
[26] G. F. Bickerstaff, in Immobilization of Enzymes and Cells, ed. G. F. Bickerstaff, Humana Press, New Jersey, 1997, pp. 1–11.
[27] R. F. Taylor, in Protein Immobilization, ed. R. F. Taylor, Marcel Dekker, New York, 1991, pp. 139–160.
[28] P. Gemeiner, in Enzyme Engineering, ed. P. Gemeiner, Ellis Horwood, New York, 1992, pp. 13–119.

Table 1 *Some support materials for biocatalyst immobilization*

Inorganic	Organic	Biological
Aluminium oxide	Polyethylene	Cellulose
Nickel oxide	Polystyrene	Dextran
Stainless steel	Polyacrylate	Agarose
Porous glass	Nylon	Starch
Porous silica	Polyacrylamide	Alginate
Activated carbon	Polymethacrylate	Carrageenan
Diatomaceous earth	Polypyrrole	Chitin
Iron oxide	Polyaniline	Bone
Titanium oxide	Polyphenol	Chitosan
Pumice stone	Polyester	Collagen, gelatin
Zirconium oxide	Polyvinyl alcohol	Liposome
Vanadium oxide		Cells (yeast, red blood)

- Chemical features are basic considerations, and include hydrophilicity (water binding by the support), inertness towards the biocatalyst, and available functional groups for modification. Regeneration and re-use of the support may be important.
- The stability characteristics of the support material may influence storage, residual enzyme activity, cell productivity, regeneration of biocatalyst activity, maintenance of cell viability, and mechanical integrity of the support.
- A resistance to bacterial/fungal attack, disruption by chemical/pH/ temperature/organic solvent, cellular defence mechanisms (antibodies/killer cells) may be required if the biocatalyst will operate in a complex environment.
- Operational safety considerations such as biocompatibility (immune response), toxicity of component or residual chemical reagents may be important if the support will come into direct or indirect contact with food, biological fluids, *etc.*
- There may be health and safety concerns with the support itself or support modification procedures, and may be important for process workers, end product users, and immobilized biocatalyst GRAS (generally recognized as safe) requirements for FDA approval (for food/pharmaceutical/medical applications).
- Key economic concerns may be raised concerning availability and cost of the support material, and/or chemicals, special equipment/ reagents, technical skill required, environmental impact, industrial scale chemical preparation, feasibility for scale-up, continuous processing, effective working life, re-useable support, contamination (enzyme/cell-free product).

- Overall reaction characteristics influenced by the support will include flow rate, enzyme/cell loading, catalytic productivity, reaction kinetics, side reactions, multiple enzyme and or cell systems, batch or continuous operation system, reactor type, and diffusion limitations on mass transfer of cofactors/substrates/products.

3.1.1 Next Generation of Support Material. A range of support materials are becoming available that will add significant new possibilities for immobilization of biocatalysts. Developments in the computer microelectronics industry have advanced the manufacture of nanostructure materials, and utilization of such materials in biotechnology is likely to lead to novel immobilized biocatalysts.[29] Polymeric microspheres (nanoparticles) have been produced with sizes rangeing from 50 nm to 10 000 nm, and are used to covalently bind enzymes, antibodies and proteins.[30] Advantages of these particles have been observed in the development of biosensors. Further indications are that nanoparticles may be engineered to provide cell-specific delivery of therapeutic agents such as enzymes, proteins and drugs after injection into the bloodstream.[31] As the secrets unfold from natural biocatalytic structures such as the cellulosome and the proteosome, it is envisaged that nanobiotechnology will enable design and production of biomolecular nanostructure arrays of sophisticated biocatalysts for the next generation of biotransformations as depicted in Figure 2.

3.2 Choice of Immobilization Procedure

There are five principal methods for immobilization of biocatalysts: adsorption, covalent binding, entrapment, encapsulation and crosslinking (Figure 4), and the relative merits of each are discussed below.

3.2.1 Adsorption. Immobilization by adsorption is the simplest method, and involves reversible non-covalent interactions between biocatalyst and support material (Figure 4).[32] The forces involved are mostly electrostatic such as van der Waals forces, and ionic and hydrogen bonding interactions, although hydrophobic bonding can be significant. These forces are very weak, but sufficiently large in number to provide reasonable binding. For example, it is known that yeast cells have a surface chemistry that is substantially negatively charged so that use of a positively charged support will enable immobilization. Existing

[29] Y. Dolitzky, S. Sturchak, B. Nizan, B.-A. Sela, and S. Margel, *Anal. Biochem.*, 1994, **220**, 257.
[30] J.-M. Laval, J. Chopineau, and D. Thomas, *Trends Biotechnol.*, 1995, **13**, 474.
[31] S. S. Davis, *Trends Biotechnol.*, 1997, **15**, 217.
[32] J. Woodward, Immobilized Cells and Enzymes: A Practical Approach, IRL Press, 1985.

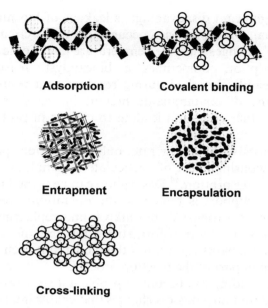

Figure 4 *Principal methods of biocatalyst immobilization*

surface chemistry on the biocatalyst and on the support is utilized for adsorption, so chemical activation or modification of the support is not required, and little damage is inflicted on biocatalysts in this method of immobilization.[33] The procedure consists of mixing together the biocatalyst and a support with adsorption properties, under suitable conditions of pH, ionic strength, *etc.*, for a period of incubation, followed by collection of the immobilized biocatalyst, and extensive washing to remove non-bound biocatalyst.

Advantages
- Little or no damage to the biocatalyst.
- Simple, cheap and quick to obtain immobilization.
- No chemical changes required for the support or biocatalyst.
- Easily reversed to allow regeneration with fresh biocatalyst.

Disadvantages
- Leakage of biocatalyst from the support/contamination of product.
- Non-specific binding of other material to the support.
- Overloading of the support with biocatalyst.
- Steric hindrance of the biocatalyst by the support.

[33] W. H. Scouten, *Methods Enzymol.*, 1987, **135**, 19.

The most significant disadvantage is leakage of biocatalyst from the support material. Desorption can occur under many circumstances, and environmental changes in pH, temperature and ionic strength will promote desorption. Sometimes a biocatalyst, reasonably firmly adsorbed, is readily desorbed during reaction as a result of substrate binding, binding of contaminants present in the substrate, product production or other condition leading to change in protein conformation.

Physical factors such as flow rate, bubble agitation, particle-particle abrasion and scouring effect of particulate materials on reactor vessel walls can lead to desorption.[34] Desorption can be turned to advantage if regeneration of support is built into the operational regime to allow rapid expulsion of 'exhausted' biocatalyst and replacement with fresh biocatalyst. In some biotransformations the use of crude substrate containing contaminants can poison the biocatalyst such that regeneration is an essential part of the reaction process.

Non-specific binding can become a problem if substrate, product and or residual contaminants possess charges that enable interaction with the support. This can lead to diffusion limitations and reaction kinetics problems, with consequent alteration in enzyme kinetic parameters (see Section 4). Furthermore, binding of hydrogen ions to the support material can result in an altered pH microenvironment around the support with consequent shift in pH optimum (1–2 pH units) of the biocatalyst, and this may be detrimental to enzymes with precise pH requirements. Unless carefully controlled, overloading the support with excess biocatalyst can lead to low catalytic activity, and the absence of a suitable spacer between the enzyme molecule and the support can produce problems of steric hindrance (see Section 4).

3.2.2 Covalent Binding.   This method of immobilization involves the formation of a covalent bond between the biocatalyst and a support material (Figure 4). The bond is normally formed between functional groups present on the surface of the support and functional groups belonging to amino acid residues on the surface of the enzyme. A number of amino acid functional groups are suitable for participation in covalent bond formation. Those most often involved are: the amino group (NH_2) of lysine or arginine; the carboxyl group (CO_2H) of aspartic acid or glutamic acid; the hydroxyl group (OH) of serine or threonine, and the thiol group (SH) of cysteine.[3]

Many varied support materials are available for covalent binding, and

[34] M. P. J. Kierstan and M. P. Coughlan, in Protein Immobilization, ed. R. F. Taylor, Marcel Dekker, New York, 1991, pp. 13–71.

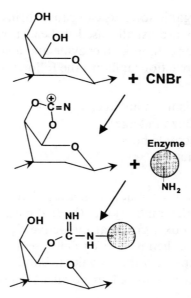

Figure 5 *Covalent immobilization showing activation of a carbohydrate support material with CNBr, and coupling of an enzyme with free amino group to the activated support to form a covalent isourea bond*

the extensive range of supports available reflects the fact that no ideal support exists (see Table 1). Therefore, the advantages and disadvantages of a given support must be taken into account when considering possible procedures for a given enzyme immobilization.

As indicated above, many factors may influence the selection of a particular support, and research work has shown that hydrophilicity is the most important factor for maintaining enzyme activity in a support environment. Consequently polysaccharide polymers, which are very hydrophilic, are popular support materials for enzyme immobilization.[35] For example, cellulose, dextran (trade name Sephadex), starch and agarose (trade name Sepharose) are used for enzyme immobilization. The sugar residues in these polymers contain hydroxyl groups, which are ideal functional groups for chemical activation to provide covalent bond formation (Figure 5). Also, hydroxyl groups form hydrogen bonds with water molecules, and thereby create an aqueous (hydrophilic) environment in the support.

However, polysaccharide supports are susceptible to microbial/fungal

[35] M. Paterson and J. F. Kennedy, in Immobilization of Enzymes and Cells, ed. G. F. Bickerstaff, Humana Press, New Jersey, 1997, pp. 153–165.

disintegration, and organic solvents can cause shrinkage of polysaccharide gels. The supports are usually used in bead form. There are many reaction procedures for coupling an enzyme and a support in a covalent bond, however, most reactions fall into the following categories:

- Formation of an isourea linkage.
- Formation of a diazo linkage.
- Formation of a peptide bond.
- An alkylation reaction.

It is important to choose a reaction method that will not inactivate the enzyme by reacting with amino acids at the enzyme active site. Thus, if an enzyme employs a carboxyl group at the active site for participation in catalysis (see Figure 1), then it is wise to choose a reaction method that involves amino groups for the covalent bond with the support.[36] Basically, two steps are involved in covalent binding of enzymes to support materials.

First, functional groups on the support material are activated by a specific reagent and, second, the enzyme is added in a coupling reaction to form a covalent bond with the support material. Normally the activation reaction is designed to make the functional groups on the support strongly electrophilic (electron deficient). In the coupling reaction, these groups will react with strong nucleophiles (electron donating), such as the amino (NH_2) functional groups of certain amino acids on the surface of the enzyme, to form a covalent bond. It is possible to vary the coupling conditions to allow single- or multiple-point attachment (see Figure 4), and influence the stability of the immobilized enzyme (see Section 4).

Cyanogen bromide (CNBr) can be used to activate the hydroxyl functional groups in polysaccharide support materials (see Figure 5). In this method the enzyme and support are joined *via* an isourea linkage. In carbodiimide activation the support material should have a carboxyl (CO_2H) functional group, and the enzyme and support are joined *via* a peptide bond. If the support material contains an aromatic amino functional group, it can be diazotized using nitrous acid. Subsequent addition of enzyme leads to the formation of a diazo linkage between the reactive diazo group on the support and the ring structure of an aromatic amino acid such as tyrosine.

It is important to recognize that no method of immobilization is

---

[36] M. F. Cardosi, in Immobilization of Enzymes and Cells, ed. G. F. Bickerstaff, Humana Press, New Jersey, 1997, pp. 217–227.

restricted to a particular type of support material, and that an extremely large number of permutations are possible between methods of immobilization and support material. This is made possible by chemical modification of normal functional groups on a support material to produce a range of derivatives containing different functional groups. For example, the normal functional group in cellulose is the hydroxyl groups (OH) of glucose, and chemical modification of these groups have produced a range of cellulose derivatives such as AE-cellulose (aminoethyl), CM-cellulose (carboxymethyl) and DEAE-cellulose (diethylaminoethyl). Thus chemical modification increases the range of immobilization methods that can be used for a given support material.[37]

3.2.3 Entrapment. Immobilization by entrapment differs from adsorption and covalent binding in that enzyme molecules are free in solution, but restricted in movement by the lattice structure of a gel (Figure 4). The porosity of the gel lattice is controlled to ensure that the structure is tight enough to prevent leakage of enzyme or cells, but at the same time allow free movement of substrate and product. Inevitably the support will act as a barrier to mass transfer and, although this can have serious implications for reaction kinetics, it can have useful advantages as harmful cells, proteins, and enzymes are prevented from interaction with the immobilized biocatalyst. There are several major methods of entrapment:

- Ionotropic gelation with multivalent cations (*e.g.* alginate).
- Temperature-induced gelation (*e.g.* agarose, gelatin).
- Organic polymerization by chemical/photochemical reaction (*e.g.* polyacrylamide).
- Precipitation from an immiscible solvent (*e.g.* polystyrene).

Entrapment can be achieved by mixing an enzyme with a polyionic polymer material, and then cross-linking the polymer with multivalent cations in an ion-exchange reaction to form a lattice structure that traps the biocatalyst (ionotropic gelation).[38] Temperature change is a simple method of gelation by phase transition using 1–4% solutions of agarose or gelatin, however the gels formed are soft and unstable. A significant development in this area has been the introduction of κ-carrageenan polymers, which can form gels by ionotropic gelation and by tempera-

[37] J. M. Guisan, V. Rodriguez, C. M. Rosell, G. Soler, A. Bastida, R. M. Blanco, R. Fernandez-Lafuente, and E. Garcia-Junceda, in Immobilization of Enzymes and Cells, ed. G. F. Bickerstaff, Humana Press, New Jersey, 1997, pp. 289–298.

[38] J. E. Fraser and G. F. Bickerstaff, in Immobilization of Enzymes and Cells, ed. G. F. Bickerstaff, Humana Press, New Jersey, 1997, pp. 61–66.

ture-induced phase transition, and this has introduced a greater degree of flexibility in gelation systems for immobilization.

Alternatively, it is possible to mix the enzyme with chemical monomers that are then polymerized to form a cross-linked polymeric network, trapping the enzyme in the interstitial spaces of the lattice. The latter method is more widely used, and a number of acrylic monomers are available for the formation of hydrophilic co-polymers. For example, acrylamide monomer is polymerized to form polyacrylamide, and methyl acrylate is polymerized to form polymethacrylate. In addition to the monomer, a cross-linking agent is added during polymerisation to form cross-linkages between the polymer chains and help to create a three-dimensional network lattice. The pore size of the gel and its mechanical properties are determined by the relative amounts of monomer and cross-linking agent. It is therefore possible to vary these concentrations to influence the porosity of the lattice structure. The formed polymer may be broken-up into particles of a desired size or polymerization can be arranged to form beads of defined size to suit particular reactor requirements. Precipitation occurs by phase separation rather than by chemical reaction, but does bring the biocatalyst into contact with a water-miscible organic solvent; most biocatalysts are not tolerant of such solvents. Thus this method is limited to highly stable or previously stabilized enzymes or non-living cells.

3.2.4 Encapsulation. Encapsulation of enzymes and/or cells can be achieved by enveloping the biological components within various forms of semi-permeable membrane (Figure 4). This process is similar to entrapment in that the enzymes/cells are free in solution, but restricted in space. Large proteins or enzymes cannot pass out of or into the capsule, but small substrates and products can pass freely across the semi-permeable membrane. Many materials have been used to construct microcapsules varying from 10 μm to 100 μm in diameter, and nylon membranes and cellulose nitrate membranes have proved popular.[39]

The problems associated with diffusion of substrates in and products out are more acute, and may result in rupture of the membrane if products from a reaction accumulate rapidly. A further problem may arise if the immobilized biocatalyst particle has a density fairly similar to that of the bulk solution and therefore floats. This can cause process problems and may require reassessment of reactor configuration, flow dynamics, *etc.*

It is also possible to use biological cells as capsules, and a notable

[39] C. K. Colton, *Trends Biotechnol.*, 1996, **14**, 158.

example of this is the use of erythrocytes (red blood cells). The membrane of the erythrocyte is normally only permeable to small molecules. However, when erythrocytes are placed in a hypotonic solution they swell up, stretching the cell membrane and substantially increasing the permeability. In this condition, erythrocyte proteins diffuse out of the cell and enzymes can diffuse into the cell. Returning the swollen erythrocytes to an isotonic solution enables the cell membrane to return to its normal state, and the enzymes trapped inside the cell do not leak out.

3.2.5 Cross-linking. This type of immobilization is support-free and involves joining the cells (or the enzymes) to each other to form a large three-dimensional complex structure, and can be achieved by chemical or by physical methods (Figure 4). Chemical methods of cross-linking normally involve covalent bond formation between the cells by means of a bi- or multifunctional reagent such as glutaraldehyde and toluene diisocyanate.[40] However, the toxicity of such reagents is a limiting factor in applying this method to living cells and many enzymes. Both albumin and gelatin have been used to provide additional protein molecules as spacers to minimize the close proximity problems that can be caused by cross-linking a single enzyme.

Physical cross-linking of cells by flocculation is well known in biotechnology industry and leads to high cell densities. Flocculating agents such as polyamines, polyethyleneimine, polystyrene sulfonates and various phosphates have been used extensively and are well characterized. Cross-linking is rarely used as the only means of immobilization because the absence of mechanical properties and poor stability are severe limitations. Cross-linking is most often used to enhance other methods of immobilization, normally by reducing cell leakage in other systems.

A recent and potentially important development is the production of cross-linked enzyme crystals (CLEC).[41] Production of enzyme crystals has long been seen as the final step in enzyme purification procedures as it signifies high purity. The principal use of enzyme crystals *per se* has been in X-ray crystallography to study enzyme structures, and the recent realization that enzyme crystals were catalytically active has led to the development of CLECs.

Indications are that CLEC have superior stability characteristics, and may enable novel enzyme-catalysed biotransformations to occur in

[40] M. Koudelka-Hep, N. F. de Rooij and D. Strike, in Immobilization of Enzymes and Cells, ed. G. F. Bickerstaff, Humana Press, New Jersey, 1997, pp. 83–85.
[41] A. L. Margolin, *Trends Biotechnol.*, 1996, **14**, 223.

difficult environments such as in organic solvents, in the gas phase, or in supercritical fluids. Enzyme crystals might be more easily incorporated into microelectronic devices such as CHEMFET (chemically sensitive field effect transistor) and ISFET (ion selective field effect transistor) to produce novel bio-chips for the next generation of new biosensors.

4 PROPERTIES OF IMMOBILIZED BIOCATALYSTS

When a biocatalyst is immobilized its fundamental characteristics are usually changed in one way or another, and the change may be a drawback or an improvement. The nature of the alteration depends on the inherent properties of the biocatalyst and additional characteristics imposed by the support material on the biocatalyst, substrate and product.[42]

It is very difficult to quantify these properties and characteristics given the diversity of biocatalysts, support materials and methods of immobilization. Consequently, it has proved impossible to completely predict what effect a particular immobilization will have on an enzyme or the reaction that it catalyses, and the only recourse is to evaluate a number of methods to discover the system that provides the greatest positive improvement for the application under consideration.[43]

4.1 Stability.   The two most important properties that may be changed by immobilization are enzyme stability and catalytic activity. Stability is defined as an ability to resist alteration, and in the context of biocatalyst stability it is important to distinguish several different mechanisms for loss of stability.

- Inactivation by heat.
- Disruption by chemicals.
- Digestion by proteases or cells.
- Inactivation by change in pH.
- Loss of catalytic activity during storage.
- Loss of catalytic activity due to process operations.

The various types are not necessarily interdependent, and an observed increase in heat stability does not indicate that there will be a corresponding increase in storage stability or operational stability. Although immobilization does not guarantee an improvement in stability, it is

[42] D. S. Clark, *Trends Biotechnol.*, 1994, **12**, 439.
[43] J. Rudge and G. F. Bickerstaff, *Biochem. Soc. Trans.*,1984, **12**, 311.

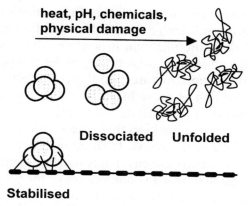

Figure 6 *Stabilization of an enzyme against protein denaturation by multi-point attachment to a support material*

widely recognized that it does represent a strategy, which can be used as a means of developing more stable enzyme preparations.[44]

Generally it is found that covalent immobilization is more effective than the other methods of immobilization at improving enzyme resistance to heat, chemical disruption and pH changes. Disruptants normally induce loss of catalytic activity by causing a considerable alteration in the protein structure of an enzyme (Figure 6). In particular, disruptants disperse the many non-covalent bonds responsible for holding an enzyme polypeptide chain in its highly specific shape or conformation, thus causing the polypeptide chain to unfold with consequent loss of active site structure and catalytic activity. Given that unfolding is associated with loss of activity, it is probable that multi-point attachment of an enzyme polypeptide chain to a support material provides extra rigidity to the folded protein chain, and therefore greater resistance to protein unfolding (Figure 6).

4.2 Catalytic activity. Immobilization almost invariably changes the catalytic activity of an enzyme, and this is clearly reflected in alterations in the characteristic kinetic constants of the enzyme catalysed reaction. In particular, the maximum reaction velocity (V_{max}) obtained with an immobilized enzyme is usually lower than that obtained with the corresponding soluble enzyme under the same reaction conditions. The Michaelis constant (K_m), which reflects the affinity that the enzyme has for its substrate, is usually changed upon immobilization, indicating that the binding of substrate to the active site has been altered. Four principal factors influence the catalytic activity of immobilized enzymes, namely (i) conformation, (ii) steric, (iii) micro-environment, and (iv) diffusion.

[44] J. Toher, A. M. Kelly and G. F. Bickerstaff, *Biochem. Soc. Trans.*, 1990, **18**, 313.

The conformation of an enzyme refers to the particular shape adopted by the polypeptide chain, which is essential for maintaining the active site structure (Figure 1). Immobilization procedures that involve modification or interaction with amino acid residues on the polypeptide chain can sometimes disturb protein structure, and thereby affect enzyme activity. Covalent immobilization is most likely to cause an alteration in the protein conformation of an enzyme. A steric problem arises if the enzyme is immobilized in a position that causes the active site to be less accessible to the substrate molecules. In solution, a free enzyme molecule is surrounded by a homogeneous micro-environment in which the enzyme is fully integrated with all components of the solution.

Immobilization creates a heterogeneous micro-environment consisting of two phases, *i.e.* the immobilized enzyme and the bulk of the solution from which the immobilized enzyme is separated (Figure 3). Therefore, all components of the reaction, substrate, products, activators, ions, *etc.*, are partitioned between the immobilized enzyme phase and the bulk solution phase. This problem can significantly alter the characteristics of an enzyme reaction even if the enzyme molecule itself is not changed by immobilization. The support material may influence the partitioning effect. If the support material attracts the substrate then this can improve the catalytic activity. Reaction rate is also reduced by diffusion restriction. As the substrate is consumed, more substrate must diffuse into the enzyme from the bulk solution, and product must diffuse away from the active site. This is normally a problem for all forms of immobilized enzymes, but particularly so for encapsulated enzymes.

Diffusional limitations may be divided into two types, external diffusion restriction and internal diffusion restriction. The external type refers to a zone or barrier that surrounds the support material, called the Nernst layer. Substrate molecules can diffuse into this layer by normal convection, and by a passive molecular diffusion. If substrate molecules pass through this layer slowly, then this may limit the rate of enzyme reaction. External diffusion restriction can be improved by speeding up the flow of solvent over and through the immobilized enzyme by increasing the stirring rate.

Internal diffusion restrictions are due to a diffusion limitation inside the immobilized enzyme preparation. In this case diffusion of substrate molecules occurs by a passive molecular mechanism only, which may be more difficult to overcome if it is a seriously limiting factor. The overall rate of diffusion is markedly influenced by the method of immobilization. Covalent and adsorption procedures cause less diffusion limitation than do entrapment and encapsulation procedures.

Table 2 *Some applications of immobilized biocatalysts*

Biotransformation	Biocatalyst	Immobilized Biocatalyst
Food industry		
Glucose → fructose	Bacterial cells	*Arthrobacter* cells/polyelectrolyte complex
Glucose → fructose	Glucose isomerase	Enzyme/polystyrene beads
Lactose → glucose + fructose	Lactase enzyme	Enzyme/alginate beads
Aspartame biosynthesis	Thermolysin	Enzyme/hydrophobic beads
Starch → maltose	Alpha amylase	Enzyme/hollow fibres
Coagulation of casein in milk	Chymosin	Calf stomach crude extract/ chitosan gel
Coagulation of casein in milk	Chymosin	*Aspergillus* recombinant enzyme/ alginate beads
Pharmaceutical industry		
6-APA → ampicillin	Bacterial cells	*Bacillus magaterium*/DEAE-cellulose
Glucose → penicillin G	Bacterial cells	*Penicillium chrysogenum*/ polyacrylamide gel
Digitoxin → digoxin	Plant cells	*Digitalis lanata*/alginate beads
Human gamma interferon	Animal cells	Chinese hamster ovary/alginate
L-Dopa biosynthesis	Tyrosinase	Enzyme/nylon 6,6 mesh
Tetracycline biosynthesis	Bacterial cells	*Streptomyces aureofaciens*/ carrageenan
Medicine		
Cancer therapy	Neuraminidase	Enzyme/gelatin membrane
Cystic fibrosis lung decongestion	DNAase	Enzyme/dextran
Phenylketonuria	Phenylalanine hydroxylase	Enzyme/agarose
Acatalasaemia	Catalase	Enzyme/microspheres
Treatment of burns and ulcers	Collagenase	Enzyme/cotton fibres
Drug detoxification (overdose)	Cytochromes P450	Enzymes/hollow fibres
Pancreatic insufficiency	Digestion enzymes	Enzyme extract/enteric-coated micropsheres
Adenosine deaminase deficiency	Adenosine deaminase	Enzyme/polyethylene glycol
Biosensors		
Serum cholesterol level	Cholesterol esterase	Enzyme/nylon membrane
Serum glucose level	Glucose dehydrogenase	Enzyme/collagen membrane
Serum alcohol level	Alcohol oxidase	Enzyme/cross-linked polymer
Diamine level in foodstuff	Diamine oxidase	Enzyme/cross-linked with glutaraldehyde
NADH level	NADH oxidase	Enzyme/Immobilon AV membrane
Glucuronide-drug levels	Animal cell microsomes	Microsomes/organic membrane

(continued overleaf)

Table 2   *continued*

Biotransformation	Biocatalyst	Immobilized Biocatalyst
Biotechnology		
Urocanic acid biosynthesis	Bacterial cells	*Achromobacter liquidum*/ polyacrylamide gel
Coenzyme A biosynthesis	Bacterial cells	*Brevibacterium ammoniagenes*/ cellophane
Hydrolysis of palm oil	Lipase	Enzyme/zeolyte type Y
Nitrate removal (groundwater)	Bacterial cells	*Scenedesmus obliquus*/polyvinyl foam
Clarification of fruit juice	Pectinesterase	Enzyme/agarose
Amino acid biosynthesis	Aminoacylase	Enzyme/aluminium oxide
Cellulose hydrolysis	Cellulosome complex	Multi-enzyme complex/calcium phosphate gel

5   APPLICATIONS

Immobilization technology is now firmly established[45] among the tools used by scientists to improve the effectiveness of biocatalysts for biotransformations in industry, analysis and medicine.[46] Consequently there are many examples[47] of applications that use immobilized bio-catalysts, and further information can been obtained from reviews,[48,49] and books.[2,32] A few examples are given in Table 2 that indicate the wide range of applications and associated immobilized biocatalysts. The future for immobilized biocatalysts is promising, and exciting developments in this field will support the expected expansion and progress of biotechnology in various industries and medicine in the 21st century. New biocatalysts such as ribozymes and abzymes will expand the range of novel biocatalysis, and immobilized multienzyme complexes will mimic biochemical pathways to offer complex biotransformations and produce more sophisticated products.

[45] M. K. Turner, *Trends Biotechnol.*, 1995, **13**, 253.
[46] W. D. Crabb and C. Mitchinson, *Trends Biotechnol.*, 1997, **15**, 349.
[47] C. P. Champagne, C. Lacroix and I. Sodini-Gallot, *Crit. Rev. Biotechnol.*, 1994, **14**, 109.
[48] M. D. Trevan and A. L. Mak, *Trends Biotechnol.*, 1988, **6**, 68.
[49] A. Wiseman, *J. Chem. Tech. Biotechnol.*, 1993, **56**.

CHAPTER 17

Downstream Processing: Protein Extraction and Purification

MIKE D. SCAWEN and P. M. HAMMOND

1 INTRODUCTION

Enzymes are employed as diagnostic reagents in clinical chemistry, as catalysts in industrial processes and increasingly as therapeutic agents in chemotherapy. Recent advances in molecular genetics have resulted in an increased awareness of the importance of protein recovery and purification.

Over the years, the increasing use of microorganisms as a source of proteins, particularly enzymes, has led to improved efficiency in production and a more reproducible product. The great majority of enzymes in industrial use are extracellular proteins from organisms like *Aspergillus* sp. and *Bacillus* sp., and include α-amylase, β-glucanase, cellulase, dextranase, proteases and glucoamylase. Many of these are still produced from the original, wild-type strains of microorganism. However, in the production of proteins for use in the fields of clinical diagnosis and for therapeutic applications, genetic and protein engineering play an ever increasing role. Recombinant DNA technology, besides permitting vast improvements in yields, has allowed the transfer of genetic material from animal to bacterial hosts. In this way, those proteins once available in minute amounts from animal tissues can now be produced in virtually limitless quantities from easily grown bacteria. One such example is human growth hormone; this was once produced in tiny amounts from human pituitaries, until it was recognized to present a potential risk to the patient from a contaminating prion which has been implicated in Creutzfeld–Jacob syndrome. Human growth hormone is now produced in far greater quantities from the bacterium *Escherichia coli,* and is completely free from the unwanted prion.

Enzymes or proteins produced by microorganisms may be intracellular, periplasmic or secreted into the culture medium. For extracellular enzymes, the degree of purification required is often minimal, as the final product is intended for industrial use and does not have to be of high purity. Such large-scale processes may yield tonnes of protein product.

Many other enzymes are produced in more complex intracellular mixtures and represent a greater challenge to the protein purification scientist. Those produced for therapeutic use must attain very high and exacting standards of purity and to achieve this it may be necessary to develop complex purification protocols. Again, recombinant DNA technology has been able to help in this area. In the case of therapeutic proteins, there are often advantages if the protein of interest can be secreted, either into the periplasm or into the medium. This results in an enormous reduction in the level of contaminating proteins and other macromolecules, such that pure product can often be obtained in only two or three steps of purification. Choice of a suitable expression system can result in higher yields in the fermenter, thereby improving the specific activity of the starting material. It is also possible to add groups to a protein which aid its purification by conferring specific properties and subsequently removing these groups when they are no longer required.

In designing a large-scale purification process the number of steps, and the recovery of product at each step can have a major effect on the overall yield of the product, as is shown in Figure 1. The typical recovery of a chromatographic step is between 80% and 90%, so a complex purification requiring many steps may have an overall yield as low as 10% of the starting material. This may not matter for a laboratory-scale purification, but for large-scale production it is important to optimize the overall process from expression system or fermentation to final polishing step, so as to minimize the number of purification steps required.

2 CELL DISRUPTION

There are three main methods for the release of intracellular proteins from microorganisms: enzymic, chemical or physical. Not all of the techniques available are suitable for use on a large scale. Perhaps the main example is sonication, which is frequently the method of choice for the small-scale release of proteins. On a large scale it is difficult to transmit the necessary power to a large volume of suspension and to remove the heat generated.

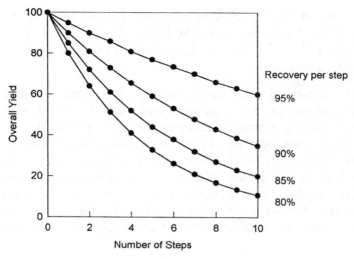

Figure 1 *The effect of recovery per step on overall recovery*

2.1 Enzymic Methods of Cell Disruption

Lysozyme, an enzyme produced commercially from hen egg white, hydrolyses beta-1,4-glycosidic bonds in the mucopeptide of bacterial cell walls. Gram-positive bacteria, which depend on cell wall mucopeptides for rigidity are most susceptible, but final rupture of the cell wall often depends upon the osmotic effects of the suspending buffer once the wall has been digested. In Gram-negative bacteria lysis is rarely achieved by the use of lysozyme alone, but the addition of EDTA to chelate metal ions will normally result in lysis. Although gentle, this technique is rarely used for the large-scale extraction of bacterial enzymes, perhaps due to the relatively high cost of lysozyme and the possibility of introducing contaminants. It has been used for the large-scale release of an aryl acylamidase from *Pseudomonas fluorescens*.[1]

2.2 Chemical Methods of Cell Lysis

2.2.1 Alkali. Treatment with alkali has been used with considerable success in small- and large-scale extraction of bacterial proteins. For example, the therapeutic enzyme, L-asparaginase, can be released from *Erwinia chrysanthemi* by exposing the cells to pH values between 11.0

[1] P. M. Hammond, C. P. Price, and M. D. Scawen, *Eur. J. Biochem.*, 1983, **132**, 651.

and 12.5 for 20 min.[2] The success of this method relies on the alkali stability of the desired product. The high pH may inactivate proteases, and the method is of value for the combined inactivation and lysis of rDNA microorganisms.

2.2.2 Detergents. Detergents, either ionic, for example, sodium lauryl sulphate, sodium cholate (anionic) and cetyl trimethyl ammonium bromide (cationic), or non-ionic, for example Triton X-100 or X-450, or Tween, have been used to aid cell lysis, often in combination with lysozyme. Ionic detergents are more reactive than non-ionic detergents, and can lead to the denaturation of many proteins. The presence of detergents can also affect subsequent purification steps, in particular salt precipitation. This can be overcome by the use of ion exchange chromatography or ultrafiltration, but obviously introduces additional steps.

2.3 Physical Methods of Cell Lysis

2.3.1 Osmotic Shock. Osmotic shock can be used for the release of enzymes and proteins from the periplasmic space of a number of Gram-negative bacteria. The method involves washing the cells in buffer solution to free them from growth medium, and then suspending them in 20% buffered sucrose. After being allowed to equilibrate, the cells are harvested and rapidly resuspended in water at about 4°C. Only about 4–8% of the total bacterial protein is released by osmotic shock, and if the required enzyme is located in the periplasmic region it can produce a 14- to 20-fold increase in purification compared with other extraction techniques. A major disadvantage of osmotic shock is the large increase in volume which occurs.

2.3.2 Grinding with Abrasives. Initially this technique was restricted to the grinding of cell pastes in a mortar with an abrasive powder, such as glass, alumina or kieselguhr. It has since been developed and mechanized using machines originally developed for the wet grinding and dispersion of pigments in the printing and paint industries. A typical product, the Dynomill (W. A. Bachofen, Switzerland) can be used to release proteins from a wide variety of microorganisms. It consists of a chamber containing glass beads and a number of fixed and rotating impeller discs. The cell suspension is pumped through the chamber, and the rapid agitation is sufficient to break even the toughest of bacteria. The disintegration

[2] T. Atkinson, B. J. Capel, and R. F. Sherwood, in 'Safety in Industrial Microbiology and Biotechnology.', ed. C. H. Collins and A. J. Beale, Butterworth, Oxford, 1992, p.161.

chamber must be cooled to remove the heat which is generated. A laboratory-scale model, with a 600 ml chamber can process up to 5 kg bacteria per hour, and production scale models are available with chambers of up to 250 litre capacity.

Many factors influence the rates of cell breakage, such as the size and concentration of the glass beads, the type, concentration and age of the cells, chemical pre-treatment, the agitator speed, the flow rate through the chamber, the temperature, and the arrangement of the agitator discs, and these have been investigated for yeasts[3] and bacteria.[3-5] This type of cell disrupter has the advantage that it can be readily mounted in an enclosed cabinet when pathogenic or rDNA organisms are to be broken.

2.3.3 Solid Shear. Methods of cell disruption employing solid shear have long been used on a small scale. It involves the extrusion of frozen cell material through a narrow orifice at high pressure and an outlet temperatures of about –20°C. It has found little application on an industrial scale, due to limitations on the amount of material which can be processed.

2.3.4 Liquid Shear. Liquid shear is the principle choice for the large-scale disruption of microbial cells, finding widespread application in both industrial processes and in research. It is particularly useful for the disruption of bacteria and yeast.[6]

As with solid shear, the cells are passed through a restricted orifice under high pressure, this time in a liquid suspension. For smaller scale work, a French Press is used. Larger scale work usually employs a homogenizer of the type developed for emulsification in the dairy industry. A temperature increase of at least 10°C in a single pass is not uncommon and it is necessary to pre-cool the cell suspension before homogenization. The liquid shear homogenizer is normally operated at wet cell concentrations of about 20%.

For large-scale work the Manton–Gaulin homogenizer (APV Ltd., Crawley, UK) is the most frequently used. It consists of a positive displacement piston pump with a restricted outlet valve, which can be adjusted to give the required operating pressure, up to 95 MPa. The smallest Manton–Gaulin homogenizer, the 15M-8TA, has a throughput of about 50 l h^{-1} at a pressure of 55 MPa. A larger version, the MC-4, has a throughput of about 300 l h^{-1}, again at a pressure of 55 MPa.

The rate of cell breakage and of protein release is dependent on a

[3] F. Marffy and M. R. Kula, *Biotechnol. Bioeng.*, 1974, **16**, 623.
[4] J. R. Woodrow and A. V. Quirk, *Enzyme Microb. Technol.*, 1982, **24**, 385.
[5] S. T. Harrison, J. S. Dennis, and H. A. Chase, *Bioseparation*, 1991, **2**, 95.
[6] D. Foster, *Bio/Technology*, 1992, **10**, 1539.

number of factors, including cell type, fermentation conditions, concentration and pre-treatment, such as freezing, as it is often observed that microbial cells break more easily if they have first been frozen. It has also been found that the presence of inclusion bodies makes *E. coli* cells more easily broken.[7]

The rate of protein release from yeast cells can be described by the empirical first order rate equation:

$$\text{Log}\,(R_m/R_m - R) = K\,n\,P^\alpha \tag{1}$$

Where: R_m is the theoretical maximum amount of soluble protein to be released, R is the actual amount of protein released, K is a temperature dependent constant, n is the number of passes, P is the operational back pressure, and α is a constant depending on the organism.

The value of the exponent α varies with the organism; for yeast it was found to be 2.9.[8], and for *E. coli* it is about 2.0. A similar first order equation can be used to describe the rate of release of proteins from other organisms, although the value of the exponent varies.[9] There are many examples of the use of the Manton–Gaulin homogenizers for the large-scale disruption of microbial cells. Beta-galactosidase has been released from *E. coli*,[10] and carboxypeptidase from *Pseudomonas* spp.[11] A large number of enzymes have been isolated from the thermophilic bacterium *Bacillus stearothermophilus*, including glycerokinase[12] and a glucose-specific hexokinase.[13] For a reliable process the conditions for cell breakage must be carefully optimized, as variations in the degree of cell breakage and protein release can have significant effects on subsequent purification steps.

3 INITIAL PURIFICATION

3.1 Debris Removal

Following cell disruption, the first step in the purification of an intracellular enzyme is the removal of cell debris. The separation of solids from liquids is a key operation in enzyme isolation, and is normally accomplished by centrifugation or filtration. Many protocols

[7] A. P. Middelberg, B. K. O'Neill, and D. L. Bogle, *Biotechnol. Bioeng.*, 1991, **38**, 363.
[8] M. Follows, P. J. Hetherington, and M. Lilly, *Biotechnol. Bioeng.*, 1971, **13**, 549.
[9] C. R. Engler and C. W. Robinson, *Biotechnol. Bioeng.*, 1981, **23**, 765.
[10] J. J. Higgins, D. J. Lewis, W. Daly, F. G. Mosqueira, P. Dunnill, and M. D. Lilly, *Biotechnol. Bioeng.*, 1987, **20**, 159.
[11] R. F. Sherwood, R. G. Melton, S. M. Alwan, and P. Hughes, *Eur. J. Biochem.*, 1985, **148**, 447.
[12] P. M. Hammond, T. Atkinson, and M. D. Scawen, *J. Chromatogr.*, 1986, **366**, 79.
[13] C. R. Goward, T. Atkinson, and M. D. Scawen, *J. Chromatogr.*, 1986, **369**, 235.

also include the addition of small quantities of DNase at this point, to break up long DNA chains which can cause the extract to become gelatinous.

3.2 Batch Centrifuges

Batch centrifuges are available with capacities ranging from less than 1 ml up to several litres, and capable of applying a relative centrifugal force of up to $100\,000 \times g$ (gravitational constant). However, for the removal of bacterial cells, cell debris and protein precipitates, fields up to $20\,000 \times g$ are adequate. Many centrifuges of this type, suitable for intermediate-scale preparations, are available.

3.3 Continuous-flow Centrifugation

Because of the large volumes of liquid which need to be handled at the beginning of a large-scale enzyme purification, it is preferable to use a continuous flow centrifuge to remove particulate matter. Three main types of centrifuge are available; the hollow bowl centrifuge, the disc or multi-chamber bowl centrifuge, and the basket centrifuge. Hollow bowl centrifuges have a tubular rotor which provides a long flow path for the extract, which is pumped in at the bottom and flows upwards through the bowl. Particulate matter is thrown to the side of the bowl, and the clarified extract moves up and out of the bowl into a collecting vessel. As centrifugation proceeds, the effective diameter of the bowl decreases, so reducing the settling path and the centrifugal force which can be applied. The ease with which the bowl can be changed, and the possibility of using a liner to aid sediment recovery has contributed to the popularity of this type of centrifuge. The flow rate must be determined empirically as it will vary from one type of extract to another, but a rate of about $60\,\mathrm{l\,h^{-1}}$ is generally satisfactory for the larger machines. Centrifuges of this type are produced by Pennwalt Ltd. (Camberley, Surrey, UK) and by Carl Padberg GmbH (Lahr, Germany).

Disc centrifuges provide an excellent means of clarifying crude extracts, and in many cases the sediment may be discharged without interrupting the centrifugation process. The bowl contains a series of discs around a central cone. As the extract enters, particulate matter is thrown outwards, impinging on the coned discs and sedimented matter collects on the bowl wall. This provides a constant flow path, so there is little loss of centrifugal efficiency during operation. A disadvantage of these centrifuges is that some loss of product may be experienced during the discharge process.

The rotors of instruments which do not have the facility to discharge sediment during operation are tedious to clean and this again may result in loss of product if the solids are required. These centrifuges achieve an RCF (rotational centrifugal force) of about $8000 \times g$ and have a capacity of up to 20 kg of sediment. As with hollow bowl centrifuges, the correct flow rate must be determined empirically. A variation of the disc-type centrifuge is the multi-chambered bowl centrifuge, in which the bowl is divided by vertically mounted cylinders into a number of interconnected chambers. The feed passes through each chamber from the centre outwards, before leaving the centrifuge. This type of arrangement also ensures a short and constant settling path as the bowl fills, and is easier to dismantle and clean than the disc-type of centrifuge.

Typical centrifuges of these types are produced by De Laval Separator Co. (New York, USA), and by Westfalia Separator Ltd. (Wolverton, UK).

A problem suffered by all types of centrifugation when applied on an industrial scale to enzyme recovery is that many homogenates produce wet sloppy precipitates, which reduces the efficiency of centrifugation.[14] The degree of clarification achieved with an industrial continuous flow centrifuge is never as great as that obtained with a laboratory centrifuge, and subsequent steps may be necessary to achieve the desired degree of clarification. One approach to alleviating this problem is to add a coarse microgranular cellulose anion exchanger (cell debris remover (CDR), Whatman) which binds cell debris and increases its density such that it is removed more efficiently by low speed centrifugation.

3.4 Basket Centrifuges

These are designed to operate at much lower g forces, perhaps only 1000 rev min^{-1}, and are basically centrifugal filters. The bowl is perforated and is normally lined with a filter cloth. The main use of these centrifuges is to collect large particulate material; in the context of enzyme purification this usually means ion exchange materials which have been used for the batch adsorption of the desired protein. Examples of such centrifuges are available from Carl Padberg GmbH (Lahr, Germany).

3.5 Membrane Filtration

Filtration is an alternative method of clarifying cell extracts. However, microbial broths and extracts tend to be gelatinous in nature, and are

[14] M. Hoare, P. Dunnill, and D. J. Bell, *Ann. N. Y. Acad. Sci.*, 1983, **413**, 254.

difficult to filter by traditional methods, unless very large filter areas are employed.

This can be overcome by using tangential or cross-flow filtration. In this method the extract flows at right-angles to the direction of filtration, and the use of a high flow rate tends to reduce fouling by a self-scouring action, although this action must be balanced against the possibility of losses due to shear effects. Membranes with an asymmetric, anisotropic pore structure are less prone to blockage than isotropic membranes, and have been used for the large-scale recovery of L-asparaginase from *Erwinia chrysanthemi*. A 1 m^2 membrane assembly was used to harvest the cells from 100 litre of culture fluid in 2.5 h, when the solids concentration in the retentate increased from 0.55% to 22% dry weight. This same membrane assembly was then used to clarify the extract obtained by the alkali lysis of these bacteria. These data indicated that to harvest the cells from 500 litre culture in 2.5 hours would require 7.5 m^2 of membrane, and that the costs compared favourably with the costs of centrifugation.[15] Membranes suitable for cross-flow filtration are available as spirally wound cartridges, which offer the same surface area as flat membranes, but in a more compact space.

Because of the limitations of large-scale centrifugation the two techniques are often combined to ensure that the extract is sufficiently clear for subsequent chromatography.

4 AQUEOUS TWO-PHASE SEPARATION

An alternative to centrifugation or filtration is aqueous two-phase separation. Aqueous two-phase systems are typically created by mixing solutions of polyethylene glycol and dextran or polyethylene glycol and salts such as potassium phosphate or ammonium sulphate to form two immiscible phases. Proteins and cellular debris show differential solubility between the two phases, so that the technique can be used both for the separation of proteins from cellular debris and for the partitioning of enzymes during protein purification. The precise partitioning of a protein depends on parameters such as its molecular weight and charge, the concentration and molecular weight of the polymers, the temperature, pH and ionic strength of the mixture and the presence of polyvalent salts such as phosphate or sulfate.[16–18] The optimal conditions required

[15] M. S. Le and T. Atkinson, *Process Biochem.*, 1985, **20**, 26.
[16] H. Walter, D. E. Brooks, and D. Fisher, 'Theory, Methods, Uses and Applications to Biotechnology: Partitioning in Aqueous Two Phase Systems', Academic Press, New York, 1985.
[17] N. L. Abbot, D. Blankschtein, and T. A. Hatton, *Bioseparation*, 1990, **1**, 191.
[18] F. Tjerneld and G. Johansson, *Bioseparation*, 1990, **1**, 255.

for a particular protein are found empirically. Although the conditions required to achieve satisfactory separation can often be precisely defined, the mechanism of partitioning is not fully understood.[19]

The phases can be separated in a settling tank, but a more efficient and rapid separation can usually be achieved by centrifugation. Since it is easier to separate liquids of different density than solids from liquids on the large scale, this approach can be used to advantage in large-scale enzyme purification. Although the relatively low cost of the polyethylene glycol–salt system makes it attractive for large-scale use, the more generally useful polyethylene glycol–dextran system can also have economic advantages in comparison with other purification methods, despite the high cost of purified dextran, providing the total processing costs are evaluated.[20] Its use is not restricted to materials of microbial origin, and the method has been successfully used to isolate materials from both plant[21] and animal[22] sources, including human alpha-L-antitrypsin expressed in transgenic milk. In this case the relatively high starting purity of the protein meant that after a single two phase separation the desired protein was 73% pure.[23]

Aqueous two phase separation can be adapted to offer biospecific partitioning by attaching ligands to the polymers in order to alter the partitioning of a protein.[17,24] All phase-forming polymers can have ligands attached covalently to them, and a wide range of such ligands have been investigated. Because of their simple coupling chemistry, the reactive dyes have frequently been used as ligands.[25] The power of the technique was demonstrated by the 58-fold purification of yeast phosphofructokinase that could be obtained in two steps using Ciba-cron Blue F-3GA immobilized on polyethylene glycol.[26] Besides the reactive dyes, a number of other ligands have been investigated. These include cofactors, such as the pyridine nucleotides used successfully in the affinity partitioning of a number of dehydrogenases.

On the process scale, affinity partition has been used for the purification of formate dehydrogenase from 10 kg quantities of the yeast

[19] J. Huddleston, A. Veide, K. Kohler, J. Flanagan, S. Enfors, and A. Lyddiatt, *Trends Biotechnol.*, 1991, **9**, 381.
[20] K. H. Kroner, H. Hustedt, and M. R. Kula, *Process Biochem.*, 1984, **19**, 170.
[21] H. Vilter, *Bioseparation*, 1990, **1**, 283.
[22] M. J. Boland, *Bioseparation*, 1990, **1**, 293.
[23] D. P. Harris, A. T. Andrews, G. Wright, D. L. Pyle, and J. A. Asenjo, *Bioseparation*, 1997, **7**, 31.
[24] G. Johansson and F. Tjerneld, in 'Highly Selective Separations in Biotechnology', ed. G. Street, Blackie Academic and Professional, London, 1994, p.55.
[25] G. Kopperschlager, *Methods Enzymol.*, 1994, **228**, 121.
[26] G. Johansson, G. Koperschlager, and P. A. Albertsson, *Eur. J. Biochem.*, 1983, **131**, 589.

Candida bodinii, using the triazine dye, Procion Red HE-3B, immobilized on polyethylene glycol.[27]

Although aqueous two-phase separation is a method which can easily be scaled up to a manufacturing level, it does not appear to be often used for industrial-scale purifications.

5 PRECIPITATION

5.1 Ammonium Sulfate

Salting out of proteins has been employed for many years, and fulfils the dual purposes of purification and concentration. The most commonly used salt is ammonium sulfate, because of its high solubility, lack of toxicity towards most enzymes, and low cost.

The precipitation of a protein by salt depends on a number of factors: pH, temperature, protein concentration, and the salt used.[28] The protein concentration is particularly important when scaling-up, because most large-scale purifications are carried out at higher protein concentrations than laboratory-scale purifications. This can have a dramatic effect on the concentration of salt needed to precipitate a given protein.

5.2 Organic Solvents

The addition of organic solvents to aqueous solutions reduces the solubility of proteins by reducing the dielectric constant of the medium. Various organic solvents have been used for the precipitation of proteins, with ethanol, acetone, and propan-2-ol being the most important. Because proteins are denatured by organic solvents it is necessary to work at temperatures below 0°C.

Because of their flammable nature, requiring flameproof equipment to be used, and high cost, coupled with a low selectivity, organic solvents are not often used in large-scale enzyme purification. The one notable exception is in the blood processing field, where ethanol precipitation is the major method for the purification of albumin; indeed it has been developed into a highly automated, computer-controlled system.[29]

[27] M. R. Kula, in 'Extraction and Purification of Enzymes. Applied Biochehmistry and Bioengineering', ed. L. B. Wingard, E. Katchalski-Katzir, and L. Goldstein, Academic Press, New York, 1979, p.71.

[28] M. C. Dixon and E. C. Webb, 'The Enzymes', Longmans, London, 1979.

[29] P. Foster and J. G. Watt, in 'Methods of Plasma Fractionation', ed. J. Curling, Academic Press, New York, 1980, p.17.

5.3   High Molecular Weight Polymers

Other organic precipitants which can be used for the fractionation of proteins are water-soluble polymers like polyethylene glycol. This has the advantage of being non-toxic, non-flammable and not denaturing to proteins. It is mainly used in the blood processing field.

5.4   Heat Precipitation

When a protein is sufficiently robust, heat treatment can provide a high degree of purification as an initial step. In a large-scale example, 55% of unwanted proteins were removed in a single step by heating an *E. coli* extract containing recombinant Staphylococcal Protein A at 80° for 10 min.[30] If the recombinant protein to be purified is from a thermophilic organism the results of an initial heat treatment can be even more dramatic: when an *E. coli* extract containing recombinant malate dehydrogenase from *Thermus aquaticus* is heated to 80° for 20 min the enzyme in the supernatant is about 90% homogeneous.[31]

6   CHROMATOGRAPHY

The purification of proteins by chromatography has been a standard laboratory practice for many years. These same chromatographic techniques can equally well be applied to the isolation of much larger quantities of protein, although the order in which they are used must be considered carefully.

For the purification of high value/low volume products, typically therapeutic or diagnostic proteins, chromatography is the most widely used method. Chromatography is the only method with the required selectivity to purify a single protein from a complex mixture of proteins to a final purity of greater than 95%.

6.1   Scale-up and Quality Management

In analytical chromatography, as well as in many laboratory-scale applications, the quantity of sample to be applied is small, and the overall aim is to achieve the maximum number of peaks or to produce a small amount of highly purified protein. The flow rates used are low, as

[30] K. A. Philip, P. M. Hammond, and G. W. Jack, *Ann. N. Y. Acad. Sci.*, 1990, **613**, 863.
[31] R. M. Alldread, D. J. Nicholls, T. K. Sundaram, M. D. Scawen, and T. Atkinson, *Gene*, 1992, **114**, 139.

ultimate resolution is of greater importance than throughput. In contrast, the aim in preparative chromatography is to purify the maximum amount of protein in the minimum of time. The flow rates used are high to increase throughput, but care must be taken to ensure adequate resolution is maintained.

The greatest resolution and throughput will be given by the use of small particles at high flow rates. Unfortunately, this combination results in high pressures, which can only be reduced by using a lower flow rate. Small particles are generally more expensive than larger ones, and high pressure equipment is more expensive than low pressure equipment. For these reasons, process-scale chromatography often uses larger particles at the highest flow rate compatible with the matrix, the resolution required and the chromatography equipment. The limiting factor is often the diffusion of macromolecules into and out of the pores in the matrix.

The scale-up of a chromatographic separation is in principle simple, as chromatography theory shows that the column diameter has little effect on resolution, so any increase in scale can be accomplished by increasing the column diameter; in practice the scaling-up of a chromatographic separation can present a challenging problem.[32]

Before scaling-up it is important to ensure that the purification process is fully understood and optimized on the laboratory scale. The initial screening for adsorptive techniques can be conveniently carried out by the batch addition of different adsorbents under various conditions of pH and ionic strength, or by using small 1 ml columns containing various matrices. Suitable chromatographic methods can then be developed using either conventional, low pressure equipment, or by using higher pressure, high performance equipment. It is now possible to largely automate this screening and optimization process by using equipment that can be programmed to run columns repeatedly under different conditions.

Following this optimization process, the next step is to increase the sample load by a factor of between 10 and 20. The bed height of the column should be kept constant and the surface area increased in proportion to the sample load. The packing material should be the same, or at least have similar characteristics. The linear flow rate, the ionic strength, and the pH must all be held constant. If gradient elution is used the ratio of gradient volume to column volume must be the same. It is important that the protein load per unit column volume is held constant, rather than sample volume, as large-scale production extracts

[32] G. K. Sofer and L. E. Nystrom, 'Process Chromatography: A Practical Guide', Academic Press, New York, 1989.

normally contain a higher concentration of protein than the corresponding laboratory-scale extracts.

In principle this scaling-up process can be repeated until the desired
scale of operation is reached. However, there are limitations. The
maximum column diameter is governed by those available, and is
currently around 120 cm, which for the bed heights of 15 cm to 20 cm
commonly used for ion exchange chromatography is equivalent to 170–
220 litres of gel. As the column diameter increases so does the cost, as
well as the difficulty of ensuring even loading of the sample over the
entire surface.

One of the prime considerations in scale-up and design of large-scale
purification protocols is the intended use of the final product. Following
the above guidelines, it should be possible to scale-up most purifications
to a scale suitable for manufacturing purposes. If the product is for
industrial or research use, this may be all that is required. However, if the
protein product is intended for therapeutic use, there are other factors
besides the mechanics of scale-up to take into account. There is also a
requirement to demonstrate that any materials which have come into
contact with the product during manufacture do not have a residual
presence in the final product. There needs to be a strict control over raw
materials, including the chromatography matrices used, to ensure that
they meet stringent criteria in terms of purity and acceptability. This in
turn may influence the choice of method. If the protein is to be used for
therapeutic purposes it will also be necessary to ensure the efficient
removal of pyrogens.

In quality management, it is necessary to distinguish between contaminants deriving from the biological system or process raw materials
and those introduced incidentally or by design during processing. It is
especially necessary to have suitable analytical methods for quantifying
those impurities which may still be present in the final product.
Contaminants introduced during processing can be wide-ranging in
nature. Before the deliberate introduction of any component(s) during a
purification protocol, it is necessary to ensure that their presence can be
adequately monitored. It may also be necessary to demonstrate that they
can be satisfactorily removed (using a positive removal step) at a later
stage. All of these factors may have an influence on the design of the
manufacturing protocol. For example, affinity chromatography may
involve a biological ligand; in such an instance, it is necessary to show
that there is no measurable leakage of ligand into the product. Practically
all affinity matrices leach ligand to some degree, and the purification
scientist may therefore be faced with proving that the level of leakage is
not significant. Analysis to a certain level of sensitivity may not be

sufficient alone. It may merely reflect the fact that the level of leakage is below the sensitivity of the detection methodology; no detection does not equal no leakage of ligand!

Standards of purity must be extremely high, and this may rule out certain approaches to scale-up. The demonstration of purity is an important consideration of quality management and has led to the need for complex strategies for the purity analysis of proteins, not least where such protein pharmaceuticals have been produced by recombinant DNA technology. The development of analytical methods such as capillary electrophoresis[33] and mass spectrometry[34] for the rapid structural characterisation of such products is therefore a fundamental requirement in both research laboratories and in commercial production.

6.2 Method Selection

All of the available chromatographic techniques, gel filtration, ion exchange, hydrophobic interaction, affinity, immunoaffinity and chromatofocusing, can be used for the large-scale isolation of proteins. Although all these techniques can be used, there are limitations to each technique, which must be taken into account when designing a process, as shown in Table 1. It is preferable to be able to proceed from one step to the next with the minimum of alteration to the conditions. Therefore, steps which concentrate the product, like ion exchange, hydrophobic or affinity chromatography, should precede steps which cause a dilution, like gel filtration. Hydrophobic interaction chromatography can conveniently follow an ion exchange step with the minimum change in buffer, because most proteins bind more strongly to a hydrophobic support at high ionic strength. Apart from its use as a means of exchanging buffers, gel filtration is probably best used as a final polishing step, when the volume of product is low. Affinity chromatography has provided a uniquely powerful method for the purification of proteins on the laboratory scale. Although it is used on a preparative scale the number of published examples are few, partly due to the problems of using expensive and delicate ligands on a large scale.

6.3 Selection of Matrix

Perhaps the most important decision to be taken when designing a large-scale purification process concerns the type of chromatography matrix to

[33] V. R. Anicetti, B. A. Keyt, and W. S. Hancock, *Trends Biotechnol.*, 1989, **7**, 342.
[34] M. Geisow, *Biologicals*, 1993, **21**, 125.

Table 1   *Chromatographic techniques for the large-scale purification of proteins*

Molecular property exploited	Chromatography type	Characteristics	Application
Size	Gel filtration	Resolution is moderate for fractionation. Good for buffer exchange. Capacity limited by volume of sample. Speed is slow for fractionation. Fast for buffer exchange	Fractionation is best left to later stages of a purification. Very useful for final polishing of product. Buffer exchange can be used at any time, although sample volume may be a limitation
Charge	Ion exchange	Resolution can be high. Capacity is high, and not limited by sample volume. Speed can be very high, depending on matrix.	Often most effective at early stages in fractionation, when large volumes have to be handled.
Polarity	Hydrophobic interaction	Resolution is good. Capacity is very high, and not limited by sample volume. Speed is high.	Can be applied at any stage, but is most useful when the ionic strength is high, after salt precipitation or ion exchange.
Biological affinity	Affinity	Resolution is very high. Capacity can be high, though may be low, depending on ligand. Speed is high.	Can be used at any stage, but is not normally recommended at early stage.

be used at each step. A matrix which is to be used for large-scale chromatography should be hydrophilic, macroporous, rigid, spherical, chemically stable (yet easily derivatized) inert, and reuseable.

No single matrix can completely satisfy all of these criteria, and it is not surprising that many different types of matrix are available, as shown in Table 2. In many cases these are available in a variety of derivatized forms suitable for the different types of chromatography. All of the different types of matrix listed have advantages and disadvantages which must be taken into account, as shown in Table 3.

Some gels, such as those based on agarose or cellulose are natural products; others, such as those based on cross-linked dextrans or agarose are modified natural products; yet others, such as those based on polyacrylamide, polyhydroxethyl methacrylate or polystyrene are

Table 2 *Examples of base matrices for large-scale chromatography. This list is not meant to be exhaustive, but is intended to give some idea of the range of materials available*

Matrix type	Example	Manufacturer
Cross-linked dextran	Sephadex	[a]
Cross-linked polyacrylamide	BioGel-P	[b]
Agarose	Sepharose	[a]
	BioGel-A	[b]
Cross-linked agarose	Sepharose-CL	[a]
	Sepharose-HP	[a]
	Sepharose-FF	[a]
Polyacrylamide–dextran composite	Sephacryl	[a]
Dextran–agarose composite	Superdex	[a]
Methacrylate copolymer	Macro-Prep	[b]
Cellulose	DE-52, CM-52	[c]
	Sephacel	[a]
Rigid organic polymers	Monobeads	[a]
	Bio-Beads	[b]
	TSK-PW	[d]
	POROS	[e]
Porous silica	Zorbax	[f]
	TSK-SW	[f]

[a] Amersham Pharmacia Biotech, Uppsala, Sweden; [b] Bio-Rad Laboratories, Hercules, CA, USA; [c] Whatman Ltd., Maidstone, Kent, UK; [d] Toso Haas, Montgomeryville, PA, USA; [e] PE Biosystems, Foster City, CA, USA; [f] Rockland Technologies, Wilmington, DE, USA.

wholly synthetic. The gels can be further classified as macroporous, such as the agarose or cellulose gels, or microporous, such as the cross-linked dextran or polyacrylamide gels. The macroporous gels are most useful for ion exchange or affinity chromatography or for the size fractionation of very large molecules, such as viruses, large proteins or glycoproteins. The microporous gels are most useful for the size fractionation of the majority of proteins. The rigidity of the gels varies widely. The earliest gels, based on cross-linked dextran, cellulose, polyacrylamide or agarose were very soft and not readily suited to large-scale chromatography. These were replaced by highly cross-linked, yet macroporous gels, based on agarose or on composites, such as polyacrylamide and agarose, are much more rigid and far more suited to large-scale chromatography. In addition to their improved rigidity, these newer gels are available in a smaller and more controlled particle size, ensuring that comparable resolution is obtained at the higher flow rates which are possible. The hydrophilic, polymeric gels, for example Monobeads or Superose (Amersham Pharmacia Biotech) or the TSK-PW range (TosoHaas), are

Table 3 *Properties of base chromatographic matrices*

Matrix type	Porosity	Non-specific adsorption	Rigidity	Stability	Ease of derivatization	Relative cost
Cross-linked dextran	Low	Low	Low	High	Good	Low/medium
Cross-linked polyacrylamide	Low	Low	Low	High	Good	Low/medium
Agarose	High	Low	Low	Poor	High	Medium
Cross-linked agarose	High	Low	Medium	High	High	Medium/high
Polyacrylamide–dextran composite	High	Medium	Medium	High	Good	Medium
Dextran–agarose composite	Medium	Low	Medium	High	Good	Medium/high
Methacrylate copolymer	High	Low	Medium	High	Good	Medium
Cellulose	Medium	High	Low	Good	Good	Low
Rigid organic polymers	Low to high	Low/medium	High	High	High	High
Porous silica	Low to medium	High	High	Poor to good	Poor	High

truly high performance gels, in that they are available as 10 μm particles that are able to withstand pressures of 3–10 MPa.

The truly inorganic materials, silica or porous glass, although ideal in terms of their rigidity and availability as small particles, are not suitable for the majority of biological applications. The hydrophobic nature of their surfaces often means that proteins are bound irreversibly, or can only be eluted under denaturing conditions. In addition, silica particles offer a high resistance to the flow of aqueous solvents and are unstable at pH values > 8.

6.4 Gel Filtration

In gel filtration, separation is based on molecular size. The stationary phase consists of porous beads surrounded by a mobile solvent phase. When the sample is applied, the molecules in the mixture partition between the pores in the beads and the solvent. Large molecules are unable to enter the pores and so pass through the interstitial spaces and elute first. Smaller molecules, which can enter the pores are eluted later, in decreasing order of size.

The total volume of a column can be represented by:

$$V_t = V_o + V_i + V_m \tag{2}$$

Where: V_t is the total volume of the column, V_o is the volume of solvent external to the particles, V_i is the volume of solvent that occupies the interior of the particles, and V_m is the volume occupied by the matrix itself.

The elution volume of a protein can therefore vary between V_o for one which cannot enter the pores in the gel, and V_i for one which is able to fully enter the pores in the gel. Thus it is possible to calculate an effective partition coefficient, K_{av}, which can vary between zero and unity:

$$K_{av} = (V_e - V_o)/(V_t - V_o) \tag{3}$$

Where: V_e is the elution volume of the solute, V_o is the void volume of the column, and V_t is the total volume of the column.

For globular proteins it has been shown empirically that the value of K_{av} is inversely proportionally to the logarithm of the relative molecular mass.

It is essential that there should be no interaction between the matrix and the solute, therefore the ideal gel filtration medium should be totally inert. For maximum capacity it should also be rigid and highly porous.

For large-scale working, rigidity is perhaps most important as it determines the highest flow rate that can be obtained.

The traditional gel filtration materials were based on cross-linked dextran (Sephadex), or polyacrylamide (BioGel P). These materials are sufficiently inert but, in the porosities suitable for the fractionation of most proteins, are too soft for easy application on the large scale. For this reason most applications of gel filtration were limited to desalting, using the low porosity, but rigid gels like Sephadex G-25 and G-50.

More rigid gel types based on a variety of materials have been introduced, and these are more suited to large-scale application. Examples include Sephacryl (Amersham Pharmacia Biotech), based on dextran and polyacrylamide; Superdex (Amersham Pharmacia Biotech), based on dextran and agarose; Superose (Amersham Pharmacia Biotech), based on a highly cross-linked agarose. All of these materials are available in particle sizes which are smaller than their traditional counterparts, so that resolution is retained at the higher flow rates permitted by their increased rigidity. A brief history of the development of gel filtration is given by Porath,[35] who was closely involved in the early developments of the technique.

6.5 Ion Exchange Chromatography

Traditionally, ion exchange media for the fractionation of proteins were based on microgranular cellulose substituted with various charged groups, such as diethylaminoethyl (DEAE) and carboxymethyl (CM), as shown in Table 4. Cellulose ion exchangers are more suited to batch-type operations at the early stages of a process. For example, in the purification of L-asparaginase from *Erwinia chrysanthemi* a six-fold purification and 100-fold reduction in volume can be achieved by batch adsorption and elution from CM-cellulose. For routine use in large-scale columns cellulose cannot support the highest flow rates, and suffers the disadvantage that its volume changes with pH or ionic strength, making it difficult to regenerate without unpacking the column, a laborious procedure. This volume change is less marked with the cross-linked, beaded forms which can be regenerated *in situ*.

Historically the next development was the introduction of ion exchange groups into Sephadex G-25 or G-50 (Amersham Pharmacia Biotech). Those based on Sephadex G-25 were rigid, but had a low capacity for most proteins. Those based on Sephadex G-50 had a high capacity for proteins, but showed large changes in volume with variation

[35] J. Porath, *J. Protein Chem.*, 1997, **16**, 463.

Table 4 *Ion exchange ligands*

	Ligand
Cation exchange	
Carboxymethyl (CM)	$-O \cdot CH_2 \cdot COO^-$
Sulfoethyl (SE)	$-O \cdot CH_2 \cdot CH_2 \cdot SO_3^-$
Sulfopropyl (SP)	$-O \cdot CH_2 \cdot CH_2 \cdot CH_2 \cdot SO_3^-$
Sulfonate (S)	$-O \cdot SO_3^-$
Anion exchange	
Diethylaminoethyl (DEAE)	$-O \cdot CH_2 \cdot CH_2 \cdot N \cdot (C_2H_5)_2$
Quaternaryaminoethyl (QAE)	$-O \cdot CH_2 \cdot CH_2 \cdot N^+ \cdot (C_2H_5)_2 \cdot (CH_2 \cdot CHOH \cdot CH_3)$
Quaternary amine (Q)	$-O \cdot CH_2 \cdot N^+ \cdot (CH_3)_2$

in pH or ionic strength, making them unsuitable for large-scale ion exchange chromatography.

These materials have been superseded by ion exchange media based on cross-linked agarose or other macroporous synthetic gels, such as Fractogel (Merck), giving a range of ion exchange materials which are both rigid and of high capacity. The particles are small and spherical, supporting high flow rates with good resolution, and do not deform under the pressures normally encountered in large-scale chromatography. Because of their rigidity and chemical stability, all of these materials can be regenerated in the column with NaOH solutions.

A typical example of large-scale ion exchange chromatography, using gradient elution from a 40 litre column of DEAE Sepharose is the purification of glycerokinase from 20 kg *Bacillus stearothermophilus* cell paste. The column was 80 × 25 cm, and was eluted with a 200 litre linear gradient of increasing phosphate concentration.[36] Recombinant *Erwinia* asparaginase has been purified by large-scale ion exchange chromatography on S-Sepharose.[37] The enzyme from 16 kg of bacterial cell paste was adsorbed onto a 30 litre column at a flow rate of 600 1 h^{-1} and eluted with a linear gradient of increasing salt concentration.

6.6 Affinity Chromatography

Affinity chromatography can provide perhaps the most elegant method for the purification of a protein from a complex mixture. Although used extensively as a laboratory technique, it is less frequently used for industrial-scale purification. One of the most frequently cited reasons

[36] M. D. Scawen, P. M. Hammond, M. J. Comer, and T. Atkinson, *Anal. Biochem.*, 1983, **132**, 413.
[37] C. R. Goward, G. B. Stevens, R. Tattersall, and T. Atkinson, *Bioseparation*, 1992, **2**, 335.

for this is the difficulty of demonstrating that no ligand has leached into the final product.

Affinity chromatography relies on the interaction of a protein with an immobilized ligand. A ligand can be either specific for a particular protein, for example, a substrate, substrate analogue, inhibitor or an antibody. Alternatively it may be able to interact with a variety of proteins, for example, AMP, ADP, NAD, dyes, hydrocarbon chains, or immobilized metal ions. Affinity chromatography using immobilized nucleotides is little used for process-scale purification because of their instability, expense, low capacity and the difficulties of coupling them to a support matrix.

Immobilized dyes have been used for the large-scale purification of many enzymes, where they offer the advantages of cheapness, ready coupling to a support matrix, stability and high capacity.[38,39] The reactive dyes are often anthroquinone-like structures and were originally thought to bind proteins by interacting with the nucleotide-binding domain of dehydrogenases. As so many different types of protein have been isolated by dye affinity chromatography, the precise interaction must be more variable, and is in most cases unknown. The reactive dyes were originally made for the textile dyeing industry, and are not ideally suited for the purification of proteins. Recently, the synthesis of ligands, loosely based on the structure of Cibacron Blue has been described, which are better able to mimic the interaction between a natural cofactor and a protein. One such series of biomimetic dyes, in which a spacer group was inserted between the anthroquinone rings and the rest of the structure, and the means of attachment to the support could be varied had a three- to four-fold higher affinity for alcohol dehydrogenase.[40] A more recent example of a synthetic biomimetic ligand is given by the synthesis of a protein A mimic based around the triazine ring framework found in the reactive dyes.[41]

Another variant of affinity chromatography is immobilized metal affinity chromatography. Immobilized metal ions, such as Cu^{2+}, Zn^{2+} or Ni^{2+}, can be used to separate proteins. This separation depends on the interaction between the metal ion and histidine residues on the surface of the protein. The metal ion is immobilized by chelation onto an iminodiacetate group attached to a suitable matrix, usually agarose. Bound proteins can be eluted with a competing ligand, such as imidazole.

[38] C. V. Stead, *Bioseparation*, 1991, **2**, 129.
[39] N. Garg, I. Y. Galaev, and B. Mattiasson, *J. Molec. Recognit.*, 1996, **9**, 259.
[40] C. R. Lowe, S. J. Burton, N. P. Burton, W. K. Alderton, J. M. Pitts, and J. A. Thomas, *Trends Biotechnol.*, 1992, **10**, 442.
[41] R. Li, V. Dowd, D. J. Stewart, S. J. Burton, and C. R. Lowe, *Nat. Biotechnol.*, 1998, **16**, 190.

A recent large-scale application of metal affinity chromatography is given by the purification of human factor IX from plasma, where the method allowed the removal of the solvent–detergent reagents added to inactivate viruses.[42]

Perhaps the most highly selective ligand is a monoclonal antibody, since this has been selected to recognize a single epitope on the surface of a protein. Because of this immunoaffinity chromatography can be used to selectively purify the correctly folded form of protein. This can be important when the protein is expressed as an inclusion body which requires solubilization, oxidation and refolding, as such techniques are never 100% efficient. An example is the purification of recombinant human CD4 on the gram scale using a monoclonal antibody directed against conformational epitopes of the protein.[43]

Affinity chromatography has seen an upsurge in interest recently following the introduction of genetically engineered affinity tails into proteins to facilitate purification by immobilized metal ion chromatography.[44]

The criteria for the selection of a matrix for affinity chromatography are similar to those for ion exchange chromatography. As a result the most commonly used matrices are macroporous, such as agarose. The matrix must be activated chemically to enable the ligand to be covalently coupled. A wide range of methods are available,[45] but one of the most generally used reagents is cyanogen bromide. A less commonly used coupling agent, which provides a very stable linkage is divinyl sulfone; it has been used for the large-scale immobilization of IgG onto Sepharose for the purification of Protein A.[30] For those who do not wish to activate their own matrix a range of activated matrices are available from the major suppliers of chromatography media.

Methods for the elution of a bound protein may be either specific, using, for example, a substrate or cofactor, or non-specific, using, for example, salt or a change in pH. Non-specific elution is normally used with highly selective adsorbents which bind only one component; with a less selective adsorbent specific elution may give a greater degree of purification. Examples of available methods include affinity elution with substrate or free ligand, change in pH or ionic strength, addition of a chaotropic or denaturing agent (*e.g.* KSCN, urea, guanidine–HCl),

[42] P. A. Feldman, P. I. Bradbury, J. D. Williams, G. E. Sims, J. W. McPhee, M. A. Pinnell, L. Harris, G. I. Crombie, and D. R. Evans, *Blood Coagul. Fibrinolysis*, 1994, **5**, 939.
[43] P. A. Wells, B. Beiderman, R. L. Garlick, S. B. Lyle, J. P. Martin, J. T. Herberg, H. F. Meyer, S. L. Henderson, and F. M. Eckenrode, *Biotechnol. Appl. Biochem.*, 1993, **18**, 341.
[44] C. E. Glatz and C. F. Ford, *Appl. Biochem. Biotechnol.*, 1995, **54**, 173.
[45] M. Wilchek, T. Miron, and J. Kohn, *Methods Enzymol.*, 1984, **104**, 3.

change in solvent polarity, by the addition of an organic modifier, such as ethylene glycol.

The mildest method of elution is preferred, but must usually be found empirically, and in many cases cost considerations may be important. Although affinity steps can be used in batch or in columns, elution is best carried out in columns.

6.7 Hydrophobic Interaction Chromatography

Hydrophobic interaction chromatography was first developed following the observation that proteins were unexpectedly retained on affinity gels containing hydrocarbon spacer arms. This concept was extended and families of adsorbents prepared using an homologous series of hydrocarbon chains over the range C_2 to C_{10}[46] although in practice most proteins can be purified using agarose substituted with phenyl or octyl groups.

Hydrophobic interactions are strongest at high ionic strength, so adsorption can often be conveniently performed after salt precipitation or ion exchange chromatography, with no change in the salt concentration of the sample. Bound proteins can be eluted by altering the solvent pH, ionic strength or by use of a chaotropic agent or an organic modifier such as ethylene glycol.

The various ions can be arranged in a series, depending on whether they promote hydrophobic interactions (salting-out effect) or disrupt the structure of water (chaotropic effect) and lead to a weakening of the hydrophobic interaction, as shown in Table 5. Those ions which promote the hydrophobic interaction, such as sulfate or phosphate, are useful in promoting binding. Those that are increasingly chaotropic are useful for strongly eluting proteins, which cannot be eluted by decreasing the ionic strength.

Chromatography on columns of Phenyl-Sepharose has been used for the purification of aryl acylamidase from *Pseudomonas fluorescens*. The enzyme from 2 kg bacteria was eluted from an ion exchange column in 0.3 M phosphate buffer, pH 7.6, and applied directly to a 500 ml column of Phenyl-Sepharose in the same buffer. The enzyme was eluted by using a decreasing gradient from 0.1 M to 0.01 M Tris-HCl at pH 7.6.[47] Recombinant human superoxide dismutase from 100 litres of *E. coli* fermentation broth has been purified on a 20 litre column of Phenyl-Sepharose. The enzyme was adsorbed after fractionation with 60%

[46] S. Shaltiel, *Methods Enzymol.*, 1984, **104**, 69.
[47] P. M. Hammond, C. P. Price, and M. D. Scawen, *Eur. J. Biochem.*, 1983, **132**, 651.

Table 5 *Effect of ions on hydrophobic interactions*

	Increasing salting-out effect								
Cations	Ba^{2+} Ca^{2+} Mg^{2+} Li^+ Cs^+ Na^+ K^+ Rb^+ NH_4^+								
	Increasing chaotropic effect								
Anions	PO_4^{3-} SO_4^{3-} CH_3COO^- Cl^- Br^- NO_3^- ClO_4^- I^- SCN^-								

saturation ammonium sulfate, and eluted with a decreasing linear gradient of ammonium sulfate from 60% saturation to zero.[48]

6.8 High Performance Chromatographic Techniques

One of the potentially most significant advances in chromatography has been the development of matrices of small particle size capable of high resolution and operation under relatively high pressures. Originally developed for the separation of small organic molecules soluble in non-aqueous solvents, the technique has rapidly developed into a form suitable for the separation of proteins and enzymes in aqueous solvents, using all of the normally available chromatographic methods. However, the majority of reported applications have been confined to the laboratory scale.

High performance matrices are highly efficient because of their small particle size, (3–50 μm). Because of this small particle size, high pressures are needed to generate good flow rates and very rigid particles are necessary. Two approaches have been taken to solve this problem. Silica-based matrices are sufficiently rigid, and can be extensively modified with monochloro- or monoalkoxysilanes to give a hydrophilic surface which can be further derivatized. Silica has the disadvantage of being unstable at pH values above pH 8, but this has been partially overcome either by using a polymer coating to reduce the availability of the inorganic particles to the solvent, or more specifically by surface stabilization with zirconium. The second approach has been the development of rigid, cross-linked polymeric supports such as Monobeads (Amersham Pharmacia Biotech) or TSK-PW (TosoHaas).

These true high performance matrices have been followed by higher performance derivatives of conventional packing materials, which have smaller particle sizes, for use in what is termed 'medium performance

---

[48] K. Vorauer, M. Skias, A. Trkola, P. Schulz, and A. Jungbauer, *J. Chromatogr.*, 1992, **625**, 33.

liquid chromatography' (MPLC). Pharmacia have introduced Sepharose HR ion exchangers and a gel filtration matrix, Superdex, which have particle sizes of about 35 μm. These matrices have a particle size and a performance which is in between HPLC matrices (3–20 μm) and conventional, low performance materials (> 100 μm). Although initially more expensive, the higher flow rates and reduced processing times can make high performance chromatography and attractive proposition for large-scale protein purification.

6.9 Perfusion Chromatography

Another approach to the problem of flow rates and pressure in high performance separations has been the introduction of perfusion chromatography.[49] This technique differs from conventional chromatography in that the particles have large, interconnecting pores which pass through the particles, with smaller diffusive pores branching off them. This combination of pore sizes allows rapid access of molecules to the high surface area diffusive pores within the particles. The very short diffusive paths within the particles and the rapid bulk transport through the perfusive pores makes separation and capacity largely independent of flow rate. The particles have a low resistance to flow, and are able to give high resolution performance using either HPLC-type systems or conventional pumping systems. The particles are composed of an organic polymer, and are available in ion exchange, hydrophobic and reverse phase derivatives. They can be used at flow rates which are some 10-fold higher than those used for conventional high performance matrices; a 1 ml column can give good resolution at 10 ml min^{-1}. The material is commercially available under the trade name POROS (PE Biosystems, Foster City, CA, USA). A potential advantage of the perfusion approach is the use of small columns which are operated at high flow rates in a cycling mode. For example, if a perfusion column can operate at 10 times the flow rate of a conventional column, a perfusion column one-tenth the size can process the same amount of material over 10 cycles as the conventional column can over one cycle. This approach has advantages in terms of the column and pumping hardware, all of which can be smaller and less expensive. Several examples of the application of perfusion chromatography are given by Fulton *et al.*[50]

[49] N. B. Afeyan, S. P. Fulton, N. F. Gordon, I. Mazsaroff, L. Varady, and F. E. Regnier, *Bio/Technology*, 1990, **8**, 203.
[50] S. P. Fulton, A. J. Shahidi, N. F. Gordon, and N. B. Afeyan, *Bio/Technology*, 1992, **10**, 635.

6.10 Expanded Bed Adsorption

One of the problems in large-scale protein purification has always been the initial clarification of the extract, as large-scale centrifuges are both expensive and inefficient. One approach to solve this problem is aqueous two-phase separation, but this has limited application, and may cause problems in subsequent steps. Adsorption chromatography is a favoured early step, but conventional packed bed columns require perfectly clear extracts to prevent blocking of the column. What is required is a technique that allows the proteins to be adsorbed from a unclarified extract. Batch adsorption followed by washing in a stirred tank can be used, but is very inefficient. To overcome these problems fluidized bed adsorption has been used for many years in the antibiotic industry for the recovery of low molecular weight compounds from fermentation broths. Despite this the successful use of fluidized or expanded beds for the recovery of proteins has only recently been reported.

When liquid is pumped through a bed of adsorbent beads which are not constrained by an upper flow adapter the bed will expand and spaces will develop between the beads. The degree of expansion is mainly dependent on the size and density of the beads and the flow velocity, density and viscosity of the mobile phase.

Matrices designed for use in conventional packed columns are only slightly denser than water; in consequence they give the desired two- to three-fold bed expansion at linear flow velocities of 10 to 30 cm h^{-1}, which are too low for practical application.[51] Denser composite materials, such as dextran–silica, agarose–Kieselguhr or cellulose–TiO$_2$ have been used, as have perfluorocarbon polymers. All of these materials give a two- to three-fold bed expansion at linear flow velocities of 100–300 cm h^{-1}, but suffer either from low capacity for proteins or are unstable at the high pH values necessary for sanitization. More recently an agarose–quartz composite has been introduced which appears to offer both reasonable capacity and pH stability.[52] This material is commercially available under the name Streamline from Amersham Pharmacia Biotech, in both ion exchange and affinity derivatives.

The columns for expanded bed adsorption have a moveable upper flow adapter and a lower adapter which will retain the packing, give an even distribution over the bed and allow the passage of particulate material without blocking. In typical application flow is upwards for loading and washing to remove residual particulates. For the highest

[51] H. A. Chase, *Trends Biotechnol.*, 1994, **12**, 296.
[52] R. Hjorth, *Trends Biotechnol.*, 1997, **15**, 250.

efficiency of elution the bed is allowed to settle, the upper flow adapter is lowered and the resulting column eluted conventionally by downward flow.

Several applications of expanded bed processes on a pilot scale are given by Hjorth.[52] The starting materials include *E. coli*, yeasts and mammalian cell cultures. In general the purifications achieved are less than those obtained by conventional chromatography, but this disadvantage must be set against the advantages gained by avoiding centrifugation and filtration.

6.11 Membrane Chromatography

Another potential solution to the problem of maintaining high flow rates and productivity is to use membranes rather than packed columns. Membrane technology is well advanced for filtration and ultrafiltration of fermentation broths and for concentration or buffer exchange of protein solutions. Similar membranes can be derivatized with ion exchange or affinity ligands, and used for adsorption chromatography. The ligands are attached to the through pores of the membrane, where mass transport is mainly by convective flow, thus reducing the limitations imposed on conventional matrices by diffusion through the pores. Because a membrane can support very high flow rates with low pressure drops the time required to complete a chromatographic cycle is reduced, leading to increased throughput.[53] The enzyme glucose-6-phosphate dehydrogenase from 400 g yeast was purified on a total of 230 cm^2 of affinity membrane derivatized with Cibacron Blue.[54]

6.12 Maintenance of Column Packing Materials

The effective maintenance of the matrix in a large-scale column is vital, so as to ensure both the integrity of the product and the longevity of the matrix. There are three main contributors to fouling of the gel bed: particulate matter in the sample, material which is non-specifically adsorbed and microbial contamination.

The first of these can be dealt with by passing all solutions through filters with a pore size between 5 μm and 10 μm. Because large-scale extracts are inevitably heavily contaminated with particulate matter, it is preferable to pass the sample and buffers through separate filters. Non-specifically bound material can be removed by washing the gel with a

[53] J. Thommes and M. R. Kula, *Biotechnol. Prog.*, 1995, **11**, 357.
[54] B. Champluvier and M. R. Kula, *Biotechnol. Bioeng.*, 1992, **40**, 33.

variety of agents, either singly or in combination, such as 2 M NaCl, up to 6 M urea, non-ionic detergents, such as 1% Triton X-100, or up to 1 M NaOH. Most of the cross-linked matrices are stable in the presence of alkali, but the manufacturers information should always be consulted.

In the case of many affinity media the problem is more complex because of the limited stability of many ligands. Some, such as hydrophobic and dye ligands are very stable, and can be subjected to the same harsh conditions used for ion exchange materials. Other ligands, such as nucleotides, lectins, or antibodies are much less stable, and washing procedures are limited to the use of high concentrations of salts, or perhaps alternating high and low pH, for example, pH 8.5 and 4.5. For these reasons it is often best to apply affinity chromatography at a late stage in a purification, when the worst of the contaminants have been removed. One possible exception to this is in the purification of products from mammalian cell culture, where the medium is very clean in comparison to bacterial extracts.

The prevention and removal of bacterial contamination, or the pyrogens that can result from such contamination, can also be achieved by treating the column with 0.5 M to 1 M NaOH, providing the matrix is sufficiently stable. Many gels can be autoclaved, but although this can sterilize it does not destroy all pyrogens and cannot be carried out without first removing the gel from the column, and in addition does not have any cleaning action.

For the long-term prevention of microbial contamination, during storage, for example, some form of antibacterial agent must be employed. In laboratory columns it is common practice to use agents like sodium azide or merthiolate, but these are unsuitable for use on a large scale, particularly when proteins intended for therapeutic use are being purified. Suitable preservatives are 0.1 M NaOH or 25% ethanol, both of which can be readily removed from the column when required.

6.13 Equipment for Large-scale Chromatography

Columns for large-scale chromatography should be constructed so as to have the minimum dead volume above and below their packing, and the end pieces should be designed so as to ensure an even distribution of sample over the entire surface area of the column, which may be $10\,000\ \text{cm}^2$ or more.

Large-scale columns of glass or plastic construction are available from Amicon (Stonehouse, Gloucestershire, UK), Amersham Pharmacia Biotech (Uppsala, Sweden), BioRad Laboratories (Hercules, CA, USA) and Whatman (Maidstone, Kent, UK). Amicon and Pharmacia also manufacture short, sectional, stack columns, either for adsorptive techniques, or for use with soft gels which cannot be used in long columns but where a long column is required. This type of configuration has the advantages that not all of the column need be used at one time and that if one section becomes contaminated it can be removed without disrupting the remainder of the bed.

The Pharmacia 'stack' column is 37 cm in diameter, with a volume of 16 litres. Up to 10 of these columns can be connected in series, giving a column of 160-litre capacity which has most of the flow characteristics of only one-tenth of that capacity. Amicon manufacture a range of sectional columns from 25 cm to 44 cm diameter, with capacities ranging from 10 to 30 litres, and which have an adjustable end-piece to compensate for changes in bed height. Both of these manufacturers also supply columns in stainless steel. These columns have the disadvantage of being opaque, but in the industrial situation offer the advantages of robustness and ease of cleaning and sterilization.

The packing of large columns needs to be carried out with care so as to avoid stratifying particles of differing sizes or creating cavities within the bed. The recommended procedure is to use an extension piece on the column and to fill it with a gel slurry containing the correct amount of gel, at a concentration of about 40% settled gel by volume in a solution containing 0.5 M NaCl. The solvent is then removed with a pump at a linear flow rate of about 5 cm min^{-1} until the top of the gel bed has dropped just below the top of the column. At this point the column outlet is closed, the extension removed and the top fitting attached as quickly as possible. Cellulose media can also be packed in this way, although both Amicon and Whatman recommend using a slurry packing technique, in which the column is filled with buffer, and a slurry consisting of about 25% settled gel pumped into the column at a high flow rate, until the column is filled with packing. Care must be taken during this operation not to exceed the maximum pressure limits of the column or packing.

Pumps used for process chromatography should be reliable, resistant to corrosion and able to operate at variable flow rates with minimum pulsation. They should not generate excessive heat or shear, and there must be no risk of contamination of the process stream with lubricants or seal materials. For the purification of therapeutic proteins a sanitary design is also important. The peristaltic pumps normally used in the laboratory are less widely used in process chromatography, because of the risk of loss of product if the tubing should split, although they do have the advantage that the pumping mechanism is separated from the liquid stream. The alternative is to use a lobe rotor type pump, fitted with a frequency controlled motor, for the control of flow rate. These pumps are readily available in a sanitary design, and cover a wide range of flow rates.

6.14 Control and Automation

Chromatographic separations can be readily automated, using either specialized microprocessor-based controllers, or microcomputers with suitable programs and interfaces, to operate valves at preset points in the process. Using such equipment it is possible to monitor the column effluent for pH, conductivity, flow rate and pressure. The flow rate should be held constant by means of a feedback control between the pump and the flow meter, which will compensate for variations in the flow rate during the process. It is also convenient to be able to alter the flow rate during the process, perhaps high during equilibration, sample load and washing, and low during elution and regeneration. Pressure transducers can be placed both before and after the column and be set to shut the system down should the pressures exceed the maximum and minimum values allowed. Sensors which can detect the presence of air in the liquid stream are also available, and these can be placed immediately before the column to protect it from running dry by shutting the system down should air enter for any reason. The column eluate should be passed through an ultraviolet monitor to detect protein, and this can be used to initiate the collection of product when the absorbance reaches a threshold value.

7 ULTRAFILTRATION

Ultrafiltration has become a standard laboratory technique for the concentration of protein solutions under very mild conditions. It can be used as an alternative to dialysis or gel filtration for desalting or buffer exchange. By using affinity precipitants to increase the molecular weight of the desired protein it can also be used as means of purification.

Ultrafiltration units are available as either stirred cells with a flat membrane, or as hollow fibres. These fibres have similar characteristics to the flat sheets, but for large-scale processing give a much larger surface area for a given volume. For pilot-scale operation, units are available with up to 6.4 m^2 of membrane area, which have ultrafiltration rates of up to 200 l h^{-1}, depending on the protein concentration. Much larger units are available which have ultrafiltration rates of several hundred litres per hour, making this method applicable to almost any scale of operation.

8 DESIGN OF PROTEINS FOR PURIFICATION

Upstream factors can have a major impact on the development of enzyme and protein purification regimes, and recombinant DNA technology has had a major impact on protein purification.[44] By fusing the gene of interest to an efficient promoter sequence a heterologous protein can be expressed in a host organism at 10% to 40% of the total soluble protein of the cell. This should be compared to the expression of many natural proteins, which may only constitute 0.01% to 4% of the total protein of the cell. As a result the subsequent purification of the protein is simplified. For proteins which are expressed in a soluble form, genetic techniques can be used to direct the newly synthesized protein into the periplasm of the cell, or even into the culture medium. This can increase the stability of the expressed protein, since only two of the eight known proteases of *E. coli* are wholly in the periplasm, and simplify the purification, because only about 8% of all *E. coli* proteins are periplasmic.

8.1 Inclusion Bodies

Such high levels of expression can lead to the protein being produced as dense, insoluble granules called inclusion bodies. These have been observed with many recombinant proteins, including urogastrone, interleukin-2, prochymosin and interferons. After cell disruption, such granules can be sedimented at a relatively low RCF, to yield insoluble material containing over 50% of the desired protein.

The reasons for the formation of inclusion bodies are not understood, since not all proteins expressed at high level form inclusion bodies. The host cell may also play an important role, because human growth hormone, which forms inclusion bodies in some *E. coli* strains, is freely soluble, with correctly formed disulfide bridges in *E. coli* RV308. The precise structure of the recombinant protein can also affect the formation of inclusion bodies. An extensive study using recombinant human interferon gamma showed that just a few amino acid changes could affect the transition between soluble and insoluble expression of the protein in *E. coli*.[55]

The solubilization and subsequent refolding of inclusion bodies represents a major challenge. Commonly the inclusion bodies are solubilized in urea or guanidinium chloride, often at high pH values, and sometimes with the addition of detergent. Once solubilized the

---

[55] R. Wetzel, L. J. Perry, and C. Veilleux, *Bio/Technology*, 1991, **9**, 731.

protein must be allowed to refold into the native conformation. In many cases simple dilution of the solubilized extract into a suitable buffer is sufficient. If the protein contains disulfide bonds it is often necessary to include oxidized and reduced glutathione to provide a suitable environment to encourage their correct formation.[56] Sometimes it is necessary to add a co-solvent, such as polyethylene glycol, or a detergent, such as Triton X-100, Tween 20 or Zwittergent 3-16. In some cases solubilized proteins can be very difficult to refold. Under these circumstances it has been observed that the addition of chaperonins can aid the refolding process. Chaperonins are proteins, which are thought to be involved in ensuring correct folding of proteins *in vivo*; with the availability of recombinant chaperonins they have been used to catalyse the correct refolding of proteins *in vitro*.[57]

8.2 Affinity Tails

Another example of genetic design to aid protein purification is the concept of affinity tails. The gene for the protein of interest is fused to a DNA sequence that codes for some amino acid sequence which will simplify the purification of the protein, by modifying its properties in a predictable manner.[58,59] One of the first examples was the genetic fusion of several arginine residues to the *C*-terminus of urogastrone. This made an unusually basic protein, which was strongly bound to a cationic ion exchange matrix. Since this group of matrices binds only about 10% of all cellular proteins, a large purification can be obtained on elution. The polyarginine tail was then removed by use of immobilized carboxypeptidase A; re-chromatography of the liberated recombinant protein on the same matrix results in a major change in its elution position, but not in the elution position of the contaminants.[60] Several examples of affinity fusion systems are shown in Table 6.

A major problem with affinity fusions for purification is the successful removal of the both the affinity tail and the reagents used. This problem may be partially solved by the genetic introduction of highly specific points for protease cleavage or of acid-labile bonds at the junction of the desired recombinant protein and the affinity tail. Many methods for cleavage have been suggested, as shown in Table 7. Those that require strongly acid conditions are only suitable for proteins which are stable

[56] A. Mukhopadhyay, *Adv. Biochem. Eng. Biotechnol.*, 1997, **56**, 61.
[57] J. G. Thomas, A. Ayling, and F. Baneyx, *Appl. Biochem. Biotechnol.*, 1997, **66**, 197.
[58] H. M. Sassenfeld, *Trends Biotechnol.*, 1990, **8**, 88.
[59] R. F. Sherwood, *Trends Biotechnol.*, 1991, **9**, 1.
[60] H. M. Sassenfeld and S. J. Brewer, *Bio/Technology*, 1984, **2**, 76.

Table 6   *Examples of protein purification methods using affinity tails*

Affinity tail	Ligand/matrix	Binding conditions	Elution conditions
Oligo arginine[a]	S-Sepharose	pH 4–8	NaCl gradient
Oligo histidine[b, c]	Iminodiacetate–Sepharose (Ni^{2+})	pH 7–8 ± guanidinium chloride	Imidazole buffer or decreasing pH gradient ± guanidinium chloride
Flag[TM] antigenic peptide[d]	Anti-Flag antibody–Sepharose	0.15 M NaCl, 1 mM $CaCl_2$, pH 7.8	10 mM EDTA, pH 7.4
Beta-galactosidase[e]	TPEG-Sepharose	1.6 M NaCl, pH 7	0.1 M sodium borate
Chloramphenicol acetyl transferase[f]	p-aminochloramphenicol–Sepharose	0.3 M NaCl, pH 7.8	5 mM chloramphenicol
Protein A[g, h]	IgG–Sepharose	pH 7.6	0.5 M acetic acid
Glutathione-S-transferase[i,j]	Glutathione–Sepharose	pH 7.3	Glutathione

[a] S. J. Brewer and H. M. Sassenfeld, *Trends. Biotechnol.*, 1985, **3**, 119.
[b] F. H. Arnold, *Bio/Technology*, 1991, **9**, 151.
[c] M. W. Van Dyke, M. Sirito, and M. Sawadogo, *Gene*, 1992, **111**, 99.
[d] T. P. Hopp, K. S. Prickett, V. L. Price, R. T Libby, C. J. March, D. L. Urdal, and P. J. Conlon, *Bio/Technology*, 1988, **6**, 1204.
[e] A. Ullman, *Gene*, 1984, **29**, 27.
[f] J. Robben, G. Massie, E. Bosmans, B. Wellens, and G. Volckaert, *Gene*, 1993, **126**, 109.
[g] T. Moks, L. Abrahmsen, B. Osterlof, S. Josephson, S. Ostling, S.-O. Enfors, I. Persson, I., B. Nilsson, and M. Uhlen, *Bio/Technology*, 1987, **5**, 379.
[h] P. A. Nygren, M. Eliasson, L. Abrahmsen, M. Uhlen, and E. Palmcrantz, *J. Molec. Recognition*, 1988, **1**, 69
[i] S. Sankar and A. G. Porter, *J. Virol.*, 1991, **65**, 2993.
[j] D. B. Smith and K. S. Johnson, *Gene*, 1988, **67**, 31.

Table 7 *Examples of methods for removing affinity tails*

Linker sequence	Cleavage method	Conditions
↓ –Asn–Gly–[a]	Hydroxylamine	pH 9, 45°
↓ –Asp–Pro–[b]	Acid	10% acetic acid, 55°
↓ –Met–Xxx–[c]	CNBr	70% formic acid, 20°
↓ –Xxx–(Arg)$_n$ [d]	Carboxypeptidase B	pH 8, 37°
↓ –Xxx–(His)$_n$ [e]	Carboxypeptidase A	pH 8, 37°
↓ –Gly–Val–Arg–Gly–Pro–Arg–Xxx–[f]	Thrombin	pH 7–8, 37°
↓ –Ile–Glu–Gly–Arg–Xxx–[g]	Factor X$_a$	pH 7–8, 37°
↓ –Asp–Asp–Asp–Lys–Xxx–[h]	Enterokinase	pH 8, 37°
↓ –Leu–Glu–Val–Leu–Phe–Gln–Gly–Pro–[i]	PreCission Protease	pH 7, 5°

↓ indicates the bond cleaved; Xxx indicates any amino acid.
[a] T. Moks, L. Abrahmsen, E. Holmgren, M. Bilich, A. Olsson, M. Uhlen, G. Pohl, C. Sterky, and H. Hultberg, *Biochemistry*, 1987, **26**, 5239.
[b] B. Nilsson, L. Abrahmsen, and M. Uhlen, *EMBO J.*, 1985, **4**, 1075.
[c] B. Hammarberg, P.-A. Nygren, E. Holmgren, A. Elmblad, M. Tally, U. Hellman, T. Moks, and M. Uhlen, *Proc. Natl. Acad. Sci. USA.*, 1985, **86**, 4367.
[d] H. M, Sassenfeld. and S. J Brewer, *Bio/Technology*. 1984, **2**, 76.
[e] M. C. Smith, T. C. Furman, T. D. Ingolia, and C. Pidgeon, *J. Biol. Chem.*, 1988, **263**, 7211.
[f] J. A. Knott, C. A. Sullivan, and A. Weston, *Eur. J. Biochem.*, 1988, **174**, 405.
[g] J. Shine, I. Fettes, N. C. Y. Lan, J .L. Roberts, and J. D. Baxter, *Nature*, 1980, **285**, 456.
[h] T. P. Hopp, K. S. Prickett, V. L. Price, R. T. Libby, C. J. March, D. L. Urdal, and P. J. Conlon, Bio/Technology, 1988, 6, 1204.
[i] P. A. Walker, L. E. Leong, P. W. P. Ng, S. H. Tan, S. Walker, D. Murphy, and A. G. Porter, *Bio/Technology*, 1994, **12** 601.

under such conditions. Specific proteases are preferred because milder conditions can be employed. The main proteases used are thrombin, enterokinase and Factor X$_a$. These all act at 37°C and may cleave at sites within the protein of interest, either because of a lack of specificity or from contaminating proteases. In addition, subsequent chromatography steps are required to remove the cleaved affinity tail and the protease.

A recent development in this area has been the use of a recombinant form of the protease 3C from human rhinovirus. This is small in size (20 kDa) and has a very restricted specificity, as shown in Table 6. It has been expressed as recombinant protein fused to glutathione-*S*-transferase. This offers several advantages. The protease itself can be purified by affinity chromatography on glutathione–Sepharose. If the target protein is expressed as a fusion with glutathione-*S*-transferase it can be purified on a column of glutathione–Sepharose. After treatment with the protease the target protein can be separated from both the glutathione-*S*-transferase tail and the protease by passage through a second glutathione–Sepharose column. Alternatively, cleavage can take place on the first glutathione-Sepharose column by adding the protease to the column buffer, so avoiding the need for a second column.[61] This protease is commercially available under the name PreCission protease from Amersham Pharmacia Biotech.

9 FUTURE TRENDS

The future of large-scale protein purification is assured, if only because of the increasing number of biopharmaceuticals that will be coming onto the market, as many of these will be required in very large amounts at a high degree of purity. The value of such products will mean that high performance techniques will be used if the process is applicable. It is likely that the availability of recombinant proteases will overcome many of the problems associated with affinity tailing of proteins.

[61] P. A. Walker, L. EC. Leong, P. W. P. Ng, S. H. Tan, S. Walker, D. Murphy, and A. G. Porter, *Bio/Technology*, 1994, **12**, 601.

CHAPTER 18

Monoclonal Antibodies

CHRISTOPHER J. DEAN

1 INTRODUCTION

Antibodies have been an important tool of the biologist, biochemist and immunologist for decades, but it was the development of hybridoma technology by Köhler and Milstein[1] and the ability to produce mono-specific antibodies (monoclonal antibodies or mAbs) that has revolutionized our ability to approach a vast range of problems in biomedical research. When an animal meets a challenge from a foreign antigen such as a virus, bacterium or deliberately injected foreign protein, phagocytic cells of the immune system such as macrophages engulf the antigen as a first line defence mechanism. Digestion products of the foreign antigen are then displayed at the surface of the phagocytic cell and attract the attention of specialized lymphocytes derived either from the bone marrow (B-cells or antibody producing cells) or thymus (T-cells). Both T- and B-cells have specific receptors at their surface which, if they can interact with antigen, may lead to proliferation of the B-cells. In this way an antibody response can be generated in the lymph nodes of the infected animal which may involve many B-cells which recognize different components of the foreign antigen. This 'polyclonal' response leads to the generation of many different antibodies that recognize the original antigen. Repeated immunization ('hyper-immunization') with, for example, a protein antigen, leads to the selection of B-cells producing high affinity antibodies in the lymph nodes (affinity maturation). Sera from these hyperimmune animals contain a number of antibodies specific for antigen (polyclonal) and while this may be fine for many uses it precludes analysis of the individual epitopes involved. It is the ability to capture the

[1] G. Köhler and C. Milstein, *Nature (London)*, 1975, **256**, 495.

individual antibody secreting B-cells as hybridomas that has influenced our approach to, and understanding of, many problems in biomedical research. In this chapter we discuss the 'classical' method for producing and testing monoclonal antibodies by hybridoma formation and discuss their use in biomedical research. More recent developments based on the use of recombinant gene technology are also described including the use of mice bearing human immunoglobulin transgenes and the preparation of libraries of immunoglobulin genes with their products displayed on bacteriophage.

2   ANTIBODY STRUCTURE

The basic antibody is a four chain structure (Figure 1) where a light chain is linked *via* a disulfide bond to a heavy chain to give an antigen binding unit (Fab) and two of these structures are linked *via* their heavy chains by one or more disulfide bonds in the so-called 'hinge region'. The latter allows flexibility of the two Fabs and can assist in binding of antigen. Each light and heavy chain consists of two or four domains respectively, the first of which at the 5' end shows considerable variation from one antibody to another (V_H and V_L) and constitutes the antigen combining site. The variable portions of the heavy and light chains are each derived from gene sequences which on processing contribute the variable (V), diversity (D) and joining (J) regions and from which the three complementarity determining regions (CDRs 1–3) in each heavy and light chain

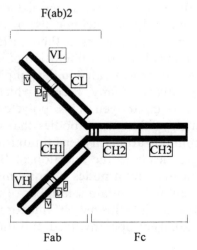

Figure 1   *Schematic representation of immunoglobulin G showing the domain structure*

are constructed. The remaining domains known as the constant regions show much less variability but are important in defining the different classes of antibody (IgA, IgG, IgM, *etc.*) which relate to functions of the antibodies such as complement activation and recruitment of host immune effector cells *via* their Fc (fraction crystallizable) receptors. The third and fourth constant domains of the heavy chains are important for the latter functions and are known as the Fc region. Because the biological functions of the antibodies are dependent on their antibody class or isotype it may be important to look for such characteristics in the monoclonal antibodies that are generated. The requirements for making hybridomas will be discussed and examples of mAbs prepared against an isolated protein antigen (human immunodeficiency I virus envelope glycoprotein) or against cell surface receptors (*e.g.* the receptor for epidermal growth factor (EGFR) and the related product of the c-*erb*B-2 oncogene) indicated.

3   PREPARATION OF HYBRIDOMAS BY SOMATIC CELL FUSION

3.1   Principle of the Technology

The method is essentially straightforward, namely, fuse a B-cell producing the desired antibody with a continuously growing myeloma cell-line and plate onto a selective medium containing hypoxanthine, aminopterin and thymine (HAT). The myeloma cell-line has lost the enzyme hypoxanthine guanosine phosphoribosyl tranferase (HGPRT) and is unable to synthezise DNA in the presence of the nucleotides and the folic acid inhibitor aminopterin. The B-cell partner does not survive for long in culture whereas the hybridoma acquires the HGPRT gene from the B-cell partner and can survive and proliferate in HAT-containing medium. By cloning the fusion products, hybridomas secreting a single antibody species can be obtained. The steps in this process are illustrated diagrammatically in Figure 2. Detailed protocols for the preparation of hybridomas and testing of the monoclonal antibodies are given in references 2–4.

[2] G. Galfré and C. Milstein, 1981, *'Methods in Enzymology, Vol. 73, Immunochemical Techniques'* ed. J.J. Langone, and H. Van Vunakis, Academic Press, New York, pp.3–46.
[3] C. J. Dean, in, *'Methods in Molecular Biology, Vol. 80, Immunochemical Protocols'*, Humana Press Inc., Totowa, N. J., 1998, pp.23–37.
[4] E. Harlow and D. P. Lane, *'Antibodies: A Laboratory Manual'*, Cold Spring Harbor Laboratory, Cold Spring Harbor, NY, 1988.

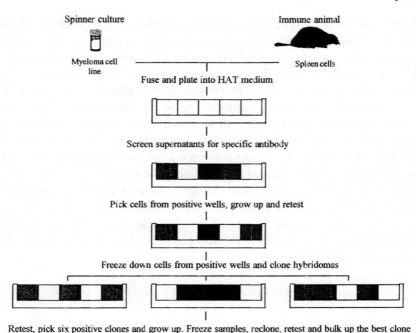

Spinner culture   Immune animal

Myeloma cell ——————————————————— Spleen cells
line
Fuse and plate into HAT medium

Screen supernatants for specific antibody

Pick cells from positive wells, grow up and retest

Freeze down cells from positive wells and clone hybridomas

Retest, pick six positive clones and grow up. Freeze samples, reclone, retest and bulk up the best clone

Figure 2 *Production of monoclonal antibodies by somatic cell fusion*

3.2 Choice of Myeloma Cell-line

The original procedure describing hybridoma formation[1] utilized mineral oil-induced plasmacytomas (MOPC) from BALB/c mice which had been selected in 8-azaguanine so that they lacked the enzyme hypoxanthine guanosine phosphoribosyl transferase (HGPRT). These cells are unable to utilize the salvage pathway for DNA synthesis and consequently die in media containing aminopterin or methotrexate. Today, the majority of hybridomas are prepared using a number of different derivatives of such mouse myelomas (Table 1). Rats of the LOU/wsl strain spontaneously develop ileocaecal plasmacytomas and several of these have been developed for use in hybridoma production (Table 1). An advantage of rat × rat hybridomas is that they are genetically more stable and their products are often of higher affinity for the target antigen. Ideally the myeloma cells used for fusion should not produce or secrete their own heavy and light chains which will increase the variability of the secreted immunoglobulin and a number of cell-lines have been produced with this goal in mind.

Table 1 *Rodent myeloma cell-lines commonly used for hybridoma production*

Name	Species/origin	Ig expression	Reference
Mouse lines			
P3-X63/Ag8	BALB/C mouse	IgG1 (K)	1
NS1/1.Ag 4.1	BALB/C mouse	K chain (non-secreted)	3
X63/Ag 8.653	BALB/C mouse	None	4
Sp2/0	Hybrid Sp2	None	5
NS0/1	NS1/1.Ag 4.1	None	6
Rat lines			
Y3-Ag 1.2.3	LOU/wsl rat	K chain	7
IR984F	LOU/wsl rat	None	8

3.3 Choice of Host for Production of Immune B-cells

Since the original myeloma cell-lines were derived from BALB/c mice, spleens from these animals have been used routinely as the source of immune lymphoid cells. However, it is possible to generate hetero-hybridomas by using immune B-cells from other animals and fusing them with mouse myeloma cell-lines; a number of successful fusions have been reported (Table 1). An advantage of using rats as the source of immune lymphocytes is that not only are the number of B-cells obtained from the spleen some four-fold greater than from the spleen of a mouse, but also that it is possible to isolate B-cells from immune lymph nodes and so widen the choice of antibody isotype obtained.[10] Also non-secreting hetero-hybridomas which subsequently have been rendered resistant to 8-azaguanine, have been used as fusion partners in certain cases.[11,12] The reason for making monoclonal reagents from sheep or rabbits is that it is expected that the affinity of the antibodies will be substantially greater than those obtained from mice. Hybridoma formation requires the successful fusion of an immune B cell with its myeloma partner and the capacity of the B-cell to undergo fusion is short lived and the fully differentiated plasma cells do not undergo fusion. Indeed, the ability to capture antibody genes from all antibody producing cells is one

[5] G. Köhler, S. C. Howe and C. Milstein, *Eur. J. Immunol.*, 1976, **6**, 292.
[6] J. F. Kearney, A. Radbruch, B. Liesegang and K. Rajewsky, *J. Immunol.*, 1979, **123**, 1548.
[7] M. Shulman, C. D. Wilde and G. Köhler, *Nature (London)*, 1978, **276**, 269.
[8] G. Galfré, C. Milstein and B. W. Wright, *Nature (London)*, 1979, **277**, 131.
[9] H. Bazin, (1982) in *Protides of Biological Fluids, 29th Colloquium* (Peeters, H., ed.) Pergamon, New York, pp.615–618.
[10] C. J. Dean, J. M. Styles, L. A. Gyure, J. Peppard, S. M. Hobbs, E. Jackson and J. G. Hall, *Clin. Exp. Immunol.*, 1984, **57**, 358–364.
[11] D. V. Anderson, E. M. Tucker, J. R. Powell and P. Porter, *Vet. Immunol. Immunopathol.*, 1987, **15**, 223–237.
[12] J. N. Flynn, G. D. Harkiss and J. Hopkins, *J. Immunol. Meth.*, 1989, **121**, 237–246.

advantage of using recombinant gene technology and this is discussed in Section 4.

3.4   Immunogen and Route of Immunization

Many different types of immunogen can be used, *e.g.* whole cells, macromolecules such as precipitated DNA, purified proteins, glyco-proteins and carbohydrates, recombinant proteins expressed in bacteria, yeast, insect or Chinese hamster ovary cells. Where the recombinant proteins are normally glycosylated, the carbohydrate side chains added will be those of the species in which they are expressed. Such alterations may show differential immunogenicity compared with the original glycoprotein, *e.g.* if it is of human origin. Peptides based on derived cDNA sequences have been widely used to produce polyclonal antisera in rabbits, but they are often poorly immunogenic in mice and rats even when coupled to larger proteins bearing T-cell reactive epitopes such a keyhole limpet haemocyanin or ovalbumen.

It is not necessary to have purified materials for immunization although it will increase the chances of success with poorly immunogenic molecules. Partially purified detergent extracts of cells expressing high levels of the protein of interest have been used successfully. Indeed, intact live cells, expressing high levels of receptors at their cell surface have proven to be the method of choice for the preparation of monoclonal antibodies against certain receptors, *e.g.* that for epidermal growth factor (EGFR) and the product of the EGFR-related c-*erb*B-2 proto-oncogene.[13,14] An important feature of using cells expressing these glycoproteins is that the conformation of the receptor is retained since it is known that, for example, the affinity of the EGF receptor for ligand is an order of magnitude less following isolation in non-ionic detergent.

To obtain the highest level of antibody response, peptides, proteins and other soluble macromolecules are usually emulsified with an adju-vant to give a relatively stable emulsion which acts as a slow release agent. The most usually employed adjuvant is Freund's which contains the tumour promoter croton oil together with heat-killed mycobacteria which act as attractants for phagocytic antigen-presenting cells such as macrophages (Freund's complete adjuvant, FCA). It is usual to give the initial immunization in FCA then to follow this at weekly to monthly intervals with the protein emulsified in Freund's incomplete adjuvant (FIA) which lacks the mycobacteria. Use of this adjuvant can lead to

[13] H. Modjtahedi, J. M. Styles and C. J. Dean, *Br. J. Cancer*, 1993, **67**, 247–253.
[14] J. M. Styles, S. Harrison, B. A. Gusterson and C. J. Dean, *Int. J. Cancer*, 1990, **45**, 320–324.

ulceration at the site of injection and for this reason less damaging adjuvants have been sought; these include precipitates made by admixture with alum. Intact, live cells do not require mixing with adjuvant and are usually injected in medium lacking serum or phosphate-buffered saline (PBS) at pH 7.4.

Initially, the animals are anaesthetized and bled to obtain a baseline serum sample (from the tail vein in mice or the jugular vein in rats). The animals are then immunized usually at five sites (4 × subcutaneously and 1 × intraperitoneally) with antigen in FCA or live cells in PBS or medium. The animals are re-immunized at intervals of 1 week to 1 month and blood samples are taken just prior to immunization. Serum samples are assessed for the appearance of specific antibody and when the titre is judged to be sufficient the animals are given a final immunization, and then killed and the spleens removed 3–4 days later. In rats it is also possible to immunize the animals *via* their Peyer's patches, immune tissues that lie along the small intestine. The Peyer's patches drain into the mesenteric nodes which have been found to be an excellent source of antibody-producing cells for hybridoma production.[9] In many countries the use of animals for experimental purposes is rightly under strict control and the procedures outlined above can only be performed by licence-holding, trained personnel.

3.5 Preparation of Myeloma Cell-line and Host Immune Lymphocytes for Fusion

It is important that the myeloma cell-line is growing well prior to its use and while static cultures grown in flasks can be suitable with some lines (*e.g.* Sp2/0), it is usual to grow them in spinner culture. With the rat myeloma Y3, which has been widely used for preparation of rat × rat hybridomas, growth in spinner culture is essential. The myelomas should be kept in exponential growth prior to fusion (in Dulbecco's modified Eagles medium or RPMI 1640) then spun down, washed twice in serum-free medium and resuspended in serum-free medium at $1–2 \times 10^7$ cells ml^{-1} just before use.

Immune animals are killed humanely, swabbed with ethanol where an incision is to be made, then the spleens or mesenteric lymph nodes are removed by blunt dissection and placed in sterile growth medium without serum. The lymphoid tissue is disaggregated by forcing through a stainless steel mesh (sterile tea strainer) using a sterile spoon-headed spatula (dipped in alcohol and flamed) and the cells collected by centrifugation. After three washes in serum-free medium the cells are resuspended at $5 \times 10^7–10^8$ lymphocytes ml^{-1}. Some workers prefer to remove erythrocytes before fusion and they can be lysed by the addition of NH_4Cl.

3.6   Hybridoma Formation by Somatic Cell Fusion

Although a number of procedures have been used to induce cell fusion (*e.g.* sendai virus, electrofusion) the majority of hybridomas have been generated by the addition of polyethylene glycol to bring about fusion of the cell membranes of the myeloma and B-cell. The fusion partners are mixed to give about 10^8 splenic lymphocytes and 10^7 myeloma cells (mouse) or 5×10^7 myeloma cells (rat) and then pelleted by centrifugation in a 10 ml round-bottom tube. After removal of the supernatant the pellet is loosened by tapping the tube then 1 ml of polyethylene glycol 1500 (50% in medium) is stirred slowly into the pellet over 1–2 min to induce cell fusion. The cell suspension is then diluted with medium alone over a further 1–5 min and then diluted into medium (DMEM or RPMI) containing 20% foetal calf serum, hypoxanthine, aminopterin and thymine (HAT selection medium) and plated into multiwell plates. Not all batches of foetal calf serum support the growth of newly formed hybridomas and it is worth testing batches for this property or use commercial batches of FCS which have been shown to support growth. With mouse fusions, it is usual to prepare thymocyte feeders from the spleen donor which are added to the HAT medium, while with rats X-irradiated rat fibroblasts[3] derived by trypsinization of the xiphisternae (terminus of sternum) are used. The fusion mixture is plated as 2 ml aliquots into four 24-well plates, although some other workers prefer to use multiple 96-well plates to ease selection of positive hybridomas. The cultures are incubated at 37°C in an atmosphere of 5% CO_2 and examined after 7–10 days for the presence of hybridoma colonies. If necessary the medium is exchanged for fresh HAT selection medium (*e.g.* 96-well plates) and incubation continued until it is judged to be time for screening of the culture supernatants for a specific antibody.

3.7   Screening Hybridoma Culture Supernatants

Perhaps the most important part of hybridoma production is the test(s) used to identify specific antibody in culture supernatants. Tests which can be performed quickly and routinely are essential since some of the hybridomas have generation times of under 12 h. Consequently long-term biological assays are not the first method of choice. Where conjugated peptides, proteins or other macromolecules are available, these can be coated onto 96-well plates and used directly to assay for specific mAb (monoclonal antibody). The procedure is illustrated in Figure 3. Adherent cells can be used in the same way but it may be necessary to cool them to 4°C to prevent internalization of antibodies from the cell surface. In practice, samples of culture supernatant are added to duplicate wells and incubated for one hour. After washing three times, bound antibody is revealed using a secondary antibody (anti-mouse or rat F(ab)$_2$) conjugated to 125-iodine (radioimmunoassay, RIA) or an enzyme such as alkaline phosphatase or horseradish peroxidase (enzyme-linked immunosorbant assay, ELISA). After incubation for one hour the plates are rewashed and then the 125-iodine label counted in a gamma counter or, in the case of the ELISA, specific substrate is added and the intensity of the colour of the product read in a spectrophotometer. The use of polyvinyl chloride (PVC) plates for the radioimmunoassays greatly assists subsequent counting since the plates can be cut into individual wells with scissors. When confluent cell monolayers are used for screening they may be assayed directly in multiwell plate readers such as the Titertek (ELISA) or solubilized in ionic detergent (*e.g.* 1% sodium dodecyl sarkosinate in 0.5 M NaOH) and the supernatant removed for counting (RIA).

Colonies are picked from positive wells of the fusion culture and plated into mediun containing hypoxanthine and thymine (HT medium). When the picked colonies have grown up they are retested to determine

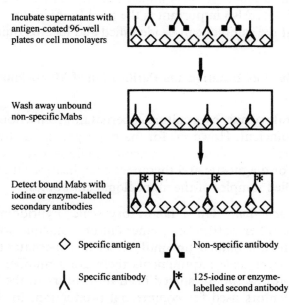

Incubate supernatants with antigen-coated 96-well plates or cell monolayers

Wash away unbound non-specific Mabs

Detect bound Mabs with iodine or enzyme-labelled secondary antibodies

◇ Specific antigen   Λ Non-specific antibody

Λ Specific antibody   Λ* 125-iodine or enzyme-labelled second antibody

Figure 3 *Screening for specific antibody using the indirect method*

which clone is secreting specific mAb and those cells are expanded and frozen down in liquid nitrogen (we use 5% dimethyl sulfoxide–95% foetal calf serum for freezing down all our cells). When only 4 × 24-well plates are used there may be more than one specific antibody producing clone per well. The selected cells are then cloned twice (see Section 3.8).

3.8 Cloning Hybridomas

By definition, a monoclonal antibody is derived from one specific B-cell. To ensure that the hybridomas are producing mono-specific mAb it is necessary to dilute cells from an antibody-producing colony such that the colonies developed are derived from individual hybridomas. This can be done in two ways. First, cells are counted, then diluted in medium and < 50 cells plated into a 96-well plate. The cells are allowed to grow and then the wells are checked for the presence of colonies and only those with a single colony per well are re-tested for specific antibody. Cells from six of the best producing colonies are picked, grown up and frozen down in liquid nitrogen. The best of this first cloning is then re-cloned using the same procedure and the six best clones bulked up and frozen down. Alternatively, hybridomas may be plated into soft agar so that cells from individual colonies can be picked and tested. It is clear from the above description that it is easy to fill up the liquid nitrogen bank with hybridoma samples and it is imperative that unwanted stocks are disposed of quickly. Also important is the need to maintain a computer database so that the samples can be identified and located quickly.

3.9 Bulk Production, Isolation and Purification of Monoclonal Antibodies

For many laboratory purposes culture supernatant from a doubly cloned hybridoma is sufficient. However, for many purposes, *e.g.* the preparation of immunoaffinity columns, conjugation to fluorochromes, isotopes or enzymes or for diagnostic and therapeutic applications it is necessary to prepare purified samples of the monoclonal antibodies.

3.9.1 Bulk Production. On a laboratory scale, hybridomas can be grown in flasks, roller bottles or spinner culture. Starting with a small seed lot the cells are bulked up until sufficient supernatant has been collected. For larger-scale requirements there are a number of devices from hollow fibre reactors to stirred bead fermentors to the huge 2000 litre airlift fermentors used for commercial production. In the laboratory, the cells are weaned onto media containing low concentrations

(2%) of foetal calf serum before bulking up as this reduces the amount of non-specific protein in the culture supernatants. For commercial production, serum-free media are used, notably if the antibodies are to be used for diagnostic or therapeutic purposes in patients.

An interesting recent development has been the use of transgenic plants for antibody production. Here the genes for the immunoglobulin heavy and light chains (see Section 5) are transferred into plant cells either by insertion into an infective agent such as the T vector of the crown gall bacterium (*Agrobacterium tumefascians*) followed by infection of the plant cells in culture with the transformed bacterium,[15] or alternatively, the DNA coding for the antibody is coated onto gold particles and, using microprojectile bombardment, the particles 'fired' into plant cells derived from tobacco leaves or wheat embryos.[16] In the latter case the antibody can be isolated from the transgenic crushed wheat seed and it has been suggested that a field of transgenic wheat is a cheap source of starting material for the preparation of specific monoclonal antibody.[17]

3.9.2 Isolation and Purification. While commercially produced supernatants may contain 1 gm or more of specific mAb per litre, the concentrations are usually much lower in laboratory-scale production. For this reason procedures to concentrate the mAb containing supernatants are first carried out. One method is to precipitate the immunoglobulin fraction with ammonium sulfate followed by centrifuging off the precipitate and redissolving it in water. Another method is to concentrate the supernatants by diafiltration, although the disadvantage of this method is that all the proteins remain in the supernatant and no change in specific activity is obtained. A good way of producing a 10-fold increase in antibody concentration exists where serum-free media (low protein) have been used. Alternatively, the supernatants may be passed through affinity columns containing antigen, anti-immunoglobulin or the bacterial Ig-binding proteins such as Protein A, G or L. For laboratory purposes, an initial precipitation with ammonium sulfate (to 45% of saturation) is the method of choice, followed by purification by ion exchange chromatography, *e.g.* passage through a cation exchanger such as Whatman DE52 cellulose and/or by affinity chromatography.[3,4]

It is important to determine the relative purity of the mAb obtained, particlarly if it is to be conjugated to fluorochromes and enzymes or

[15] A.C. Hiatt, R. Cafferkey and K. Bowdish, *Nature (London)*, 1989, **342**, 76.
[16] T. M. Klein, E. C. Harper, Z. Svab, J. C. Sanford, M. E. Fromm and P. Maliga, *Proc. Natl. Acad. Sci. USA*, 1988, **85**, 8502.
[17] P. Drake, E. Stoger, L. Nicholdon, P. Christen and J. K.-C. Ma, in *'Monoclonal Antibodies: A Practical Approach'*, (ed. P. Shepherd and C. Dean) Oxford University Press, Oxford, in press.

radiolabelled with ^{125}I, since the presence of contaminants may interfere with the use of the conjugates. The simplest procedure is to run a sample on denaturing and non-denaturing SDS-PAGE and stain with Coomassie Blue, where the intact molecule should run at about 160–180 kDa whereas on the denaturing gels two bands should be seen at 20–25 kDa (light chain) and 40–50 kDa (heavy chain). Gross protein contamination should be readily visible. It is also important to make sure that the purified mAb has the same antigen-binding characteristics of the antibody in culture supernatants.

The derivation of mAbs with radioisotopes, fluorochromes, enzymes, *etc.*, is beyond the scope of this chapter and the interested reader is referred to references 4 and 18 for further information on these aspects.

4 EXAMPLES OF THE PREPARATION OF RAT MONOCLONAL ANTIBODIES WHICH HAVE BEEN USED TO INVESTIGATE THE STRUCTURAL AND FUNCTIONAL PROPERTIES OF MACROMOLECULES

Antibodies can be generated against discrete antigenic sequences, *e.g.* a sequence of amino acids (continuous epitope) or a discontinuous epitope formed either from two adjacent amino acid chains or a conformation formed by folding of the protein molecule. Continuous epitopes can be recognized by antibody even when denatured (*e.g.* Western blots) or present in formalin-fixed paraffin-embedded sections. Conformational epitopes are often important in biological interactions and their activity is frequently lost when the three-dimensional structure is destroyed, *e.g.* by fixation or denaturation in strong ionic detergents. Under these conditions the mAbs no longer bind.

To illustrate how it is possible to generate mAbs which can be used to investigate protein structure and function we will consider first the results of some early studies on the envelope glycoprotein (gp120) of the Type 1 human immunodeficiency virus (HIV I) using rat monoclonal antibodies. Then we shall discuss briefly the preparation, evaluation and use of anti-receptor antibodies for use in diagnostic and therapeutic application in medicine

4.1 HIV I gp120

The Type 1 human immunodeficiency virus is the major cause of acquired immunodeficiency disease syndrome (AIDS) in the Western

[18] J. D. Pound, (Ed.) *'Methods in Molecular Biology, Vol. 80, Immunochemical Protocols'*, Humana Press, Totowa, New Jersey, 1998.

world. The envelope glycoprotein (gp120) is involved in the binding of the human immunodeficiency virus I to the CD4 antigen on the surface of T-cells and in man autologous antibodies directed against gp120 can block infectivity. The *env* gene encodes a 160 kDa precursor protein (gp160) which is processed into the mature external surface gp120 and the transmembrane gp41 glycoproteins. The gp120 from one of the variants of HIV I has been cloned and expressed in Chinese hamster cells and purified as rgp120 and this was used to generate a series of rat mAbs for use in structural and functional studies.

CBH/cbi rats were immunized three times via their Peyer's patches with rgp120 until tests on plates coated with this antigen showed that good titres of antibodies were present in sera obtained from the rats. Using lymph node cells a series of fusions were carried out which generated a total of 62 mAbs that bound to rgp120-coated plates.[19-22] A number of the mAbs were selected for further investigation since they bound to non-overlapping epitopes on HIV I gp120 and of these, some were found to neutralise infective virus.[18] Of the latter, some were found to block infectivity by interfering with the binding of gp120 to the CD4 receptor, *e.g.* mAb 39/13g which defined a conserved conformational epitope while others acted at a different site.[19] Further investigation showed that some antibodies bound to linear determinants (*e.g.* mAb 38.1a, amino acids 430–447[20]) whereas others (*e.g.* mAb 8/19b) bound to conformational determinants involving the first and third constant regions (C1 and C3) or within the third variable region (V3 loop).[21] Of the 62 mAbs, 31 were found to recognize linear determinants within the C1, V2, V3, C4 and C5 regions and 31 bound to conformation dependent epitopes.[21] In this way it has been possible to build up a picture of the immunogenic regions of the viral envelope glycoprotein based on its cDNA-derived amino acid sequence and to investigate the involvement of the different regions in virus activity.

4.2 mAbs to Growth Factor Receptors

The receptor for epidermal growth factor (EGFR) is expressed by a number of normal dividing tissues, in particular skin and squamous epithelium, but the receptor is over-expressed by certain tumour cells, notably carcinomas of squamous cell origin.[23] The related receptor

[19] J. Cordell, J. P. Moore, C. J. Dean, P. J. Klasse, R. A. Weiss and J. A. McKeating, *Virology*, 1991, **185**, 72.
[20] J. A. McKeating, J. Cordell, C. J. Dean and P. Balfe, *Virology*, 1992, **191**, 732.
[21] J. A. McKeating, C. Shotton, J. Cordell, *et al.*, *J. Virol.*, 1993, **67**, 4932.
[22] J. A. McKeating, C. Shotton, S. Jeffs, *et al.*, *Immunol. Lett.*, 1996, **51**, 101.
[23] H. Modjtahedi and C. Dean, *Int. J. Oncol.*, 1994, **4**, 277.

which is a product of the c-*erb*B-2 proto-oncogene is expressed during embryogenesis and by adenocarcinomas in the adult. Patients with tumours expressing high levels of this receptor have been found to have a poor prognosis.[24]

To prepare hybridomas, CBH/cbi rats were immunized three times at monthly intervals *via* their Peyer's patches with about 5×10^6 cells known to express high levels of the respective receptor, *i.e.* MDA-MB 468 (EGFR) or BT 474 (c-*erb*B-2). For the EGFR, mAbs were required that would block the interaction of the specific growth factors EGF and TGFα and so prevent proliferation of the tumour cells. With this in mind the antibodies were selected on the basis that they inhibited the binding of ^{125}I-labelled EGF. No ligands were available that bound to the c-*erb*B-2 receptor and so an alternative approach was made using cell-lines that expressed high (BT 474), intermediate (MDA-MB 361) or low (MCF7) levels of receptor.[25] To determine if specific antibodies were elicited following immunization dilutions of pre-immune and immune sera were tested for binding to monolayers of cells expressing *erb*B-2 or that inhibited the binding of ^{125}I-EGF to monolayers of the human bladder carcinoma EJ which expresses moderate levels of the EGFR. Three days after the last immunization the animals were killed, the mesenteric nodes harvested and the cells fused with the Y3 rat myeloma using PEG 1500 and the mixture plated into four 24-well plates containing X-irradiated rat fibroblasts as feeders. Ten to twelve days following fusion, supernatants were tested for antibodies that blocked the binding of ^{125}I-EGF to EJ cells[26] or gave the required differential binding to *erb*B-2 expressing cells.[24] Re-testing showed that positive colonies were present and these were picked, grown up and cloned twice by limiting dilution. At each stage samples of cells were frozen down in liquid nitrogen for long-term storage.

To characterize the mAbs obtained, they were grown up and purified from culture supernatant by precipitation with ammonium sulfate followed by ion exchange chromatography on Whatman DE52 cellulose. The purified antibodies were labelled with 125-iodine in turn and those binding to the EGFR were competed with each other for binding to EJ cells. In this way it was possible to show that each antibody fell into one of four distinct groups (epitope clusters A–D). Likewise, the mAbs to c-*erb*B-2 were found to bind to one of five distinct epitopes. In no case was there cross-reactivity between the EGFR-specific and *erb*B-2 specific mAbs, despite the homologies between the two receptors. Next, the

[24] N. E. Hynes and D. F. Stern, *Biochim. Biophys Acta*, 1994, **1198**, 165.
[25] J. M. Styles, S. Harrison, B. A. Gusterson and C. J. Dean, *Int. J. Cancer*, 1990, **45**, 320.
[26] H. Modjtahedi, J. M. Styles and C. J. Dean, *Br J. Cancer*, 1993, **67**, 247.

isotype (subclass) of the antibodies was determined using a capture assay employing mouse anti-rat isotype specific mAbs then detecting bound rat mAb with specific ^{125}I-sheep anti-rat F(ab)$_2$. The class or subclass of an antibody is an important indicator of its potential function or site of action, *e.g.* rat IgG2b antibodies can activate, complement and recruit host immune effector cells while IgA antibodies are usually present and active in secretory fluids. When the mAbs to the EGFR were tested for biological activity those directed against epitopes C and D were found to be the most effective inhibitors of ligand binding and receptor activation and they prevented the growth in culture and *in vivo* of the EGFR over-expressing human head and neck carcinoma HN5.[27] One of these antibodies (ICR62) is currently undergoing clinical evaluation.[28] Of the rat mAbs to c-*erb*B-2 one (ICR12) was outstanding and has been used in the clinic to target breast cancer. In pre-clinical testing, it has been found to be ideal for use in antibody-directed, enzyme pro-drug therapy (ADEPT[29]).

4.3 Monoclonal Antibodies for Clinical Application

The ability to generate unlimited quantities of antigenically defined monoclonal antibodies led to the expectation that their use would revolutionize therapeutic treatments. While there have been a number of successful applications notably in the field of transplantation,[30,31] their possible limitations soon became apparent. Some of the rodent antibodies proved to be highly immunogenic, inducing a human anti-mouse antibody response (HAMA) and when they were given repeatedly to patients, the injected foreign proteins were rapidly removed from circulation. This was due partly to the fact that rodents and humans use different amino acid sequences to build their combining sites, but largely due to the highly immunogenic nature of the murine Fc. A number of strategies involving the use of recombinant gene technology were then developed to try and overcome this problem namely

  (i) Making chimaeric antibodies by joining mouse variable regions segments (Fv) to a human Fc.

[27] H. Modjtahedi, S. Eccles, G. Box, S. Styles and C. J. Dean, *Br. J. Cancer*, 1993, **67**, 254.
[28] H. Modjtahedi, T. Hickish, M. Nicolson, *et al.*, *Br. J. Cancer*, 1996, **73**, 228.
[29] S. A. Eccles, W. J. Court, G. A. Box, C. J. Dean, R. J. Melton and C. J. Springer, *Cancer Res.*, 1994, **54**, 5171.
[30] G. Goldstein, J. Schindler, H. Tsai, *et al.*, *N. Eng. J. Med.*, 1985, **313**, 337.
[31] H. Waldmann and S. Cobbold, *Immunol. Today*, 1993, **14**, 247.

(ii) Splicing the complementarity determining regions (CDRs) into a human framework to reduce overall immunogenicity.

(iii) Constructing yeast artificial chromosomes (YACs) containing human germ-line immunoglobulin genes and using these to prepare transgenic mice whose own immunoglobulin genes have been inactivated.

(iv) Preparing libraries of human immunoglobulin genes isolated as single chain variable fragments (scFvs) from peripheral B-cells of healthy individuals and expressing them on the surface of bacteriophage for selection.

Of these strategies the first two have received considerable attention and a number of such antibodies have seen clinical use.[32] The generation of transgenic mice producing human antibodies[33] is attractive because it means that conventional hybridoma technology can be used to generate human monoclonal antibodies and high affinity antibodies should be generated by affinity maturation in the mouse lymph nodes. However, it should be remembered that the glycosylation patterns will still be of murine origin. Finally, great store has been placed on the preparation and use of libraries of human recombinant immunoglobulins prepared as scFvs or Fabs then expressed at the surface of bacteriophage and these are discussed in more detail in Section 5.

5   GENERATION OF MONOCLONAL ANTIBODIES USING RECOMBINANT GENE TECHNOLOGY

5.1   Isolation of Immunoglobulin Variable Region Genes and Expression on the Surface of Bacteriophage

With the discovery of thermostable DNA polymerases and the development of the polymerase chain reaction (PCR) it has become a relatively simple matter to isolate and clone immunoglobuin genes from immune or non-immune B-cells and hybridomas. The procedures rely on the body of information concerning the DNA sequences of rodent and human immunoglobulins (Kabat Database[34]) which has facilitated the design of forward and backward primers needed for amplification of the DNA sequences by PCR.[35] The usual starting material is mRNA rather

[32] W. J. Harris and C. Cunningham, 'Antibody Therapeutics', R. G. Landes, Austin, TX, 1995

[33] M. J. Mendez, L. L. Green, J. R. E. Corvalan, et al., Nature Genetics, 1997, 15, 146.

[34] E. A. Kabat, T. T. Wu, H. M. Perry, K. S. Gottesman and C. Foeller, 'Sequences of Proteins of Immunological Interest', 5th Ed., US Deptartment of Health and Human Services, Bethesda, 1991.

[35] G. Winter and C. Milstein, Nature (London), 1991, 349, 293.

than DNA since the former contains only the coding exons; the introns having been spliced out during transcription. Most attention has been focused on isolation and cloning of the immunoglobulin variable regions of the heavy (V_H) or light chains (V_L) which contain the complementarity determining regions (CDRs) involved in antigen binding. By linking the products of the (V_H) and (V_L) genes single chain variable fragments (scFv) can be generated, the protein products of which have the capacity to bind to antigen. These scFv constructs can be inserted into phagemid vectors for display at the surface of filamentous bacteriophage[36] coupled to a minor coat protein (gene 3 protein) present at one end of the phage.[37] Libraries of human antibody genes have been generated in this way from genes isolated from B-cells in the peripheral blood of both healthy[38] and HIV infected[39] individuals. An outline of the procedures involved is presented in Figure 4.

5.1.1 Isolation of mRNA for V_H and V_L and Generation of cDNA. RNA is particularly sensitive to degradation by RNases and it is therefore important to wear gloves during handling and ensure that all equipment and solutions are free of this enzyme; autoclave virgin plastic tubes, pipette tips, *etc.*, and treat water and glassware with diethyl pyrocarbonate and then autoclave before use. Total RNA is isolated using the one step procedure of Chomczynski and Sacchi[40] where spleen cells or hybridomas are homognized in 4 M guanidine thyocyanate then mixed with phenol to give a monophasic solution. Chloroform is then added and phase separation results in the RNA remaining in the aqueous phase while DNA and proteins are at the interface or in the organic phase. The RNA is then precipitated from the aqueous phase with isopropanol and washed with 75% ethanol. Messenger RNA (mRNA) contains a poly A tail and for this reason it can be isolated from the total RNA by binding to deoxythymine cellulose (dT-cellulose). After washing with high salt buffer, the mRNA is eluted with low salt buffer. Complementary DNA (cDNA) is synthesized using the mRNA as template by the addition of suitable forward primers (which anneal 3' of the coding sequences for the variable regions), the four deoxynucleotide triphosphates (dNTPs) and avian myeloblastosis virus (AMV) reverse transcriptase.

[36] G. P. Smith, *Science*, 1985, **228**, 1315.
[37] J. McCafferty, A. D. Griffiths, G. Winter and D. J. Chiswell, *Nature (London)*, 1990, **348**, 552.
[38] J. D. Marks, H. R. Hoogenboom, T. R. Bonnert, J. McCafferty, A. D. Griffiths and G. Winter, *J. Mol. Biol.*, 1991, **222**, 581.
[39] D. R. Burton, C. F. Barbas III, M. A. A. Persson, S. Koenig, R. M. Chanock and R. A. Lerner, *Proc. Natl. Acad. Sci. USA*, 1991, **88**, 10134.
[40] P. Chomczynski and N. Sacchi, *Anal. Biochem.*, 1987, **162**, 156.

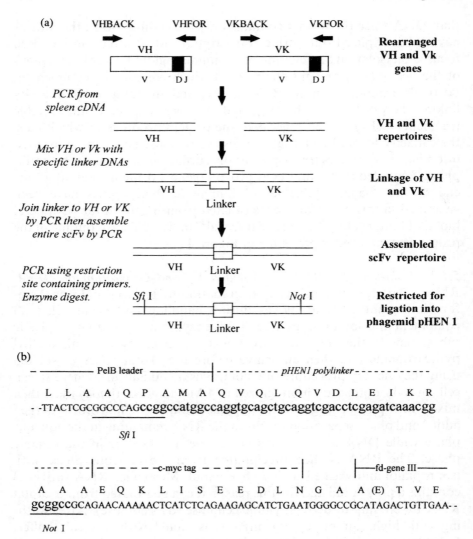

Figure 4   *Preparation of a single chain Fv library from cDNA isolated from immune spleen cells. (a) Amplification of VH and VK gene families, linkage to form scFv library and restriction for insertion into phagemid pHEN1. (b) Sequence of pHEN1 cloning site showing the polylinker that is excised and replaced by the restricted scFv*

5.1.2   PCR Amplification of cDNAs for Antibody V_H and V_L. PCR (ploymerase chain reaction) amplification makes use of the fact that the thermostable DNA polymerases can be subjected to repeated rounds of high temperature (95°C) without losing substantial activity. The PCR reaction mixture contains the cDNA sample (derived from many different mRNAs), suitable back and forward primers, the four dNTPs and DNA polymerase (Figure 4a). The sequences of the primers used (usually 21-mer oligonucleotides) depend on the origin of the immunoglobulin cDNAs, *i.e.* whether they are of human, rat or mouse origin. Back primers (5'–3') may be chosen from sequences at the beginning of the first framework region of the (V_H) or (V_L) or within the leader sequences. Forward primers (3'–5') are usually selected from the constant domains (C_{H1} or C_L) just 3' of the V_H or V_L regions. PCR amplification is carried out for example, by cycling from 95°C to denature the DNA and primers then annealing of the primers with the DNA strands at 50–60°C, followed by polymerization (DNA extension) at 70°C. Using a PCR cycler the rounds of denaturation, annealing and extension are continued automatically until sufficient DNA has been synthesized. The DNA samples are then electrophoresed through agarose containing ethidium bromide, the heavy and light chain bands visualized by ultraviolet light are cut out and the DNA isolated and purified. Many of the steps that are carried out in recombinant gene technology make use of kits supplied by companies such as Amersham Pharmacia Biotech.

5.1.3   Linking of V_H and V_L to give scFv.   The linker used is a gene coding for the peptide (Gly_4–Ser)$_3$ and the oligonucleotides coding for the peptide are amplified using primers to create complementarity between the amplified linker sequence and the relevant heavy or light chain primers. Where a gene library is being constructed from say spleen B-cells, a number of different linkers need to be prepared with ends that have complementary to the relevent heavy and light chain primers. The linkers are then joined to the H- or L-chain Fv gene fragments by PCR amplification then the linker assembled H- and L-chain joined by a second round of PCR amplification to give scFv antibody fragments. A further round of amplification is then carried out using primers containing restriction sites for the enzymes used to insert the fragments into a phagemid.

5.1.4   Insertion of scFv into Phagemid Vector.   Phagemid vectors such as pHEN-1[41,42] contain both *Escherichia coli* and phage origins of replication in addition to an ampicillin resistance gene. The phagemid allows the cloning of Sfi I/Not I restriction enzyme fragments of the scFv antibody genes (Figure 4b) in fusion with the gene 3 protein (pIII) of filamentous bacteriophage fd.[36,37] The transcription of antibody–pIII

[41] J. Viera and J. Messing, *Meth. Enzymol.*, 1987, **153**, 3.
[42] H. R. Hoogenboom, A. D. Griffiths, K. S. Johnson, D. J. Chiswell, P. Hudson and G. Winter, *Nucl. Acids Res.*, 1991, **19**, 4133.

fusions is driven from the inducible LacZ promoter and the fusion protein targeted to the periplasm by means of the pelB leader. In addition, there is an in-frame fusion of a c-myc epitope tag so that soluble scFv antibodies can be detected. These phagemid vectors require rescuing by helper phage which provide proteins for replication and packaging.

To insert the scFv genes into the phagemid vector, the vector is cut with the same two restriction enzymes used on the scFv preparation. The restriction-digested antibody scFv genes are ligated into the cut phagemid vector using T4 DNA ligase and the ligation mixtures are then electroporated into competent *E. coli* TG1 to make the phage library.

5.1.5   Expression of scFv on the Surface of Bacteriophage.   Samples of the phage library in *E. coli* TG1 are grown up in glucose containing media and helper phage (*e.g.* M13KO7) is added to superinfect the bacteria. The superinfected cells are then spun down, and grown overnight in glucose-free medium containing ampicillin and kanamycin to first, induce the packaging of phagemid expressing scFv in association with the gene III protein and second, induce phage formation. Intact bacteria are pelleted by centrifugation and the phage are precipitated from the supernatants with polyethylene glycol. This purified and concentrated phage stock can be used in a first round selection on antigen.

5.1.6   Screening Phage Display Libraries of Immunoglobulin Genes.
Phage displaying specific antibodies (*e.g.* scFv) are selected by binding to and elution from specific antigen. Where purified antigen is available, the most convenient method is to use antigen coated tubes (*e.g.* Nunc, 75 × 12 mm maxisorb immunotubes) that have been blocked with phosphate-buffered saline (PBS) containing 3% skimmed milk to reduce non-specific binding of phage. Phage is added to the tubes in the same buffer and incubated for 1–2 h at room temperature. After rigorous washing the bound phage is eluted with 100 mM triethylamine and the eluate neutralized with 1 M Tris-HCl at pH 7.4. TG1 cells are then infected with the selected phage which is rescued with helper phage and grown up to provide phage for the second round of selection on antigen-coated tubes. In this way, several rounds of selection of the antibody-bearing phage can be carried out to increase the specificity of the antibodies obtained. However, if a diverse population of antibodies is required it is better to limit the number of rounds in this initial selection.

Other methods for selection of specifically binding phage antibodies include the use of biotin-conjugated antigen which, after binding to phage antibody, is mixed with streptavidin-coated magnetic beads and the specific phage-containing complexes are removed from the suspension with a magnet, washed and the phage eluted. Other methods include those described earlier for mAbs produced by conventional cell fusion. However, it is not easy to select phage antibodies by direct binding to cells because of the number of antigens expressed at the cell surface or intracellularly; further information can be found in reference 43.

5.1.7 Preparation of Soluble scFv. Soluble scFv that can be used for different applications as immunological reagents (*e.g.* immunocytochemistry, ELISA, western blotting, *etc.*) can be generated by targeting the scFv to the bacterial periplasm and extracting the antibody fragments. Glucose-containing media are inoculated with an individual colony of the clone of interest generated from the initial screening tests and exponentially growing cells obtained. The cells are spun down resuspended in glucose-free media containing ampicillin and the inducer isopropyl β,δ-thiogalactopyranosde (IPTG). After 4 h at 30°C the cells are centrifuged down and then extracted for 15 min at 4°C by resuspending in Tris-HCl at pH 8.0 containing 20% sucrose and 1 mM EDTA. The cell debris is spun off and the supernatant containing scFv retained and used directly.

5.1.8 Screening Supernatants Containing Soluble scFv. The scFvs express the myc tag and this can be exploited in their detection using 9E10l, a mouse mAb recognizing this determinant[44] as the second reagent. Supernatants can be screened in the same way as conventionally produced mAbs using antigen-coated plates or cell monolayers. Scfv with specificity for a number of antigens have been produced in this way.[45] At the present time, great importance is placed on the possibility that by generating phage libraries of human immunoglobulin genes, scFv with specificities for human target antigens can be isolated which will form the building blocks for human antibodies that can be used in the detection and treatment of human disease.

6 MONOCLONAL ANTIBODIES IN BIOMEDICAL RESEARCH

mAbs are essential reagents for the isolation, identification and cellular localization of specific gene products and for assisting in the determina-

[43] J. Osbourne, in 'Monoclonal Antibodies: A Practical Approach', ed. P. Shepherd and C. Dean Oxford University Press, Oxford (in press).
[44] S. Munro and H. R. B. Pelham, *Cell*, 1986, **46**, 291.
[45] A. Nissim, H. R. Hoogenboom, I. M. Tomlinson, G. Flynn, C. Midgley, D. Lane and G. Winter, *EMBO J.*, 1994, **13**, 692.

tion of their macromolecular structure. Although the ability to clone and sequence specific genes has revolutionized our understanding of gene structure and function, it is the facility to make mAbs against the recombinant proteins or peptides based on protein sequences derived from cDNA clones that has made an outstanding contribution to our understanding of cell biology. Whether it is in structural studies by light or electron microscopy, specific protein isolation by immunoprecipitation or affinity chromatography or analysis by electrophoretic and other biochemical techniques, monoclonal antibodies are essential tools. It is not possible to discuss in detail the many applications of mAbs in basic research. Instead, we focus briefly on their diagnostic and therapeutic application in medicine.

mAbs against cytoplasmic and cell surface markers have had an impact not only in basic studies but also in offering new strategies for therapy. We have already referred to the over-expression of the receptor for epidermal growth factor (EGF) and the related receptor, a product of the c-*erb*B-2 proto-oncogene, which have been found to be indicators of poor prognosis in patients bearing squamous cell carcinomas or adenomas.[22,23] Also, the many cell surface markers that have been discovered on leukocytes are used both for identifying malignancies of haemopoetic origin and for determining the type of treatment to be given.

These cell surface molecules are also expressed by normal leukocytes and constitute receptors, adhesion molecules, enzymes and so forth. A systematic approach was made with the 'cluster of differentiation' (CD) designation for human leukocytes that were identified with mAbs.[46] mAbs were submitted to workshops and, as a result of these collaborative studies, they were then placed into groups on the basis of their fluorescent staining patterns for different leukocyte populations. In this classification, the antigens are given a CD number. In most cases, the biochemical nature of the molecule was not known initially, but the antibodies were used subsequently to isolate the antigens from cells and to identify recombinant proteins in expression libraries. This has enabled the elucidation of the amino acid sequence and potential structure of these molecules. The very large database on the leukocyte antigens is summarized in reference 47.

Although many of the CD antigens are expressed on cells of widely differing origins, the expression of others is confined to leukocytes. For

[46] A. Bernard, 'Leucocyte Typing. Human Leukocyte Differentiation Antigens Detected by Monoclonal Antibodies: Specification, Classification, Nomenclature', Springer-Verlag, Berlin, 1984.
[47] A. N. Barclay, M. H. Brown, A. Law, A. J. McKnight, M. Tomlenson and P. Anton vander Mewe, 'The Leukocyte Antigen Facts Book', Academic Press, London, 1997.

example, some CD antigens delineate particular subsets of leukocytes with important functions in the immune system, such a CD4, expressed by T-helper cells, and CD8, expressed by cytotoxic T-cells. In the last two decades it has proved possible to dissect the interactions between T-cells and antigen-presenting cells to determine how T-cells function in the generation of both humoral and cellular-immune responses. The results of these investigations have shown that exogenous antigens, taken up and processed by macrophages, dendritic or B-cells are presented at the cell surface as peptides associated with class II histocompatibility molecules to the CD4-positive T-helper cells, whereas peptides derived from endogenously derived proteins are presented at the surface of all cells in association with class I histocompatibility molecules.[48] The latter complexes are recognized by the receptors on the CD8-positive cytotoxic T-cells. In this way, normal or abnormal (*e.g.* viral products) constituents of the cytoplamic compartment are presented for examination by the CD8-positive T-cells. Elucidation of the molecular events involved in these processes has relied substantially on the use of specific mAbs directed not only against the interacting cell surface molecules but also against intracellular components of signal transduction and other biochemical pathways that are activated during the immune response.

7 MONOCLONAL ANTIBODIES IN THE DIAGNOSIS AND TREATMENT OF DISEASE

We have referred already to the important role that monoclonal antibodies play in the diagnosis of disease and they are vital tools for the pathologist. The therapeutic application of antibodies has a long history and mAbs were hailed as the magic bullets to specifically target cancer and other diseases. One of the success stories is in organ transplantation where the use of antibodies to T-cell determinants such as the anti-CD3 murine antibody OKT3 has helped to prevent rejection of cadaveric kidney grafts.[49] Less successful, so far, has been the use of mAbs to treat human cancers. This is due in part to the problem that all new experimental procedures are tried on patients that have failed conventional treatments such as surgery, radiotherapy or chemotherapy and have a large tumour burden. Also, some of the earlier mAbs were of relatively low affinity and cross-reacted with normal tissues and this may have reduced their localization to tumours in patients. For a recent review of mAbs that either have been or are undergoing clinical evaluation, see reference 32.

[48] K. L. Rock, *Immunol. Today*, 1996, **17**, 131.
[49] G. Golstein, J. Schindler, H. Tsai, *et al.*, *N. Engl. J. Med.*, 1993, **313**, 337.

There have been some notable successes, however, including the use of 131-iodine labelled antibody B1 (anti-CD20) in the treatment of B-cell lymphoma,[50] 90-yttrium labelled HMFG1 (anti-PEM) for adjuvant therapy of ovarian cancer[51] and unconjugated 17–1A for the treatment of colorectal cancer.[52] In the latter, an extensive and ongoing study involving 189 patients given adjuvant therapy, has yielded some promising results. Following surgical resection of the colon, the patients were treated over a period of 5 months with a total of 900 mg of antibody 17-1A, which recognizes a 37–40 kDa cell surface glycoprotein. Despite HAMA responses in 80% of the patients, the treatment reduced the overall 5-year death rate by 30% and the recurrence rate by 27%. Interestingly, it appeared that this positive result was due to the prevention of outgrowth, or elimination, of small metastatic foci of tumour cells. The treatment appeared to be less effective where more extensive tumour deposits remained following surgery.

We referred earlier to the potential immunogenicity of rodent mAbs. When they are given repeatedly to patients undergoing treatment, the HAMA response can result in rapid loss of circulating antibody. The development of high affinity antibodies of low immunogenicity should overcome this problem and improve the effectiveness of the mAbs. While the lower immunogenicity of 'humanized' mouse antibodies can reduce HAMA responses substantially, it is thought that fully human antibodies generated by recombinant gene technology will overcome this problem.

[50] O. W. Press, J. F. Eary, F. R. Appelbaum, *et al.*, *N. Engl. J. Med.*, 1993, **329**, 1219.
[51] V. Hird, A. Maraveyas, D. Snook, *et al.*, *Br. J. Cancer*, 1993, **68**, 403.
[52] G. Reithmüller, E. Schneider-Gadicke, G. Schlimok, *et al.*, *Lancet*, 1994, **343**, 1177.

CHAPTER 19

Biosensors

MARTIN F. CHAPLIN

1 INTRODUCTION

Biosensors are analytical devices that convert biological actions into electrical signals in order to quantify them.[1-3] They make use of the specificity of biological processes; enzymes for their substrates or other ligands, antibodies for their antigens, lectins for carbohydrates and nucleic acids for their complementary sequences or antibodies. The parts of a typical biosensor are shown in Figure 1. The biological reaction usually takes place in close contact with the electrical transducer, here shown as a 'black-box'. This intimate arrangement ensures that most of the biological reaction is detected. The resultant electrical signal is compared with a reference signal that is usually produced by a similar system without the biologically active material. The difference between these two signals, which optimally is proportional to the material being analysed (*i.e.* the analyte), is amplified, processed and displayed or recorded.

The primary advantage of using biologically active molecules as part of a biosensor is due to their high specificity and, hence, high discriminatory power. Thus, they are generally able to detect particular molecular species from within complex mixtures of other materials of similar structure that may be present at comparable or substantially higher concentrations. Often, samples can be analysed with little or no prior clean-up. In this aspect they show distinct advantages over most 'traditional' analytical methods; for example, colorimetric assays like the Lowry assay for proteins.

[1] J. D. Newman and A. D. F. Turner, *Essays Biochem.*, 1992, **27**, 147.
[2] A. E. G. Cass, 'Biosensors, A Practical Approach', Oxford University Press, Oxford, 1990.
[3] A. P. F. Turner (ed.) 'Advances in Biosensors', Vols 1 (1991), 2 (1992) and 3 (1995), with ed. Y. M. Yevdokimov, JAI Press Inc., Greenwich, NY.

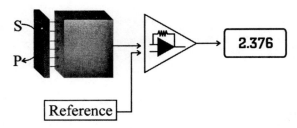

Figure 1   *The functional units of a biosensor*

Biosensors may serve a number of analytical purposes. In some applications, for example in clinical diagnosis, it is important only to determine whether the analyte is above or below some pre-determined threshold, whereas in process control there often needs to be a continual and precise feedback of the level of analyte present. In the former case, the biosensor must be designed to give the minimum number of false positives and, more importantly, false negatives. In the latter case it is generally the response to changes in the analyte that is more important than its absolute accuracy. Minimizing false negatives is often more important in clinical analyses than minimizing false positives as, in the former case a diagnosis may be missed and therapy delayed, whilst in the latter case further tests may show the error. Other biosensor analytical applications may require accuracy over a wide analytical range.

Biosensors must show advantages over the use of the free biocatalyst, which possesses equal specificity and discriminatory power, if they are to be acceptable. Their main advantages usually involve their reusability and rapid response or the reduced need for sample pre-treatment. Repetitive re-use of the same biologically active sensing material generally ensures that similar samples give similar responses as any need for accurate aliquoting of such biological materials to contain precisely similar activity is avoided. This circumvents the possibility of introducing errors by inaccurate pipetting or dilution and is a necessary prerequisite for automated and on-line monitoring. Repetitive and reagentless methodology offers considerable savings in terms of reagent costs, so reducing the cost per assay. In addition the increased operator time per assay and the associated higher skill required for 'traditional' assay methods also involves a cost penalty which in many cases may be greater than that due to the reagents. These advantages must be sufficient to encourage the high investment necessary for the development of a biosensor and the purchase price to the end-user. Table 1 lists a number of important attributes that a successful biosensor may be expected to possess. In any particular case only some of these may be achievable.

Table 1   *The properties required of a successful biosensor*

Required property	Achievable with ease
Specificity	Yes
Discrimination	Yes
Repeatability	Yes
Precision	Yes
Safe	Yes
Accuracy	Yes, as easily calibrated
Appropriate sensitivity	Yes, except in trace analysis
Fast response	Yes, usually
Miniaturizable	Yes, generally
Small sample volumes	Yes, generally
Temperature independence	Yes, may be electronically compensated
Low production costs	Yes, if mass-produced
Reliability	Yes, generally
Marketable	Difficult, due to competing methodology
Drift-free	Difficult but possible
Continuous use	Yes, for short periods (days)
Robust	No, generally need careful handling
Stability	No, except on storage (months) or in the short term (weeks)
Sterilizable	No, except on initial storage
Autoclavable	Not presently achievable

The different types of biosensors have their own advantages and disadvantages[4] that are summarized in Table 2. Not all types of biosensor have been used in commercially viable products. Apart from the important related area of colorimetric test strips, the most important commercial biosensors are electrochemical. There are amperometric biosensors available for glucose, lactate, glycerol, ethanol, lactose, L-amino acids, cholesterol, fish freshness and microorganisms. Potentiometric devices have been marketed for glucose, other low molecular weight carbohydrates and alcohols. Immunosensors have been marketed utilizing surface plasmon resonance devices and piezocrystals.

The market share of the application areas for biosensors is shown in Table 3. Only a few years ago, market research projections for the growth of the biosensor industry predicted expansion at an impressively fast pace of almost 50% per year. These were over-optimistic, but the industry sales as a whole were predicted to be worth about £500 million in the year 2000 and there are currently over a thousand scientific papers, concerning biosensors, published each year. It is now accepted that there are substantial and investment-intensive difficulties involved in producing such robust and reliable commercial analytical devices that are able

[4] J. H. T. Luong, C. A. Groom, and K. B. Male, *Biosensors Bioelectronics*, 1991, **6**, 547.

Table 2 *The properties of the various biosensor configurations*

Biosensor	Amperometric	Conductimetric	Potentiometric	Piezoelectric	Calorimetric	Optical
Cost	Low	Lowest	Low	Medium	High	High
Reliability	High	Medium	Medium	Varies	High	Medium
Complexity	Medium	Low	Low	Low	Very high	High
Selectivity	High	Low	High	Low	Low	High
Sensitivity	High	Medium	Medium	Low	Medium	Medium
Speed of response	Medium	Medium	Slow	Fast	Slow	Medium
Applicability	Medium	Small	Medium	Narrow	Wide	Medium
Present usage	Highest	Smallest	High	Small	Very small	Medium
Future prospects	High	Medium	High	Medium	Medium	High

Table 3 *The market share of biosensor application areas*

Application area	Market share (%)
Clinical diagnostics	
Glucose	85
Lactate and others	4
Research	4
Pharmaceuticals	3
Environmental	2
Food	2
Robotics, defence and others	< 1

to operate under authentic real-life conditions, even where a novel and highly promising research device has been produced.[5] The cost of the sensor is often of importance and considerable effort has been expended in the production of disposable devices using cheap integrated chip technology; disposable technology offering opportunities for continued and increased revenue. Such methodology has opened up the possibility for having a number of different sensors on one device, allowing multi-parameter assays. A realistic target for the future is a density of a million sensors per square centimetre.

By far the largest biosensor application area is in clinical diagnostics. This includes monitoring of critical metabolites during surgery. The major target markets are concerned with use within the physician's office, the casualty department of a hospital and in home diagnosis. These application areas are potentially very wide. The use of rapid biosensor techniques in doctors' surgeries avoids the need for expensive and, most importantly, time-consuming testing at central clinical laboratories. Thus diagnosis and treatment may start during the first visit of a patient. This removes the need to wait for a return visit after the clinical tests have been completed elsewhere and allowing time, perhaps, for the patient's condition to deteriorate somewhat. Also there is less likelihood of the sample being mishandled or contaminated. Centralized clinical analytical facilities remain a necessity due to the need, in many cases, for multiple different analyses on the same sample and difficulties such as regulatory compliance and quality assurance. Legislation imposes stringent quality assurance and control standards on clinical analyses. This makes it much more expensive to bring novel clinical biosensors to the market place today than in the past but generally permits those biosensors that are already established, giving them a distinct competitive edge.

[5] D. Griffiths and G. Hall, *Trends Biotechnol.*, 1993, **11**, 122.

Home diagnosis is an area that is being opened up by, for example, pregnancy and ovulation test kits. Clearly there are risks and problems involved with their more widespread use but many people prefer to use them as indications for whether a trip to the doctor's surgery is really necessary or not. As counselling may be necessary in some potential home diagnosis applications (for example, cancer and AIDS) controversy exists over their development.

One of the major potential uses for biosensors is for *in vivo* applications such as for the control of diabetes.[6] The purpose here is to continuously monitor the levels of metabolites so that corrective action can be employed when necessary. Clearly such biosensors must be biocompatible and miniaturized so that they are implantable. In addition they should be reagentless, the reaction being controlled only by the presence of the metabolite and the stabilized bioreagent. The signal generated must be stable over the period of interest. At the present time such biosensors have a relatively short lifespan of a few days, at most, due mainly to problems which arise from the body's response.

Industrial analysis involves food,[7,8] cosmetics and fermentation process control and quality control and monitoring. The defence industry is interested in detectors for explosives, nerve gases and microbial toxins. Environmental uses of biosensors are mainly in areas of pollution control. A typical application might be to detect parts per million of particular molecular species such as an industrial toxin within the highly complex mixtures produced as process effluent.

2 THE BIOLOGICAL REACTION

An important factor in most biosensor configurations is the sensing surface. This normally consists of a thin layer of biologically active material in intimate contact with the electronic transducer. In some cases the biological material may be covalently or non-covalently attached to the surface but often in electrochemical biosensors it forms part of a thin membrane covering the sensing surface. Generally the conversion of the biological process into an electronic signal is most efficient where the distance between the place where the biological reaction takes place and the place where the electronic transduction takes place is minimal. In addition, it is important for the retention of biological activity that the biological material is not lost into analyte solutions. The immobilization

[6] J. Pickup, *Trends Biotechnol.*, 1993, **11**, 285.
[7] G. Wagner and G. G. Guilbault (ed.), 'Food Biosensor Analysis', Marcel Dekker, Inc. New York, 1994.
[8] J. H. T. Luong, P. Bouvrette and K. B. Male, *Trends Biotechnol.*, 1997, **15**, 369.

Table 4 *Examples of biosensor immobilization methods*

Physical entrapment
  Biologically active material held next to the sensing surface by a semi-permeable membrane that prevents it from escaping to the bulk phase but allows passage of the analyte. Sometimes the membrane can be made such that it increases the specificity of the sensing or reduces unwanted side-reactions. This is a simple and inexpensive method.

Non-covalent binding
  Adsorption of enzymes to a porous carbon electrode. This may suffer from a gradual leaching of the enzyme to the bulk phase.

Covalent binding
  Treatment of the biosensor surface with 3-aminopropyltriethoxysilane followed by coupling of biologically active material to the reactive amino groups remaining on the cross-linked siloxane surface. Proteins may be attached by use of carbodiimides, which form amide links between amines and carboxylic acids. Such methods permanently attach the biological material but are difficult to reproduce exactly and often cause a large reduction in activity.

Membrane entrapment
  Crosslink proteins with glutaraldehyde within a cellulose or nylon supporting net.

technology for holding the biocatalyst in place is extensive.[9] The various methods of immobilization are summarized in Table 4. Much current research is directed at stabilizing enzymes for use in biosensors. At suitable pH, polyelectrolytes such as diethylaminoethyldextran wrap around enzymes so restricting their movement and reducing their tendency to denature. Polyelectrolytes in combination with polyalcohols, such as sorbitol and lactitol, have been shown to considerably stabilize enzymes against thermal inactivation for use in biosensors.

3 THEORY

In the absence of diffusion effects (see later) most biological reactions can be described in terms of saturation kinetics:

$$\text{Biological material} + \text{Analyte} \rightleftharpoons \text{Bound analyte}$$

The bound analyte then gives rise to the biological action, so generating the electronic response. This electronic response varies with the biological response which, in turn, varies with the concentration of the bound analyte. Apart from the logarithmic relationship in potentiometric

[9] M. F. Chaplin and C. Bucke, 'Enzyme Technology', Cambridge University Press, Cambridge, UK, 1990.

biosensors, the biological and electronic responses are often proportional.

$$\text{Bound analyte} \rightarrow \text{Biological response} \rightarrow \text{Electronic response}$$

$$\text{Electronic response} = \frac{(\text{Maximum electronic response possible}) \times (\text{Analyte concentration})}{(\text{Half-saturation constant}) + (\text{Analyte concentration})}$$

where the half-saturation constant is equal to the analyte concentration which gives rise to half the maximum electronic response possible (Figure 2). The response is linear, to within 95%, at analyte concentrations up to about a twentieth of the half-saturation constant. A biosensor may be used over a wider, non-linear range if it has compensatory electronics.

This relationship holds for both reacting (*e.g.* enzymes) and non-reacting (*e.g.* immunosorption) processes. Where the analyte reacts as part of the biological response (*e.g.* during a biocatalytic reaction utilizing enzyme(s) and/or microbial cells), an additional factor is the diffusion of the analyte from the bulk of the solution to the reactive surface.[9] If this rate of diffusion is less than the rate at which the analyte

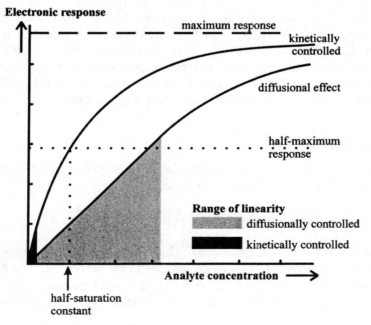

Figure 2   *The range of response of a biosensor under biocatalytic kinetic and diffusional control*

would otherwise react, there follows a reduction in the local concentration of analyte undergoing reaction. The rate of diffusion increases as the concentration gradient increases.

Rate of diffusion = (Diffusivity constant) × (Analyte concentration gradient)

where the analyte concentration gradient is given by the difference between the bulk analyte concentration and the local (microenvironmental) analyte concentration on the sensing surface of the biosensor divided by the distance through which the analyte must diffuse.

As most biocatalytic biosensor configurations utilize a membrane-entrapped biocatalyst this concentration gradient depends not only on the analyte concentration in the bulk and within the membrane but also on the membrane's thickness. The thicker the membrane, the greater the diffusive distance from the bulk of the solution to the distal sensing surface of the biosensor and the greater the amount of biocatalyst encountered. Both effects increase the likelihood that the overall reaction will be controlled by diffusion. Thus, such biosensors can be designed to be under diffusional or kinetic control by varying the membrane thickness. When the rate of analyte diffusion is slower than the rate at which the biocatalyst can react, the electronic response decreases due to the lower level of analyte available for reaction. A steady state is rapidly established when the rate of arrival of the analyte equals its rate of reaction. This steady state condition may be determined wholly by the rate of diffusion (diffusional control) or wholly by the rate of reaction (kinetic control), or by an intermediate dependency. Where the reaction rate depends solely on the rate of diffusive flux of the analyte, this determines the electronic response.

Electronic response ∝ Rate of diffusive flux

As the rate of diffusion depends on the bulk concentration of the analyte, this electronic response is linearly related to the bulk analyte concentration and, most importantly and intriguingly, is independent of the properties of the enzyme. Thus, the biosensor is linear over a much wider range of substrate concentrations (see Figure 2) and relatively independent of changes in the pH and temperature of the biocatalytic membrane, so long as the system remains diffusion controlled. It should be noticed, however, that under these conditions the response is reduced relative to a system containing the same amount of biocatalyst but not diffusionally limited. Maximum sensitivity to analyte concentration would be accomplished by the utilization of thin membranes containing a high biocatalyst activity and a well-stirred analyte solution. The overall kinetics of most biosensor configurations are difficult to predict. They depend on the diffusivities in the bulk phase and within the biocatalytic volume, the nature, porosity and physical properties of any membrane, the intrinsic biocatalytic kinetics, the electronic transduction process and kinetics, the way in which the analyte is presented, and on other non-specific factors. Generally, such overall kinetics are determined experimentally using the complete biosensor and, hence, it is very important that the biosensor configuration is reproducible.

In biosensors utilizing binding only, such as immunosensors, the major problem encountered is non-specific absorption that blocks the binding sites. There is need to minimize this and maximize the specific binding. As binding is an equilibrium process, high sensitivity necessitates a very high affinity between the analyte and the sensor surface.

4   ELECTROCHEMICAL METHODS

Electrochemical biosensors are generally fairly simple devices. There are three types utilizing electrical current, potential or resistive changes:

(i) Amperometric biosensors which determine the electric current associated with the electrons involved in redox processes.
(ii) Potentiometric biosensors which use ion selective electrodes to determine changes in the concentration of chosen ions (*e.g.* hydrogen ions).
(iii) Conductimetric biosensors which determine conductance changes associated with changes in the ionic environment.

There has been much progress in miniaturizing these devices using microfabrication technologies developed by the electronics industry. These include the use of screen-printing and the deposition of nanolitre volumes of enzymes using advanced ink-jet printing and conducting inks.

4.1   Amperometric Biosensors

Enzyme-catalysed redox reactions can form the basis of a major class of biosensors if the flux of redox electrons can be determined.[10] Normally a constant potential is applied between two electrodes and the current, due to the electrode reaction, determined. The potential is held relative to

[10] A. Heller, *Curr. Opin. Biotechnol.*, 1996, **7**, 50.

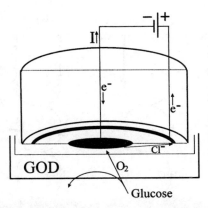

Figure 3 *Amperometric glucose biosensor based on the oxygen electrode utilizing glucose oxidase (GOD)*

that a reference electrode and is chosen such that small variations do not affect the rate of the electrode reaction. The first and simplest biosensor was based on this principle. It was for the determination of glucose and made use of the Clark oxygen electrode.[11] Figure 3 shows a section through such a simple amperometric biosensor. A potential of -0.6 V is applied between the central platinum cathode and the surrounding silver/silver chloride reference electrode (the anode). Dissolved molecular oxygen at the platinum cathode is reduced and the circuit is completed by means of the saturated KCl solution. Only oxygen can be reduced at the cathode due to its covering by a thin Teflon or polypropylene membrane through which the oxygen can diffuse but which acts as a barrier to other electroactive species.

Pt cathode reaction (-0.6 V) $O_2 + 4H^+ + 4e^- \rightarrow 2H_2O$
Anode reaction $4Ag^0 + 4Cl^- \rightarrow 4AgCl + 4e^-$

The biocatalyst is retained next to the electrode by means of a membrane, which is permeable only to low molecular weight molecules including the reactants and products.

Glucose may be determined by the reduction in the dissolved oxygen concentration when the redox reaction, catalysed by glucose oxidase, occurs (Figure 4a).

$$\text{Glucose} + O_2 \xrightarrow{\text{glucose oxidase}} \delta\text{-Gluconolactone} + H_2O_2$$

[11] L. C. Clark and C. H. Lyons, *Ann. N.Y. Acad. Sci.*, 1962, **102**, 29.

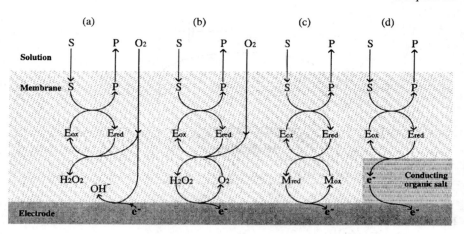

Figure 4 *The redox mechanisms for various amperometric biosensor configurations*

It is fortunate that this useful enzyme is also one of the most stable oxidoreductases found. Conditions can be chosen such that the rate at which oxygen is lost from the biocatalyst-containing compartment is proportional to the bulk glucose concentration. Other oxidases can be used in this biosensor configuration and may be immobilized as part of a membrane by treatment of the dissolved enzyme(s), together with a diluent protein, with glutaraldehyde on a cellulose or nylon support. An alternative method of determining the rate of reaction is to detect the hydrogen peroxide produced directly by reversing the polarity of the electrodes (Figure 4b). This is the principle used in Yellow Springs analysers. The use of a covering, such as the highly anionic Nafion^{TM} membrane, prevents electroactive anions, such as ascorbate, reaching the electrode without restricting glucose. Internally a cellulose acetate membrane replaces the Teflon membrane to allow passage of the hydrogen peroxide. This arrangement has a higher sensitivity than that utilizing an oxygen electrode but, in the absence of highly selective membranes, is more prone to interference at the electrode surface.

Cathode reaction $\qquad$ $2AgCl + 2e^- \rightarrow 2Ag^0 + 2Cl^-$

Pt anode reaction ($+0.6$ V) $\qquad$ $H_2O_2 \rightarrow O_2 + 2H^+ + 2e^-$

These electrodes can be developed further for the determination of substrates for which no direct oxidase enzyme exists. Thus sucrose can be determined by placing an invertase layer over the top of the glucose oxidase membrane in order to produce glucose (from the sucrose) which can then be determined. Interference from glucose in the sample can be

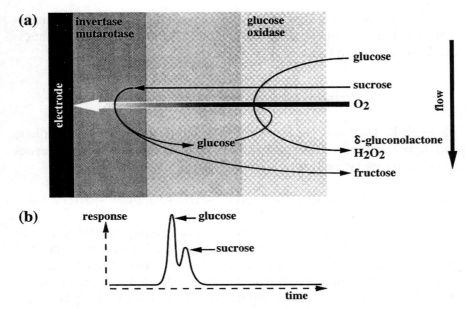

Figure 5 *A kinetically controlled anti-interference membrane for sucrose. The sample is presented as a rapid pulse of material in the flowing stream. Use of phosphate buffer, which catalyses mutarotation, removes the need for the mutarotase enzyme*

minimized by including a thin anti-interference layer of glucose oxidase and peroxidase over the top of both layers, which removes the glucose without significantly reducing the oxygen diffusion to the electrode. An alternative approach to assay sucrose and glucose together[12] makes use of the lag period in the response due to the necessary inversion of the sucrose that delays its response relative to the glucose (Figure 5).

Many biochemicals may be analysed by similar biosensors. One interesting example is the biosensor for determining the artificial sweetener aspartame in soft drinks where three enzymes are necessary in the sensing membrane in order to produce the H_2O_2 for the electrode reaction.

$$\text{aspartame} + H_2O \xrightarrow{\text{peptidase}} \text{L-aspartate} + \text{L-phenylalanine methyl ester}$$

$$\text{L-aspartate} + \alpha\text{-ketoglutarate} \xrightarrow{\text{aspartate aminotransferase}} \text{L-glutamate} + \text{oxaloacetate}$$

$$\text{L-glutamate} + O_2 + H_2O \xrightarrow{\text{glutamate oxidase}} \alpha\text{-ketoglutarate} + NH_4^+ + H_2O_2$$

[12] E. Watanabe, M. Takagi, S. Takei, M. Hoshi, and C. Shu-gui, *Biotechnol. Bioeng.*, 1991, **38**, 99.

Fish freshness can be determined using a similar concept; the nucleotides in fish change due to a series of reactions after death. Fish freshness can be quantified in terms of its K value:

$$K = \frac{(HxR + Hx) \times 100}{(ATP + ADP + AMP + IMP + HxR + Hx)}$$

where HxR, IMP and Hx represent inosine, inosine-5′-monophosphate and hypoxanthine, respectively. After fish die their ATP undergoes catabolic degradation through a series of reactions outlined below:

$$ATP \rightarrow ADP \rightarrow AMP \rightarrow IMP \rightarrow HxR \rightarrow Hx \rightarrow Xanthine \rightarrow Uric\ acid$$

The accumulation of the intermediates inosine and hypoxanthine relative to the nucleotides is an indicator of how long the fish has been dead and its handling and storage conditions, and hence its freshness. A commercialized fish freshness biosensor has been devised which utilizes a triacylcellulose membrane containing immobilized nucleoside phosphorylase and xanthine oxidase over an oxygen electrode.

$$\text{inosine} + \text{phosphate} \xrightarrow{\text{nucleoside phosphorylase}} \text{hypoxanthine} + \text{ribose phosphate}$$

$$\text{hypoxanthine} + O_2 \xrightarrow{\text{xanthine oxidase}} \text{xanthine} + H_2O_2$$

$$\text{xanthine} + O_2 \xrightarrow{\text{xanthine oxidase}} \text{uric acid} + H_2O_2$$

The electrode may be used to determine the reduction in oxygen[13] or the increase in hydrogen peroxide. The inosine content may be determined after the hypoxanthine content by the addition of the necessary phosphate. The nucleotides can be determined using the same electrode and sample, subsequent to the addition of nucleotidase and adenosine deaminase. Typically, K values below 20 show the fish is very fresh and may be eaten raw. Fish with a K value between 20 and 40 must be cooked but those with a K value above 40 are not fit for human consumption. Critical K values vary amongst species but can be a reliable indicator applicable to frozen, smoked or fish stored under modified atmospheres. Clearly a relatively simple probe that accurately and reproducibly determines fish freshness has significant economic importance to the fish industry. In its absence freshness is determined completely subjectively by inspection.

Although such biosensors are easy to produce they do suffer from

[13] S. S. Hu and C. C. Liu, *Electroanalysis*, 1997, **9**, 1174.

some significant drawbacks. The reaction is dependent on the concentration of molecular oxygen which precludes its use in oxygen deprived environments, such as *in vivo*. Also the potential used is sufficient to cause other redox processes to occur, such as ascorbate oxidation/ reduction, which may interfere with the analyses. Much research has been undertaken on the development of substances that can replace oxygen in these reactions.[14] Generally oxidases are far more specific for the oxidizable reactant than they are for molecular oxygen itself, as the oxidant. Many other materials can act as the oxidant. The optimal properties of such materials include fast electron transfer rates, the ability to be easily regenerated by an electrode reaction and retention within the biocatalytic membrane. In addition, they should not react with other molecules, including molecular oxygen, that may be present. Many such oxidants, now called mediators, have been developed. Their redox reactions, which together transfer the electrons from the substrate to the electrode so producing the electrical signal, are summarized in Figure 6.

The mediated biosensor reaction consists of three redox processes:

$$\text{substrate}_{(reduced)} + \text{enzyme}_{(oxidized)} \xrightarrow{\text{enzyme reaction}} \text{product}_{(oxidized)} + \text{enzyme}_{(reduced)}$$

$$\text{enzyme}_{(reduced)} + \text{mediator}_{(oxidized)} \xrightarrow{\text{enzyme reaction}} \text{enzyme}_{(oxidized)} + \text{mediator}_{(reduced)}$$

$$\text{mediator}_{(reduced)} \xrightarrow{\text{electrode reaction}} \text{mediator}_{(oxidized)} + e^-$$

When a steady state response has been obtained the rates of all of these processes and the rate of the diffusive flux in must be equal (see Figure 4c). Any of these, or a combination, may be the controlling factor. The level of response is, therefore, difficult to predict.

A blood-glucose biosensor, for the control of diabetes, has been built and marketed based on this mediated system.[15] The sensing area is a single-use disposable electrode, produced by screen-printing onto a plastic strip, and consists of an Ag/AgCl reference electrode and a carbon working electrode containing glucose oxidase and a derivatized ferrocene mediator. Both electrodes are covered with wide mesh hydrophilic gauze to enable even spreading of a blood drop sample and to prevent localized cooling effects due to uneven evaporation. The electro-

[14] F. W. Scheller, F. Schubert, B. Neumann, D. Pfeiffer, R. Hintsche, I. Dransfeld, U. Wollenberger, R. Renneberg, A. Warsinke, G. Johansson, M. Skoog, X. Yang, V. Bogdanovskaya, A. Bückmann and S. Y. Zaitsev, *Biosensors Bioelectronics*, 1991, **6**, 245.
[15] P. I. Hilditch and M. J. Green, *Analyst*, 1991, **116**, 1217.

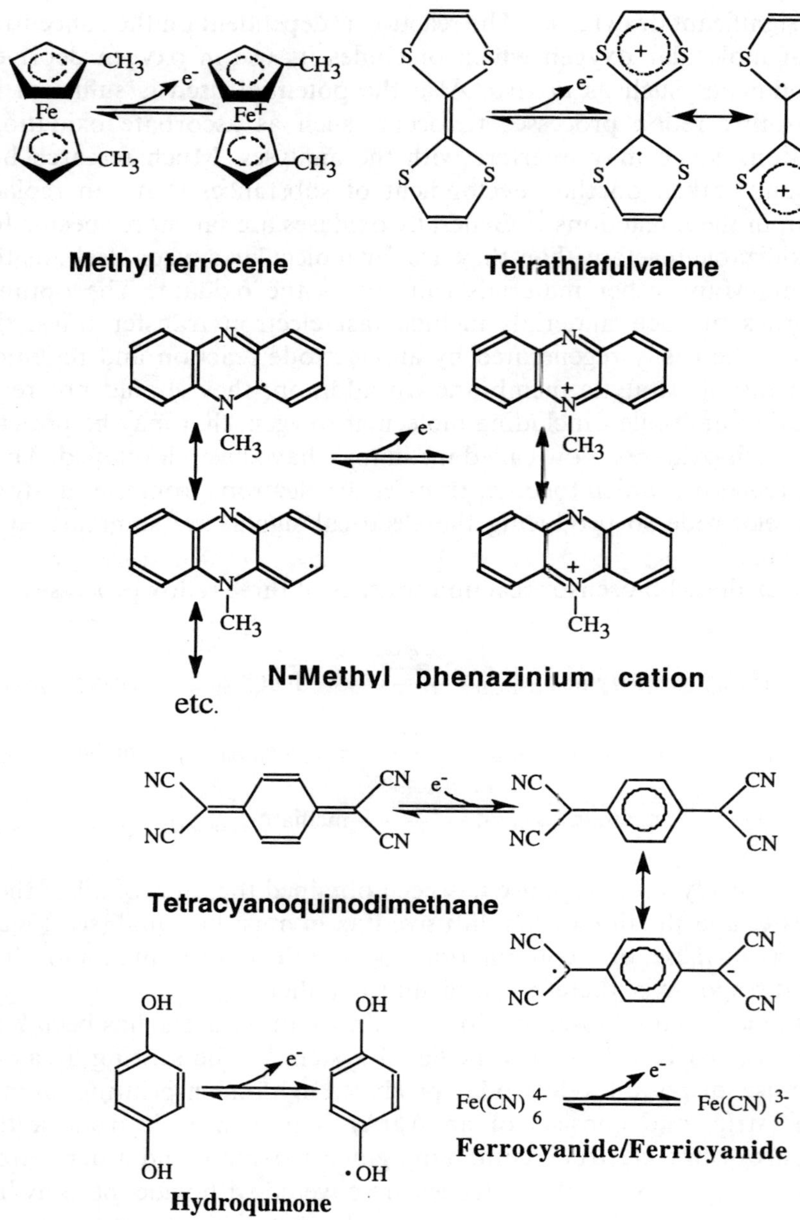

Methyl ferrocene

Tetrathiafulvalene

N-Methyl phenazinium cation

etc.

Tetracyanoquinodimethane

Hydroquinone

$$Fe(CN)_6^{4-} \rightleftharpoons Fe(CN)_6^{3-}$$

Ferrocyanide/Ferricyanide

Figure 6 *The redox reactions of amperometric biosensor mediators. Tetracyanoqinodimethane acts as a partial electron acceptor whereas ferrocene, tetrathiafulvalene and N-methyl phenazinium can all act as partial electron donors. Hydroquinone and ferricyanide are soluble mediators* des are kept dry until use and have a shelf life of 6 months when sealed in aluminium foil. They can detect glucose concentrations of 2–25 mM in single drops of blood and display the result within 30 s. Such biosensors are currently used by hundreds of thousands of people with diabetes in more than 50 countries worldwide. Although the subject of much research,[6] the use of subcutaneous *in vivo* continuous glucose sensing using electrode-based biosensors has not overcome the difficulties caused by the body's response and poor patient acceptability. New non-biosensor technologies are now competing,[16] involving near-infrared light which can penetrate tissue and detect glucose in the blood, albeit at low sensitivity and in the face of almost overwhelming interference.

When an oxidase is unable to react rapidly enough with available mediators, horseradish peroxidase, which rapidly reacts with ferrocene mediators, can be included with the enzyme. This catalyses the reduction of the hydrogen peroxide produced by the oxidase and consequent oxidation of the mediator. In this case the mediator is acting as an electron donor rather than acceptor. The oxidized mediator then can be rapidly reduced at the electrode at moderate redox potential.

$$\text{mediator}_{(reduced)} + \tfrac{1}{2}H_2O_2 + H^+ \xrightarrow{\text{peroxidase}} \text{mediator}_{(oxidized)} + H_2O$$

$$\text{mediator}_{(oxidized)} + e^- \xrightarrow{\text{electrode reaction}} \text{mediator}_{(reduced)}$$

A major advance in the development of micro-amperometric biosensors came with the discovery that pyrrole can undergo electrochemical oxidative polymerization (Figure 7) under conditions mild enough to entrap enzymes and mediators at the electrode surface without denaturation. A membrane, entrapping the biocatalyst and mediator, can be formed at the surface of even extremely small electrodes by polymerizing pyrrole in the present of biocatalyst. The polypyrrole tightly adheres to platinum, gold or carbon electrodes. This allows silicon chip microfabrication methods to be used and for many different sensors to be laid down on the same chip (Figure 8).

Another advance has been the use of conducting organic salts on the electrode. These may allow the direct transfer of electrons from the reduced enzyme to the electrode without the use of any (other) mediator (Figure 4d). Conducting organic salts consist of a mixture of two types of planar aromatic molecules, electron donors and electron acceptors (see Figure 6), which partially exchange their electrons. These molecules form segregated stacks, containing either the donor or acceptor molecules,

[16] M. A. Arnold, *Curr. Opin. Biotech.*, 1996, **7**, 46.

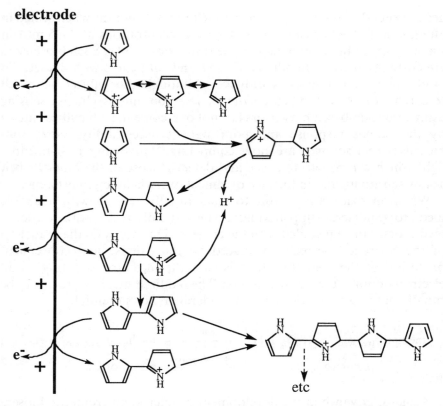

Figure 7 *The mechanism for the electrochemical oxidative polymerization of pyrrole*

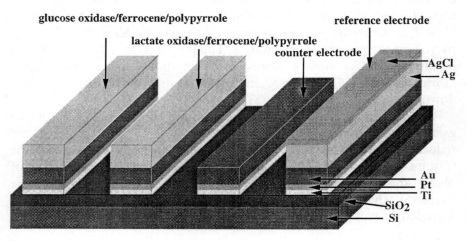

Figure 8 *A combined microelectrode for glucose and lactate*

with some of the electrons from the donors being transferred to the acceptors. These electrons, which have been partially transferred, are mobile up and down the stacks giving the organic crystals a high conductivity. There must not be a total electron transfer between the donor and acceptor molecules or the crystal becomes an insulator through lack of electron mobility. These electrodes give the somewhat misleading appearance of direct electron transfer to the electrode. As both the components of the organic salts, in the appropriate redox state, are able to mediate the reaction, it is highly probable that these electrodes are behaving as a highly insoluble mediator prevented from large-scale leakage by electrostatic effects.

4.2 Potentiometric Biosensors

Changes in ionic concentrations are easily determined by use of ion-selective electrodes. This forms the basis for potentiometric biosensors. Many biocatalysed reactions involve charged species, each of which will absorb or release hydrogen ions according to their pK_a and the pH of the environment. This allows a relatively simple electronic transduction using the commonest ion selective electrodes, the pH electrode. Table 5 shows some biocatalytic reactions that can be utilized in potentiometric biosensors. Potentiometric biosensors can be miniaturized by the use of field effect transistors (FET).

Table 5 *Biocatalytic reactions that can be used with ion-selective electrode biosensors*

Electrode	Reactions
Hydrogen ion	
Penicillin	penicillin $\xrightarrow{\text{penicillinase}}$ penicilloic acid $+ H^+$
Lipid	triacylglycerol $\xrightarrow{\text{lipase}}$ glycerol $+$ fatty acids $+ H^+$
Urea	$H_2NCONH_2 + H_2O + 2H^+ \xrightarrow{\text{urease (pH 6)}} 2NH_4^+ + CO_2$
Ammonia	
L-Phenylalanine	L-phenylalanine $\xrightarrow{\text{phenylalanine ammonia-lyase}} NH_4^+ + trans\text{-cinnamate}$
L-Asparagine	L-asparagine $+ H_2O \xrightarrow{\text{asparaginase}} NH_4^+ + $ L-aspartate
Adenosine	adenosine $+ H_2O + H^+ \xrightarrow{\text{adenosine deaminase}} NH_4^+ + $ inosine
Creatine	creatine $+ H_2O \xrightarrow{\text{creatinase}} H_2NCONH_2 + $ sarcosine
	$H_2NCONH_2 + 3H_2O \xrightarrow{\text{urease (pH 7)}} 2NH_4^+ + HCO_3^- + OH^-$
Iodide	
Peroxide	$H_2O_2 + 2I^- + 2H^+ \xrightarrow{\text{peroxidase}} 2H_2O + I_2$

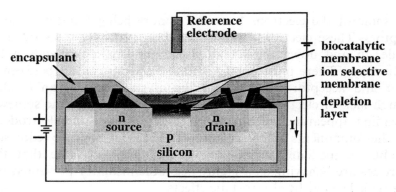

Figure 9   *An FET-based potentiometric biosensor*

Ion-selective field effect transistors (ISFET) are low cost devices that are in mass production. Figure 9 shows a diagrammatic cross-section through an npn hydrogen ion responsive ISFET with a biocatalytic membrane covering the, approximately 0.025 mm^2, ion selective membrane. The build-up of positive charge on this surface (the gate) repels the positive holes in the p-type silicon causing a depletion layer and allowing the current to flow. The reference electrode is usually an identical ISFET without any biocatalytic membrane. A major practical problem with the manufacture of such enzyme-linked FETs (ENFETs) is protection of the silicon from contamination by the solution, hence the covering of waterproof encapsulant. Because of their small size, they only require minute amounts of biological material and can be fabricated in a form whereby they can determine several analytes simultaneously. A further advantage is that they have a more rapid response rate when compared with the larger sluggish ion-selective electrode devices. The enzyme can be immobilized to the silicon nitride gate using polyvinyl butyral deposited by solvent evaporation and cross-linked with glutar-aldehyde. Some fabrication problems still exist, however, and these are currently being addressed. In particular, they need on-chip temperature compensation.

A potentiometric biosensor for fish freshness similar in biochemical principle to that described earlier has been developed, which determines the pH changes associated with the reactions, utilizing an amorphous silicon field effect transistor.[17] Potentiometric biosensors for DNA have been developed which use anti-DNA monoclonal antibodies conjugated with urease. DNA is bound to a membrane, placed on the electrode, and quantified by the change in pH on addition of urea (see Table 5).

[17] M. Gotoh, E. Tamiya, A. Seki, I. Shimizu, and I. Karube, *Anal. Lett.*, 1988, **21**, 1785.

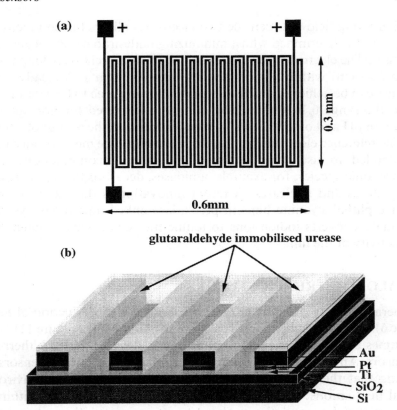

Figure 10 *Parts of a conductimetric biosensor electrode arrangement. (a) Top view, (b) cross-sectional view. The tracks are about 5000 nm wide and the thickness of the various layers are approximately; SiO₂ 550 nm, Ti 100 nm, Pt 100 nm, Au 2000 nm*

4.3 Conductimetric Biosensors

Many biological processes involve changes in the concentrations of ionic species. Such changes can be utilized by biosensors that detect changes in electrical conductivity. A typical example of such a biosensor is the urea sensor, utilizing immobilized urease,[18] and used as a monitor during renal surgery and dialysis (Figure 10). The reaction gives rise to a large change in ionic concentration making this type of biosensor particularly attractive for monitoring urea concentrations.

$$NH_2CONH_2 + 3H_2O \xrightarrow{\text{electrode reaction}} 2NH_4^+ + HCO_3^- + OH^-$$

[18] N. F. Sheppard, D. J. Mears, and A. Guiseppielie, *Biosensors Bioelectronics*, 1996, **11**, 967.

An alternating field between the two electrodes allows the conductivity changes to be determined whilst minimizing undesirable electrochemical processes. The electrodes are interdigitated to give a relatively long track length (~ 1 cm) within a small sensing area (0.2 mm^2). A steady state response can be achieved in a few seconds allowing urea to be determined within the range 0.1–10 mM. The output is corrected for non-specific changes in pH and other factors by comparison with the output of a non-enzymic reference electrode pair on the same chip. The method can easily be extended to use other enzymes and enzyme combinations that produce ionic species; for example, amidases, decarboxylases, esterases, phosphatases and nucleases. A recent innovation is the use of iodine-sensitive phthalocyanine films in peroxidase-linked immunoassays; the peroxidase converts iodide ions to iodine molecules which change the conductivity of the film.

5 CALORIMETRIC BIOSENSORS

A general property of many enzyme reactions is the production of heat (Table 6). This forms the basis of calorimetric biosensors (Figure 11),[19,20] sometimes also called enzyme thermistors or thermometric or thermal biosensors. An important factor in the manufacture of such biosensors is assuring that the resulting heat changes are not affected by environmental fluctuations. For this reason, the reaction is confined within a heat-insulated box and the analyte stream is passed through a heat

Table 6 *Exothermic reactions used in calorimetric biosensors*

Analyte	Reaction	Biocatalyst
Antigens	ELISA	Catalase/antibody
Ascorbic acid	Oxidation	Ascorbate oxidase
Cholesterol	Oxidation	Cholesterol oxidase
Ethanol	Oxidation	Alcohol oxidase
Glucose	Oxidation	Glucose oxidase
Glycerol	Catabolism	*Gluconobacter oxydans* cells
Hydrogen peroxide	Redox	Catalase
Lactate	Oxidation	Lactate oxidase
Penicillin G	Hydrolysis	β-Lactamase
Pyruvate	Reduction	Yeast lactate dehydrogenase
Oxalic acid	Oxidation	Oxalate oxidase
Urea	Hydrolysis	Urease
Uric acid	Oxidation	Uricase

[19] P. Bataillard. *Trends Anal. Chem.*, 1993, **12**, 387.
[20] K. Mosbach, *Biosensors Bioelectronics*, 1991, **6**, 179.

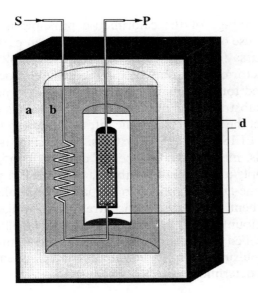

Figure 11 *Sectional view through a calorimetric biosensor; (a) insulated box, (b) heat exchanger, aluminium cylinder, (c) biocatalytic packed bed reactor, (d) matched thermistors*

exchanger. The reaction takes place within a small packed bed reactor and the difference in temperature between the incoming analyte and the product stream is determined by matched thermistors. This difference is only a fraction of a degree centigrade but temperatures can be resolved down to 0.0001°C. Although such devices use relatively more enzyme than other biosensors, they are versatile and free from optical interference.

Calorimetric devices can be cumbersome devices that have had only limited commercial success when compared to the other types of biosensors. One advantage that they have over other biosensor configurations is the ease with which a number of reactions can be linked together within the reactor. This allows the possibility of utilizing recycling reactions by co-immobilizing other enzymes and introducing other reactants into the analyte stream. Thus the sensitivity to lactate can be increased by more than an order of magnitude by co-immobilizing lactate dehydrogenase with the lactate oxidase allowing substrate recycling and, effectively reacting the lactate analyte hundreds of times. Of course, any pyruvate in the sample interferes in this assay.

$$\text{lactate} + O_2 \xrightarrow{\text{lactate oxidase}} \text{pyruvate} + H_2O_2$$

$$\text{pyruvate} + \text{NADH} + H^+ \xrightarrow{\text{lactate dehydrogenase}} \text{lactate} + \text{NAD}^+$$

A major advantage of the calorimetric principle is that it can be extended to the use of whole viable cells and to form part of an enzyme-linked immunoassay (ELISA) system. Immobilized viable cells within the packed bed reactor can not only be used to achieve bioconversions but may also be used for environmental monitoring when presented with a metabolizable substrate. The presence of toxic materials in the substrate stream will affect the general metabolic rate so indicating their presence.

Calorimetric ELISA (TELISA) systems, in a similar manner to other ELISA methods, may have a number of different configurations. One method is to apply a mixture of unlabelled antigen (analyte) and a fixed amount of enzyme-labelled antigen to the packed bed column containing an immunosorbent. Increased concentration of unlabelled antigen increases the amount bound in competition to the labelled antigen. The amount of labelled antigen remaining in the column can then be determined by pulsing the substrate, for the labelling enzyme, through the column and determining the heat produced.

6 PIEZOELECTRIC BIOSENSORS

The piezoelectric effect is due to some crystals containing positive and negative charges that separate when the crystal is subjected to a stress, causing the establishment of an electric field. As a consequence, if this crystal is subjected to an electric field it will deform. An oscillating electric field of a resonant frequency will cause the crystal to vibrate with a characteristic frequency dependent on its composition and thickness as well as the way it has been cut. As this resonant frequency varies when molecules absorb to the crystal surface, a piezoelectric crystal may form the basis of a biosensor. Even small changes in resonant frequencies are easy to determine accurately by modern electronic techniques. Differences in mass, adsorbed to the sensing surface, even as small as a nanogram per square centimetre can be measured. Changes in frequency are generally determined relative to a similarly treated reference crystal without the active biological material. A biosensor for cocaine in the gas phase may be made by attaching cocaine antibodies to the surface of a piezoelectric crystal. This biosensor changes frequency by about 50 Hz for a one part per billion atmospheric cocaine sample and can be reused on flushing for a few seconds with clean air. The relative humidity of the air is important as if it is too low, the response is less sensitive and if it is too high the piezoelectric effect may disappear altogether. Cocaine in solution can be determined after drying such biosensors.[21]

[21] B. S. Attili and A. A. Suleiman, *Microchem. J.*, 1996, **54**, 174.

Enzymes with gaseous substrates or inhibitors can also be attached to such crystals as has been proved by the production of a biosensor for formaldehyde incorporating formaldehyde dehydrogenase and organophosphorus insecticides incorporating acetylcholinesterase, respectively.

One of the drawbacks preventing the more widespread use of piezoelectric biosensors is the difficulty in using them to determine analytes in solution. The frequency of a piezoelectric crystal depends on the liquid's viscosity, density and specific conductivity and, under unfavourable conditions, the crystal may cease to oscillate completely. There is also a marked effect of temperature due to its effect on viscosity. The binding of material to the crystal surface may be masked by other intermolecular effects at the surface and bulk viscosity changes consequent upon even quite small concentration differences. There is also the strong possibility of interference due to non-specific binding.

Antibody–antigen binding can be determined by measuring the frequency changes in air after drying the crystal. Such procedures, although sensitive, are difficult to reproduce repetitively, as the antibody layer may be partially lost when the antigen is removed. However, one-shot biosensors have been developed using this principle for the detection of several food contaminants such as enterobacteria.[22]

Piezoelectricity is also utilized in surface acoustic wave (SAW) devices where a set of interdigitated electrodes is microfabricated at each end of a rectangular quartz plate (Figure 12). Binding of molecules to the surface affects the propagating wave, generated at one end, such that its frequency is reduced before reception at the other.

7 OPTICAL BIOSENSORS

Optical biosensors (also called optodes) are currently generating considerable interest, particularly with respect to the use of fibre optics and optoelectronic transducers. These allow the safe non-electrical remote sensing of materials in hazardous or sensitive (*i.e. in vivo*) environments. An advantage of optical biosensors is that no reference sensor is needed, and a comparative signal is generally easily generated using the same light source as the sampling sensor. A simple example of this is the fibre optic lactate sensor (Figure 13) which senses changes in molecular oxygen concentrations by determining its quenching of a fluorescent dye.

$$O_2 + \text{lactate} \xrightarrow{\text{lactate monooxygenase}} CO_2 + \text{acetate} + H_2O$$

[22] M. Plomer, G. G. Guilbault, and B. Hock, *Enzyme Microb. Technol.*, 1992, **14**, 230.

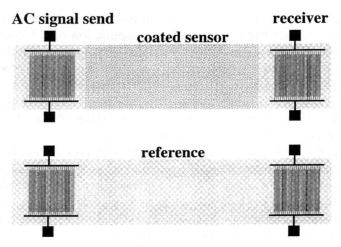

Figure 12   *A surface acoustic wave (SAW) biosensor*

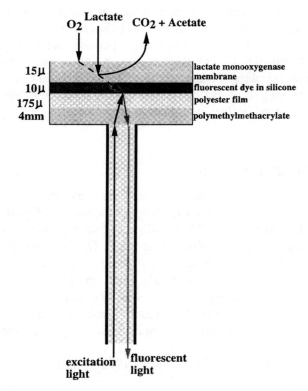

Figure 13   *A fibre optic lactate biosensor*

The presence of oxygen quenches (reduces) the amount of fluorescence generated by the dyed film. An increase in lactate concentration reduces the oxygen concentration reaching the dyed film so alleviating the quenching and consequentially causing an increase in the fluorescence output.

Simple colorimetric changes can be monitored in some biosensor configurations. A lecithin biosensor has been developed containing phospholipase D, choline oxidase and bromothymol blue. The change in pH, due to the formation of betaine, causes a change in the bromothymol blue absorbance at 622 nm.[23] Gas phase reactions can also be monitored.[24] For example, alcohol vapour can be detected by the colour change of a dry dispersion of alcohol oxidase and peroxidase plus the redox dye 2,6-dichloroindophenol.

One of the most widely established biosensor technologies is the low technology single-use colorimetric assay based on a paper pad impregnated with reagents. This industry revolves mainly round blood and urine analysis with test strips costing only a few pence. A particularly important use for these colorimetric test strips is in the monitoring of whole-blood glucose in the control of diabetes. In this case the strips contain glucose oxidase and horseradish peroxidase together with a chromogen (*e.g.* *o*-toluidine) which changes colour when oxidized by the peroxidase catalysed reaction with the hydrogen peroxide produced by the aerobic oxidation of glucose.

$$\text{chromogen(2H)} + H_2O_2 \xrightarrow{\text{peroxidase}} \text{dye} + 2H_2O$$

The colour produced can be evaluated by visual comparison to a test chart or by the use of a portable reflectance meter. Many test strips incorporate anti-interference layers to produce more reproducible and accurate assays.

It is possible to link up luminescent reactions to biosensors, as light output is a relatively easy phenomenon to transduce to an electronic output. Thus the reaction involving immobilized (or free) luciferase can be used to detect the ATP released by the lysis of microorganisms.[25]

$$\text{luciferin} + \text{ATP} + O_2 \xrightarrow{\text{luciferase}}$$
$$\text{oxyluciferin} + CO_2 + \text{AMP} + \text{pyrophosphate} + \text{light}$$

[23] V. P. Kotsira and Y. D. Clonis, *J. Agric. Food Chem.*, 1998, **46**, 3389.
[24] S. Lamare and M. D. Legoy, *Trends Biotechnol.*, 1993, **11**, 413.
[25] M. F. Chaplin, in 'Physical Methods for Microorganisms Detection', ed. W. M. Nelson, CRC Press, Inc., Boca Raton, USA, 1991 p. 81.

This allows the rapid detection of urinary infections by detecting the microbial content of urine samples.

7.1   Evanescent Wave Biosensors

A light beam will be totally reflected when it strikes an interface between two transparent media, from the side with the higher refractive index at angles of incidence (θ) greater than the critical angle (Figure 14a). This is the principle that allows transparent fibres to be used as optical waveguides. At the point of reflection an electromagnetic field is induced which penetrates into the medium with the lower refractive index; usually air or water. This field is called the evanescent wave and it rapidly decays exponentially with the penetration distance and generally has effectively disappeared within a few hundred nanometres. The exact depth of penetration depends on the refractive indices and the wavelength of the light and can be controlled by the angle of incidence. The evanescent wave may interact with the medium and the resultant electromagnetic

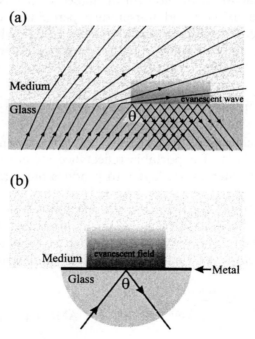

Figure 14   *Production of (a) an evanescent wave (b) surface plasmon resonance. At acute enough angles of incidence the light is totally internally reflected at the glass surface in (a) an evanescent wave extends from this surface into the air or water medium. This process is amplified in (b) by the presence of the thin metal film*

field may be coupled back into the higher refractive index medium (usually glass) by essentially the reverse process. This gives rise to changes in the light emitted down the wave-guide. Thus, it can be used to detect changes occurring in the liquid medium.

Various effects, due to biological sensing processes, can be determined including changes in absorption, optical activity, fluorescence and luminescence. Because of the small degree of penetration, this system is particularly sensitive to biological processes in the immediate vicinity of the surface and independent of any bulk processes or changes. It can even be used for the continuous monitoring of apparently opaque solutions.

This biosensor configuration is particularly suitable for immunoassays as there is no need to separate bulk components since the wave only penetrates as far as the antibody–antigen complex. Surface-bound fluorophores may be excited by the evanescent wave and the excited light output detected after it is coupled back into the fibre (Figure 15) effectively by the reverse of the process causing the evanescent wave. Sensors can be fabricated which measure oxidase substrates using the principle of quenching of fluorescence by molecular oxygen as described earlier. Another advantage of only sensing a surface reaction less than a micron thick is that the volume of analyte needed may be very small indeed.

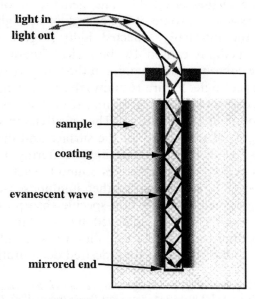

Figure 15 *The principle behind evanescent wave immunosensor. The light output is reduced by absorption within the evanescent wave*

Protein A, an important immunoglobulin-binding protein from *Staphylococcus aureus*, has been determined by this method using a plastic optical fibre coated with its antibody. Detection was by the fluorescence of a fluorescein-bound anti-Protein A immunoglobulin which was subsequently bound, sandwiching the Protein A.[26]

7.2 Surface Plasmon Resonance

The evanescent field generated by the total internal reflection of monochromatic, plane polarized light within a fibre optic or prism may be utilized in a different type of optical biosensor by means of the phenomena of surface plasmon resonance (SPR).[27] If the surface of the glass is covered with a very thin layer of metal (usually pure gold, silver or palladium just a nanometre or so thick), the electrons at the surface may be caused to oscillate in resonance with the photons. This generates a surface plasmon wave and amplifies the evanescent field on the far side of the metal (Figure 14b). If the metal layer is thin enough to allow penetration of the evanescent field to the opposite surface, the effect is critically dependent on the 100 nm or so of medium adjacent to the metal. This effect occurs only when the light is at a specific angle of incidence dependent on its frequency, the thickness of the metal layer, and the refractive index of the medium immediately adjacent the metal surface within the evanescent field. The generation of this surface plasmon resonance adsorbs some of the energy of the light so reducing the intensity of the internally reflected light (Figure 16). Changes occurring in the medium caused by biological interactions may be followed using the consequential changes in the intensity of the reflected light or the resonance angle. Figure 16 shows the change in the resonance angle of a human chorionic gonadotrophin (hCG) biosensor on binding hCG to surface bound hCG antibody.[28] The sensitivity in such devices is limited by the degree of uniformity of the surface and the bound layer and the more-sensitive devices minimise light scattering. Under optimal conditions just 20 or 30 protein molecules bound to each square micron of surface may be detected. As with other immunosensors, the main problem occurring in such devices is non-specific absorption.

The biological sensing can be achieved by attaching the bioactive molecule to the medium-side of the metal film. Physical adsorption may be used but, because this may lead to undesired denaturation and weak

[26] Y. H. Chang, T. C. Chang, E. F. Kao, and C. Chou, *Biosci. Biotech. Biochem.*, 1996, **60**, 1571.

[27] B. Liedberg, C. Nylander, and I. Lundstrom, *Biosensors Bioelectronics*, 1995, **10**, R1.

[28] J. W. Attridge, P. B. Daniels, J. K. Deacon, G. A. Robinson, and G. P. Davidson, *Biosensors Bioelectronics*, 1991, **6**, 201.

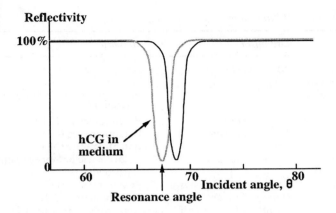

Figure 16 *The change in absorption due to surface plasmon resonance*

binding, covalent binding is often preferred. Gold films can be coated with a monolayer of long-chain 1,ω-hydroxyalkyl thiols, which are copolymerized to a flexible un-crosslinked carboxymethylated dextran gel enabling the subsequent binding of bioactive molecules. This flat plate system, marketed as BIAcore[TM] by Pharmacia Biosensor AB,[29] allows the detection of parts per million of protein antigen where the appropriate antibody is bound to the gel. Fisons Applied Sensor Technology has marketed a similar device, IAsys[TM]. Typical analyses need less than 50 μl and take times of 5 – 10 min. Similarly, biosensors for DNA detection can be constructed by attaching a DNA or RNA probe to the metal surface when as little as a few femtograms of complementary DNA or RNA can be detected[30] and, as a bonus, the rate of hybridization may be determined. Such biosensors retain the advantages of the use of evanescent fields as described earlier. They can be also be used to investigate the kinetics of the binding and dissociation processes.[31] In spite of the relatively low costs possible for producing biosensing surfaces, the present high cost of the instrumentation is restricting the developments in this area.

8 WHOLE CELL BIOSENSORS

As biocatalysts, whole microbial cells can offer some advantages over pure enzymes when used in biosensors.[32] Generally, microbial cells are

[29] M. Raghavan, P. Bjorkman, P. J. Bjorkman, P. Hughes, and H. Hughes, *Structure*, 1995, **3** 331.
[30] T. Schwarz, D. Yeung, E. Hawkins, P. Heaney, and A. McDougall, *Trends Biotechnol.*, 1991, **9**, 339.
[31] M. Fivash, E. M. Towler, and R. J. Fisher, *Curr. Opinion Biotechnol.*, 1998, **9**, 370.
[32] Y. I. Korpan and A. V. Elskaya, *Biochemistry-Moscow*, 1995, **60**, 1517.

Table 7   *Whole cell biosensors*

Analyte	Organism	Biosensor
Ammonia	*Nitrosomonas* sp.	Amperometric (O_2)
Biological oxygen demand (BOD)	many	Amperometric (O_2/mediated), polarographic (O_2) or potentiometric (FET/H_2)
Cysteine	*Proteus morganii*	Potentiometric (H_2S)
Glutamate	*Escherichia coli*	Potentiometric (CO_2)
Glutamine	*Sarcina flava*	Potentiometric (NH_3)
Herbicides	Cyanobacteria	Calorimetric or amperometric (mediated)
Nicotinic acid	*Lactobacillus arabinosus*	Potentiometric (H^+)
Sulphate	*Desulfovibrio desulfuricans*	Potentiometric (SO_3^-)
Thiamine	*Lactobacillus fermenti*	Amperometric (mediated)

cheaper, have longer active lifetimes and are less sensitive to inhibition, pH and temperature variations than the isolated enzymes. Against these advantages, such devices usually offer longer response and recovery times, a lower selectivity and they are prone to biocatalytic leakage. They are particularly useful where multistep or coenzyme-requiring reactions are necessary. The microbial cells may be viable or dead. The advantage of using viable cells is that the sensor may possess a self-repair capability but this must be balanced against the gentler conditions necessary for use and problems that might occur due to membrane permeability. Different types of whole cell biosensors are shown in Table 7.

Biochemical oxygen demand (BOD) biosensors most often use a single selected microbial species. Although rapid, linear and reproducible, they give different results to conventional testing, which involves incubation over 5 days and reflects the varying metabolism of a mixed microbial population. Thermophilic organisms have been used in such a biosensor for use in hot wastewater.

9   IMMUNOSENSORS

Most biosensor configurations may be used as immunosensors[33] (Table 8) and some of these have been mentioned earlier. Figure 17 shows some of the configurations possible. Direct binding of the antigen to immobilized antibody (Figure 17a) or antigen–antibody sandwich (Figure 17b) may be detected using piezoelectric or SPR devices,[34] as can antibody release due to free antigen (Figure 17c).

---

[33] E. Gizeli and C. R. Lowe, *Curr. Opin. Biotechnol.*, 1996, **7**, 66.
[34] C. L. Morgan, D. J. Newman, and C. P. Price, *Clin. Chem.*, 1996, **42**, 193.

Table 8 *A selection of immunosensors*

Analyte	Sensing method	Biosensor
Cocaine	Antibody/Protein A	Piezoelectric
Human chorionic gonadotrophin	Antibody/catalase	Amperometric (O_2)
	Antibody	SPR
Hepatitis B surface antigen	Antibody/peroxidase	Potentiometric (I^-)
Insulin	Antibody/catalase	Amperometric (O_2)
T2 toxin	Antibody	Evanescent wave
Trinitrotoluene (TNT)	Antibody/labelled antigen	Fluorescence

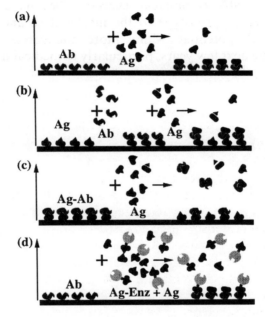

Figure 17 *Different configurations for biosensor immunoassays: (a) antigen binding to immobilized antibody, (b) immobilized antigen binding antibody which binds free second antigen, (c) antibody bound to immobilized antigen partially released by competing free antigen, and (d) immobilized antibody binding free antigen and enzyme-labelled antigen in competition*

Binding of enzyme-linked antigen (Figure 17d) or antibody can form the basis of all types of immunosensors but has proved particularly useful in calorimetric and amperometric devices. The amount of enzyme activity bound in these immunosensors is dependent on the relative concentrations of the competing labelled and unlabelled ligands and so it can be used to determine the concentration of unknown antigen concentrations.

The main problems involved in developing immunosensors centre on non-specific binding and incomplete reversibility of the antigen–antibody reaction both of which reduce the active area, and hence sensitivity, on repetitive assay. Single-use biosensing membranes are a way round this but they necessitate strict quality control during production.

10 CONCLUSION

Biosensors form an interesting and varied part of biotechnology. They have been applied to solve a number of analytical problems and some have achieved notable commercial success. They have been slow to evolve from research prototype to the marketplace and have not yet reached their full potential. Many more commercial products are expected over the next few years, particularly in medical diagnostics.

Subject Index

555